Adaptive Reuse of Cultural Heritage

Luigi Fusco Girard • Antonia Gravagnuolo
Editors

Adaptive Reuse of Cultural Heritage

Circular Business, Financial and Governance Models

Springer

Editors
Luigi Fusco Girard
Emeritus Professor at University of Naples Federico II; CNR IRISS, National Research Council, Institute for Research on Innovation and Services for Development
Naples, Italy

Antonia Gravagnuolo
CNR ISPC Institute of Heritage Science, National Research Council
Naples, Italy

ISBN 978-3-031-67627-7 ISBN 978-3-031-67628-4 (eBook)
https://doi.org/10.1007/978-3-031-67628-4

This work was supported by Consiglio Nazionale delle Ricerche

This Springer imprint is published by the registered company Springer Nature Switzerland AG
The registered company address is: Gewerbestrasse 11, 6330 Cham, Switzerland

To Marilori Cataldi Fusco Girard

Preface

Cultural heritage is a driver for sustainable development in cities. As an economic and cultural asset, it boosts economic growth, enhances urban livability and contributes to environmental adaptability. In addition, the reuse of abandoned and underused cultural heritage and landscapes is a practical substitute to demolition, bypassing the wasteful processes of demolition and new construction prolonging the cultural heritage lifespan. Adaptive reuse of cultural heritage can thus be instrumental to circularize the flows of energy, raw materials, and human and cultural capital, and hence, it plays a significant role in the transition towards circular economy. Complementary to its environmental benefit, adaptive reuse brings forth substantial economic, social and cultural advantages by reusing historic buildings, sites and landscapes attached meanings and values by a wide range of citizens and actors.

The existing governance structures and operational systems concerning reuse of cultural heritage and landscapes are still highly limited in the involvement of relevant stakeholders to the decision making process. Regulatory and planning tools are not flexible enough to allow sustainable and circular transformation processes, and are restricted in the financial resources and funding arrangements that mostly rely on public funds. Therefore, to turn cultural heritage and landscapes into a resource, instead of a cost for the collectivity, the structures of authority, institutions and financial arrangements should be adjusted. This adjustment needs to ensure larger stakeholders' involvement in decision-making, to attract private investments and to facilitate cooperation between community actors, public institutions, property owners, informal users and producers. In addition, to manage change, flexible, transparent and inclusive tools are required, thus leveraging the potential of cultural heritage to foster adaptive reuse practices.

This timely book thus aims to address this gap in existing knowledge from a circular economy and sustainable development perspective, and to introduce innovative economic, environmental and governance models and evaluation tools tested and validated for adaptive reuse within the "CLIC—Circular models Leveraging Investments in Cultural Heritage Adaptive Reuse" project funded by the European Horizon 2020 Research & Innovation Action Program. The CLIC project is a transdisciplinary research project bringing together expertise from disciplines such as heritage studies, regeneration and urban development, business management, economics, ecology and social sciences. This research responds to the European Societal Challenge 5 “Climate action, environment, resource efficiency and raw materials”, aimed at achieving resource efficient and climate change resilient economy and society through systemic innovation. The CLIC project also aims to unlock public and private investments in solutions for a more resource-efficient, greener and more competitive economy as a key part of smart, inclusive and sustainable growth strategy for Europe and worldwide.

In this context, this book comes to fore as a fundamental key reading for scholars, professionals and policy makers, towards demonstrating how the adaptive reuse of cultural heritage, in a systemic perspective, has the potential to stimulate growth, sustainable development, social regeneration, welfare, jobs, income and livability of urban/territorial settings: to implement the circular economy model. It also provides innovative models and a circular toolkit for financing, reusing and managing cultural heritage based on research outcomes and implementation of experimental models in four pilot European territories covered as case studies.

Naples, Italy

Luigi Fusco Girard

Antonia Gravagnuolo

Contents

Chapter 1
Introduction: CLIC EU Horizon2020 R&I Project: Circular Models Leveraging Investments in Cultural Heritage Adaptive Reuse

Luigi Fusco Girard and Antonia Gravagnuolo

This volume harvests the key results of on the Horizon 2020 CLIC research project on "Circular economy models to leverage investments in cultural heritage adaptive reuse". This multidisciplinary and trans-disciplinary research was developed between 2017 and 2022 engaging hundreds of citizens and stakeholders in four pilot European cities to experiment innovative approaches for the adaptive reuse and management of cultural heritage in the perspective of the circular economy.

The theoretical background of the research is grounded on the ecological economics, focusing on circular conservation, regeneration and care of cultural heritage as action aimed at environmental protection and inter/intra generational justice. Ecological economics, which dates back to the 80s with scholars such as inter alia Herman Daly (1968, 1980), Robert Costanza (1989), Robert Ayres (1969), Georgescu-Roegen (1986), Milan Zeleny (1981) is a field of evolutionary economics, concerned with the dynamic of complex adaptive systems.

Also the contribution of the human/social economy from the pioneer Sismondi to M. Gandhi to A. Etzioni to E. F. Schumaker to Amartya Sen (1970) have been considered in CLIC as a source of the humanistic approach, beyond market and state.

The urgent social and ecological challenges of our times require innovative approaches. The above regards also the cultural heritage sector to exploit its potential as a true driver of sustainable growth and long-term prosperity for people, planet and places. Today, it is not only required to identify alternative business models able to leverage financial re-sources needed for heritage conservation. Truly innovative business models can only come as a consequence of a radical ecological shift in societal values, which influence personal behaviours and lifestyle and open up to a

L. Fusco Girard
Emeritus Professor at University of Naples Federico II; CNR IRISS, National Research Council, Institute for Research on Innovation and Services for Development, Naples, Italy

A. Gravagnuolo (✉)
CNR ISPC Institute of Heritage Science, National Research Council, Naples, Italy
e-mail: antonia.gravagnuolo@cnr.it

L. Fusco Girard, A. Gravagnuolo (eds.), *Adaptive Reuse of Cultural Heritage*,
https://doi.org/10.1007/978-3-031-67628-4_1

new economic model –a circular economy model—that prioritizes care for people and the planet together with individual benefit, including ensuring equal opportunities to present and future generations ("no one is left behind", UN Agenda 2030 for Sustainable Development). Instead of being less profitable, the new evolutionary, ecological and symbiotic economics based on nature circularity principles revealed a great potential of long term success in the case studies explored in the CLIC research, suggesting the need of re-thinking heritage management models towards collaborative, community-based, nature-based and adaptive approaches. The true key of success in the implementation of the circular economy can be identified in the capacity of citizens and stakeholders to collaborate and cooperate, trusting each other towards shared objectives, thus exploring synergies instead of (only the) competitive schemes and learning how to act as ecosystems instead of individual subjects, overcoming the self-centred utilitarian culture, towards a civic culture.

Global challenges as climate change, poverty and growing inequalities cannot be addressed without cooperation within and between communities, which however starts at micro level within local communities and organisations changing approaches towards the collective care of the "common good" (being that ecological resources, cultural heritage, health or social justice).

A particular attention has been devoted to the cultural dimension and to cultural benefits of the reuse, as well as to operational tools.

The CLIC research project aimed at exploring whether and how the needed radical shift in values can be implemented through cultural heritage adaptive reuse, stimulating the development of new contemporary meanings of heritage through active conservation. Adaptive reuse can ensure that cultural heritage continues to "live" for present and future generations, because it ensures use-values in an indefinite time span, thus preserving its "intrinsic value".

Six key concepts underlie the CLIC project in the search of innovative models:

- Circularization of processes (the circular economy)
- Cultural heritage and landscape as commons goods
- The landscape approach, as promoting a systemic unifying perspective
- Creative hybridization between old and new values
- The growing role of social / solidarity economy and third sector actors
- The notion of well-being, well living, and beauty in the city-territory system as a key aspect of sustainability.

Four European cities were involved as pilot experimentations to implement the circular adaptive reuse of cultural heritage for the regeneration of abandoned sites: Salerno in Italy, Rijeka in Croatia, post-industrial villages in the rural region of Västra Götaland in Sweden, and the city of Amsterdam through the civil society organisation Pakhuis de Zwijger.

This book collects the results of these pilots' experimentations, developing participatory Local Action Plans for the circular adaptive reuse of cultural heritage and detailed reuse projects linked to circular business models to turn "dead" abandoned and underused heritage resources into "living systems".

The innovative ecological and evolutionary perspective of the CLIC research in the field of heritage conservation requires the development of suitable evaluation models and tools, which reflects also the need of engaging local communities in the co-design and co-development of adaptive reuse projects, which objectives need to be recognised and shared to build the "heritage communities" strongly evoked in the Faro Convention (Council of Europe, 2005). This heritage community should be characterized by the capacity to link young and aged generation.

New evaluation tools need to be adapted in order to adhere to the dynamic evolution of communities' needs and preferences. Multidimensional and multicriteria evaluation tools (Nijkamp 1977) to support participatory decision-making of cultural heritage adaptive reuse in the perspective of circularity were developed within the CLIC project and tested in the diverse case studies presented in this volume. The CLIC dynamic, co-evolutive and spiral evaluation model based on multidimensional criteria and indicators of circularity represents one of the results of the research, allowing the identification, evaluation and interpretation of alternative adaptive reuse solutions, towards the identification of the more "satisfying" project. Cities and regions involved in the research developed new strategies and projects for the recovery and reuse of abandoned heritage through the support of structured, dynamic and participatory evaluation processes.

Based on the analysis of more than one hundred of case studies, circular governance models for cultural heritage adaptive reuse were identified and tested/adapted to the pilot areas of the CLIC project. Third sector and community based initiatives emerged as particularly relevant for cultural heritage, leading to specific recommendations towards enhanced policies for heritage conservation. The role of third sector actors, such as community cooperatives, foundations, civil society organisations, NGOs is projected to become more and more relevant in the heritage field. Indeed, third sector actors aim to reach social objectives while generating new jobs and revenues, which can be employed for the maintenance and recovery of cultural heritage over time, in a virtuous circle of "care", opposed to the vicious circle of abandonment.

Finally, financing models for circular adaptive reuse of cultural heritage were explored at both theoretical and practical level, integrating economic, cultural, social and environmental impacts assessment towards viable business and financing solutions "blended" with societal benefit. A particular attention was paid to the impact of cultural heritage adaptive reuse on local communities, exploring how heritage regeneration can contribute to mutual trust, cohesion and inclusion at local level.

The CLIC research introduced the circular adaptive reuse of cultural heritage as entrepreneurial activity through the Startup Competition for contributing to the development of specific entrepreneurial skills and capacities in the heritage sector, leveraging investments for heritage-based business projects able to generate new jobs and also positive societal impacts.

At international and policy level, the CLIC project contributed to the New European Bauhaus initiative and contributed to a certain extent to shaping the new Horizon Europe framework, which was developed taking into account the results and further research needs expressed through this research. New research will need to continue exploring the role of cultural heritage adaptive reuse for the implementation of the circular economy to reach the objectives related to climate change mitigation and adaptation, natural resources regeneration, as well as European cohesion, inclusion and cultural identity. This volume aims to provide a valuable pathway and a blueprint for the integration of circular business, financing and governance models in cultural heritage conservation, reuse, valorisation and regeneration over the next challenging times, stimulating specific new "CLIC researches at national levels".

The book is organized in six parts: (1) theoretical foundation; (2) experimentation and tools for the adaptive reuse of cultural heritage in the perspective of the circular economy model, focusing on evaluation tools; (3) barriers, bottlenecks and new governance; (4) circular business models and financial instruments; (5) enhancing creativity in the cultural heritage entrepreneurial sector; (6) conclusions and guidelines.

1 The First Part Develops a Common Framework Starting from Robust Theoretical Foundations in the Field of Cultural Heritage Conservation

The adaptive re-use of cultural heritage is presented as the entry point for the circular economy strategy in cities and regions. A structured review of scientific literature in cultural heritage adaptive reuse completes this overview on theoretical foundations and opens to the analysis of best practices of cultural heritage adaptive reuse (ARCH) in a circular economy perspective, revealing and discussing differences and similarities between European countries, territorial scales, cultural heritage typologies, cultural, social and regulatory contexts and environmental conditions.

Specific principles for the success of the circular adaptive reuse of cultural heritage have been identified and proposed.

First of all, **Chap. 2** by Luigi Fusco Girard "The circular 'human-centred' adaptive reuse of cultural heritage: theoretical foundations" provides the theoretical foundations that allow to identify the adaptive reuse of abandoned cultural heritage as driver of circular economy implementation at urban and territorial level, leveraging the potential of its "intrinsic value" for the generation of new use values in a circular perspective. This introduction includes the historic context of economics of cultural heritage conservation and proposes specific new principles for the success of the adaptive reuse through transferring natural ecology organisational system into the human social ecology.

The following **Chap. 3** by Marco Acri, Xavier Greffe and Saša Dobričić "Intangible Matters: cultural heritage as a driver for circular economy" recognizes that decades of linear economy have contributed to the weakening of the embedded intangible dimension in the production and preservation of cultural heritage, with consequent loss of tangible communicative competence of things. As a rule, the production of cultural heritage and it reuse practice rely on the reactivation of spatial and temporal residuals, considering obsolescence as a value and not as a lack, thus opening new imaginary for circular economy. This contribution proposes a new experimental register of actions recognizable in care, maintenance, repair, tailor maid, proximity and re-design, which is capable to nourish the circular loop through a constant driver of acceptable changes.

Chapter 4 by Saša Dobričić, Marco Acri, Jukka Jokilehto "The Intimate Circular Dimension of Heritage Conservation and Historic Urban Landscape" highlights the spectrum of possible connections between cultural heritage adaptive reuse and circular economy, displaying the incredible potentials for a sustainable regeneration of European urbanised areas. Since its very beginning, however, it seemed clear that regardless of centuries of debates on the concept of conservation, the understanding on the basic notions were very different from country to country, letting potentially erroneous openings in the conservation practice, provoked by often superficial interpretations of reality as well as pressures from short term global market waves. Based on an extensive literature review on heritage and historic urban landscape preservation, this contribution concentrates on the very intimate historic relationship between cultural heritage production and preservation and circular economy, doing parallelisms between the past and the present. In particular, the Chapter addresses the new and old paradigms in preservation, with an important emphasis on the concepts of authenticity and integrity, that must be seen as reference guiding principles when dealing with adaptive reuse, conservation and restoration, but also with cultural and historic urban landscape, with the scope to set clear fundamentals for new planning and governance frameworks.

2 The Second Part of the Book Focuses on Practices, Experimentation and Tools for the Adaptive Reuse of Cultural Heritage in the Perspective of the Circular Economy Model, Exploring Particularly Evaluation Tools

Chapter 5 by Antonia Gravagnuolo, Gabriella Monteleone and Luigi Fusco Girard "In search of a circular model for cultural heritage adaptive reuse: building evidence-base", describes how the CLIC framework for circular adaptive reuse of cultural heritage was tested through the assessment of European adaptive reuse cases, to identify the elements of success in diverse contexts according to specific and multi-dimensional circularity criteria. The Chapter present the methodology and tools used in CLIC to collect, organise, analyse and interpret relevant data on European cultural heritage adaptive reuse practices, to identify good practices, success factors and barriers towards the implementation of the circular model for heritage reuse and regeneration. The CLIC Survey rationale and structure is described, followed by the Knowledge and Information Hub (KIH) platform features, data analysis and assessment. A number of case studies is provided as an overview of the relevant information collected and how it was used to advance knowledge on successful adaptive reuse cases.

The **Annex** of this chapter, developed by Gabriella Monteleone, focuses on the CLIC Knowledge and Information Hub platform and describes how innovative data management system developed by FacilityLive helped to provide clear and transparent data and to manage them in innovative ways, enhancing opportunities for collaboration/cooperation between diverse stakeholders and to increase funding for cultural heritage. Data on cultural heritage in European countries are not easily available and new user-friendly data management systems can be a valid contribution able to support complex choices and evaluations. The Annex focuses on creating an impact in the market of information technology by providing methodologies, expertise, a novel patented technology and an enabling platform for heritage-related data management. The ambition is to use and further develop highly innovative data and information technology to enable cooperation between multiple stakeholders and inform their investment and management choices.

Chapter 6 by Natale Carlo Lauro, Antonia Gravagnuolo, Luigi Fusco Girard, Immacolata Vellecco and Maurizio Lauro on "A statistical model representation and analysis of cultural heritage adaptive reuse practices based on latent variables for circularity assessment" presents the results of the statistical analysis conducted on the CLIC survey database on existing cultural heritage adaptive reuse practices, through which the key "building blocks" of circularity were identified. Through the statistical analysis of the large database with 126 practical cases, the CLIC theoretical model of circular adaptive reuse of cultural heritage was operationalised, identifying the critical elements/criteria of success that become fundamental references for ex-ante evaluations in the design phase of the adaptive reuse intervention.

Chapter 7 by Francesca Buglione, Antonia Gravagnuolo, Mariarosaria Angrisano, Silvia Iodice, Martina Bosone, Pasquale De Toro and Luigi Fusco Girard deals with "Understanding best practices of cultural heritage adaptive reuse in the perspective of the circular economy: in-depth assessment of case studies". The analysis of best practices informed the CLIC evaluation framework of circular adaptive reuse of cultural heritage as they represented useful examples of how circularity goals are reached through diverse interventions. The "ideal" circular model for cultural heritage adaptive reuse should include the achievement of all goals (cultural, environmental, social, economic), however these best practices were important to contribute to the identification of specific circularity criteria for evaluation of future projects/interventions. This contribution present the methodology implemented to select and analyse the case studies, and the results obtained through direct interviews conducted with heritage sites/buildings managers. A summary of lessons learned and the key success factor identified in the case studies supported the identification of specific circularity criteria and indicators for ex-ante evaluation.

Chapter 8 by Antonia Gravagnuolo, Martina Bosone and Luigi Fusco Girard "The CLIC multidimensional impacts assessment framework: criteria and indicators for circular 'human-centred' adaptive reuse of cultural heritage" develops an evaluation framework able to support the ex-ante and ex post assessment of for circular "human-centred" adaptive reuse of cultural heritage. This contribution identifies multidimensional evaluation tools, as criteria and indicators, according to the CLIC theoretical framework of circular adaptive reuse of cultural heritage, starting from the analysis of previous studies and ex-post evaluation of adaptive reuse practices, to structure a comprehensive operational framework for ex-ante evaluation and participatory decision-support in the perspective of circularity. The multidimensional indicators about impacts of cultural heritage are quanti-qualitative indicators and they refer to different scales from the single building to the urban scale. This Chapter highlights the novelty of the CLIC approach in the evaluation and the perspectives for the implementation of the proposed framework.

Chapter 9 by Simona Panaro, Salvatore Greco and Alessio Ishizaka "A multicriteria and multi-scalar Decision Support System to implement circular economy in cultural heritage adaptive reuse" presents a Decision Support System including the methodology and the multicriteria evaluation model. The Chapter highlights the flexibility of the proposed tool and the possibility of involving diverse stakeholders in a step-by-step process to support transparent and informed decision making at different stages of the adaptive reuse process. The theoretical and methodological part offers some practical suggestions for the implementation of the tool in two CLIC pilot cities.

The following **Chap. 10** by Agata Żbikowska, Magdalena Roszczyńska-Kurasińska, Bartosz Ślosarski, Anna Domaradzka "Social Sustainability Framework for measuring socio-cultural impacts of cultural heritage adaptive reuse" is focused on revising the existing theoretical frameworks as well as wellbeing and sustainability indicators. The aim is to propose valid tools for assessing the socio-cultural impact of adaptive reuse investments in the cultural heritage field. It starts with a macro level review of the existing and widely used measures of wellbeing, to look for potential indicators related to cultural heritage. Further examining the specificity of adaptive re-use, the focus is on relation between adaptive reuse projects and local communities, exploring the potential features of sustainable community that could be assessed.

Then, **Chap. 11** by Anna Domaradzka and Magdalena Roszczyńska-Kurasińska "Grassroots Social Sustainability Framework and Intrinsic Value measures for cultural heritage adaptive reuse projects" is focused on the results of pilot implementation of grassroot social sustainability framework (GSS) to study specific cases of adaptive reuse projects. The framework is based on the ideas described in detail in previous Chapters of the book. Here, the results of three pilot studies are provided, to illustrate how concepts of social sustainability and intrinsic value can be linked to create a hands-on methodological GSS framework. Its main use is to enable the analysis of social surroundings of cultural heritage sites as well as qualities of the site itself, to identify opportunities and threats for achieving long-term grassroots sustainability.

Chapter 12 by Christer Gustafsson and Jermina Stanojev "Multidimensional Spillovers of Cultural Heritage through Regional Development and Circular Economy" addresses predominately spillovers and non-use values which are not immediately related to the use of cultural heritage, but may generate larger benefits to local systems in terms of increase of human and relational capital. It presents a research on the links between global challenges, regional development and spillover effects of conservation and adaptive reuse of cultural heritage through the concept of circular economy and smart specialisations and how they interconnect, highlighting the generation of multidimensional benefits of cultural heritage adaptive reuse under the circularity perspective.

Finally, **Chap. 13** by Antonia Gravagnuolo, Martina Bosone, Serena Micheletti, Mariarosaria Angrisano and Luigi Fusco Girard concludes this Part with the experimentation of the CLIC evaluation methodology and tools in the pilot city of Salerno, in Italy: "Towards participatory, dynamic, co-evolutionary evaluation for circular adaptive reuse of cultural heritage: the experimentation of Salerno Circular City of Health". The methodology based on participatory co-design of circular adaptive reuse solutions was experimented, supported by multidimensional and multicriteria evaluation tools. The Chapter presents in particular the CLIC dynamic and co-evolutionary evaluation methodology for participatory circular adaptive reuse of cultural heritage. Through the CLIC project experimentation, the pilot city of Salerno developed a large adaptive reuse project of abandoned heritage buildings, localized in an area that has been considered a "urban waste". The aim is the regeneration of the historic urban site, recovering tangible and intangible cultural heritage, and generating positive impacts as jobs creation, enhanced accessibility, safety and quality of life in the urban area, together with regeneration of natural resources and the environment. The project of the "Circular City of Health" regenerates the most important intangible heritage of Salerno: the ancient Medical School inspired by Hippocrates health and medicine principles, which was developed in Salerno over centuries, contributing to the city's development and attractiveness. Today, the Salernitan Medical School is recognized as an intangible heritage that can have a key role in the city's development, stimulating the adaptive reuse of tangible, but also intangible heritage resources in the city.

3 The Third Part Is Focused on Barriers, Bottlenecks and the Need of a New Governance

Chapter 14 by Deniz Ikiz Kaya, Nadia Pintossi and Caroline A.M. Koot "Adaptive reuse of cultural heritage: barrier assessment and policy-related recommendations" focuses on the barriers and bottlenecks of adaptive reuse of cultural heritage as a driver for the transition towards the circular economy. Adaptive reuse can be regarded unviable due to a number predominant challenges identified as part of this study, which include governance-related issues arising from lack of collaboration and participation, financial constraints, and social and legislative barriers. This Chapter identifies the barriers to adaptive re-use, and examines the solutions and instruments to tackle them and to assess related multi-level policy enablers that support adaptive reuse practices based on their usefulness and feasibility at varying local contexts.

The following **Chap. 15** is proposed by Cristina Garzillo, Allison Wildman, Intza Balenciaga and Ane Izulain on "A New Approach for Cultural Heritage Adaptive Reuse: Circular Governance". It describes the circular governance models for adaptive reuse of cultural heritage and their identification. These models, collectively called Custodian Governance Models, are the Public Custodian, the Community Custodian and the Private Custodian for the Common Good. This Chapter also illustrates the Heritage Innovation Partnership model and the process that brought the pilots participating in such partnerships to develop local action plans as blueprints for adaptive reuse of cultural heritage. Each pilot city/region of the CLIC project is covered. The Boxes include a short introduction of the case studies in CLIC cities/region, discuss their existing administrative, financial and governance structures, and present their individual local action plans for further replication in other cities and regions.

4 The Fourth Part Focuses on Circular Business Models and Financing Instruments for Cultural Heritage Adaptive Reuse

This section is introduced by **Chap. 16** by Ruba Saleh and Christian Ost "Circular business model for cultural heritage adaptive reuse. An iterative journey". It focuses on the Circular Business Model (CBM) for cultural heritage adaptive reuse, conceived by the authors as a co-design process during which stakeholders in CLIC Pilots proposed reuse ideas/solutions to specific heritage assets in relation to their territorial needs and available resources. During the co-design workshops, participants tested the CBM desirability; identified partnerships, users and beneficiaries and made sure that the social, environmental and economic impacts are sustainable. The adapted canvas builds on an iterative process starting with a documentation and analysis of the existing economic landscape and perceptions mapping and ends up with a tailored CBM for a specific asset. The CBM represents a concrete innovative tool aimed at synergizing urban conservation and the circular agenda while fulfilling the philosophy of human-centred cities.

Chapter 17 by Immacolata Vellecco and Assunta Martone on "Business Models for Cultural Heritage Adaptive Reuse" investigates 34 case studies of cultural heritage adaptive reuse. The analysis mainly focuses on the pattern of "original use"/ "new use", linking the cases to the four pillars of sustainable conservation (economic, social, environmental, cultural), also highlighting the different type of actors involved (public, private, social) and their role in supporting the reuse project. As the typologies of building were considered relevant in the decision and success of the reuse projects, the analysis focuses on coastal buildings (lighthouses), religious buildings (monasteries and churches), forts and castles, industrial buildings, and minor heritage buildings. Furthermore, some rural and urban cases highlight connections between reuse initiatives, new functions of the buildings and local (contextual) needs. Although this research design by multiple case studies prevents findings generalizability to different industries, contexts, or countries, it is still appropriate to study the early state of art of circular business models in the adaptive reuse of cultural heritage both in the theory testing and in the theory building view. Some cases of reuse result aiming at eco-innovation and fit the concept of circular economy as efficient use of natural resources, but economic sustainability is now receiving much more attention than in the past. Nevertheless, the majority of reuse projects lay on a wider concept of circularity, based on the sustainable development paradigm and targeting social and cultural goals, in the view of use cultural heritage for people and next generations. Conclusions suggest two manifold streams of policy intervention: to support local communities in the phase of reuse decision, making them aware that the value proposition is at the core of the decision, constraining the alignment of all the elements of the business model, on the different levels of sustainability; and to consider cultural heritage and its reuse as an engine to produce culture and to reinforce young people education with a richer endowment of concepts and abilities, thus enabling them to face what is "new".

The following **Chap. 18** by Gillian Foster "Key Learnings from the "Kultur-Token" Sustainable Business Modelling Case Study" suggest and proposed insights and practical recommendations derived from the sustainable business modelling study of a non-adaptive reuse project in Vienna. The Kultur-Token is a unique valorisation of cultural and cultural heritage assets to encourage low-carbon mobility and reduce carbon dioxide emissions. To derive outcomes for CLIC, the current article explores parallels between CLIC and the Kultur-Token through a retrospective discussion of the case study's research philosophy, goals, and methods. The paper summarizes key lessons learned and recommendations for sustainable business modelling in the context of circular adaptive reuse of cultural heritage projects.

Chapter 19 by Tracy Pickerill "Investment Leverage for Adaptive Reuse of Cultural Heritage" deals with cultural heritage adaptive reuse investment strategies. They involve long term investment horizons, that necessitate the integration of sustainable and circular finance. This Chapter presents an integrated panoptic toolkit of financial (grant, tax, debt & equity) and non-financial (regulation, real estate, risk mitigation, risk performance, capacity building, impact metric and digital network) instruments designed to leverage capital investment and engender collaborative partnerships to encourage private investment capital to flow to grass-roots adaptive reuse activities, including: adaptive reuse, refurbishment and energy retrofit of cultural built heritage; protection and management of natural eco-systems; socio-cultural community enterprise activities. What is new, is a framework for the deployment of blended "hybrid" instruments, pooled within diverse multidisciplinary collaborative fund structures, to encourage intentional and measurable impact investment returns.

Moreover, **Chap. 20** by Aliona Lupu and Ivo Allegro "Circular financing mechanisms for Adaptive Reuse of Cultural Heritage" addresses financing cultural heritage as a "common good". Cultural heritage traditionally receives funding from public sector, primarily for the purpose of its preservation and without the aim to exploit it economically. In a period of increasing pressure on public budgets, this approach activates the spiral of increasingly inadequate investments because of scarce available resources in the hands of the public decision-maker. Different scales of investment, adaptability of assets for new uses, attractiveness of urban areas, as well as ownership and governance structures, determine the adoption of different financing mechanisms for the cultural heritage valorisation. This Chapter describes the three financing mechanisms developed under the CLIC project specifically relevant for the adaptive reuse of cultural heritage. The proposal of the financing mechanisms aims at providing an intense impulse towards sustainable financing of circular adaptive reuse of cultural heritage to preserve the cultural shared value for people and achieve the Sustainable Development Goals (SDGs) locally. An opportunity in the deployment of the mentioned financing mechanisms can come from the synergies between private patient capitals and European, national and regional public funds and, in particular, through new exploitation ways of ESIFs (European Structural Investment Funds).

5 The Fifth Part Questions Whether it Is Possible to Incentivise Creativity in the Cultural Heritage Entrepreneurial Sector

Chapter 21 by Antonia Gravagnuolo, Aliona Lupu, Jermina Stanojev and Valeria Catanese on "Heritage-led entrepreneurial ecosystems: skills and role of startups and innovation" highlights the results of the CLIC Startup Competition describing its approach and methodology, and presenting some startups taking part within three main topics: Circular Tourism, Circular and creative cities and regions, Circular creative industries and social innovation. These startups participated in a mentoring programme in which they developed their circular business models and related business plans to grow and scale-up as next generation circular startups in the heritage sector.

6 In the Sixth Part Conclusions are Drafted Based on the Results Presented

In **Chap. 22** Luigi Fusco Girard and Antonia Gravagnuolo provide critical reflections on the CLIC project results. A set of policy-related guidelines is offered to tackle the challenges of a circular reuse in order to support the adaptive reuse of cultural heritage as entry point for circular "human-centred" cities and regions. Particularly, authors conclude that adaptive reuse is too often interpreted only instrumentally as an investment to secure new employment. Today, however, this approach must be integrated with the need to improve ecosystems health and people's well-being/quality of life, as the adaptive reuse of cultural heritage can become a precious opportunity to implement the circular economy in cities and regions, towards the circular and sustainable transition in all sectors.

From the various contributions proposed, both general reflections and concrete operational suggestions emerge, which, when put together, provide an answer to the question of "why the adaptive reuse of cultural heritage through the circular model" and how to implement it in abandoned and degraded landscapes, agricultural areas, churches, industrial landscapes. Through the circular model applied at territorial level, depopulated villages challenged by ageing could find a new life, becoming villages characterized by high quality of life, of density of human relationships, proximity to lakes, rivers, landscapes, mountains, fortifications, etc., building a culture-nature network which can have as a result the increase of wellbeing, wealth and a new balance between human communities and Nature.

References

Ayres RU, Kneese AV (1969) Production, consumption & externalities. Am Econ Rev 59:282–296
Costanza R (1989) What is ecological economics? Ecol Econ 1(1):1–7
Daly H (1968) On economics as a life science. J Polit Econ 76(3):392–406
Daly H (1980) Economics, ecology, ethics: essays toward a steady-state economy. W H Freeman & Co.
Georgescu-Roegen N (1986) "The entropy law and the economic process in retrospect" (PDF). East Econ J 12(1)
Nijkamp P (1977) Theory and application of environmental economics. North-Holland, Amsterdam
Sen A (1970) Collective choice and social welfare. Holden Day, San Francisco
Zeleny M (1981) Autopoiesis: a-theory of living organization. Elsevier North Holland Inc, New York

Part I
Theoretical Foundations

Introduction

The first section of the book develops a common framework starting from robust theoretical foundations of multicriteria evaluation in the field of cultural heritage conservation. The adaptive re-use of cultural heritage is presented as the entry point for the circular economy strategy in cities and regions. A structured review of scientific literature in cultural heritage adaptive reuse completes this overview on theoretical foundations and opens to the analysis of best practices of cultural heritage adaptive reuse (ARCH) in a circular economy perspective, revealing and discussing differences and similarities between European countries, territorial scales, cultural heritage typologies, cultural, social and regulatory contexts and environmental conditions.

Chapter 2
The Circular "Human-Centred" Adaptive Reuse of Cultural Heritage: Theoretical Foundations

Luigi Fusco Girard

1 Introduction: The CLIC Research Project

1.1 The Horizon 2020 Call

The Horizon 2020 call SC5-22-2017 required to identify *"innovative financing, business and governance models for adaptive re-use of cultural heritage"*. Four kinds of cultural heritage assets were considered in the call: dismissed religious heritage, abandoned industrial buildings, farm buildings no longer used for agriculture and cultural landscape.

This chapter indicates the approach, the general perspectives, some specific outcomes of the CLIC research in reacting to the above issues and also some very general Recommendations.

The CLIC general goal was to identify evaluation processes as key tools to test, implement, validate and share innovative "circular" financing, business and governance models for systemic adaptive reuse of cultural heritage and landscape, demonstrating the economic, social, environmental convenience, in terms of long lasting economic, cultural and environmental wealth. Eleven specific objectives have been identified (CLIC project—Circular models Leveraging Investments in Cultural heritage adaptive reuse 2017). The ambition of CLIC was to propose operational tools for contributing to the implementation of Agenda 2030 goals (United Nations 2015), of New Urban Agenda goals (United Nations 2017), of the European Amsterdam Pact (European Union 2016), of the New Leipzig Charter (Informal Ministerial Meeting on Urban Matters 2020), of European Commission (2021b) and, more in particular, for improving urban regeneration strategies, assuming the cultural heritage as the entry point of the circular economy model.

L. Fusco Girard (✉)
Emeritus Professor at University of Naples Federico II; CNR IRISS, National Research Council, Institute for Research on Innovation and Services for Development, Naples, Italy

L. Fusco Girard, A. Gravagnuolo (eds.), *Adaptive Reuse of Cultural Heritage*,
https://doi.org/10.1007/978-3-031-67628-4_2

The climate change is considered as the key challenge today: as a threat for the health and well-being of human beings also of future generations and of the Mother Earth. Adaptive reuse of cultural heritage is interpreted in this perspective as an opportunity to move towards a more sustainable, resilient, inclusive, beautiful future.

Current adaptive reuse practices are not sustainable because in general they are not faced and implemented as self-sustainable ones.

For example, the flow of energy coming from the movement of the air, water, sun is understated. A key issues of climate change and growing pollution is the issue of conventional fossil energy which is consumed in the economic/urban systems. It has relevant impacts on people well-being. Our well-being/well living depends on the kind of the energy that is employed (WHO 2020).

Energy here is considered as a key issue that was substantially ignored in the current economic models.

But the energy is not only the material one. The intangible form of energy is represented by creativity, new ideas, values: by the culture.

The choice of new uses/functions in CLIC are less linked to the traditional tourism sector or to the residential reuse of heritage assets (Fusco Girard 2018) and are much more linked to contribute to implement the circular city, through place-making: with new functions, more linked to creative/cultural activities and cultural industries and waste reuse/recycle processes.

Thus, CLIC research is oriented to suggest Recommendations about innovative governance, financing, business models for the adaptive reuse of cultural heritage coherent with the European Green Deal (European Commission 2019a), implementing sustainability as self-sustainability.

1.2 Finding the Roots of the CLIC Research in the Thought of Three Masters: Schumpeter, Geddes, Genovesi

The proposals of CLIC research can be recognized in their roots into the history of the economic thinking: into the proposals of (in particular) Patrick Geddes, Joseph Schumpeter and Antonio Genovesi that have been re-read and actualized. Schumpeter and Geddes have in common the evolutionary approach and the distance from the orthodox mainstream economics. They introduced a more flexible, functional, resilient point of view useful for the adaptive reuse of the heritage. Geddes stressed the importance of "nature-led solutions", anticipating adaptive and mitigation measures of our time.

It is due to Schumpeter in particular the emphasis on the evolutionary dynamic that should characterize all proposals from economics to governance with the need for continuous creative innovations in terms of new sources, processes, organizational forms (Schumpeter 1934). It is still due to Schumpeter the importance attributed here to the cultural dimension in the sense that the orthodox economy generated a specific way of thinking/reasoning: a particular rationality and culture (Schumpeter

1943). In the proposed dynamic transitional perspective new solutions should be identified and implemented trough continuous tentative and errors and thus identifying new more and more creative solutions.

In the era of climate change, Antonio Genovesi (Genovesi 1765) should be given credit for recognizing at a very early stage emphasizing the relevance of climate in the economic production (from agriculture to the arts of textiles, metallurgy, etc.) and above all the importance of "mutual confidence" (Genovesi 1765, Chapter VIII).

The centrality of social connections was recalled many times.

The foundation of this connections was represented by the "mutual trust of one in another" (Genovesi 1765, Chapter X). The more cohesive a society is, the greater is its social well-being, i.e. its 'common happiness' and its economic wealth.

Common to both Geddes's (Geddes 1915b, Chapters V, VI) and Genovesi's vision is the reflection that economy is the instrument and not the end: it is the mean to promote the happiness of society and especially striking is the emphasis on cooperation in relation to competition (Genovesi 1765, Chapter X) which is a common message of these scholars.

Cooperation between human beings represents Genovesi's "mutual aid" and Geddes' "mutual partnership", capable of creating centripetal/attractive/integrative force fields. Genovesi's view has many points in common with Geddes' perspective also about the centrality of the educational function.

In other terms the CLIC research recognizes Geddes as able to anticipate more than a century ago many of the concepts examined and developed in the last decades. These ideas include in particular the urban regeneration of historic districts, the "recover of the **spirit of the city**" (Chap. 17), the "conservative surgery" (avoiding demolitions),[1] the key role of participation of inhabitants, the nature-led approach, the key notion of symbiosis coming from biology, the importance of cooperation (able to balance competition), the role of aesthetic quality of the environment. More precisely, it is due to a reinterpretation of Geddes the CLIC proposal for the assumption of the ecological perspective to see the reality and thus to interpret the conservation of cultural heritage, as well as the importance attributed to the "mutually helpful partnership" that occurs in the many symbioses that the natural world offers in its multiple experiences of life 'together with'. It is a form of cooperation as a source of mutual agreement that allows for a reinterpretation and reshape of the Darwinian proposal based only on competition. The plant world is the foundation or the engine of life and therefore stands as the most important common good, essential for rebuilding a desirable development. It is his insight that plants are the source of life on earth and the foundation of human survival itself. In particular, from his reflections emerges the need to learn from nature how to survive in the current context of global change/warming. From nature comes human well-being/health, but also utility and beauty.

[1] Today we avoid demolitions to reduce the CO_2 production. See: (Historic England 2020).

On the other hand, one cannot fail to recognize Geddes as the first promoter of the circular reuse model when he suggests an urban regeneration that (by avoiding demolitions) enhances the city's historical cultural memory.

We also owe it to Geddes to reflect on the "civic energy" that can be regenerated starting with the redevelopment of the cultural heritage, thus regenerating the "spirit of the city".

1.3 Toward "Re-placing" the City/Territory System Through the Lens of Bioecology

We are an hyperconnected society. But we are discovering that we are more and more dis-connecting our life, cities, from the networks of Mother Nature. We are changing the evolutionary dynamic of the nature ecosystems. Climate change impacts regard many key sectors of current economy.[2]

In CLIC research the attention is to re-connect as much as possible our city systems with the Earth patterns, starting from the reuse of heritage, recognizing significant relationships with the Earth pattern. And also, in the sometimes, to re-connect people in the fragmented society, reducing growing inequalities and poverty in its multidimensional forms. In other words the CLIC proposal is to promote the ecological transition together with the "humanistic transition", starting from the regeneration of the "spirit of the city".

Thus, which theoretical framework about relationships between adaptive reuse of cultural heritage and local development in the perspective of the circular economy and the circular city?

The adaptive reuse is read through the **lens of bio-ecology**. The "intrinsic" characteristic of the bio-ecological systems is their **regenerative capacity, grounded on their circular organization and their symbiotic capacity** (in identifying adaptive and cooperative behaviours). This bio-ecological interpretation of the adaptive reuse has implications in terms of proposal of an evolutionary approach, attentive to modifications by environment, and of key principles of adaptive reuse: on the new governance approaches and tools, on new business and new financial models. The adaptive reuse is interpreted in relation also to **the humanization challenge** (see United Nations 2016) towards a reduction of poverty in its multidimensional forms, linking **people, community and places** through lived relations of people and the emotional and affective attachment that connects a community to a space: a space of permanence in a constantly evolutionary changing environment. The **integration of the bio-ecological and humanistic approach** is proposed in CLIC **through the circular economy model** (from which new governance, business and financial models can be identified).

[2] See Stern Review, 2006.

The adaptive reuse of the cultural assets is a key element for a **creative place-making**, for implementing the circular and human-centred paradigm valorizing the "spirit of the cities" (Geddes 1915a).

The adaptive reuse interpreted in an **evolutionary context** requires strong innovations to become effective and to allow the needed complex management (Schumpeter 1943). Ancient signs of human creativity should be combined with innovative tools for humanizing urban spaces.

Data and information together with digital innovative technologies are essential for implementing and managing the circular organization model in the space/territory. Cultural heritage is a robust connective infrastructure if integrated with digital/ICT infrastructure. But information and data are not the key of sustainability. ***Culture is the critical resource for implementing sustainable development***, and thus the circular model. Culture and community are interconnected. They drive development strategies (ICOMOS Delhi Declaration on Heritage and Democracy 2017). Thus, the circular economic model is proposed here not only to better reuse/recycle materials, but also as a bearer of soft, immaterial, intangible components and values.

2 The General Characteristics of the CLIC Research

According to the above, CLIC research offers first of all a **cultural framework** about the promotion of a "circular" inclusion in time and in space, evoking the contribution of H. Daly (Daly 1996; Daly and Farley 2003), K. Boulding (Boulding 1973, 2013), F. Capra (1989), E. Schumacher (Schumacher 1973), R. Costanza (1992) and also of A. Sen (2003) and I. Serageldin (1999) who have stressed the goal of humanizing the economic development: not the profit but the well-being of people is the core of the economic development.

Summing up, the specific characteristics of CLIC research in interpreting the adaptive reuse of cultural heritage between the bio-ecology and the human-centred approach (Fig. 2.1) are:

- The assumption of the **bio-ecological** paradigm to reconnect in a circular way the adapted built environment with nature life networks and thus to face the climate change/global warming challenge (Fig. 2.1);
- The assumption of the **human-centred paradigm** to reconnect the human beings with each other and also with future generations, reducing distances/inequalities (thus interpreting the climate change challenge integrated with the social challenge). The human-centred paradigm offer a specific direction/perspective/sense to the change. The overarching goal becomes the well-being of people: its happiness. Happiness is strictly linked to the density of relationships. The humanization project is a project which unifies, integrating multiplicity into a systemic uniqueness. Approaches and tools assessment really change assuming the human-centred perspective. Participation in the choices is really deepened,

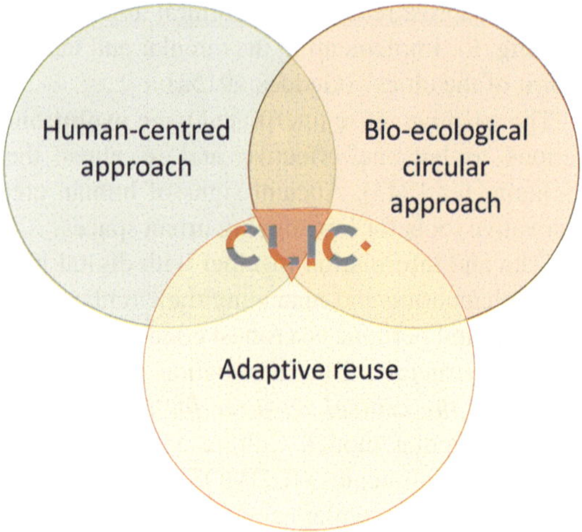

Fig. 2.1 The CLIC general approach

also with the identification of perception indicators for evaluating subjective variations in well-being (Kahneman and Krueger 2006; Domaradzka et al., in this volume (Chap. 11)).

- The assumption of a key role recognized to **culture** (as the more important human production) and to **cultural conditions for implementing sustainable development.** The role of culture is fundamental. As in nature a resource degenerates and is lost if it is not regenerated, in our society a cultural value (as trust, respect of rules, inclusion, etc.) is going to be lost if it is not regenerated (Morin 1990). A culture that recognizes the weight of the relationships is the engine for the implementation of the circular models. Thus, it emerges the need of a **culture local strategic plan** to avoid the risk of an entropy crisis (coming from inside of our society due to the loss of sense/meanings/shared directions).

From the above, a cultural framework (or rather a cultural model), is offered from which to deduce new models of business, financing, and governance. In the context of the European New Green Deal (aimed at addressing climate change and triggering ecological modernization in the European reality for its sustainable development), it is introduced a third "pole" between economy and ecology: the ***cultural*** pole.

Interpreting the **adaptive reuse in the circular and human-centred** perspectives means to reconnect cultural heritage (in the space and also in the time dimensions) with the territory and its community, assuming an **evolutionary perspective/ approach** (characterized by notions of complexity, metabolism, entropy). From the integration of the ecological paradigm with the human-centred paradigm through the circular model, it derives that the outcomes of CLIC regard specific tools not only in technical terms. The outcomes of CLIC are also in the immaterial/cultural dimension, considering the heritage asset also as the immaterial city infrastructure for regenerating new loops.

Fig. 2.2 Adaptive reuse of cultural heritage in the bio-ecological circular organisation

Thus, more in particular, some specific **common characteristics** connecting the research contributions of CLIC research are:

1. **The adaptive reuse in a bio-ecological perspective.** This means the assumption of the ecological paradigm in interpreting the adaptive capacity (as the creative power of nature that evolves in its dynamics through exploratory tentatives). They are *remembered* by the ecosystem memory. As nature adapts and *remembers*, also circular adaptive reuse should be characterized by adaptive, evolving capacity to a changing context and attention to its memory. The image of the tree/forest and its circular metabolism is the reference model: as a tree/forest is «generous» of spill overs, also adaptive reuse should do the same. As the tree/forest is characterized by a circular set of regenerative processes, also the adaptive reuse should be characterized by a circular logic in its functioning (Fig. 2.2).
2. **The assumption of the human-centred paradigm**. Culture is the production "for excellence" of human beings (Greffe 2002). CLIC introduces the cultural condition for sustainable development that integrates the economic/ecological conditions. The circular economy model is interpreted also in soft terms: offering relational and cooperative synergistic culture able to promote development and citizenship.

This means not only to stress the key role of **creativity and innovation capacity** as engine of a new spatial shape and new architecture. But they are key elements in planning and designing the adaptive reuse and also in management. Innovations improve metabolism, reducing entropy and multiplying benefits. On the other side, the attention is to avoid material and immaterial forms of waste, including the human/social capital waste.

An example of a waste of human capital is the elderly population which is becoming an increasing percentage in the European Union (which is the continent with the most elderly people in the world).

Adaptive reuse of cultural heritage should refer to the ability to connect people by avoiding all forms of isolation and loneliness. An example is the adaptive re-use of religious heritage which con become the "centre of gravity" in different urban neighbourhoods for social/cultural functions.

3. **The assumption of the evolutionary perspective/paradigm** coming from the notion of adaptation. This Darwinian "enlarged" paradigm (first of all due to Patrick Geddes) has been assumed more recently by Robert Ayres (1994) who introduced this terminology and, some years before, also to J. Schumpeter (1943) and others.

The evolutionary paradigm characterizes the whole CLIC research. This evolutionary perspective should be recommended in the planning and design of adaptive reuse; in business models; in financial tools; in governance models; in evaluation tools, knowledge and culture. In the Evolutionary Economics the attention shifts from market price and values and market cost to quantitative/qualitative multidimensional impacts and quantitative/qualitative metabolism, also in their circular mutual causation.

Evolutionary Economics, starting from the dynamics of complex systems, recognizes that all economic systems are dissipative in the sense that they "import" energy internally and "export" entropy externally (Georgescu-Roegen 1971, 1976).

Evolutionary Economics (and in particular Ecological Economics) links the evolution of the economic system to technological innovations, thus stressing the attention to the node of **energy** and of **metabolism,** that are underestimated in current economics.[3]

Patrik Geddes (Geddes 1911a) interpreting the **bio-ecological dimension** of the dynamics of the urban system together with the role of symbioses and of mutual cooperation in natural living systems, anticipated the approach of **Nicholas**

[3]The evolutionary perspective should be assumed when:

- it is needed to introduce innovations into a certain heritage site/context in order to make it not only more vital but also more long-lasting;
- there are circular processes of cause and effect between economic, ecological and social factors;
- the entropic nature of all economic processes is recognised;
- the importance of community values is recognised also in the time;
- it is recognised that the preferences of the subjects for whom transformation is envisaged are not already given, but need to be oriented/constructed (the climate change crisis being the most glaring demonstration of the failure of the orthodox economics);
- it is recognised that the object of analysis is the evolutionary dynamics of complex and adaptive systems;

The above makes it essential to abandon the typical approach of mainstream economics (orthodox economics) and move in the direction of evolutionary economics, of which ecological economics is one of the most significant interpretation.

Georgescu-Roegen (Georgescu-Roegen 1971, 1976), founder of **bio-economics and critic of mainstream economics** (as this ignored the relationship between economics and ecology). This relationship is a key characteristic of circular economy.

3 Culture and Cultural Values

3.1 The Role of Culture for Proposing Meanings to Face the Neo-humanistic Challenge

Cultural values represent key non-material values: they are the foundation of any vision for the transformation/development project. The city culture shapes its specific identity, its soul. The city culture shapes also the adaptive reuse at micro and macro scale: it gives form to the new participated governance, to adaptation capacity, to cooperative attitude.

Another reason of this emphasis here on culture is due to the fact that the real challenge in our time is ***within/inside*** people. There is an inner challenge, in the way of thinking that should be modified and "enriched" in a systemic and critical perspective. The current culture (grounded on the individual self-interest and on the short horizon) is exactly opposite to the culture of sustainability (based on the "US" and on the long term). This culture can promote a sustainable, resilient, inclusive and safe city (Agenda 2030, SDG no. 11).

There is first and foremost a **cultural challenge to be faced**. The current culture, through an exaggerated confidence in technology, has become the engine of alteration of the evolutionary dynamics of nature: the degradation of the environment, the loss of biodiversity, the global warming. But it has also produced social fragmentation and isolation, growth of inequality and marginalization.

This ***cultural foundation*** refers to the common roots of the European cultural tradition, which can be interpreted, in synthesis, as the search and ***proposal of meaning for a more human life***, being linked to a specific vision of man and life: man as the end of the economy, and not as a mean; the well-being and the dignity of every human person (regardless of any belonging) as the source of his rights and as the goal of development.

In designing a sustainable future, the roots of the European culture should be firmly taken into account.

The common European values are fundamental to face the technological challenge in the perspective of humanization, of the "neo-humanistic transition" in the digital era. These values should not be lost now that the extraordinary technological development provides powerful innovations, from artificial intelligence to the internet of things to robotization etc. that can be used for human promotion but also for the exact opposite of "human": that is, for any form of manipulation and control/submission by the various forms of economic and political power.

It is necessary not to lose these values in the era of economic globalization, in which the relationship between population and its history should be strengthened.

The CLIC project interprets the re-use, preservation and enhancement of cultural heritage as a source of regeneration of common memory, able to generate new values and new creativity: a matrix of creative mediation which opens to the future vision on the basis of memory/past. That is, cultural heritage is considered as the entry point for the proposal of the neo-humanistic transition in the time of the digital revolution.

The heritage of cultural assets represents the signs of the creative activity of past generations: it expresses a way of feeling the life. It can help to transform the *status quo* with new meanings. It represents not only an economic entry point, but also and above all a cultural entry point. In short, its "transformative force/capacity"[4] is not only a consequence of economic, social, and environmental benefits, but first and foremost of cultural benefits (often neglected because they are "intangible").

3.2 The Circular Economy Model in Integrating the Neo-humanistic and Ecological Paradigms

The ecological transition, which is configured as the greatest transformation ever achieved in the history of human civilization, is absolutely essential. But it will produce a series of human and social costs that must be anticipated and foreseen for their mitigation. Some new activities will be imposed, while others will become obsolete with respect to the new paradigm of sustainability. Therefore, they are destined to disappear, with all the consequent negative impacts on employment: with significant human and social costs.

Considering all the risks in terms of work that it will entail, and the existing general scene, the ecological transition should be carried out from the perspective of the ***human scale of development***, thus combining and integrating the ecological and the humanistic dimensions through culture: **through the culture of circularity**. That is, the culture of the virtuous circuits of integration.

To achieve an effective ecological transition, it should be necessary to promote a "community spirit", generating responsibility in people's behaviour.

The circular economic model integrates the two paradigms. It not only proposes a new urban metabolism, reducing and reusing waste, minimizing impacts and recycling materials and giving them a longer life, also through a "second hand market". This is of great importance today. But above all, it is based on and promotes a culture of cooperation/collaboration/symbiosis with other subjects, with nature and with future generations: in the spatial and temporal dimension.

The model of **circular economy**, as it has been realized in the experiences of Industrial Ecology/Symbiosis, has proved to be capable of ensuring a profit to the

[4] New Leipzig Charter; Ausbel J. H. 1989.

company, but also a profit to society (new jobs) and to the environment (less climate-altering and polluting impacts). This model is configured as more suitable to meet the economic and ecological conditions of sustainability. But it is also able to satisfy the cultural conditions of sustainable development: the ability to regenerate cultural values. The rate should be at least equivalent to that of their consumption (by current capitalistic economy).

The culture of the circular model opens to create links in the space (between different subjects and between people and nature) and in time (between today and yesterday; between today and tomorrow generations).

The integration of the neo-humanistic paradigm with the ecological paradigm is fundamental in shaping the transformation of our society: towards an *"ecological and humanistic transition"*. It requires not only green tools but also cultural processes. They are related to the cultural challenge, and thus to culture as the specific product of excellence of human beings, through which interpersonal relationships and also their relationships with nature are shaped. Human and cultural dimensions are closely linked. Culture shapes (and reshapes) the worldview, the interpretation of reality and behaviours in relation to nature and in relation to others. Cultural values such as cooperative, collaborative values are becoming increasingly important in our fragmented society. **Transformation of mindset** is at the core of everything for growth and success for "reacting" but also to became "pro-active", as well as for the creation of positive social/environmental impacts.

The neo-humanistic paradigm is linked to the ability to re-produce and re-share cultural values such as cooperation, collaboration, coordination, which can reduce social fragmentation, towards a long-term future. They should be able to reshape business, market choices, investment decisions.

Certainly, the **ecological transition**, characterized by an "integrated system of renewable energy", offers a strong contribution to the conservation and regeneration of natural resources, with a drastic reduction of waste, which are recycled/regenerated as much as possible, and transformed into resources (thus reducing the amount of extraction from the natural ecosystem) and making sure that the outputs can return to the natural ecosystem as much as possible. Certainly, the use of renewable energy sources is essential to the de-carbonisation of current economy, as is the planting of new green areas etc. The ecological transition is required to be able to implement the "city of symbiosis between humans and nature", nature being recognized as the most important city infrastructure. Nature should inspire us to face new challenges.

But this is not enough. It is necessary to introduce a shared **new rationality** to face new complex choices: a rationality different from that of the homo-oeconomicus (Daly 1996; Fusco Girard 1987). A **relational and multidimensional rationality** that refuses optimization (maximization and minimization) but is interested in the search of balanced solutions, able to creatively combine also conflicting needs/objectives (Simon 1959; Nijkamp 1980).

The "**human-centred and ecological/circular approach**" here evoked is inspired by the circular model. Circular processes are the common element. On its turn, it reshapes the development project of the city **towards a circular project**

that unites, generating and multiplying relationships and links between subjects, in space and time: between human beings, between young and old people of this and future generations, but also between people and nature (Mother Earth) and between people and the past, the history. The memory of the past contributes to shape the notion of "we", thus reshaping the notion of "I". In conclusion, the circular economy is here proposed not only for reducing the consumption of natural and man-made forms of capital, but also in relation to intangible forms of capital (human and social), underlining its capability to generate social ties through reciprocal bonds and reciprocity practices.

3.3 The "Regenerative Re-use" of Cultural Heritage/ Landscape

A re-generation/re-use of cultural/landscape assets (in which Europe is particularly rich), because of its multiple transversal dimensions, is proposed in CLIC as an entry point for the implementation of the **circular city**.

The circular city is the spatial/territorial aspect of the circular economy.

The valorisation of cultural heritage—at certain conditions—strengthens and promotes, in its turn, the culture of collective memory in a circular process in the time dimension. Thus, it can stimulate the regeneration of the values of inclusion, solidarity, integration. These values are not "given," but must be regenerated as quickly as they are consumed.

The regeneration of cultural values is essential to reshape the value of freedom in the perspective of responsibility.

In effects, the circular economy model is proposed here not only as a model able to reuse materials, row materials and energy, but also as the bearer of a **new culture**: an inclusive/relational culture that is strengthened by reusing cultural heritage (emphasizing inclusive values in the temporal dimension and not only in the spatial one).

No effective adaptive reuse, and more in general, no ecological transition or no new symbiotic circular ecosystem will be characterized by effectiveness if it is not able to re-generate values such as inclusion, solidarity, responsibility, the ability to care for others and for nature. No Next Generation Plan to improve resilience and ensure recovery will be implemented effectively if the preconditions for all of the above are not re-generated: namely, the interpersonal and institutional ***trust***. If a real project of regeneration of the ***culture of trust*** is not promoted.

In conclusion, adaptive reuse should be interpreted also in the above cultural perspective.

For implementing the circular human-centred paradigm, it is necessary to build and spread a new "culture", characterized above all by a **long-term horizon**, by the recognition of **intrinsic and not merely instrumental values**, by **critical knowledge and the ability to evaluate** as the foundation of the culture of responsibility. The "horizon of the city" and in particular of the "circular city" allows to promote the overcoming of forms of radical hyper individualism, which transforms legitimate rights into particularistic selfishness. Circular city should promote **trust,** *i.e.*, social capital, as the necessary glue of society for its development.

The foundation of the **circular economic model** is represented by cooperation. Cooperation is grounded on ***trust***. From the trust, that finds its origin in the search of the truth, of the transparency, of the respect of the rules and of the civil virtues, it springs the ability of collaboration, cooperation, synergy/symbiosis. Therefore, not only social benefits but also economic ones arise. Locke (1663) already underlined how trust represents the bond of society. Antonio Genovesi and the Neapolitan School of the eighteenth century considered trust as the foundation of resilience, of economic development and of the humanization of society itself. They converge with the Geddes intuitions about the key role of symbioses for improving the productivity.

The "adaptive reuse" becomes, therefore, part of a more general **"cultural strategy"**. It is necessary to regenerate the physical assets but, at the same time, also to regenerate the link between man and man, between man and nature/ecosystems (Rees 1997, n.d.): to rethink behaviours, lifestyles and choices. Through adaptive reuse of cultural heritage new "circular heritage ecosystems" should be implemented, able to produce positive integrated impacts.

3.4 The Key Role of Culture: The Cultural Conditions of Sustainable Development

From the above it is possible to deduce more clearly the cultural condition for implementing the circular/sustainable development.

The existing economy has been described as "devouring natural resources, socially divisive, and environmentally hostile" (European Commission 2019b). The economic organization of the capitalist model is also a devourer of cultural values.

It should be noted that Schumpeter (1943) already pointed out that the economy, with its rationality of ends/means, has become the matrix of logic, in the sense that economic rationality and rationality in general tend to coincide: the economy has offered the current logic, that is, the "rational" way of reasoning and behaving/choosing. But Schumpeter also stressed that the capitalist process erodes its own institutional structure and its founding culture.

Max Weber, sometime later, in analysing "the spirit of capitalism", had focused on the relationship between cultural values and economic phenomena (for example, the economic rationalism), emphasizing the role of ethical values (starting with trust) in economic dynamics (Weber 1991, 2019).

More recently, Hollis substantially confirmed how economic development erodes the cultural values it needs, making the "value" of trust increasingly fragile, as Fukuyama also pointed out (Fukuyama 1995; Hollis 1998).

The economic and ecological conditions of sustainability are well known: it is necessary that the speed of extraction of resources from nature does not exceed the speed of their re-generation and that the speed of production of waste discharged into the ecosystem does not exceed the speed of absorption of waste.

The identification of the *cultural condition of sustainable development* integrates the above two ecological conditions. The teaching of nature shows that a resource that is not capable of regeneration ends up dissolving and disappearing (Morin 1990). This statement is transferred and re-proposed from the field of tangible resources to the field of intangible resources, such as the cultural values of trust, respect, cooperation, co-responsibility.

Thus, in analogy, values of solidarity, cooperation, and inclusion are not already "done", but they have to be re-generated with the same speed of their consumption by the market capitalistic economy, in analogy with the functioning of the natural ecosystems. If they are not re-generated, they de-generate and are lost (Morin 1990).

How? In which way?

Circular economy and heritage valorisation/reuse contribute to re-generate values and not only tangible/material components.

The above opens the analysis of industrial and urban metabolism, fundamental in the implementation of the circular model in a new perspective. Also, the analysis of *intangible/cultural components* should be considered: trust as a fundamental value today increasingly scarce but for this reason increasingly important. From the re-generation and dissemination of this value follows the ability to improve the effectiveness of investments. The above appears particularly topical in the time of National Resilience and Recovery Plans, for the ecological modernization of EU society.

The circular model satisfies the ecological and also the cultural conditions for sustainability. It not only proposes a new urban tangible metabolism. It also reflects, is grounded and promotes a new culture: a culture of cooperation/collaboration/symbiosis in the space and in the time dimensions with other subjects, with Mother nature, and with future generations. The linked valorisation of cultural heritage reinforces and promotes on its turn the culture of collective memory in a circular process in the time dimension.

Thus, the circular human-centred approach recognizes a particular attention to the cultural dimension, as a key dimension for implementing sustainability. Cultural dimension is not only interpreted as the material cultural heritage valorisation, but also in terms of changing the capacity of the current worldview, mindset, way of life.

4 Adaptive Reuse Between Ecological Paradigm (Nature) and Culture (Human-Centred) Paradigm: The Implications

4.1 Current Definitions and the Circular Adaptive Reuse Notion

Figures 2.2 and 2.3 resume in graphical terms the above points assumed in CLIC research about the perspective of previous paragraphs.

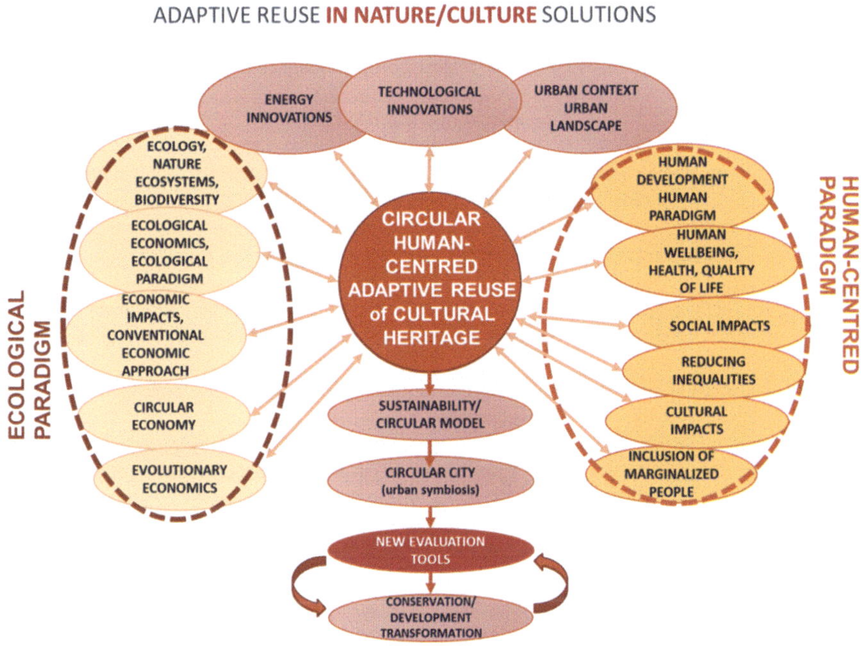

Fig. 2.3 Circular and human-centred adaptive reuse

The circular logic is theoretically "intrinsic" into the notion of adaptive reuse (because it is interested to extend the time life of an asset, avoiding all forms of waste), also if the conventional current definitions of adaptive reuse do not incorporate this **circular logic**.

For example, it is not included in Douglas's definition "any building work and intervention aimed at changing its capacity, function or performance to adjust, reuse or upgrade a building to suit new conditions or requirements" (Douglas 2006). The above is more and more strange if the adaptive reuse regards cultural heritage, considering that cultural heritage "is not just a monument: it is a **non-removable evolving resource**, supporting identity, memory and sense of place". Nor it is incorporated in the Paul Getty Research Institute (2017) (Normandin and Macdonald 2013): reuse of "the conversion of outmoded or unused structures, such as buildings and objects of historic value, to new uses or application in new contexts".

One reason is that, in current interpretation, the emphasis is focused on the design step, and not also in the dynamic management process (see Fig. 2.4: linear adaptive reuse). A second reason is that the reciprocal relationships between the cultural site/asset and its dynamic environment is not underlined.

A third reason is that adaptive reuse is not interpreted through the dynamic lens of bioecology.

The consequence is that practically the adaptive reuse is interpreted and managed in a linear (Fig. 2.4) and not in a circular perspective.

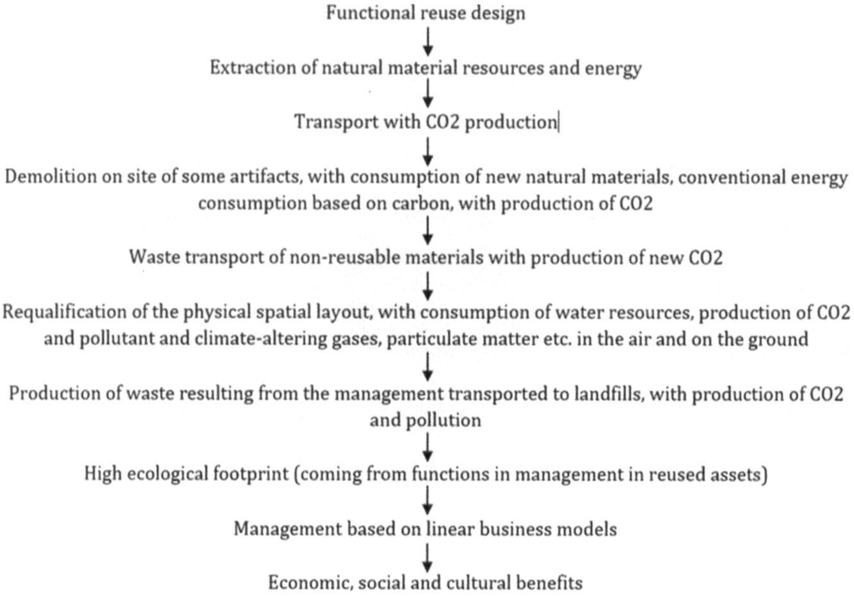

Fig. 2.4 The linear adaptive reuse

4.2 The Adaptive Capacity of Cultural Assets Reuse Should Mimic the Dynamic Creative Power of Nature: Its Auto-poietic Capacity

Continuous adjustments and adaptation over time are required to allow an increase of the usefulness/useful life of the cultural assets: to guarantee a future as long as possible and, eventually its “eternity”. This adaptive capacity is itself a continuous circular process, linking the site with its context.

The adaptive reuse interpreted in the bio-ecological perspective (Fig. 2.2) as proposed in CLIC, becomes more and more structurally characterized by a circular process. Circularity guarantees the life in nature and also in man-made assets. Circularity is the characteristics of symbioses.

The adaptive capacity mimics the creative power of nature to evolve in its dynamics of life through exploratory attempts that are remembered by the ecosystem memory: as nature adapts and remembers, also circular adaptive reuse should be characterized by evolving capacity to react to a changing context, on the base of its memory (its “intrinsic value”) (Zeleny 1992; Fusco Girard and Vecco 2021). In other words, in the interpretation of **adaptive reuse** in a **bio-ecological perspective** (that is in the assumption of the ecological paradigm) the adaptive capacity is similar to the creative power of nature to evolve in its dynamics of life, thanks to self-regenerative potential.

In this way, adaptive reuse can really become able to transform a dead site into a living system.

The analogy with trees (or with the forest) allows to imagine an adaptive reuse that contributes as much as possible to nature regeneration through the renewable energy from the sun to lower pollutant and climate-changing concentrations, purifying the air even with the appropriate planting, generating oxygen, sequestering/reducing carbon dioxide, dust, combustion residues, mitigating heat islands, thus helping to improve the local microclimate. Water, as precious resource, is managed with care. Reuse thus contribute to the required transition toward a de-carbonized (local) economy (Figs. 2.2 and 2.6).

This circular economy adaptive reuse underlines the attention to the circular metabolism (the entry and the exit of energy and materials) and to reduce the comprehensive entropy through the co-evolution between the manmade and the natural capitals. Also, intangible forms of capital, as the human capital and the social capital, should be considered through the re generation of cultural assets. At the same time, attention is due to promote complementarities, and thus (circular) synergies and symbioses.

If the fundamental objective (or the "objective of objectives") of sustainable development is the promotion (as much as possible in the spaces and in the time) of the good living/well-being of people, as recognized in the first item of the Rio de Janeiro Agenda 21, a key step is the capacity to assess the impacts of a change on the well-being level of different subjects.

The quantitative and qualitative indicators about the quality of life/well-being become fundamental.

4.3 Adaptive Reuse and the Notion of "Change": The Key Role of Evaluation

A characteristic of adaptive reuse project is its procedure through "attempt by poiesis". There are many solutions to combine in different way pre-existences and contemporaneity/modernity, old are new meanings. Thus this procedure requires evaluation capacity. Evaluation represents a process oriented to propose "good reasons", design/planning, that is persuasive argumentations to justify a choice. This process, in the case of adaptive reuse of cultural heritage, becomes very complex because one essential characteristic of the cultural asset is the capacity to produce a flow of "intangible" benefits. Their evaluation requires to go beyond the "willingness to pay". Other characteristics are the important role of the civil society (third sector participation) in the decision-making process, the role of young people and the long-term horizon. Through the evaluation process the "vision" of the change of the status quo can be operationally implemented.

Adaptive reuse incorporates the notion of **"change"** towards a circular organization/structure. This circular model has economic, social and environmental high potentials.

The change is emphasized in the circular model in terms of reciprocal interdependences with the context and in terms of the change actions. The change should be as much as possible coherent with the existing economic, social, environmental, institutional context, so that existing performances can be enhanced. The change due to a set of **actions** reflects a dynamic and circular relationship between an asset/site and its comprehensive environment. On the other side, the change due to a set of actions, is the outcome of a specific **decision process grounded on evaluations** in an uncertain environment. Alternative solutions are identified, able to meet multiple/multidimensional objectives coming from many involved subjects, and are compared.

Balanced satisfying solution that better fits into the context is thus designed and implemented, knowing that the reciprocal relation with the context will suggest new adjustments over time. Figure 2.5 explains better this spiral process in the different steps (from bottom up towards top down):

1. identification of alternatives
2. evaluation of impacts coming from each alternative in relation to objectives (ex-ante evaluation)
3. identification of the balanced and satisfying solution

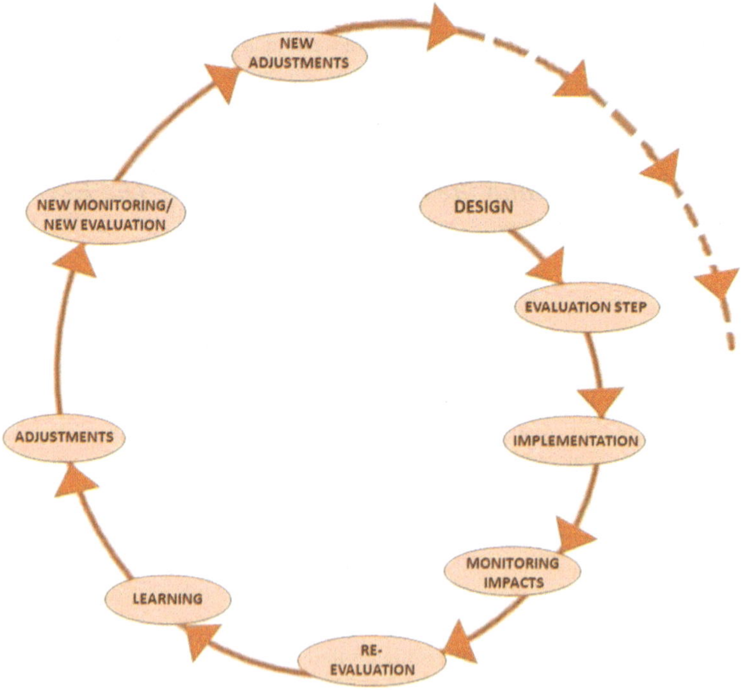

Fig. 2.5 The spiral process in adaptive reuse and monitoring evaluation

4. implementation of the chosen solution
5. monitoring of impacts (net benefits)
6. learning from impacts, for deducing performances (ex-post evaluation)
7. using this new knowledge to adapt/adjust the chosen solution for improving productivity
8. adoption of a management plan, and making new ex-ante, on-going, ex-post assessments
9. new adjustments implemented

More in details, in the micro scale, the spiral process of the reuse is showed in Fig. 2.6.

The logic of the circular reuse is strongly stressed in the circular economy model, as in CLIC assumed.

In other words, in the adaptive reuse interpreted in the circular economy, this loop starts from the reuse of natural materials, water and waste and renewable energy and **includes also the immaterial/intangible resources and values re-generation in the management strategy.**

4.4 Adaptive Reuse and the "Spiral" Evaluation Process

Adaptation and evaluation are characterized by reciprocal relationships. Adaptive reuse requires to make choices regarding the conservation of some elements as permanence, allowing the change/transformation/substitution for others, thus combining the logic of change with the logic of conservation.

Adaptive reuse of cultural heritage is thus a **complex process** because it requires **the capacity to distinguish and select essential elements from other accessory elements.** It requires to make choices between intrinsic value (the essential meaning of the heritage) and the instrumental values. It requires to manage conflict between needs, demands, preferences, values, goals (that are multidimensional and also in conflict) coming from different involved subjects. The adaptive reuse is, in any case, a very important exercise of critical thinking requiring continuous innovations and, consequently, continuous evaluation capacity.

The **evaluation process is** a key characteristic of Adaptive Reuse. **It is the core phase of any adaptive reuse process.**

The first evaluation exercise is about the general context interpretation, in its economic activities, and also ecological, social, cultural ones. It requires to identify better actions to fit the adaptive reuse in the context in order to increase the existing activities, generating an attractive and integrative force-field, reducing the "distance" between a site/activity and its general context.

The evaluation process is iterative and interactive. This regards also the Adaptive management plan, in the search of the more satisfying solution of adaptive reuse, of testing this solution, verifying the impacts through new experiments, new monitoring etc., learning from impacts, reviewing the original solution through continuous adjustments.

Fig. 2.6 Actions required in the circular human-centred adaptive reuse

The required evaluations for the implementation of the design and for the adaptive management plan are grounded on the Ecological Economics approach. Ecological Economics recognizes that a single monetary assessment, based on WTP, is useful, but not sufficient for the evaluation of all economic, socio/cultural/environmental benefits. Wich WTP of future generations or of poor people can be assessed? From values incommensurability it follows that only a democratic participated debate can integrate technical economic evaluations in a multicriteria integrated process.

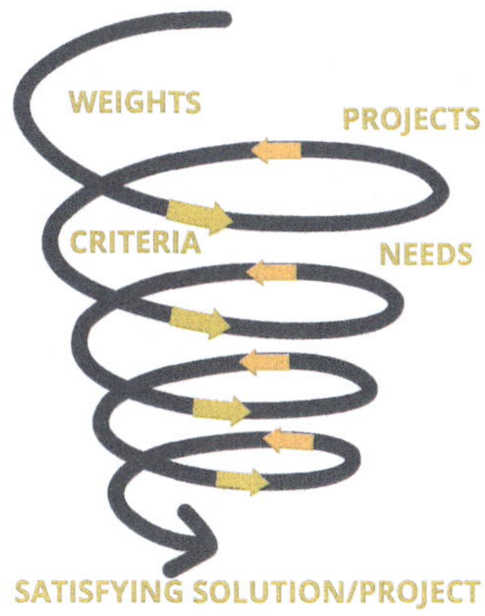

Fig. 2.7 The evolutionary evaluation spiral. (Zeleny 1993)

The dynamic approach of adaptive reuse in the perspective of the circular economy should follow a coherent dynamic/evolutionary evaluation approach. In this dynamic approach criteria, weights and alternatives are going to change in an evolutionary spiral during the entire process (Fig. 2.7), converging towards down and identifying the satisfying solutions (Zeleny 1993). Also the impact matrix should be organized in dynamic terms.

Cost benefit analysis certainly helps identify the economic impacts. But it should be integrated by other approaches, first of all to consider the flow of the energy (for example through Life Cycle Assessment LCA). Other environmental and social impacts should be assessed through specific tools (Fusco Girard and Nijkamp 1997; Nijkamp et al. 1990).

Approaches grounded on MAUT (Multi Attribute Utility Theory), specific approaches as AHP (Saaty 1984), ELECTRE (Roy 1990), REGIME (Hinloopen et al. 1983), EVAMIX (Voogd 1983), NAIADE (Munda 1995), CIE (Lichfield 1995), are useful because they help to better compare alternatives, in a multi-dimensional space deducing a priority (through iterative/interactive steps).

But the participatory and inclusive processes in decision making should be more clearly stressed to stimulate real engagement and inclusion.

One perspective is to assess the landscape as the entry point for evaluations. The aesthetic value of the landscape is perceived by all persons. It does not require a specific knowledge, being enough the "common knowledge", for entering and participating in the evaluation process. It provokes imagination, emotions, memories. It requires only qualitative indicators, being grounded on the subjective perceptions.

The TOPSIS evaluation model has been used in the CLIC concrete evaluation process in Salerno (Assari 2012; Carlsson and Zeleny 1983; Gravagnuolo et al. 2021, 2024; Lootsma 1999). Approaches grounded on LCA are required at the micro scale to check specific environmental impacts that are to be recommended and incorporated in the "circular evaluation model".

Figure 2.8 proposes an example of reuse of sites introducing six iterative/interactive steps for identifying a satisfying solution, balanced in economic, ecological, social dimensions. The final solution is identified combining and re-combining self-sustainable functions, sustained functions in new (symbiotic) relationships with the context thus including also near natural capitals. In the Fig. 2.8 two new green areas are added to better integrate the adaptive reuse of the existing cultural site into the context and to improve the multidimensional benefits, in an ecological perspective: in the perspective of a circular symbiotic "ecosystem" (see paragraph 5.3).

The evaluation process is oriented towards the adoption of ***a place-based and nature-based approach,*** that means to connect single points/areas in the space into a network of built and natural heritage: connecting them through synergies and circular/cooperative activities. Circularity means symbioses.

Evolutionary co-evaluation is a key step in the adaptive reuse, because it requires a sequence of choices in design and in the management process, for enhancing performances and productivity. The spiral is the image of the evolutionary circular evaluation approach. The image of the spiral can be interpreted as a vortex spiral, as in Fig. 2.7, but can also indicate the spiral of creative evolution: the growing spiral representing the creative dynamic processes in transforming the status quo (Fig. 2.8). Innovative digital technologies (IoT, AI, etc.) help in identifying all multidimensional impacts and thus the more satisfying solutions.

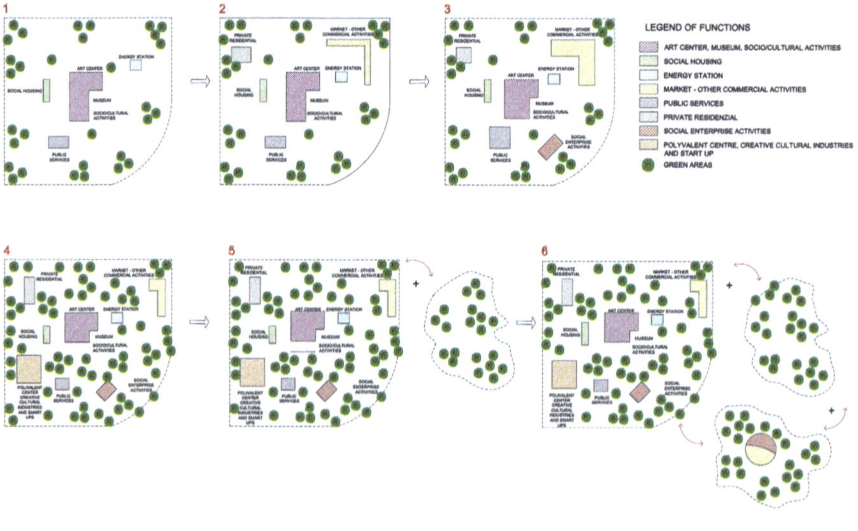

Fig. 2.8 The search of a balanced solution between self-regenerative, regenerative, symbiotic functions through iterative/interactive spiral steps

The evolutionary evaluation is a necessary process for designing the more satisfying reuse project.

It stimulates also a change of mindset. City dwellers should experience the benefits of this approach, less shaped by mainstream economics and more founded on the ecological/circular economy, starting with the definition of urban planning choices that are too often seen as the prerogative of technicians and politicians. In this way, plans/projects proposed from above can also find support from below. And above all, urban regeneration planning becomes the entry point for an "evolutionary policy" (in the perspective of regeneration of confidence between inhabitants and with institutions).

As repeatedly emphasised, here we want to stress once again that regenerative and adaptive governance integrates urban planning not only as interested in the regeneration of the *urbs* but also of the *civitas*. That is, the elaboration of urban solutions and urban management should become the entry point for the promotion of a civic culture: of a culture of active citizenship, because planning should no longer be the prerogative only of technicians and politicians. Evaluation and co-assessment processes become a central learning processes and training/education in the culture of active citizenship (Fusco Girard et al. 2023).

The spiral evaluation and participatory processes becomes a learning process which overcomes the culture of the here-and-now, i.e. the culture of general amnesia of memory, which today must be countered by the culture of circularity (between past, present and future).

Thus, the importance of these participatory evaluation processes is not only that of identifying new effective reuse solutions. The aim is above all to make the inhabitants aware of the not-linear but circular interdependencies between causes and effects about the issues at stake, and to stimulate their critical knowledge as well as their responsibility. The aim is to broaden the time horizon from the here and now that characterises current culture.

But participatory evaluation processes should be also oriented to change the structure of preferences, going against the traditional assumption by mainstream economics that preferences and their hierarchy are "given": that each subject is the best judge of its own utility and interests. In practice, the evaluation process, through the learning steps, is transformed into a primarily cultural and educational process for inhabitants, aimed at questioning existing preferences through public debate and thus developing a public spirit, collaborative/cooperative/public values.

The existing challenge of climate change and its progressive danger is precisely due to this assumption: that public institutions must accept and assume current preferences and values as "given". They should be modified and constructed (Hodgson 1995).

In this way, urban design and regenerative planning contribute also to the production of civic/civil knowledge.

From the combination of different multidimensional balance sheets, the strategy that best achieves the common good of the city can be deduced. The human-centred approach suggests a particular form of impact analysis to be added: a human centred matrix (see paragraph 2).

4.5 Adaptive Reuse and Creativity in Place-Making

Creativity is the fundamental ingredient of adaptive reuse to face continuous challenges and for managing its complexity: for managing/solving conflicts between needs, objectives, interests and values; for identifying a virtuous relationship between utility, social justice and beauty (Zeleny 1992). Collective creativity is stimulated by participation and public engagement.

The assonance of this approach to adaptive reuse with Schumpeterian Evolutionary Economic approach, characterized by the introduction of technological innovations, becomes more explicit.

Creative activities prefer in general to be localized in heritage assets. Adoption of new technologies better allow to produce actions in a coordinated/collaborative way, thus implementing the circular model. Digital technologies are key elements to be introduced not only in the energy sector but also toward the human-centred and circular paradigm implementation: to enhance the attractiveness of a place, its vibrancy and human interactions. The creativity of the adaptive reuse project is not only related to the reuse architectural project, but it is especially related to the management project. It regards the creativity of the entrepreneurs/managers. The entrepreneur decides to take on a certain investment and to run the relative risks by creating a new organization, through new solutions that transform the cultural resource into a complex of complementary systemic relations. This takes place by adapting the cultural man-made capital, elaborating a new order, a new organization, also through subsequent experimentation and thus "learning" from successes and possible failures, in the search for ever more performing solutions, through new combinations, involving also nature capital and intangible forms of capital.

Adaptive reuse is called to make choices always under conditions of increasing uncertainty, first of all because of the dynamics of demand market. This implies the need to make complex forecasts about the evolutionary dynamics of the demand for the various needs/goods/services in the territorial context where the cultural asset is localized.

On the other hand, the entrepreneur/manager is led to multiply the possible functional combinations in search of new synergies between different functions, activities and subjects that can mutually reinforce and consolidate over time, increasing the attractiveness of the "micro ecosystem" in which he operates. This means a continuous search for a new organisational architecture, which transforms the simple "collection" of component elements into a system of interdependencies through adaptive reuse: into a "circular heritage ecosystem" (Harper 1996).

Therefore, in the adaptive reuse, a continuous adaptation/reorganization/remodulation (as in living systems) is needed to transform a simple "aggregate" into a new "structure", with an innovative organization based on reciprocal and dynamic relations of complementarity, which in turn generate synergies and symbiosis grounded on learned lessons through which the reuse design is continuously re-shaped and improved through ex-ante, ongoing and ex-post evaluations.

At the end it is clear that CLIC research recognizes circular adaptive reuse as a complex process having to face with the dynamic of change, considering not only the above characteristics, but also some other elements: the position in the territorial context (in the city centre, with high potential of use value or in marginal poor areas, with no potential without connective infrastructures); the state of conservation and the intensity of transformation or adapting to new functions (with high structural costs versus low adaptive costs); the involved subjects: public, private, social. They all contribute to the complexity of effective adaptive reuse, considering multiple, multidimensional and in conflict objectives and criteria to be satisfied.

4.6 A Conflict Between Environment/Landscape and Adaptive Reuse Solutions

With the implementation of ecological transition strategies towards de-carbonization, conflicts may occur between environment/landscape and adaptive reuse: i.e. between the general context and the specific cultural assets. The environment provides a range of natural ecosystem services (food, water, fiber, biomass, clean air, microclimate, nutrients, etc.) from which it derives its particular "value" (Costanza et al. 1997). This is increasingly degraded by the various forms of pollution and climate-altering impacts that are increasingly incidental and capable of destabilizing the ecosystem cycles themselves (water, carbon, etc.).

Landscape is the human being's perception of the environment: not only of aesthetic and symbolic quality but also of meaning. It then becomes a shared vision. All the challenges of a certain territory are reflected in the landscape, from social inequalities to pollution, etc. In short, the landscape is a synthetic indicator of the health of an area, of its sustainability or unsustainability. In it one can read the winning and losing values and interests in their mutual priority.

A quality environment/landscape has a high transformative potential, because it determines attractiveness, quality of life, well-being and thus development (Council of Europe 2000).

With the implementation of the ecological transition strategy (i.e. new photovoltaic, wind, hydroelectric etc. installations) negative impacts on the aesthetic/perceptual quality of the landscape may occur. To what extent is one community prepared to give up this specific aesthetic/landscape quality in order to realize the general benefits coming from the de-carbonization processes, that are so necessary today?

Design choices require great attention and equally great creativity in changing the sources of energy to be used in the physical assets and production systems: how, how much, where, with whom, for whom to produce energy in an unconventional way. There is a need for the community, and not just for expert knowledge, to consider the complex and multidimensional impacts. We need real participation in these choices, taking a medium and long-term perspective: we need to decide together, avoiding easy simplifications. The introduction of innovative energy technologies should not be "juxtaposed" to the cultural asset, but technological innovation must become an original contribution capable of connecting old and new elements in new solutions/configurations. Integrated participative new evaluation tools are required to effectively support creativity.

5 The General Conditions for Success: The Circular Model for Adaptive Reuse

5.1 The General Issue

The general issue for the regeneration is about the question: how to transform a waste/abandoned/died area into an living and attractive system. The principles for an effective adaptive circular reuse can be summarized into: **the re-generative capacity**, **the symbiotic capacity**, **the generative capacity**.

Figure 2.9 proposes the three principles of the CLIC circular model, that are analysed more in detail in Fig. 2.10.

They are inspired by the nature economy (that is bio-ecology) by its perfect circular metabolism, by its adaptive capacity, by many symbiotic practices and regenerative/auto-poietic processes.

In this way, the circular reuse becomes able to close multidimensional loops between the cultural asset and its multidimensional context.

Many good practices can be viewed through these three principles, with different combinations between different regenerative, generative and symbiotic functions.

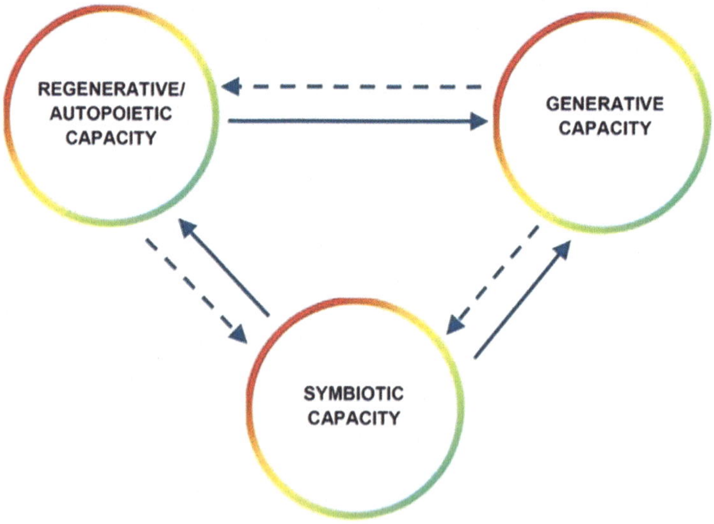

Fig. 2.9 The three principles of the CLIC circular model

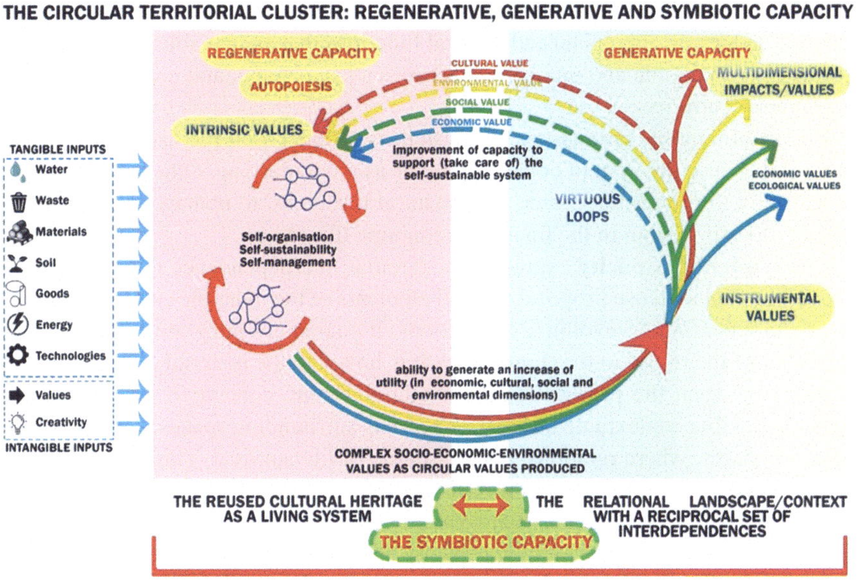

Fig. 2.10 The CLIC circular model: Tangible and intangible impacts from self-organization capacity, and loops

The "ideal" (Zeleny 1992, 1993) adaptive reuse is able to implement as much as possible the above three principles.

5.2 *The CLIC Circular Model for the Regeneration*

The above Fig. 2.10 underlines the organization of the **symbiotic/circular ecosystem/clusters**, with (many multidimensional) externalities (coming from the reuse of the cultural asset) and the relevance of symbiotic processes in the comprehensive context. It suggests that for the reuse of the heritage assets the functions should be chosen so that some of them can sustain themselves and thus they can also support some other activities. For example, in the reuse of a dismissed industrial site, some residential and commercial functions are justified for supporting social, cultural, educational, research, civic ones (not able to self-sustain themselves) coherent with the intrinsic value of the asset.

The figure distinguishes between different multidimensional impacts, character ized by feedback loops, reciprocal integrations, systemic interdependences which can transform vicious processes into virtuous ones, starting from a specific threshold.

The **re-generative capacity** is the auto-poietic capacity. It reflects the capacity to maintain over time the organizational structure of a **living system**: its identity and profile (characterized by a circular metabolism), with continuous activities for adapting through making adjustments because of degradation/decay and remaking processes. This regenerative capacity (grounded on the "intrinsic value") allows the "energy" to sustain also social and cultural functions that are not able to self-sustain. The available tangible and intangible energy is a key element and condition for this auto-poietic capacity.

The ideal tangible circular energy system is grounded on the use of renewable energies because the system could behaviour as a circular one. Other attributes are the capacity to re-generate water, other natural resources, to manage waste and the capacity to self-sustain in the financial/economic dimension.

The **symbiotic capacity** expresses the circular interdependence relationships of a specific adaptive reuse proposal with all elements of the dynamic external context, that require continuous evolutive adaptations. It guarantees integration and thus the durability of the re-use in the (long) time. It is linked to the material and immaterial relations between the physical asset and the systemic context: it guarantees the dynamic flexible contextualization of a site to its surrounding spaces, as in the natural eco-systems, where relationships are source of life/survival. Thus, the re-use of these particular (heritage) assets in-forms, shapes, re-shapes its surrounding environment (which is in its turn re-shaped and deformed).

The re-generative and symbiotic capacity also generates the capacity to produce multiple tangible and intangible impacts: the **generative capacity**.

Generative capacity is the multidimensional usefulness which an eco-system "offers" to its context (as "instrumental" values), multiplying its relationships. Positive tangible externalities are, for example: removal of air pollution, amenity of cultural natural landscape, sense of place, inclusion, education services, reduction in emission of greenhouse gas, in coherence with the priority of these goals recognized by European Union and by Agenda 2030. Also, the soil consumption is avoided, while the production of waste material is reduced. In the same time, new economic activities improve the local economy and the employment.

This circular model of reused territorial heritage asset is able to produce external effects that partially impact on the context and partially are able, in turn (in a circular perspective) to "come back" (from the context) to the ***cultural heritage ecosystem***. Some of them are economic impacts on the context that, in turn, become input again for cultural circular heritage place or ecosystem. These economic outcomes and values, in fact, can be "re-used" to support the activities included in the space/place.

The choice of multiple functions in design and management should be able to satisfy, as much as possible, the three above principles in the adaptive circular reuse.

The described circular adaptive reuse model of the cultural heritage is interpreted and managed in ecological terms, in the perspective of the Green Deal of European Union (to face the climate challenge) and of the New European Bauhaus. But also, as a way to improve the immaterial social infrastructure of the city, generating micro-communities through the management itself of the old heritage as a **common**, characterized by a specific value (an "intrinsic value", that reflects the particular spirit that has been connoting the site over centuries and millennia).

5.3 The "Circular Symbiotic Heritage Ecosystem" as Self-sustainable Model

What to do with the many signs of memory/history that are devoid of vitality today? How to promote a future for so many cultural assets, so many abandoned buildings or industrial plants or agricultural farms, or disused churches or abandoned landscapes?

The ecological transition suggests that the bio-ecological lens should be adopted in the regeneration reuse of cultural heritage. On the other hand, "adaptive" reuse is that which mimics the adaptive capacities of living organisms in light of the evolutionary dynamics of the context in which they are embedded (Zeleny 1992).

The adaptive capacity of a living organism, that is, its evolutionary capacity, depends on its self-regenerative capacity, the foundation of which is due to a circular organization of processes (the production phase is followed by the degradation phase and then the regeneration phase) (Zeleny 2021).

The adaptive reuse of cultural heritage thus finds its foundation in this circular/self-regenerative organizational structure.

The production of places as ***circular symbiotic ecosystems, localized in the heritage assets*** (in degraded areas) should be the "ideal" (Zeleny 1992) outcome of the regeneration/reuse of old un-used or died spaces, such as in abandoned industrial buildings. Here the three conditions of §5.1 and 5.2 are implemented. The reuse of cultural heritage goes as well beyond traditional tourism functions. For example, in old residential, port, industrial dismissed assets, common spaces for sharing experiences, ideas, knowledge should be multiplied, for testing new solutions, integrated with productive activities, thus attracting new skills, researchers, entrepreneurs, investments thanks to a trust environment.

In particular, the adaptive reuse/regeneration of degraded spaces in the city (starting, for example, but not only, from abandoned industrial buildings) should be incorporated more and more in the ecological perspective (that characterizes every living organism), to reconnect the built assets with the bio-ecological systems, in coherence with the evolutionary dynamics of nature. This means to become able to continuously re-shape circular city/territorial ecosystem in an integrated perspective through **nature-led solutions**, able to contribute to re-generate ecosystem services.

This ideal symbiotic/circular adaptive ecosystem assumes nature and biodiversity as the main infrastructure for the economic development.

At the same time, the nature-led solutions also contribute to the beauty of the built/natural landscape: beauty should be effectively stressed as a relevant contribution to the human-centred strategy. It affects the well-being perception of people. Beauty of the natural landscape is the outcome of the harmony of the ecosystem: of their regenerative, generative, symbiotic capacity.

The "circular heritage ecosystem" promotes a "learning community" able to continually assess and transfer good and best practices.

The "circular symbiotic heritage ecosystem" should become an example of Living Laboratory, as an incubator of circularity, able to implement the circular organization that guarantees a better metabolism, avoiding/reducing waste, negative environmental impacts, and greenhouse gas emissions. New business models are experimented, producing in the same time also a new mindset and new values: an innovative way of thinking that educate and inspire a new way to create economic wealth, closing the loops.

Thus, the circular symbiotic ecosystem becomes grounded also on a culture-led strategy for regenerating human relationships and lifestyles., and not only for generating economic wealth.

The implementation of circular city territorial ecosystems starts from the reuse of dismissed heritage assets for their transformation into "productive integrated environments": into creative places as "condensation nuclei of development" (Zeleny 1993, 2021), from which continually to learn. It becomes fundamental the capacity to continuously self-correct, that derives from the self-regenerative capacity.

The key condition is the capacity to identify some "engine functions" as self-sustainable ones, able "to sustain" other not sustainable functions, covering existing growing needs. Thus, these cultural ecosystems should try to mimick as far as possible the organisation of natural systems, transferring this organisation in the social/productive dimension, based on the circular model. Artificial intelligence helps to identify new symbiotic/circular relationships for increasing effectiveness: combining public, private and civil sectors into self-organizing symbiotic partnerships.

6 Toward Circular Governance, Business and Financial Models

All the above represents the conceptual grounds of the CLIC proposals in terms of principles of governance, business, and financial models.

The aim of CLIC is to offer some guidelines to actors, cities, public institutions, financial institutions, private bodies, social actors and cultural institutions in implementing local development through new models, grounded on the "**transformative potential**" of the heritage circular reuse.

The New Leipzig Charter (Informal Ministerial Meeting on Urban Matters 2020) and the New Territorial Agenda 2030 (European Commission 2021a) invite to recognize the key role of cultural heritage adaptive reuse in the transformative power towards the **common good** of the city (Informal Ministerial Meeting on Urban Matters 2020). The complex flow of impacts, coming from the adaptive reuse, is greater if compared to investments in other sectors (mobility, etc.) because of multiple and multidimensional impacts (in particular in the cultural dimension).

The five principles of the New Leipzig Charter (the promotion of common good as general objective of all urban policies, systemic approach on knowledge about interdependences, participation of all stakeholders, subsidiarity, production of places) should be integrated with a last sixth principle: the bio-ecological or nature-led principle, as in CLIC proposal.

CLIC research has assumed the above model in the perspective of circular economy. It requires the knowledge of all metabolic flows in adaptive reuse.

CLIC research introduces also the attention to **all** forms of metabolism and not only the material metabolism: also to the immaterial "hidden" aspects of the adaptive reuse.

This is why CLIC suggestions are linked to an interpretation of the new circular governance model, new collaborative business and financial models starting with the effort of enlarging the current world-view (that is determining the growing disconnection with natural system of life and among people). Thus, the new circular governance, business models and financial tools should reflect and promote also the regeneration of inclusion/integration values. They are not done, but they have to be regenerated with the same speed of their consumption. The notion of "*intrinsic value*" is useful in this cultural perspective: it assumes the interpretation capacity of people/participants to recognize the roots as the **essential meanings** incorporated into the heritage assets: as direction and boundary of the functional change (Fusco Girard and Vecco 2021).

6.1 Towards the Circular Human-Centred Governance

Figure 2.11 proposes the image of the "good government" in the city as in well-known fresco by Ambrogio Lorenzetti in Siena. It offers the perspective through which outcomes of CLIC research have been read in the evolutionary and human-centred perspective.[5]

Fig. 2.11 Ambrogio Lorenzetti, allegory of the good government, 13,381,339. Siena, public palace

[5] See chaper of Garzillo et al. in this book.

Today governance is facing with the ecological modernization of society, which is the greatest transformation ever imagined in history. But there is a context of growing mistrust: for example, between young people and governments (regarding the fight against climate change), between citizens and public institutions, among citizens etc. Here it is proposed a local experimental governance that takes as its entry point the valorisation of public spaces, starting with those with cultural values, to promote collaborative processes between citizens and institutions also based on self-organization.

Public government is usually considered as a tool for promoting reuse/regeneration in traditional governance. The general effect is to promote the perception of the common interest/good. Innovative governance model is related to the capacity to promote **self-organization through incentives, new regulations and specific investments.** Self-organization produces many values and in particular trust, cooperation, synergies and, thus, better functioning of public institutions and of society.

The implementation of the subsidiarity principle is a critical step in governance to regenerate values and self-organizating micro-community. Self-organization is the pre-condition for adaptive capacity and for dynamic evolution.

The subsidiarity principle (see art. 5 of Consolidated Version of the Treaty on European Union and the Treaty on the Functioning of the European Union 2012) is a key governance principle for transforming cultural, natural historic assets into common goods, to be managed through self-organization/self-management approach with specific "Pacts of Collaboration") (Rapporto Labsus 2021 2021).

Thus, cultural values and also public spirit, citizenship, responsibility are regenerated moving towards the implementation of the good governance and of the city "good life".

Subsidiarity principle of adaptive reuse stimulates the self-organization/management capacity (as many good practices confirm), thus orienting towards the achievement of the common good, and of an active civic culture. Young people and the "third sector" should be strongly engaged because they are very attentive to the ethical values of sustainability.

The adoption of the circular economy model, especially if supported by the "**culture of evaluation of results**", contributes to re-generate inclusion, cooperation, trust. Innovative governance requires innovative city planning and management. Innovative urban planning is a *"value-based planning"*, not only a reflection of the economic reductionist approach, linked to the neo-liberalism strategy. Urban planning and regeneration of old cultural assets should be reconsidered for achieving the common good of the city/territory system: to increase as much as possible the well-being of as much as possible number of inhabitants in the space and in the time.

The Fig. 2.12 shows innovative, regenerative, experimental governance that reflects the above characteristics of CLIC approach in the city/territory place-making: the ecological paradigm integrated with the human paradigm in a dynamic evolutionary context, in which innovative technologies are to be introduced. The New Leipzig Charter (Informal Ministerial Meeting on Urban Matters 2020), and also the Territorial Agenda 2030 stressed the notion of a governance oriented to "ensure the common good", and the "transformative power of the European cities", starting from the implementation of the urban heritage preservation/development.

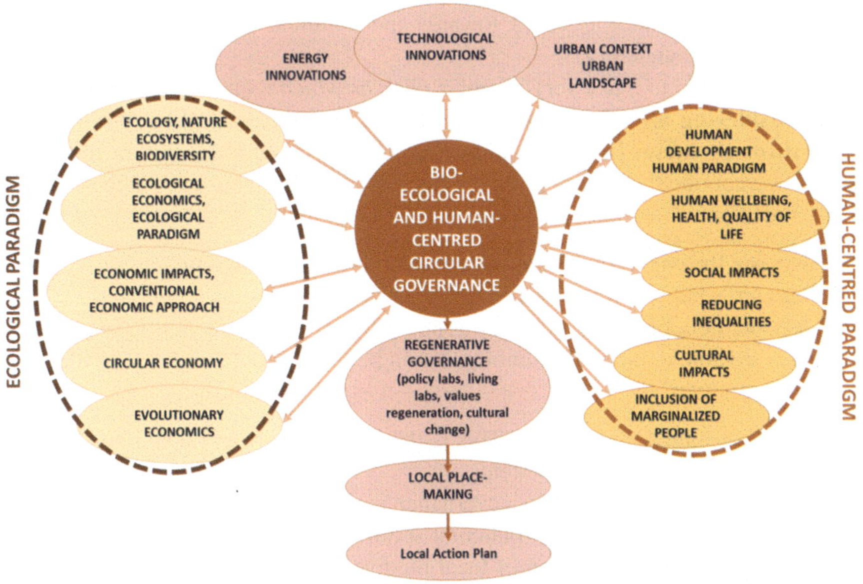

Fig. 2.12 The ground of the circular bio-ecological and human-centred governance

The CLIC research has demonstrated also through the ex-post evaluation of many good practices the potential transformative power of reusing the cultural heritage assets because of the main flows of net benefits, in relation to other investments (in mobility, waste management, etc.).

The investment in adaptive reuse employs more people with less specialization than in other sectors, contributes to the benefits of the tourism economy, promotes a "heritage communities" (Faro Convention) (Council of Europe 2005) characterized by ties to a cultural asset that can transform themselves into interpersonal relationships of collaboration and self-organization (as occurs in shared management based on the principle of subsidiarity).

The key process of the innovative circular governance is linked to the spiral co-evaluation process **from bottom up towards top down** (see §7.3, 7.4 and 8) and vice versa.

Figure 2.12 is related to the ground of the new bio-ecological, human-centred circular governance. The bio-ecological paradigm suggests first of all to introduce not only a new sixth key principle in the list proposed in the New Leipzig Charter, that are all referenced to the human-centred paradigm. To face the climate change challenge, it is needed also to introduce technical innovations in particular in energy sector (reducing impacts ageing both on causes and on the capacity to reduce the negative impacts) and in artificial intelligence.

Here we can remember the conclusions of a ETH Zurich research (Bastin, J.F. The global tree restoration potential, Science, 2019, 365) about the need to use tree replanting as a good effective strategy for CO^2 sequestration to reduce the negative impacts of global warming (European Commission 2023; Nowak et al. 2021).

The new governance oriented towards the common good requires strategic choices values grounded.

Today, we need to make very complex choices which require to **think in a new way, considering all values and not only the instrumental ones. We have to think through values (in urban planning, in landscape planning, in management)**, overcoming the current economistic approach. New strategic choices should be grounded on values, because (Keeney 1996):

- Values are the energy/force able to drive choices and behaviours
- Values are what we want to take care, because we perceive as important
- Values are not objectives, goals, criteria, attribute. Values are their ground
- Values are able to facilitate communication between people, and also to identify new alternatives

Values are consumed during time. This is the reason why they should be regenerated, as happens in natural ecosystems.

In conclusion, the new circular governance is characterized by a strong attention to promote first of all nature-led planning solutions: new urban paths, forests, soil de-permeabilization, valorisation of local biodiversity, with strong attention to the climate change issues and strong capacity to interpret adaptive reuse as a real opportunity to face global warming impacts, the capacity to integrate the challenge of climate change with challenge oriented to reduce social inequalities and marginalities. The promotion of cooperative/collaborative relationships between all different actors of the civil society, of the private subjects, of public bodies for a better management of multiple and conflictual interests and objectives contributes to promote also community and thus the humanistic transition. A key role of evaluation process is to be recognized for improving decision-making processes from bottom-up and not only from top-down, thus identifying agreed satisfying solutions.

A strong attention to the aesthetic quality of the urban environment in coherence with New European Bauhaus is needed for contributing to the ecological and humanistic transition. This aesthetic quality is achieved through multiplying symbioses as it happens in nature ecosystems. Symbiotic principle can be transferred from nature to reciprocal social relationships, improving complementarities and thus cohesion between inhabitants and communities. The beauty should be interpreted also as the product of the density of the interpersonal relationships, and not only in the aesthetic dimension.

However, the beauty of landscape is the entry point for enlarging the participation of people because it does not require a specific knowledge.

6.2 Towards Circular Regenerative and Human-Centred Business Model

Also current business models should be modified first of all recognizing the perspective to be "in transition": the business plan should be organized in a ***dynamic and evolutive*** way, to better fit/adapt in the existing changing environment, also for improving resilience (Saleh and Ost 2023).

In the circular model each firm does not act in isolation, but it searches possible collaborations with others in a systemic perspective. Digital innovative design can stimulate innovative circular business: they require circular knowledge and specific information about reuse/renovation/recycling.

In traditional business models, grounded on market prices and market costs, all natural resources (as forest, water, flora, etc.) are used. But the cost of this consumption is totally ignored because these resources are unpriced (Saleh and Ost 2023).

New adaptive circular business models, first of all, have to recognize this key mistake. Profit achieved damaging the commons (represented by natural ecosystem that supports human activities) should not still included (Porter and Kramer 2011).

Emissions into air, water and soil are considered together with waste.

Another attention is to specific evaluation process: for example, to "*life cycle assessment*" and "*metabolic flows evaluation*". The image of the spiral evolution should also characterize the new business approach with spiral evaluation processes with incremental steps.

A key aspect is the specific capacity of the new circular business model to really improve the trust of consumers and of stakeholders.

The circular business model introduces innovations on a number of points in particular: on how to recover/regenerate the natural resources used (water, materials, energy, etc.); on how to recover and recycle waste or by-products by considering them as inputs for new production processes; on how to improve the quanti-qualitative metabolism between input and output. For example, with decrease of the water (coming from the climate change) a key issue is to regenerate (potable) water after its use in the consumption and production processes, and how to re-introduce it in the city water network.

The circular business model is therefore highly integrated, being aimed at promoting new collaborative partnerships to improve competitiveness of the network in the market. One of its characteristics is that of going beyond the specialist as "vertical" approach, adopting an open or "horizontal" approach to the valorisation of possible symbiotic relationships of interdependence (in imitation of symbiosis in nature). It encourages reasoning from the perspective of building an ecosystem in which each subject perceives itself as an effective component. From the ability to collaborate/co-operate follows the ability to co-ordinate reciprocal activities as well as the promotion a network of micro-communities of enterprises based on reciprocity of behaviour, perception of co-partnership, and a shared condition of trust.

The above means also to make changes in the existing *status quo* taking care of the territory: incorporating the bio-physical space into the economic business models. Innovative business models have to be implemented not in the a-spatial dimension, but in a concrete city/territory, taking care, at the same time, also of the social environment, that is of the social space (and thus considering possible collaboration). The above elements are linked to the responsibility toward society (as Olivetti (Olivetti 1960), Bat'a introduced (Bat'a 2011). The profit is considered also in relations to the achievement of the general interest, towards the health of the whole system.

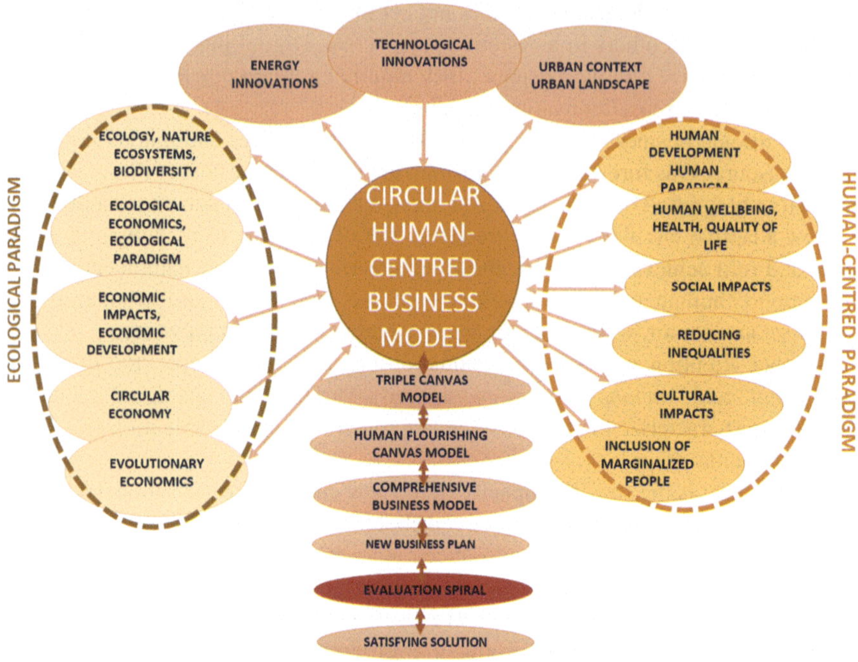

Fig. 2.13 The ground of the innovative circular business models

The attention moves towards the local/regional scale and territory (which provides wood, food, fresh air, energy, with many regulating services, habitat services and cultural services). Innovations and innovative technologies, such as AI, etc., become the possible integration for comprehensive impacts in the environmental, social and economic dimension (see Fig. 2.13): for identifying new symbioses transforming some competitive relationships into cooperative activities.

Time includes the short, the medium and the long term, during which all the potential of new relationships between the single activity and other activities/subjects should be brought into symbiosis, taking into account the three types of profit: economic/financial profit, social profit, ecological/environmental profit. Ultimately, the above implies a business model based on the interpretation of entrepreneurial activity that considers the adaptive reuse of cultural heritage towards a "living asset" as a "self-reproductive asset": which tends to become regenerative, self-organising, self-reproductive and thus capable of continuous new synergies, thanks to the adaptive capacity to change.

The "return" to nature suggests comparative measures for balancing CO_2 production (for example in terms of new planting, the restoration of natural ecosystems as well as the respect of natural ecosystem services and new symbiosis) (de Groot et al. 2010, 2012).

6.3 Towards Circular Human-Centred Financial Models

The circular economy model requires circular funding solutions for the access to implement circular projects. Local financial institutions have a key role to play for sustaining the circular model (Borrero Ochoa 2011; Smit et al. 2014; European Investment Bank 2018).

A well-known circular way to fund investments is the land value capture, through "recover" land value increases coming from a public investment, thus returning values from private to the public body (OECD 2022).

It can be used also in the reuse of cultural assets or in the regeneration of historic district (Pickerill 2021 and Chap. 19, in this volume). But the list of existing financial tools is rich.

To stimulate investments in adaptive reuse, **new financial hybrid circular instruments** have to be proposed combining market driven impacts with crowdfunding platforms with other specific tools. The Triple Bottom Line well known approach (Elkington 1994, 2018) influences new financial tools, stimulating new indicators research.

The challenge is to shape a favourable environment to stimulate private for profit and also not for profit, circular financial tools able to generate recycled financial flows. The green finance should be in particular attentive to include social subject that are in general excluded from the current financial system. It should be able to better stimulate for example reforestation, biodiversity conservation, wetlands restauration in circular heritage reuse, considering the transformative power of the finance dimension.

Some general conditions for financing an adaptive circular reuse project by a (local) financial institution can be synthetized here:

1. the effective integration of the building historic assets into the context and the natural environment (symbiotic principle);
2. the conservation of the permeability of the soil (regenerative principle);
3. the contribution to the biomass (intense trees planting integrating the reuse of cultural man-made assets);
4. the reuse of rainwater (regenerative principle);
5. the use of renewable energy (regenerative principle);
6. the use of natural light (regenerative principle);
7. the use of local materials (regenerative principle);
8. the use of recycled materials (regenerative principle);
9. the choice of functions able to sustain other functions (generative principle);
10. the number of new employed (generative principle);
11. the contribution to the production of a young and old persons community (generative principle)
12. the co-production of a citizenship culture (generative principle)

For each condition a specific set of specific and effective new indicators is necessary to be identified, also for a continuous monitoring.

Fig. 2.14 Circular human-centred financial models

But these general criteria should also be linked to the introduction of innovative technologies for improving the self-sustainability, thus minimizing the external financial supports; to the capacity to organize systemic complementarities; to the regeneration by the project not only of new employment, but also of social capital production (trust) (see Fig. 2.14).

In any case, it is important that a general environment is created to reduce speculative processes, to improve the quality of heritage conservation rules. Forms of **"revolving funds",** as preferably public "revolving circular funds", are effective for adaptive circular reuse, for financing PPP (or PPPP), for financing debt (see chapters of Pickerill (Chap. 19); Lupu and Allegro (Chap. 20), in this volume).

Many existing and effective case studies produce evident/base proposals for identifying new financial specific tools.

The premise is that a current capital market and asset pricing system have to be redesigned within a circular perspective. The place-led financial tools can vary from "impact investing" to blended finance, to PPP (or PPPP), to urban value capture tools to crowdfunding, to grants. Social enterprise should be stimulated to take part in this perspective.

The "Circular City Funding Guide"[6] provides elements on financing and funding circular projects.

[6] https://www.circularcityfundingguide.eu/

7 Some CLIC Governance Proposals Towards the "Ecological and Humanistic Transition"

In conclusion, the triple model grounded on "regenerative", generative and symbiotic principles provides the CLIC circular adaptive reuse strategy and thus desirable circular governance strategy. It shapes the design and the management of the reuse of cultural assets too. It is a guiding model for complex choices in governance, business and financial tools considering all tangible and intangible aspects toward the integration of reuse into development strategy.

The circular human-centred adaptive reuse of cultural heritage should be shaped considering the following aspects and recommendations for passing from the general principles to more operational aspects, coherent with the specific and different rules/regulations existing in the countries of European Union:

1. The **adaptive reuse of cultural heritage** should be framed in the perspective of the **circular natural ecosystems model.** This vitality is grounded on their adaptive capacity during the time: that is, on its turns, grounded on their autopoietic self-regenerative capacity.

   The adaptive reuse of cultural heritage represents an effective entry point for the circular city implementation, for contributing to improve the urban comprehensive metabolism (as many practices can confirm).
2. The adaptive reuse of cultural heritage has to be interpreted into the **ecological transition** linked to the European Green Deal. Interpreting the adaptive reuse in the perspective of the ecology/biology means to consider the adaptive reuse linked to the return of "nature" in the city: nature as the most important infrastructure of the city, able to sustain all human activities through ecosystem services. This means connecting adaptive reuse to the CO_2 capture, water management and recycling, use of renewable energy, circular management of waste. Green spaces, green surfaces and other nature-based solutions are considered fundamental in the circular city (European Commission 2023). This source of the adaptive reuse means also the ability to promote a regenerative capacity. The regenerative capacity regards not only the natural and manmade capital, but also the human-social capital. The living assets are critical elements for the successful reuse of the cultural capital: for creating a new life, avoiding the depreciation and the death.
3. All impacts of climate change should be assessed at local level for identifying their costs. Adaptive reuse reduces these costs. Mitigation plans become the general perspective of all adaptive reuse projects.

   Adaptive reuse is more effective if it is incorporated in the regeneration of urban historic districts and in the local urban planning. The Local Action Plan of reuse should move in this direction.

   The **local regenerative planning** should be grounded on the ecological dimension (on primary intrinsic ecological values) to improve the city metabolism. Each adaptive reuse should be considered as an action against the climate change impacts. Adaptive reuse should thus be coherent as much as possible with mitigation and adaptation plans and with urban regeneration projects. Specific incentives should be offered to inhabitants to check bottom up the achievement of circular model.

4. In many local planning rules there are not included focused incentives for the reuse of building materials coming from demolition. Specific incentives should be offered for sustaining the reuse at micro scale starting from the self-reuse (in the same building site) of particular materials as bricks, stones, pavements, roof tiles, iron beams, wood beams etc. on the basis of specific performances.[7]
5. The circular economy model should be interpreted not only in terms of quantitative reuse, recycle, regeneration of hard components and materials, but also considering immaterial soft ones, for really improving choices today. The circular economy understands specific cooperative/inclusive values that are to be regenerated today in coherence with the cultural condition of sustainability. Urban metabolism considers water management, waste management, energy management, etc., but also it should consider the entry and the production of immaterial infrastructure, with cultural, symbolic components and values. And first of all with a strong attention to future.
6. The adaptive reuse of cultural heritage should be integrated in the **"human development"** strategy of the city/territory. It means to frame the adaptive reuse in the strategies to reduce poverty in all its different dimensions (employment, new jobs, knowledge etc.), for the reduction of social inequalities, for inclusion—also with reference to future generations. Community building linked to heritage sites should be strongly promoted, thorough specific "collaboration pacts" (Labsus 2014) according to the **subsidiarity principle.** The **subsidiarity principle** of the Treaty of European Union[8] is the essential starting principle for a governance to improve citizenship and for implementing the **"common good",** delivering the full potential of urban areas (See the Pact of Amsterdam, 2016 and the New Leipzig Charter). The "Pacts of Collaboration"[9] (based in Italy on the Constitution Chart, art. 118) are a good example of the regulatory instrument to generate a "management community" of the heritage assets, turning them from a public good (or private good) into a "common good". These "Pacts" are characterized by and move towards a culture of trust between the subjects involved (private, public, and "people" partnerships – PPPP).
7. The adaptive reuse interpreted as the fundamental element in the **generation of "places"** suggests the transformation of public spaces into areas characterized by a particular attractiveness (visual, economic, social, civil, environmental, aesthetic). The generation of "places" should be characterized by a network perspective, linking one place to many others also of natural capital (see § 5.2).

---

[7] In general, the prices of recovered materials are not specified in the price list and in some cases the required performances imply costs that overcomes the ones of new materials. In the same time, for all reused materials the VAT should be eliminated.

[8] https://eur-lex.europa.eu/legal-content/EN/TXT/?uri=celex%3A12012M%2FTXT

[9] Labsus Organisation, "Pacts of Collaboration" (in Italian) https://www.labsus.org/cose-un-patto-di-collaborazione/

The ***circular symbiotic cultural ecosystem*** is the "ideal" project of adaptive reuse in the city regeneration (see § 5.3). The symbiosis between public, private and civil subjects should be supported through specific incentives oriented to close the loops of energy, water, natural materials, waste; and as an example from which to learn how (small) businesses, creative entrepreneurs, new start-ups, social entrepreneurs, artists can inspire a new way of working. And also, a new way of thinking: as an example of continuously evolving Living Laboratory towards the implementation of the city circular development.

8. The adaptive reuse of cultural heritage should guarantee/regenerate not only authenticity but first of all the **"spirit of the place". I**ts **"intrinsic value"**, deduced (at least) from the perception place attachment, pride, co-ownership of the place is the reference element for manging the change of uses/functions.
9. In the relational/reconnecting perspective in which to interpret the adaptive reuse, a fundamental importance is the **choice of new uses/functions** to be introduced. From the choice of functions, from their intensity and reciprocal combination, depends the new vitality of the heritage site and its capacity to generate virtuous circles, thus to **transform a site lacking vitality into an attractive/centripetal forces field**: a site characterized by the capacity to become an attractive "pole". These functions should be chosen on the base of the above three general principles (see Fig. 2.9 e 10): symbiotic self-regenerative and regenerative capacity. Regenerative principle is the ground of adaptation change, releasing the "energy" for sustaining vitality. Symbiotic principle guarantees circularity, transforming heterogeneity into systemic unity. Generative capacity guarantees the satisfaction of needs.
10. Technological innovations become of fundamental importance, especially **digital technologies** and **energy technologies**. They should be strongly supported by specific public incentives. Public institution should become able to better assess the impacts of these technologies on natural and built landscape.

    Digital technologies, through sensors, cameras, IoT, automation, AI, etc. allow to constantly monitor flows in entrance and exit, and thus the metabolism at micro level. They allow to generate data and information on environmental conditions, on health, accessibility, etc. so to help all subjects to enhance their decisions on mobility, logistics, production of goods and services. They also allow to monitor and enhance the capacity to adapt to the context and to learn. Furthermore, they allow to localize specific uses/functions in marginal areas enabling remote working, reducing in this way the marginality of many remote areas. For example, "digital villages" can reduce the depopulation processes offering new opportunities to work in little towns. Artificial intelligence can contribute (if well-oriented) to the good governance, identifying new productive/effective solutions in implementing the circular model, in searching new hidden/potential symbioses, thus transforming the competitive approach into a cooperative one, in the general interest. Digital technologies are also employed into energy technologies innovations.

Energy innovative technologies allow to reduce energy consumption, to use renewable sources of energy (also in the perspective of using green hydrogen), to filter and depurate in a botanic way polluted air. They should be combined with natural ventilation systems. Innovative materials such as photovoltaic ceramics tiles, photovoltaic roof tiles should be employed to integrate renewable energy in heritage buildings reused. Digital innovations support the recycling/reuse/regenerative economic model.

11. New governance is shaped by culture. New governance promotes new circular relationships between people and the man-made assets using technological innovations and should also be oriented to avoid the **ambivalent/negative role of digital technologies. They can promote socialization, but also de-socialization.** They reduce often the capacity to distinguish the real information from fake news, hindering the exercise to hierarchize and the recovery of memory. On one side, they allow to build communitarian relationships and to work from remote areas remaining connected to knowledge communities, but on the other side they can generate isolation and "dehumanization".
12. Cultural heritage reuse should be considered as an important booster for the creation of "real life" communities, in which the multidimensional nature of human relationships can be expressed. The "energy communities" for producing energy from renewables can be considered as a good starting step to be supported towards the connection of local community. Other communities can be stimulated through a governance oriented to implement co-planning, co-programming, co-design, co-management the heritage assets as common goods. This governance, grounded on the subsidiarity principle, stimulates self-organization capacity and public community spirit. These communities are linked to the specific heritage site, to the spirit of the place, and contribute to **"human-centred" circular cities**.
13. From the complexity of the adaptive reuse interpreted in the ecological perspective, it comes the necessity to develop specific **regulatory innovations to foster mutual relationships and community building**. Adaptive reuse, urban regeneration and citizenship are to be considered as interdependent. Adaptive reuse requires real public participation in the choices of new uses and in management, thus stimulating the birth of new communities.
14. The **circular governance of cultural heritage contributes to co-planning and co-design the special transformations, fostering trust**, which is based on transparency, truth, and thus responsibility. In this sense, the reuse is associated to the capacity of **critical analysis,** public data/discussion and evaluation. This represents another element for the **humanization of cities**, expressing the creative capacity of people and communities able to develop **dynamic, evolutionary co-evaluation, through self-organisation, self-management, self-government: thus active citizenship, (together with jobs for young people).**
15. Beauty is the general characteristic of the natural urban landscape. The beauty contributes to the quality of life. The perception of the aesthetic values does not require specific scientific attitude in the sense that all people can take part to

deliberative processes. All people can "meet" the notion of the general interests, participating in public choices. The adaptive reuse of cultural heritage should aim to valorize and re-generate the value of **"beauty"**, also in line with the New European Bauhaus approach, as a particular contribution to the humanization of cities. Beauty is the outcome of the harmony of circular flow of symbiotic interdependencies, and in its turn is able to generate new virtuous circuits.

Beauty is also the outcome of the harmony in the social dimension: of inclusion, capacity to take care, to cooperate together towards shared objectives/values.

The aesthetic quality of the physical and natural environment fosters a sense of interpersonal openness, reducing each person's potential for conflict with others, and thus promoting behaviours of respect, unselfishness.

Beauty is a useful entry point for other dimensions, including the ethical dimension and the common good achievement. The beauty of places, of public spaces, should become, more in particular, interpreted as the entry point for a ***"civil aesthetics"***, for promoting creative syntheses between personal interests and general common interests (Arena 2020).

16. In the dynamic-evolutionary perspective, the **circular business model** is characterized by its capacity to be in relationship with the territorial context/society. It has to improve adaptive capacity to make resilient the economic activity. Profit should be reinterpreted as **economic profit along with social profit, ecological profit, and also civil profit**. The enterprise is in circular relationship with the city (as in the experiences of Olivetti in Italy and Bat'a in Czech Republic) and shapes (and is shaped by) the environment. These spatial and multidimensional impacts should be recognized and assessed in the new business model, opening to a multidimensional "profit and loss" balance of the economic activity, and opening up a larger perspective of interrelations also with the third sector/social enterprise. Interactive/interactive and spiral evolutionary process characterize all the circular new business plans. It becomes a statement of the multidimensional values towards the stakeholders and the territory citizens and investors. New collaborations and symbioses can be introduced in circular businesses, also thanks to AI. A new governance should take specific and effective initiatives for incentivizing circular business.
17. The **circular financial model** for cultural heritage adaptive reuse has a significant transformation power, as green finance already can confirm (in empowering nature to enhance resilience). It should be (also) grounded on the assessment of the impacts generated by the funded/financed project, assuming an experimental and evidence-based approach. Social and impact investors, as well as the public sector, already try to identify specific measures to assess the impacts of supported projects, "blending" social and financial return on the investment. However, in light of the EU Taxonomy[10] and sustainable finance initiative at

[10] EU Taxonomy https://finance.ec.europa.eu/sustainable-finance/tools-and-standards/eu-taxonomy-sustainable-activities_en

European level, all businesses and financial initiatives should ensure that social goals are reached through finance, and follow the principle of not only doing good, but also "do not harm". Thus, the new governance should focus on this perspective, highlighting the necessity of **assessing impacts** in the ex-ante, ongoing and ex-post phase of the investment. Finally, circular financing models are aimed to **regenerate financial resources** in the territory, avoiding places "consumption" from a financial point of view—by including diverse local stakeholders and co-investors, even enlarged to the entire community through hybrid models of community enterprise, community foundations, and community finance/crowdfunding.

18. New governance, business and financial models require in any case innovative **hybrid (spiral) evaluation tools** for their implementation (see §4.3 e 4.4). All impacts (on economy, on environment, on landscape, etc.) are to be assessed through new effective performance indicators for "demonstrating" the added value of cultural heritage reuse in the circular model and its priority in relation to other investment sectors (Fusco Girard and Nocca 2019). The perspective of humanization suggests to integrate the above indicators with subjective ones, for expressing the quality of life perception (see §2), before and after the change. It suggests also to improve the engagement of all actors and beneficiary people in the choice processes. The balancing of interests and values and priorities should become very clear.
19. The benefits of the adaptive reuse are economic, financial, environmental, social, cultural, aesthetic. They should be systematically recognized and assessed including intangible impacts for contributing to the **"transformative power of the city"** (Informal Ministerial Meeting on Urban Matters 2020). The real added value of reuse is intangible. For example, it is represented by the generation of **relational goods**, by **the capacity to generate a community**. It is represented by the promotion of a **culture of cooperation, collaboration, complementarity**: of a culture of "us", and of long-time horizon. In this sense, the circular economy model proposed by CLIC is not only based on waste and materials management, but it is much larger, including also the immaterial/cultural dimension. For example, the circular economy model that inspired CLIC is related to **industrial symbioses**, which is based on a complex relational system, and not only the recycling of materials (which is one important but limited technical aspect). Circular economy evokes and underlines the long-term perspective and the relational dimension.
20. The local governance is asked to integrate the adaptive reuse of cultural assets into a **local "cultural strategy" towards the regeneration of relational values** involving schools and all educational institutions and associations. In other terms, circular model allows to link past with present with future time and not only elements in the space.
21. The circular heritage symbiotic ecosystem is the CLIC proposal for regenerating the life in a dead site. The bio-ecological lens allows to recognize that nature is characterized by a plurality of ecosystems in which every living organism lives into a community. But nature also shows a holistic behaviour grounded on

the circularity of reciprocal interdependences. The circular heritage ecosystem is conceived as a "trust ecosystem" able to produce virtuous circular processes between utility, beauty and inclusion. It should be planned, designed and managed as a living system integrated in the society and nature, linking human and social capital with the territory, imitating the nature symbiotic experiences. Its organization, going beyond traditional tourism approaches, regards first of all, but not only, the abandoned industrial building areas (see § 5.3).

22. **Religious cultural heritage** represents the typology most strongly present in UNESCO sites. It represents a sign that strongly characterizes the European landscape. With the demographic processes of decreasing and ageing and the rarefying of traditional religious practice, new uses must be identified for these properties once destined only for worship. These new use values should be consistent with the "intrinsic value" of religious heritage (Fusco Girard and Vecco 2021). It is necessary to research which new use values are the most compatible in a perspective that is as participatory as possible with the local community, i.e. with voluntary associations, civil society movements, church groups, etc. In the perspective of the circular economy, these new uses should allow the cultural heritage to continue to 'exchange' cultural, social, symbolic values, meanings. With the context in which they are inserted. It should be emphasized again that in the circular economy perspective it is not only a matter of reusing/recycling material resources but also of considering the role of intangible resources capable of regenerating links with the territory. It can also be noted that the reuse of disused religious heritage, located in inland and generally in poor areas, can be effectively integrated into a network (for example in the valorization of cultural itineraries and cultural routes).

23. The circular reuse of **cultural farm buildings no longer used for agriculture** regards in general the ones localized in the countryside and in particular in inner and marginal areas. The above requires the organization of a specific ecosystem: a self-sustainable system through the organization of systemic complementarities. This auto-poietic system has the "energy" to sustain also others components/subjects/activities through its generative capacity, in a symbiotic reciprocal relationship with the context.

    The agro-food sector is characterized by high level of water and energy consumption and by production of many waste and climate change gasses (due to transports and long chains of production/consumptions). Short loops become necessary together with the recovery of meteoric water, the use of renewable energies and local materials, recycling all waste from the agricultural production. In the agro-ecological symbiosis, in which energy for production and transportation is the biogas produced through animals' manure and field biogas, fertilizers are produced through organic matter employed from grain and vegetable production. Here the revitalizaion of man-rural areas is achieved through a new systemic organization: through reshaping the food production, closing the loops, connecting producers and consumers, promoting thus a bottom up "food community", integrating the participation of academic subjects with the citizen knowledge.[11,12] The agro-ecological symbiosis of Palopuro in Finland[13] can be considered as a good practice in this direction. Inland areas are characterized by some potential benefits: the beauty of cultural/natural landscape, rich biodiversity, pure air and water, low rents/prices, social cooperative environment, clean available renewable energies. They offer specific opportunities already recognized in the European Commission (2016) and in European Commission (2018). These documents suggested specific approaches to rural development, for attracting new activities and people, grounded on the reuse of local potential resources and on the development of nature-led, community-led and eco/spirituality-led approaches (as in the experience of Findhorn ecovillage network). However, it is today necessary to integrate these recommendations within the circular model, interpreted in terms of reuse, recycle, regeneration of all forms of waste coming from rural agriculture, agrifood production and zootechnical production. Rural regeneration strategies suggest to use all wastes for producing new forms of resources/jobs: for example, transforming waste from agriculture into new biomaterials, and so on.

24. Depopulation, migration, ageing of people, the de-localization of activities generates the decay of **cultural landscapes.** New investments in natural capital, manmade, human and social capital generate new virtuous loops between the quality of landscape and the local attractiveness. Re-activating in this perspective internal landscape subject to abandonment, depopulation and underuse means in particular to reconnect human beings to nature, and re-building the symbiotic relationships among them into a systemic perspective (Fusco Girard 2021). Nature-led solutions, supporting climate mitigation and enhancing resilience (with specific reforestation, tree planting) become one of the fundamental components of the local economy development towards the human-centred approach, interpreted here as a return to solidarity with nature. The many symbioses that we can recognize in nature inspire for the organization, planning and management of new regeneration/development projects (Fusco Girard et al. 2023).

    Nature-based solutions conserve and regenerate biodiversity, thus becoming one of the fundamental components of the new development strategies, because of their multidimensional benefits, including economic ones (Roe et al. 2021).

    Cultural Routes are also in this case a particularly interesting strategy for the development of inland landscapes based on the promotion of relational values.

[11] Helenius J, Hagolani-Albov S E, Koppelmäki Kari (2020) Co-creating Agroecological Symbioses (AES) for Sustainable Food System Networks. Front Sustain Food Syst 4:588715. https://doi.org/10.3389/fsufs.2020.588715

[12] Koppelmäki K, Lamminen M, Helenius J, Schulte R P O (2021) Smart integration of food and bioenergy production delivers on multiple ecosystem services. Food and Energy Security 10(2): 351–367. https://doi.org/10.1002/fes3.279

[13] www.palopuronsymbioosi/english/

Cooperation and collaboration at local level are key values to be stimulated and promoted through specific actions.

The success of Cultural Routes is based on landscape valorization. Landscape means natural landscape but also man-made landscape, such as towns of art, etc. This perspective of "landscape" offers an integrated, systemic/holistic, and human-centred approach to promote territorial transformations/development (UNESCO 2005, 2016): the landscape as the key complex systemic/territorial/ resource of Cultural Routes. Landscape becomes the fundamental element of identity/specificity that can make a place different from other areas, through elements of uniqueness, integrity and authenticity. This uniqueness can become a catalyst first of all for cultural tourism activities, especially creative/innovative ones.

Cultural Routes are a values repository of great importance today. They consist of a group of small villages, castles, forests, rivers, lakes, settlement fragments located (most often) in "inland areas" and thus originally characterised by isolation, abandonment and oblivion.

Putting these heterogeneous resources in mutual synergy through specific networks, within the framework of cultural routes characterized by historical, artistic and religious values, represent a development proposal that is entirely consistent with the current ecological transition strategy, because they do not determine climate-polluting impacts.

In order to realise new relational values on which the variety and therefore the identity of a cultural route depend, a suitable infrastructure strategy is needed. Light or green infrastructures represent the "backbone" structure of Cultural Routes. They are tangible and intangible network infrastructures through which a unitary strategy is realised, capable of activating new symbiosis and thus the breaking of various isolations and fragmentations.

The Cultural Routes should contribute to "making culture". That is, they must contribute to the fruition/enjoyment of culture, communication of culture, production of culture. The aesthetic dimension is the entry point for promoting culture in its aforementioned three perspectives and for enlarging participative experiences.

The specialist contributions in the next chapters make operational the above recommendations. They should be integrated and reshaped in each European Country to become operational/effective: new CLIC National researches should be improved.

8 Discussion

In the general context of CLIC research, the role that the active reuse of cultural heritage can play has been highlighted under many perspectives. It is the source of a set of benefits that can be enhanced if the adaptive reuse is interpreted on the basis of the three general principles already outlined in Sect. 5.

In the above perspective, Geddes, Genovesi and Schumpeter have "many messages" to tell us still today about the Horizon research issues and the required new governance models, in our time of climate change.

Schumpeter, promoting of "the need of change", offers directions still relevant today for adaptive reuse being structurally linked with change and the need to introduce innovations. The creativity is a key ingredient for combining ancient and new assets, meanings and values and to improve technological infrastructures and also the immaterial infrastructures. He also suggests to stress the cultural dimension.

Geddes, recognizing that "plants are the foundation of life on Earth" offers a fundamental perspective for the needed ecological transition essential to reconnect the city to nature: the health of the ecosystem is the foundation for the health and well-being of all human beings (Geddes 1911a).

Genovesi with his emphasis on "relationships" becoming social glue, that is, a force for social cohesion and thus an engine for development introduces the "humanistic transition".

For both, the well-being and the happiness represent the real goal of economics. And for both relationships are critical elements for the (city) good life.

Geddes proposed the "biological view of economics", recognizing more than a century ago that the conservation of nature "is surely the very foundation of any utilitarism…" and "….. be a main aim of higher utilitarism (or "true utilitarism"). There is no wealth but "life". "Practical economics finds its supreme end in the maintenance and evolution of humanity" while the aims of practical economics are in terms of quality of life (Geddes 1934). In many chapters he stressed the relationship between urban regeneration and citizenship through "civic movements" for the construction of a vision of future, to be cultivated through civic education (Geddes 1911a, b).

Genovesi related the economic development and the social welfare by emphasizing that this depends on the "connection" between the various social partners by laying the foundation for human development based on relationships of reciprocity: that is on cooperation.

The importance of the educational process (recognized by both Geddes and Genovesi) suggested to better underline in CLIC the centrality of the cultural and human dimensions.

Finally, today the wellbeing and the happiness represent the real goal of the economy (Costanza and Kubiszewski 2023).

The interpretation of the adaptive reuse in the perspective of Agenda 2030 invites to consider the "**circular**" adaptive reuse, being the 12th SDG the "goal of the goals" of Agenda 2030 (taken again in the new Urban Agenda). This goal introduces a new way to produce and to consume, avoiding any waste.

Dismissed cultural assets are a clear example of waste.

Certainly, the construction sector plays a key role in the operational implementation of the circular model. It should be supported by specific measures and governance as above underlined (fiscal and financial incentives, new rules in public procurement, new specific analysis about the metabolism etc.).

Many are the implications for how to evaluate investment in heritage reuse. It becomes necessary to move beyond the conventional approaches of neoclassical economics.

The circular adaptive reuse suggests specific evaluation processes, grounded on the spiral approach and on multicriteria tools, using a complex set of indicators (not only related to WTP). In this way can emerge the real comprehensive added value of the heritage reuse compared to other investments (in mobility, waste management etc.) because of many multidimensional impacts.

Through ex post evaluations and the experiences implemented in some specific areas (good practices) the added value of the cultural heritage reuse in the circular perspective has been documented.

The participative evaluation process has been considered here as the entry point towards a critical knowledge: towards a civic culture, necessary to stimulate an active role of citizens for promoting the circular model, involving in particular young people, and the third sector.

Therefore, it becomes easier to convince that the reuse of cultural heritage is an effective entry point for implementing the circular city.

In addition, through the assessment of all costs and benefits (in tangible and intangible terms) in different dimensions it becomes possible to increase the trust in local public institutions putting in relation spent money with outcomes and impacts: trust being the condition for sustaining any long-term reuse-transformation project by bottom up.

9 Conclusions

Abandoned industrial plants, disused churches, agricultural structures often reduced to the state of ruins (due to movements of people and demographic dynamics), landscapes in a state of decay are only one aspect of the loss of so many 'places', more and more *'out of time'*. How many villages are undergoing the same process under the impetus of extreme events or simply due to the transfer of population from inland (often mountainous) areas to the coast?

The question that arises is: *what can be done with so many signs of the past*, which are now real 'waste': the result of a development model based on maximum consumption here and now?

The general answer is that it is necessary to regenerate ties, relations, relationships of interdependence, through virtuous circuits/loops.

The circular development model appears in this perspective a necessary choice, being it characterized by the reduction of all forms of waste and being grounded on virtuous regenerative processes of relationships between human beings and with nature; between natural, man-made, human and social capitals.

The circular economy model and reuse of cultural assets are characterized by reciprocal interdependence, by their convergence about the long time, about the recover capacity of assets, materials, energy etc. for combatting any form of loss/refusal.

The CLIC reuse proposal is focused to re-connect our cities, the housing/industrial/commercial systems to the patterns of organic life, thus cultivating a healthy environment for the wellbeing of people. This is first of all a cultural challenge, considering that the dis-connections are grounded (at the end) on the current vision of the world/reality.

The application of this model to the adaptive reuse of cultural heritage in 'marginal conditions' offers, under certain conditions, positive answers (being itself founded on the elimination of all waste and the regeneration of relationships).

The Circular model (proposed in the SDG of Agenda 2030 number 12) is a key condition for implementing sustainability.

But today we have to recognize that the realisation of all the strategic goals of the 2030 Agenda are becoming increasingly difficult and complex. Social inequalities are multiplying and the impacts of climate change appear ever more intense and accelerated.

If the adaptive re-use is interpreted on the basis of the circular model aimed at realising "***circular cultural ecosystems***" as Living Labs, multidimensional benefits can be significantly produced and enhanced. For instance, an important role of adaptive reuse in relation to the fight against climate change is more and more evident. But also in relation to other aspects: for example, for stimulating processes of democratisation in the construction of public decisions, i.e. in choices affecting the good of all, the general good.

The importance of these ecological and social aspects in relation to historic/artistic/cultural impacts and to economic impacts themselves has been growing over time. But attention to the ecological condition of the "*common house*" cannot obscure attention to the subjects themselves who live in this "*common house*". In other words, the ecological paradigm must increasingly intersect and integrate with the humanistic paradigm. The human perspective has been a cornerstone of European culture and can also be a reference today when we proceed to implement active reuse experiences. It not only regenerates attention to the human scale of development but also invites one to question the very meaning of the human paradigm, that is linked to relationships/ties.

Circular model implementation depends on culture and, on its turn, culture/mindset promotes circularity. The circular model is consistent with the above also because it invites a virtuous, i.e. circular, relationship between past and present and therefore between present and future. In fact, the entire circular model does not only refer to the recycling/reuse/regeneration of materials and used resources but invites a general reorganisation of production and settlement processes on the basis of the model offered by nature, where all waste is eliminated in a context of metabolism that is going to be perfected over time. The adaptive reuse in the human-centred approach at micro and macro (urban) scale is essentially a culture-led challenge and not only an economic, financial, technological issue. The use of material products of culture (monuments, landscape, etc.), its selection, its adaptation is a cultural construction.

Adaptive reuse should contribute to conserve the **original soul of the city**: its particular identity, its meanings. Adaptive reuse should stimulate the capacity of the city to self-regenerate in the time and in the space, reacting to specific external stimuli.

The circular model interrelates different elements not only in space but also in time. The adaptive circular reuse of cultural assets tells us *where we come from, who we are,* and it correlates not only with nature but also with the history: with the recognition of our roots, towards the foundation of a *new humanism*. Memory and history become generators of new energy to build a desirable and sustainable and human future (Fusco Girard et al. 2023).

References

Arena G (2020) I custodi della bellezza. Prendersi cura dei beni comuni. Un patto fra cittadini e istituzioni per far ripartire l'Italia. Touring Club Italiano

Assari A (2012) Role of public participation in sustainability of historical city: usage of TOPSIS method. Indian J Sci Technol 5(3). https://doi.org/10.17485/ijst/2012/v5i3.2

Ayres RU (1994) Industrial metabolism: theory and policy. In: The greening of industrial ecosystems

Bat'a (2011) Bata Reminiscence and Resource Centre. https://www.bataheritagecentre.org.uk/

Borrero Ochoa O (2011) Betterment levy in Colombia: relevance, procedures, and social acceptability. Land Lines, April, 14–19

Boulding KE (1973) The economics of energy. Ann Am Acad Pol Soc Sci 410(1). https://doi.org/10.1177/000271627341000011

Boulding KE (2013) The economics of the coming spaceship earth. In: Environmental quality in a growing economy: essays from the Sixth RFF Forum, vol 3. https://doi.org/10.4324/9781315064147

Capra F (1989) Il Tao della fisica. Adelphi

Carlsson C, Zeleny M (1983) Multiple criteria decision making. J Oper Res Soc 34(10):1011. https://doi.org/10.2307/2580902

CLIC project – Circular models Leveraging Investments in Cultural heritage adaptive reuse. (2017). Horizon 2020 project "CLIC project – Circular models leveraging investments in cultural heritage adaptive reuse." Horizon 2020. https://www.clicproject.eu/

Costanza R (1992) Ecological economics. The science and management of sustainability. Columbia University Press

Costanza R, Kubiszewski I (2023) Urban and regional planning using the four capital framework of ecological economics. In: Fusco Girard L, Nijkamp P, Nocca F (eds) Reconnecting the city with nature and history. Towards circular regeneration strategies. Collana "Studi urbani e regionali". Franco Angeli, Milano

Costanza R, D'Arge R, De Groot R, Farber S, Grasso M, Hannon B, Limburg K, Naeem S, O'Neill RV, Paruelo J, Raskin RG, Sutton P, Van Den Belt M (1997) The value of the world's ecosystem services and natural capital. Nature 387:253–260. https://doi.org/10.1038/387253a0

Council of Europe (2000) European landscape convention. In: European treaty series – No. 176. Council of Europe Publishing. https://rm.coe.int/16807b6bc7

Council of Europe (2005) Framework convention on the value of cultural heritage for society. In: Council of Europe Treaty Series (No. 199). https://doi.org/10.1007/978-3-030-30018-0_1051

Daly HE (1996) Beyond growth. The economics of sustainable development. Beacon Press. http://pinguet.free.fr/daly1996.pdf

Daly HE, Farley J (2003) Ecological economics: principles and Applicatons. Island Press. https://indomarine.webs.com/documents/Ecological_Economics_Principles_And_Applications.pdf de Groot RS, Alkemade R, Braat L, Hein L, Willemen L (2010) Challenges in integrating the concept of ecosystem services and values in landscape planning, management and decision making. Ecol Complex 7(3):260–272. https://doi.org/10.1016/j.ecocom.2009.10.006

de Groot R, Brander L, van der Ploeg S, Costanza R, Bernard F, Braat L, Christie M, Crossman N, Ghermandi A, Hein L, Hussain S, Kumar P, McVittie A, Portela R, Rodriguez LC, ten Brink P, van Beukering P (2012) Global estimates of the value of ecosystems and their services in monetary units. Ecosyst Serv. https://doi.org/10.1016/j.ecoser.2012.07.005

Douglas J (2006) Building adaptation 2nd edition. In: Building (Issue November 1994). https://doi.org/10.1017/CBO9781107415324.004

Elkington J (1994) Towards the sustainable corporation: win-win-win business strategies for sustainable development. Calif Manag Rev 36(2):90–100. https://doi.org/10.2307/41165746

Elkington J (2018) 25 years ago I coined the phrase "triple bottom line." Here's why it's time to rethink it. Harv Bus Rev 25:2–5

European Commission (2016) Cork 2.0 Declaration "A Better Life in Rural Areas". Luxembourg, Publications Office of the European Union. https://doi.org/10.2762/618522

European Commission (2018) Bled Declaration "for a Smarter Future of the Rural Areas in EU". EU Action for Smart Villages, European Commission, Bled, Slovenia.

European Commission (2019a) The European green Deal. European Commission. https://doi.org/10.1017/CBO9781107415324.004

European Commission (2019b) The human-Centred. Opportunities for citizens through research and innovation, City. https://doi.org/10.2777/489486

European Commission (2021a) Territorial agenda 2030 – a future for all places. European Commission. https://ec.europa.eu/regional_policy/en/information/publications/brochures/2021/territorial-agenda-2030-a-future-for-all-places

European Commission (2021b) New European Bauhaus – beautiful, sustainable, together. Available at: https://new-european-bauhaus.europa.eu/index_en

European Commission (2023) Guidelines on biodiversity-friendly afforestation. Reforestation and Tree Planting. https://environment.ec.europa.eu/publications/guidelines-biodiversity-friendly-afforestation-reforestation-and-tree-planting_en

European Investment Bank (2018) The EIB Circular Economy Guide. Supporting the circular transition. May 2023. ISBN 978-92-861-5570-3. https://doi.org/10.2867/235223

European Union (2012) Consolidated version of the Treaty on European Union and the Treaty on the Functioning of the European Union, Pub. L. No. 2012/C 326/01 (2012). http://data.europa.eu/eli/treaty/teu_2012/oj

European Union (2016) Urban agenda for the EU pact of Amsterdam

Fukuyama F (1995) Trust the social virtues and the creation of prosperity. The free press

Fusco Girard L (1987) Risorse Architettoniche e Culturali: valutazioni e Strategie di Conservazione. Franco Angeli

Fusco Girard L (2018) Discourse of Professor Luigi Fusco Girard at the European parliament high-level european parliament conference "Cul-tural heritage in Europe: linking past and future." European Parliament Conference "Cultural heritage in Europe: Linking past and future." https://www.clicproject.eu/wp-content/uploads/2018/09/CLIC-Luigi-Fusco-Girard-Discourse-at-the-European-Parliament.pdf

Fusco Girard L (2021) The evolutiorary circular and human centered city: towards an ecological and humanistic "re-generation" of the current city governance. Hum Syst Manag 40(7):753–775. https://doi.org/10.3233/HSM-211218

Fusco Girard L, Nijkamp P (1997) Le valutazioni per lo sviluppo sostenibile della città e del territorio. Franco Angeli

Fusco Girard L, Nocca F (2019) Moving towards the circular economy/city model: which tools for operationalizing this model? Sustainability (Switzerland) 11(22):6253. https://doi.org/10.3390/su11226253

Fusco Girard L, Vecco M (2021) The "intrinsic value" of cultural heritage as driver for circular human-centered adaptive reuse. Sustain For 13(6):3231. https://doi.org/10.3390/su13063231

Fusco Girard L, Nijkamp P, Nocca F (2023) Reconnecting the city with nature and history. Towards circular regeneration strategies. Collana "Studi urbani e regionali". Franco Angeli, Milano
Geddes P (1911a) Chapters in modern botany. Kessinger's Lessinger Preprint
Geddes P (1911b) The civic survey of Edinburgh. Transactions of the town planning conference for the civic department, outlook tower. Edinburgh and Crosby Hall, Chelsea. https://atom.lib.strath.ac.uk/civic-survey-of-edinburgh-by-patrick-geddes
Geddes P (1915a) Cities in evolution: an introduction to the town planning movement and to the study of civics. Williams & Norgate
Geddes P (1915b) Cities in evolution: an introduction to the town planning movement and to the study of civics. In: The Geographical Journal. Williams & Norgate. https://doi.org/10.2307/1779702
Geddes P (1934) Jhon Ruskin, Economist. William Brown/Harvard University Library, Edinburgh
Genovesi A (1765) Lezioni di Economia Civile (F. Dal Degan (ed.))
Georgescu-Roegen N (1971) The entropy law and the economic process. Harvard University Press
Georgescu-Roegen N (1976) Economics and educational development. J Educ Finance 2(1):1–15. https://doi.org/https://www.jstor.org/stable/40703073
Gravagnuolo A, Bosone M, Fusco Girard L (2021) Deliverable D2.5—Methodologies for impact assessment of cultural heritage adaptive reuse. https://www.clicproject.eu/wp-content/uploads/2022/01/D2.5-Methodologies-for-impact-assessment-of-cultural-heritage-adaptive-reuse.pdf
Gravagnuolo A, Angrisano M, Bosone M, Buglione F, De Toro P, Fusco Girard L (2024) Participatory evaluation of cultural heritage adaptive reuse interventions in the circular economy perspective: a case study of historic buildings in Salerno (Italy). J Urban Manag (In Press). ISSN 2226-5856. https://doi.org/10.1016/j.jum.2023.12.002
Greffe X (2002) Culture et économie. Periferica 3:83–96. https://doi.org/10.25267/Periferica.2002.i3.06
Harper DA (1996) Entrepreneurship and the market process. Routledge, New York
Hinloopen E, Nijkamp P, Rietveld P (1983) Qualitative discrete multiple criteria choice models in regional planning. Reg Sci Urban Econ 13(1). https://doi.org/10.1016/0166-0462(83)90006-6
Historic England (2020) Heritage counts 2019 – There's no place like old homes. Re-use and Recycle to Reduce Carbon https://historicengland.org.uk/content/heritage-counts/pub/2019/hc2019-re-use-recycle-to-reduce-carbon/
Hodgson GM (1995) Economics and evolution. Marshall Studies Bulletin, 5. University of Cambridge, pp 41–50. http://www.cce.unifi.it/rivista/hodg5.htm
Hollis M (1998) Trust within reason. Cambridge University Press
ICOMOS (2017) Delhi declaration on heritage and democracy, 19th general assembly of the international ICOMOS. New Delhi, India 11–15 December 2017
Informal Ministerial Meeting on Urban Matters (2020) The New Leipzig Charter. The transformative power of cities for the common good (30 November 2020). https://ec.europa.eu/regional_policy/sources/docgener/brochure/new_leipzig_charter/new_leipzig_charter_en.pdf
Kahneman D, Krueger AB (2006) Developments in the measurement of subjective Well-being. J Econ Perspect 20(1):3–24. https://doi.org/10.1257/089533006776526030
Keeney RL (1996) Value-focused thinking. Harvard University Press
Labsus (2014) Regolamento sulla collaborazione fra cittadini e amministrazione per la cura e la rigenerazione dei beni comuni urbani. Labsus. https://www.labsus.org/2014/02/beni-comuni-un-regolamento-cittadini-attivi-piu-forti/
Lichfield N (1995) Community impact evaluation. University College Press
Locke J (1663) Essays on the law of nature. Bodleian Library, Oxford University
Lootsma FA (1999) Multi-criteria decision analysis via ratio and difference judgement Applicd optimization. In: Optimization
Morin E (1990) Introduction à la pensée complexe. ESF Publisher, Paris
Munda G (1995) Multicriteria evaluation in a fuzzy environment. Physica-Verlag HD. https://doi.org/10.1007/978-3-642-49997-5

Nijkamp P (1980) Environmental policy analysis. John Wiley and Sons, New York
Nijkamp P, Rietveld P, Voogd H (1990) In: Jorgenson DW (ed) Multicriteria evaluation in physical planning. Elsevier. https://shop.elsevier.com/books/multicriteria-evaluation-in-physical-planning/nijkamp/978-0-444-88124-3
Normandin K, Macdonald S (2013) A colloquium to advance the practice of conserving modern heritage. https://www.getty.edu/conservation/publications_resources/pdf_publications/pdf/colloquium_report.pdf
Nowak DJ, Greenfield EJ, Ellis A (2021) Climate change and urban forests: executive summary. https://d3f9k0n15ckvhe.cloudfront.net/wp-content/uploads/2021/10/FINAL_Nowak-Study_Full-.pdf
OECD (2022) Global compendium of land value capture policies. OECD. https://doi.org/10.1787/4f9559ee-en
Olivetti A (1960) Città dell'uomo. Edizioni di comunità
Pickerill T (2021) Investment leverage for adaptive reuse of cultural heritage. Sustain For 13(9):5052. https://doi.org/10.3390/su13095052
Porter ME, Kramer MR (2011) Creating shared value. Harv Bus Rev 89(nos. 1–2 (January–February 2011)): 62–77
Rapporto Labsus 2021 (2021) https://www.labsus.org/rapporto-labsus-2021/
Rees W (1997) Urban ecosystems: the human dimension. Urban Ecosyst 1:63–75. https://my.vanderbilt.edu/greencities/files/2014/08/Rees-urbanecosystems1.pdf
Rees WE (n.d.) Understanding urban ecosystems: an ecological economics perspective. In: Understanding Urban Ecosystems. Springer, pp 115–136. https://doi.org/10.1007/0-387-22615-X_8
Roe D, Turner B, Chausson A, Hemmerle E, Seddon N (2021) Investing in nature for development: do nature-based interventions deliver local development outcomes? https://www.iied.org/20206iied
Roy B (1990) The outranking approach and the foundations of Electre methods. In: Readings in multiple criteria decision aid. Springer, Berlin/Heidelberg, pp 155–183. https://doi.org/10.1007/978-3-642-75935-2_8
Saaty TL (1984) The analytic hierarchy process: decision making in complex environments. In: Quantitative assessment in arms control. Springer, pp 285–308. https://doi.org/10.1007/978-1-4613-2805-6_12
Saleh R, Ost C (2023) Innovative business model for adaptive reuse of cultural heritage in a circular economy perspective. Int J Entrep Small Bus 48(1). https://doi.org/10.1504/IJESB.2023.10053086
Schumacher EF (1973) Piccolo è bello. Uno studio di economia come se la gente contasse qualcosa. Ugo Mursia Editore
Schumpeter JA (1934) The theory of economic development. An inquiry into profits, capital, credit, interest, and the business cycle. (Harvard Ec). Harvard University Press. https://www.hup.harvard.edu/catalog.php?isbn=9780674879904
Schumpeter JA (1943) Capitalism, socialism and democracy. Routledge
Sen A (2003) Gender inequality and theories of justice. In: Women, culture, and development. https://doi.org/10.1093/0198289642.003.0011
Serageldin I (1999) Cultural heritage as public good. In: Global public goods, vol 3. Oxford University Press, New York, pp 240–263. https://doi.org/10.1093/0195130529.003.0012
Simon HA (1959) Theories of decision-making in economics. Am Econ Rev 49(3):253–283
Smit M, Baarveld M, Dewulf G (2014) Value capture and value creation in urban redevelopment: the case of built industrial heritage. Conference: AESOP 2014 – From control to co-evolution
Stern N (2006) The economics of climate change: the Stern review. Cambridge University Press, Cambridge
UNESCO (2005) Vienna Memorandum. In: Memorándum de Viena. Document WHC-05/29.COM/5

UNESCO (2016) The HUL guidebook. Managing heritage in dynamic and constantly changing urban environments a practical guide to UNESCO's recommendation on the historic urban landscape. UNESCO

United Nations (2015) Transforming our world: the 2030 agenda for sustainable development

United Nations (2016) New Urban Agenda 2030. Available at: https://habitat3.org/the-new-urban-agenda/

United Nations (2017) New Urban Agenda. In: United Nations conference on housing and sustainable urban development (habitat III). United Nations

Voogd JH (1983) Multicriterial evaluation for urban and regional planning. Pion

Weber M (1991) L'etica protestante e lo spirito del capitalismo. Rizzoli

Weber M (2019) Economy and society. A new translation. Howard University Press

WHO (2020) WHO Manifesto for a Healthy recovery from COVID-19. Prescriptions and Actionables for a healthy and green recovery. Available at: https://www.who.int/publications/m/item/who-manifesto-for-a-healthy-recoveryfrom-covid-19

Zeleny M (1992) Beauty, quality, armony. In: Human system management. World Scientific, pp 115–118

Zeleny M (1993) Alla ricerca di un equilibrio cognitivo. In: Fusco Girard L (ed) Estimo ed economia ambientale. Franco Angeli, pp 113–131

Zeleny M (2021) Entering the age of accelerated change: in search of equilibrium. Hum Syst Manag 40(1):3–14. https://doi.org/10.3233/HSM-209002

Chapter 3
Intangible Matters: Cultural Heritage as a Driver for Circular Economy

Marco Acri, M. Xavier Greffe, and Saša Dobričić

1 Introduction

The following pages stress the importance of looking at cultural heritage, including immovable and movable objects, monuments, and landscapes, underscoring the inseparability of tangible and intangible and assigning to the letter an essential role in all phases of the reuse endeavour. The experiences gained from the CLIC project, which merged cultural heritage adaptive reuse and circular economy, underscored the importance of historical investigation to understand the relevance of these concepts today. The process of heritage making has been constantly related in the past to the production of daily objects, but also works of art and monuments, by means of the basic economic—oikos—principles of efficiency and efficacy: this signified perfecting the objects for their primary use, as well as optimising the use of available materials. Considering that industrialisation and globalisation reversed the traditional productive logic depending on high costs of materials and low costs of labour, making materials easily available but flattening specialisations. In this perspective, driven by sustainability concerns, we need to re-create an alternative narrative able to reinterpret the logic of traditional local economies. In this framework, the focus will be emphasising the contribution of the intangible in the production of heritage through knowledge, tacit and explicit, taste and design, performed by the craftsmanship model as an intrinsic driver of circular economy. In the following pages, taking advantage of the lesson learnt in CLIC, intangible components in heritage are essential in circular models, and vice versa, the intrinsic circularity that characterises the cultural heritage "market". However, some conditions should be

M. Acri (✉) · S. Dobričić
Graduate School, Cultural Heritage Studies, University of Nova Gorica,
Nova Gorica, Slovenia
e-mail: sasa.dobricic@ung.si

M. X. Greffe
École d'économie de la Sorbonne, Université Paris 1 Panthéon-Sorbonne, Paris, France

L. Fusco Girard, A. Gravagnuolo (eds.), *Adaptive Reuse of Cultural Heritage*,
https://doi.org/10.1007/978-3-031-67628-4_3

set up at different levels of the circular metabolism, especially in the perspective of new self-employment and innovative business creation.

This paper is organised into two main sections: the first section is meant to reorganise the conceptual framework about tailor-made from the perspective of circular economy. The concepts of information asymmetry, tacit knowledge, experience and use, maintenance and repair, taste and vintage, as well as the second life of things, are discussed in a logical sequence. The second section is meant to stress the importance of recovering the traditional cultural heritage production processes based on skills and competences, where designers and craftsmen jointly with final users and clients do contribute to the creation of heritage in a tailor-made conceptual process. Finally, some recommendations are made, mostly pushing for new theoretical and political narratives that return to cultural heritage its constituting bone, the intangible dimension.

2 New Conceptual Framework: Readability of the Intangible

Circularity in production, particularly in the building sector, has existed for ages in the practice. Lacking any theoretical background, numerous examples of reuse of materials and building typologies, from mediaeval cities to modern age, etc., are witnessing how radically different visions of obsolescence have been driving local economies and space production (Frey 2016; Liverani 2011; Rea 1999; Ruskin and Rosenberg 1987; Settis 1996). According to the western literature, the locution "circular economy" was first used in the 1980s 'to describe a closed system of economy-environment interactions' (Murray et al. 2017, p. 372). Today the concept of circular economy is getting significant attention in many sectors, with a consequent proliferation of definitions (Temesgen et al. 2021) that mainly depend on the sector and profile that investigates or applies such concept (researchers, practitioners, business leaders, foundations). Along with a growing attention and confident expectations towards a circular economy, a more critical and cautious approach is raising and questioning whether we may really expect a change in respect to the economy that works with, rather than against, nature, or it is rather a "protective belt" that actually does not significantly affect the hard core of the mainstream economy around the growth model. In other words, it may be all about business opportunity, a "feel-good concept" in compromise with environmental and social challenges that finally maintains the imperative of economic growth (Kirchherr et al. 2017; Webster 2017).

Heritage conservation evolution is an interesting example of how a "protective belt" approach may evolve from a specialised and circumscribed doctrine, for decades quite peripheric for the mainstream modern designers, to a dynamic concept that has systematically affected the very heart of the contemporary design grammar. A highly context-value orientated and relational approach that works with, not against, cultural, social, or environmental constraints and limits. In other words, design in heritage is driven by embedded values brought by a historical ensemble of multiple residuals. As a result, operating with less rather than with more (the *less is more* for the Bauhaus director Mies van der Rohe) is not an ideology or a doctrinal matter in conservation design but a *sine qua non* condition depending on the complexity itself (of the multiple residuals). Such a lifestyle approach, when dealing with complexity, matches today with any creative design action that aspires to the 'art of building' that simultaneously co-produces the 'art of dwelling' and living. In minor vernacular architecture, as Illich reminds us, "...dwellings were never completed before occupancy, in contrast to the contemporary commodity, which decays from the day it is ready to use." (Illich 1984). Through conservation design, an historical patronal palace may be reactivated as a library, a school, or a museum, but its possibilities of reuse will not be exhausted in it: it remains a palace, a reactivated residual of the past. Good conservation design allows a circular reactivation of heritage assets as valuable residuals of the past that do not fall into the logic of permanent repurposing, and so producing and wasting.

Design with heritage also assured the maintenance of certain skilfulness typical of craftsmanship: the ability to control its own production means, tools, and knowledge combined all together with distinct negligence towards highly ambitious and revolutionary results, often and rather compensated with meticulous care of materials or attention to details. This modesty, which lacks any imperative of destruction and construction from scratch (Latour 2008), is, on the other hand, highly compensated with the intensity of thought that connotes the complexity of the context-related tailor-made approach, which might emerge as the driver of contemporary design practice. Hence, the hardware of efficiency that drives re-use and re-cycling narratives of circular economy, might gain a powerful and meaningful dimension when tailor-made design brings all its knowledge means, whether tacit or explicit, to read and interpret what might be the hidden potential (as value embedded residual) of any residual, even of the weakest one as waste.

Looking at the intangible components of cultural heritage in history and today, two important ideas should be considered in order to tackle and accommodate the needs of the circular economy from a mutual benefit perspective:

1. The fact that most of our tangible cultural heritage has been produced in market conditions where labour was low-cost while materials and logistics had high costs. Many of us still remember grandfathers straightening crooked iron nails to be reused as many times as possible. In fact, many re-use and re-cycle logics were in life as well as in business, rather a rule than an exception. In this logic, after maintenance, repair was the ultimate bulwark before replacement. On the opposite, today labour increases in value while materials and logistics are increasingly more affordable. Recalling grandfather's iron nails, today in building sites fallen ironware is not even recollected. In this contemporary logic, replacement comes often before repair, given that maintenance must eventually be self-performed.

The overall triumph of products over resources made the latter almost valueless and overexploited and the former underused and over-obsolete, with extremely expensive consequences for both environment and society. This phenomenon is at the base of the concern asking for a new re-design logic of the entire production scenario, partly embedded in circular models;

2. Cultural heritage is an expression of its context-related character in terms of materiality and shape. In the case of vernacular architecture, such a relation is evident also in the typology register. Whether major or minor heritage, the production scenario was driven by a tailor-made approach, very much depending on local constraints in terms of material availability, climate conditions, socio-cultural rules, familiar economies, and local knowledge consolidated in traditions. As an example, most vernacular architectures were produced according to the available size of beams for ceilings and architraves; in Venice, these dimensions depended on the maximum dimension of trunks that could be transported by river rafts from the dolomites, approximately 4.20 meters. Consequently, the average dimension of rooms in vernacular architecture is 4 meters, a condition that consequently affected the development of the morphological aspects of the Venetian historic urban landscape (Acri and Bonafede 2000). Similar parallelism can be done for other crafts, usually tailor made and final user specific. Remaining in Venice, for example, the gondola is tailored to the gondolier that will drive it and, as such, fully designed according to his/her size, height, and weight (Penzo 2000).

In fact, it is an achieved understanding that art, crafts, and techniques are forms of communicating about human beings and civilisations (Leroi-Gourhan 1943), telling much about the evolution of individuals and communities in a specific place. Leroi-Gourham as archaeologist, beyond anthropologist and sociologist, considered archaeological artefacts as witnesses of humans after their death and disappearance: in his perspective, arts and crafts are residuals, what remains, in under a certain viewpoint, the waste. Nevertheless, they hold an intangible component that takes them out of the usual elimination or recycling process.

Our outlook to achieve the above integration sees the intangible as a key driver for the tangible, being both inseparable and impossible to exist individually (Laroche et al. 2001). Intangible and tangible mutually feed each other in a virtuous circle that is based on a conservation approach that, as a rule, upgrades by re-designing, as many traditional practices witness. In this picture, the following pages will investigate new and old grammars of different ways of doing by displaying the interweave of knowledge, skills, aptitudes, values, and relations around the new narrative of co-designing and co-producing with the residual character of things.

2.1 The Plague of Tailor-Made: Information Asymmetry

The production of most goods since the early history of crafting has been characterised by an important factor: the information asymmetry. Information asymmetry is recognised in politics (Jackson and Morelli 2011) and economics, which can be seen in the market between the final consumer and the producer only visible through the seller (Barkley Rosser Jr 2003; Greffe 2017; Stiglitz 1975), although reputation in social media and users feed-back systems have recently scaled back the phenomenon (Wolf and Myerscough 2007). Consumers should trust the producer or its intermediary role on products in line with their expectations and out of their sphere of expertise. In crafting, such asymmetry is mostly evident in the expectation vs. the certainty in the final results, which lately may also reveal a price information asymmetry. The seller of an internationally branded pair of shoes knows more about the shoes than the buyer, who can however touch and experience the product before purchase, while the shoemaker manifests his knowledge asymmetry in the uncertainty of the final result, which is only an expectation for the client, a trust in the capacities, skills, and promises of the shoemaker. Such a divide is even broader in the case of design works (Daalhuizen et al. 2009), which may additionally increase when dealing with cultural heritage adaptive reuse in all different approaches as contemplated by the doctrine (Jokilehto 2018).

Adaptive reuse design requires absolute trust about the skills of the craftsman, which may rely either on the latter's authority or on personal relationships or networks, the so-called word of mouth. The expert's authority is expressed through years of work, awards, reputation, and examples that can be displayed through images or realised projects and objects. The trust is based on the hope to get expectations fully achieved (Grant and Fox 2007) and on the desire to have something different, tailor-made, as close as possible to one's personality, but the uncertainty of the result will remain. However, such uncertainty is mediated by the relationship craftsman-client that allows the client's constant contribution (inspections, negotiations, expectations, etc.). which turns into a co-designing, co-producing experience.

The tailor is asked to make a dress, and the possibilities in the final shape are discussed with the client, who has comfortability, aesthetic, durability, and uniqueness requirements that the craftsman can guarantee. Although the craftsman has a good certainty about the final result, uncertainty partly remains. The client has expectations that the final result will be exactly as imagined, discussed and paid. The process in between is a co-creation in negotiation. An interviewed shoemaker in Venice expressed about tailor made shoes the consciousness about their high costs, justifying this as a "product for life, that can be continuously repaired, modified, and adapted in a sort of circular process" (Grlić and Potokar 2014). A similar relationship affects heritage conservation design, with an additional player represented by the heritage asset with its tangible and intangible dimensions that influence that negotiation. The client expresses wills of many kinds, but it is up to the architect-designer to direct a set of craftsmen, experts and engineers, to channel, limit and amend the client's expectations, even by necessary impositions, providing through the information the intangible dimension at stake. It is exactly this set of multiple voices that further complicated the production scenario.

Worthy to remind that in architecture and architectural restoration, being partly allographic (Goodman and Morizot 1998), the final result depends on others' contributions and skills. This means that the author is obliged to manage his own information asymmetry with the other actors involved in the process, including the permissions authorities, the building site company, and all the participating craftsmen. Differently from music, which is purely allographic because precise indications for executions drive the infinite future interpreters of the composition, in building we cannot claim that design fully describes exact indications for execution. It is exactly this asymmetry that asks for additional translations of the design project (the autographic part) for the executors. The advantage in the past, which was almost eliminating such "building asymmetry", was on the workers and craftsmen's skills (carpenters, blacksmiths, marble cutters, etc.), as well as on the much lesser quantity of construction techniques and materials.

Nowadays, with the rise of the architectural profession as an autographic holder of design, this asymmetry needs to be reduced. To solve it, both parts, the author and the executors, reiterate the interpretation and execution logic, making the design project a circular thinking model, continuously (potentially) revised, rediscussed by multiple actors. Many may have memory of lapis sketches on walls, including measures, material indications, etc., made by the architects to drive building site workers in their performance. The work of Carlo Scarpa (Codello et al. 2009; Dal Co and Mazzariol 2006), one of the main reference figures in the history of architectural adaptive reuse, is witnessed by his drawings from the building sites, meant to explain reuse, material, and technical solutions, resulting in continuous co- and re-design practice of learning-by-doing. In the case of Historic Urban landscape, vernacular architectures, and monuments, the more we move from conservation to adaptive reuse, the higher will be the asymmetry, as well as the possibility not to accommodate users' expectations. Drastic adaptive reuse projects generate more asymmetry especially if not performed at best in all details and if not letting room for new interpretations and revisions.

By giving as established that the history of our heritage has been based on local design upgrading and local crafts continuous perfecting, it may be necessary to demonstrate that information asymmetry in cultural heritage has been increasing due to the progressing removal of local design and crafts from daily lives. Villages, towns, and cities have become less and less centres of production, transforming more into centres of consumption and/or accumulation, with a consequent impossibility for urban users to live in proximity to any production. At the latest stages of the global market economy, which we presently witness, products fully packed abroad never display their hidden secrets of making, their generating skills in time, finally extending the users' unawareness, sometimes ignorance, of their nature to the extreme. Such unawareness shows how the asymmetry of information is not merely an economic factor but also a social and cultural one because the cultural traditional values linked to the expectations (also in terms of timing) and the co-design are annihilated.

How does asymmetry reduce? The rule to reduce and eventually eliminate asymmetry is to make the parts aligned or stay equally distant from the centre, the truth, which in our debate is represented by cultural heritage. If the adaptive reuse of cultural heritage should pay attention to the combination of tangible and intangible that have an additional historic dimension in the genius loci (Norberg-Schulz 1979) and sociological dimension in the sense of place (Tuan 2011), the enhancement of knowledge of both designers and users is required. There is a recognition that knowledge is mainly individual; thus, the creation act depends on an individual set of variables, that range from passion, experience, context, ambition, and possibly the *oikos* (Klamer 2017). A specific role is played by tacit knowledge, defined as the set of information in personal knowledge that are difficult to extract from the minds of individuals and to transfer to others (Doing 2011; Nonaka and Takeuchi 1995; Polanyi 1966). A set of "innate" capabilities that are nourished through experience and/or by standing next to masters.

Why call this tacit, and how does it affect heritage and further on the circular economy? The reference to tacit knowledge as opposed to explicit knowledge has been seldom explicated through the concepts of knowing-how and knowing-that (Ryle 1945, 1949), and found end meaning in terms of a corresponding distinction between embodied knowledge and theoretical knowledge, assuming that doing things is never itself an exercise of intelligence.

Knowing-that involves consciously accessible knowledge that can be articulated and is characteristic of a person learning a skill through explicit instruction, recitation of rules, and attention to his or her movements. Someone who just graduated in architecture "knows that".

Knowing-how can be defined as the characteristic of the one who acts and makes judgements without explicitly reflecting or purely relying on the principles or rules involved. It is often associated with embodied knowledge as the type of knowledge where the body knows how to act (Tanaka 2011). Knowing-how involves more than a physical know-how; it involves knowing how to obtain desired end states, knowing what to do in order to obtain them, and knowing when to do it. Then knowing-how seems to be bound up with some variety of knowing-that (Kakiuchi and Greffe 2015). The knowing how put in parallel to the embodied knowledge is a step towards the Merleau-Ponty intuition on the "*Je peux et je pense*" (I can and I think) (Merleau-Ponty 1945) versus the Cartesian concept of "I think – Cogito Ergo Sum", recognising an additional quality in the making. An architect with building site experiences probably "knows-how".

As we have seen the tacit knowledge can be clearly seen as related to the performance of "making", but can easily be understood also in the performance of using, and this is maybe its very application.

2.2 *The Circular Experience Models: The Complex Context of Knowing How*

The tacit knowledge is embedded in the human mind through experience and jobs, which we would rather define use because it may stay in the producer and user. In arts and crafts, as much as for architecture, use and experience are essential to grasp utility, efficacy, and beauty. It is through the experience and the use that an asset gains perfection. Architecture and urban space help in locating the experience. Considered an advanced form of art for its fourth dimension—time—architecture evokes experience and memories (Zevi 1948) and thus empowers the attention of the users. We all live in architecture. We know, in a tacit form, what makes us comfortable, safe and satisfied and we know how to use it. The more a building is "usual" in terms of architectural elements, syntax, and morphology, the more it reassures users (Bond 2017; Hannah 1969; Saqer Mustafa Sqour 2018). Per extension, the more an architecture is used, the more it is understood and consolidated in the user's memories, potentially turning into knowledge. Such a use though, is not enough to achieve the tacit knowledge that enables users to capture the intangible component and the sense of place of cultural heritage. Tacit is not fully achieved since physical tangibility does not per se guarantee that the user possesses its recognisable representation, because of their lack of experience with it (Laroche et al. 2001). The full tacit component of heritage can be achieved through the experience the users will reach by getting to the sense of appreciation of the intangible dimension.

A very well-designed conservation project may empower the attention of the users and solicit their understanding of cultural heritage. The conservation project may though be carried out through different approaches, such as pure conservation, critical restoration, reconstruction and adaptive reuse (see Acri, Dobricic, Jokilehto, Chap. 4) and by adopting different techniques and materials. Any solution may raise final users' attention. However, one approach can be keener than another to express the specific qualities of that heritage, displaying better its intrinsic dimension through details, respect of materials, and redesign acts able to bring users nearer. The experience will be achieved through the users awareness, being conscious of the motivations behind redesign solutions, establishing a dialogue with that heritage and the redesign intention, and being willing to "export" the experience in other contexts, thus closing the circular intellectual process.

An important contribution to "experience making" comes from the pilot initiatives made in Rijeka within the CLIC project and named "urban seeding". This initiative, meant to provide exemplary tools and pilot actions for the establishment of a reproducible set of activities to merge built environment adaptive reuse with the principles of the circular economy, focused in the core area of the city, along its historic river, including some dismissed plots of land as well as some industrial buildings. Organised through workshop sessions involving students at their later stages of research and young professionals in different multidisciplinary teams, the urban seeding was asking for ideas for circular business and action proposals in the area, characterised by ad hoc applicability but also potential replicability and respect of the cultural heritage of the city. Participants, with a good knowing that coming from their brief fresh career, had to experience the site and others competences and project proposals for their possible future, culture oriented, circular city. The results highlighted critical assessments of the state of the art of Rijeka as a sustainable city and a true capability to highlight the intrinsic values of the historic urban landscape of the city, completely different from the ones of other reference case studies, such as Nantes or Ljubljana. Although little knowledge about the inputs from the local stakeholders was provided to participants, the requirement to experience the city areas through a specific lens made traditions and local genius loci emerge strongly, without any banal tourism ambition or perspective.

Maintenance and repair offer a concrete, applied step forward because they constitute a practical contribution to the experiential moment of the tacit knowledge. Maintaining and repairing objects, when possible, obliges users to take the position of the craftsman, of the architect, of the designer, employing their own hands in a practical work that intimately obliges a relationship with materials, techniques, tools, time, etc. The effort to deal with maintenance and reuse reconnects the users to the designer and to the craftsman, reconfiguring the creative mutual exchange in the circular redesign. When repairing a door in a traditional home, one must pay attention to details that are often overlooked, ask questions and seek answers, and use tools and materials to explore necessary options and potential improvements. Such action of repairing, maintaining, and restoring additionally develops a critical aptitude towards the strengths and weaknesses of a product and its repairability in the perspective of a sustainable personal economy. Experiences in this sense have been implemented largely in the UK by means of repair manuals (Pickles 2016; Pickles et al. 2014) in the logic of the learning by doing.

In Rijeka, the urban seeding initiative proposed the organisation of ateliers that could permit normal citizens to carry out routine house finishing maintenance work. This could have been done in the newly established public space for personal maintenance activities called REPERAI. The example taken from the discussion has been the maintenance of the traditional wooden shutters and windows, often replaced with new PCV ones. Participants' and organisers' feedbacks showed that this was a relatively simple task that didn't require any specialised knowledge or equipment and, for the most part, allowed for a thorough understanding of the objects, their creative secrets, and concepts, including maintenance and repair. This example also enabled listeners and testers to eventually even reduce the asymmetry of information on this specific maintenance action, empowering them in case of dialogue with local carpenters.

A good example of experiencing by repair and maintenance is the terrazzo floor in Venice, which was introduced as a response to the logistical constraints of the Venetian building site and the consequent need to optimise the use of resources, including the residual materials. It became very soon an exceptional solution to the constant vibrations of traditional and monumental buildings, with infinite aesthetic and practical combinations (Lazzarini 2008). The immediate understanding of end users about such flexibility and ease of application of the solution accelerated its perfecting in different spatial and socio-economic contexts. In the last decades, the lack of understanding of its intrinsic qualities, despite an affordable market cost for its maintenance, determined a fast substitution with a consequent heavy loss for the city's cultural and historic capital. In reality, the terrazzo floor can be partly maintained through normal housekeeping activities by simple oiling and wax applying and through simple repair actions, such as mortar crack filling. In the case of heavier maintenance actions, the costs are lower than substitution (average 30 eur/sqm), a factor that has not been fully understood by the newest generations.

Inspired by a successful story in CLIC, as architects, we have tried to apply some circular economy in conservation as a practice of learning by doing. The successful story is from ROTOR DC (ROTOR 2021), a cooperative that organises the reuse of construction materials based on Brussels and operated by dismantling and recovering construction materials and finishing to be resold and, more interestingly, to be re-designed for reuse. The exercise consisted in dismantling some materials in an historic Austro-Hungarian building, mainly ceramic tiles and parquet. Although the imaginary and the doctrine, the knowing-that, described this as a simple practice, in reality it involved particular care and organisation, such as understanding the potential quality and value in the market, carefully detaching elements, checking their status, cleaning the fully, disposing, and…storing them onsite or for transport. From knowing-that to knowing-how there is indeed experience, especially considering the range of possible items that potentially appear in the activity. As an important stress to be mentioned here and developed further in the following section, the experience component highlighted the need for re-design of the items, their hypothetical future reuse, an aspect that is fundamental for the business viability and the poetry of repurposing.

A little section should also be given to taste. In the tradition, the making of heritage was largely dependent on daily practices, on daily needs, on survival issues, and on the relationship with the surrounding territory. The set of tools—*outils*—available did not depend on beauty factors but heavily on the so-called utility. Aesthetics was a dimension confined to higher societal classes, where the wonders about shape and taste could derive also from a higher artistic familiarity.

Wooden tables, shoes, lamps, windows and doors, tiles, dresses, and the overall set of daily, today cultural, objects had to respond mostly to utility and durability, and for such perfected through the already mentioned production-consumption mutual relationship. The advent of industrialisation and of a globalised economy offered a much higher set of commodities, much different and at lesser costs, suddenly possible for most: aesthetics, thus beauty and taste, spread as a global driver, with roots in local specificities: "to describe what you mean by a cultured taste, you have to describe a culture" (Wittgenstein 1966), that was opening considerations on the local inherited and present influence in choice making, the space in between knowing and pleasure (Kant 1993) and the ascertained association of taste to knowledge in history to the extent of being characteristic of the sapient (Agamben 2015, p. 10). The phenomenon of taste is far from being finally assessed, but it seems one of the determining factors to feed the needs of the circular economy and reassuring inputs from the so-called "vintage" trend.

Vintage is a popularised version of the antique store, given its different, but still determinant, selecting factor. If in the antique store quality and rarity are essential selecting criteria, jointly with integrity and authenticity (Jokilehto 2019; Stovel 2007), in vintage they are taken over by quality and significance in addition to the "second hand" principle. In vintage, the aesthetic dimension, in particular shape, is essential and refers purely to individual values, with an increasing attention to quality as a nostalgic (often just) belief. The selection of vintage objects is driven by materials' quality and crafting, while the typology is suggested by external factors, such as fashion and design trends. More and more the vintage market has been fed by ad-hoc design marketing in more or less specialised magazines, TV series, and architectural digests with one-to-one effects for the circular economy: on one side, in fact, it promoted the reuse, on the other hand, it nourished cheap reproductions meant to provide supply to the scarcity for lower costs, feeding the neoclassical linear economy. What started decades ago in the weekly flea markets for true need and/or collecting hobbies, became larger and larger, developing parallel markets based on reproduction.

Also reproduction is through a known concept: Benjamin and Adorno's works underlined the authenticity, the aesthetic autonomy, through the concept of *aura* (Benjamin 1991; Adorno 2009), considered at risk for the reproducibility of the works of art, as well as for the loss of those ideas of creativity, genius, and eternity that characterised properly "auratic" art (Brandi 2000; Di Giacomo 2013). Re-production and "bad taste" were soon labelled as "in-style" phenomena, very common in furniture, namely reproducing through modern techniques iconic craft languages of the past: this was done often without falsification purposes, but simply to respond to the scarcity in the market and provide options mostly for the new bourgeoisie, alias middle class (De Fusco and Fiorino 2004; Pratt 1981). A further dimension of reproduction would have been represented by *kitsch*, defined as a true consequence of industrialisation, displacing handicrafts, and mass phenomena that could be contrasted—as a model of the culture of consumerism—through modernism and the avant-garde (Greenberg 1939), but often associated with bad taste (Dorfles 1972) as a vehicle of misleading message from a missing creative action.

The capability to grasp the aesthetics of cultural heritage is at the forefront of its preservation. As we have seen, taste is a knowledge that we do not possess, it is a residual of the experiences and acts as a transformative factor that gives new sense (semantic) to the established accumulation of experiences, and it does not establish a new order but acts on the contingency of things. As such, taste in reuse is often driven by the approach "the less is the best", as the most honest critical act to preserve the perceived authenticity. The same approach is sedimented in the conservation doctrine, namely promoting the actions that give back performance by critically and minimally eliminating the "superfluous" historic additions, as in the most optimistic circular perspective.

2.3 The Second Life of Things: Re-use and Re-design

If we consider that objects have a life cycle—or even more subsequent life cycles—we are led to consider not only their creation but also their alleged end and the problems that go with it. This consideration is not new, and the history that led from the "Chambers of Wonders" to museums is witness to it. The definition of Alain Schnapper already underlined an interest considering the case of the chambers of wonders (Schnapper 1988): "The chamber of wonders is… a microcosm … meaning a summary of the world, where objects of the earth, sea and air find place, or where objects of the three kingdoms – mineral, vegetal and animal – are displayed alongside the artefacts of man", a "collection of the whole of nature". Such chambers developed during the sixteenth and seventeenth centuries by the initiative of naturalists. Similar happened in Italy, adding a dimension of hobby, leisure, and even prestige for the sole sake of knowledge (Rivallain 2001). Elias Ashmole brought together in his chamber of curiosities in England ancient coins, books, engravings, geological and zoological specimens, including the stuffed body of the last dodo that Europe had known. During the eighteenth century, the *Encyclopedia* found in these cabinets perfect places of knowledge and reflection (Cornette 1989). The contemporary era has underlined how much "works of art function when, by stimulating a penetrating vision, a sharp perception, an alert visual intelligence, and broadened perspectives, … they participate in the organisation and reorganisation of the experience, and therefore to the shaping and re-shaping of our worlds" (Goodman and Morizot 1998, p. 105). With no doubt this statement is addressed more to works of art than to common objects, but anthropologists would go further by suggesting that objects have a life of their own, that they can "function" or experience different value regimes, and that they can thus pass from the stage of everyday life object to that of heritage object or even vice versa.

In this debate, Appadurai—who has reflected on the capitalist world as well as on pre-capitalist societies—started from the analysis of Simmel, who had underlined the limited nature of the economic approaches to the value of objects (Appadurai 1986). For Simmel, the economic value of objects does not reflect their intrinsic value but rather the opinions of market economy players: customers and suppliers (Simmel 1900). Each of these groups estimates this value according to what the object represents to them, based on different contexts. The result is a shifting average value at a given point in time, which is based on subjective and objective elements, the first playing a role on the demand side and the second more on the supply side. Thus, objects can quickly change their value regime without changing their form drastically, a fact that broadens the way in which their life cycle is considered. For Appadurai, objects circulate under these alternative regimes of value, such as in a system of "gift against gift", opposed to the competitive market system, where the "subject" must follow social constraints; hence the famous quote by Appadurai: "If from a theoretical point of view human actors give a value to objects… from a methodological point of view, objects give meaning to the subjects." (Appadurai 1986, p. 5). A theory that has been partly reconsidered by Coccia in narrating the role of objects in the city, as a moral of contemporary cities (Coccia 2014).

The result is an agency of things that Bruno Latour highlighted by using human-non-human terminology rather than object-person terminology (Latour 1991). By incorporating imperatives of a technical nature, the non-human is then an agent that leads the person to react in a given way and to build social behaviours, for example when a lock is forcing people to close the door when leaving home or when a speed bump forces them to slow down their cars in front of a school (Latour 2007, pp. 33–46). Behind an object lays down a social program that makes the object itself a social actor. As Andreas Reckwitz wrote, "the main issue then is that certain things or artefacts provide more than mere objects of knowledge and become necessary and irreplaceable components of certain social practices" (Reckwitz 2002, p. 210).

We can thus speak of a cultural biography of objects or even of a social history of objects, by broadening the vision to include several of them. Through their different possible value regimes, objects open an opportunity to change the society where we live, and accordingly Hennion and Latour wrote: "Objects do something, and first of all they do us" (Hennion and Latour 1993).

This approach to the life cycle of objects and to the succession of their possible value regimes explains the need to make objects dialogue with their contexts and to change their forms if necessary. For example, demolished, an unperforming window, is actually a "matter out of place" (Douglas 2005). As Douglas argues, it matters 'Where objects are' and not 'What they are'; in other words, its status depends on our capacity to re-arrange and re-contextualise. In the absence of such a contextualisation, the value of an object would be reduced to a value of age or nostalgia, making it fragile and exposing it to a risk of losing attention and therefore of rejection, in many cases—where not having a recognised heritage value—having a value of waste or even representing a cost to be removed as quickly as possible … whatever the social cost is.

Recalling the example of ROTOR and our attempt at material repurposing, the additional creative act after dismantling is redesigning. Architectural elements are, probably more than movable objects, needing new design given that once removed they are decontextualised, loosing intrinsic specificities. This must be said for elements that have a full function, as beams and windows and doors, but also for those elements that may have a function in relation to others, as stones or bricks. Their redesign builds their second life, which can be richer and enhanced in dignity. Nevertheless, redesigning is an intense cultural performance; thus, mistakes and misunderstandings can be easily framed in the bad taste or kitsch, as the practice often shows. "Museification" of usable items is, for example, a developing ambiguous routine.

3 Implementation: Tailor Made Redesign as Cornerstone of Adaptive Reuse

3.1 Tailor-Made as a Grammar of Redesign

Looking today at circular economy from the perspective of cultural heritage adaptive reuse, namely "intervening on and adapting", we should integrate some fundamental lessons from the past: what is the new statute of the product if it embeds its own re-designed expression as resource? If all things should be prone to re-use, then everything should be redesigned (including nature) (Latour 2008) and the world is not as produced as it is re-designed, fact that empowers the same notion of design, thus becoming the most vital part of the entire productive scenario. Moreover, if the history or life (cycle) of things matters more than their statute of products (final results), then the multiple co-existing forces of outdated and new that are embedded and in friction in heritage and its preservation practice might offer a diverse design path, a tailor-made productive scenario. In such scenario the consumer/producer exchange relationship does not necessarily maintain products and waste only as valuable (even if circular) commodities, which would risk to only "negatively affirm" instead of critically engaging with current (linear) production practice. European cities are an example of continuous redesign. Historic buildings have a history of continuous redesign. Anything has the potential to be redesigned.

If things are not produced but are rather designed, redesigned, and perpetually contextualised, then the whole process is imbued with uncertainty. The entire process, even if circular, has a completely different grammar far from being flawlessly smooth, it is impregnated with negotiation, conflict, and uncertainty. After all, co-designing with, rather than against nature, cannot be an expression of the traditional linear production rationale that self-referentially optimises itself by rationalising both the residuals (recycling) as well as the mistakes-deviations of the process (the new circular logics). In this sense, the lesson from heritage is profoundly cultural rather than natural, coming from its impossibility to be endlessly trapped in harmonious recycling loops of its contradictions. Heritage accepts its residuals as they are: there is a certain standstill that blocks the testing logic of repeatedly producing and wasting. Heritage as category recognises the residual either as irreplaceable if tangible or cyclically reproducible as embedded in tradition if intangible; both are excluded from the dictate of obsolescence. This is why design with heritage lacks grandeur of gestures, proceeds with modesty and prudence, thus avoiding both the overexploitation of the given and the embedded obsolescence of the new, typical of modern commodities (take, make, waste). Designing with heritage and its engineering of perfection is not running through the mimicking of nature's wasteless cycles but rather through a profoundly cultural approach of care, attachment, precaution, and interdependence to the residual nature of things. It is a rather unpretentious intervention, accomplished through the careful re-crafting of rough materiality and spirituality of things, where objective material constraints matter as much as subjective human ones.

> As a concept, design implies a humility that seems absent from the word *construction* or *building*. Because of its historical roots as a mere addition to the "real" practicality, study materiality and functions of daily objects, there is always some modesty in claiming to design something anew. In design there is nothing foundational. It seems to me that to say you plan to design something, does not carry the same risk of hubris as saying one is going to build something. Introducing Prometheus to some other hero of the past as a *designer* would doubtlessly have angered him. Thus, the expansion of the word *design* is an indication (a weak one to be sure) of what could be called a post Promethean theory of action. (Latour 2008, p. 3)

As a rule, the preservation practice imposes that any intervention establishes a solid relationship with the existing context. As much as tinny might be the interference of the new into the existing formal or material order, this will also go much further than refunctioning or adding something only in terms of function, efficiency, adaptation, or even taste and trend. Any design action within the realms of preservation will in its essence intertwine new relationships with the residuals of the past, passing through the intrinsic formal, material, and aesthetic assets of the given context. Any such design proceeds within the highly uncertain realm of new and existing, that reactivate the multiple co-existing forces of integrity and authenticity (Acri et al. 2019).

If we imagine at the very heart of making the integration of the existing within the extensions of new, then design and re-design matter more than the production process: its tailor-made approach comprehends multiple sets of knowledge and tacit skills, not distinguishing what is planned, calculated, defined, experienced, or imagined (Latour 2008, p. 2). Ambiguity is generally the main driver in this context: we consider the rise of ambiguity when we have multiple different representations of the same thing. If "landscape (and HUL) is as perceived by people" then we have multiple interpretations and not opinions of the same thing (CoE 2000; Kovacic et al. 2019; Kovacic and Di Felice 2019).

Exactly as circular economy is not simply economy that is circular, literally, from the perspective of complexity of skills involved in design, adaptive reuse, is not just use that is adapted, but a set of "imaginary of circularity" a social artefact that relates to broader cultural values that all together shape our vision of project and future (Kovacic et al. 2019, p. 77). It is clear that if this collective imagination of futures relies on technological innovation, economic competition, and resource security narratives, the circular imaginary might be even counterproductive, namely unable to imagine a circular society that goes beyond the circular economy.

Hence, in addition to humility and doubtfulness mentioned in the initial chapter, that characterise the action of the craftsman and challenge the contemporary designer, we have here emphasised the importance of certain embedded skilfulness. "Skill" is attached to art and craft as much as it is to design. Actually, to make a good "disegno" – drawing, to make things different, although nothing will radically change, one has to be a skilled designer. From this perspective, from objects to buildings and cities to nature itself—everything needs to be redesigned, more than produced.

> *If it is true that the present historical situation is defined by a complete disconnect between two great alternative narratives – one of emancipation, detachment, modernization, progress and mastery, and the other, completely different, of attachment, precaution, entanglement, dependence and care – then the little word "design" could offer a very important touch stone for detecting where we are heading and how well modernism (and also postmodernism) has been faring….* In other words, the more we think of ourselves as designers, the less we think of ourselves as modernizers. (Latour 2008)

3.2 *The Craftmanship as Enabler of Circular Economy in Cultural Heritage*

Knowledge, aesthetics, and design are fundamental factors intangible to favour circularity, but who gives shape to them? Considering the spreading specialisation of designers in conservation and reuse, the focus should be given on crafts. For centuries, the construction of traditional buildings and monuments depended on consolidated and innovative building and crafting languages dependent on a limited set of materials and techniques, which were extremely well practiced and managed by building site masters and craftsmen. The last century's acceleration of artificial intelligence and global industrial products reduced the need for traditional skilled workers. Nevertheless, cultural heritage adaptive reuse projects are still needing specialised works, able to either initiate or complete the re-design process.

In his work on crafts, Sennet, commenting on the intellectual work of her master Hanna Arendt (Arendt 1989), introduces the figure of the craftsman from the distinction in the pre-industrial society of *animal laborans* and *homo faber*: the former referring to a human being working as a pack animal, wondering "how", the latter as a human being building a common life, wondering "why" (Sennett 2013, p. 16). H*omo faber*, the craftsman, is who invests in the pleasure of making, as a construction of the future, who learns how to make good for self-pleasure and satisfaction, as a life rule that permits the achievement of very refined techniques guided by a perfect synchrony of mind and hand (Clark and Chalmers 1998; Menary 2010). In addition, Sennet argues that a man who knows how to govern himself, balancing autonomy and rules respect, is also a righteous man. Traditional craftsmen are those that, perfectly comfortable with traditional techniques and tools, look at the redesign work as a circular practice, thus placing before repair to replacement, the material consistency of heritage.

Repair is embedded in the concept of cultivation as further antidote to obsolescence. For example, as per craftsman repair, the gardener's cultivation deals with the domestication of life, which is done with care and attention to details by proceeding with humility and preparedness for the unpredictable events. As a matter of fact, a gardener knows very well that any tiny, even invisible being can make his opus collapse. In other words, the gardener is aware that collaborative designs include also undesirable collaborators, as well as the craftsman acknowledges the negotiation and co-production with unwanted clients. Gardener's knowledge maintains continuous relation with different cultures that are meant to grow in a common plot of land, in the same way that craftsman has to embrace the unpredicted problems that arise and do not match the models or means at disposal.

In our attempts to complete a circular practice in an historic building, we could feel the excitement of the challenge to recover items by repairing and redesigning their position in space by use of traditional practices (clearly by contemporary tools and finishes). In this building site there was an interesting moment in the process: there was an old crooked Austro-Hungarian window sash that did not close properly and became dangerous. Two local firms replied that it had to be thrown away and replaced, but first it had to be taken as a copy for reproduction (needing approximately 3 working days), asking for a compensation of 800–1000 euros. By chance we turned to a hobbyist craftsman, who said he wanted to try to straighten it, which happened after a working morning, for a cost, glass replaced, of 40 euros. This example is not meant to open nostalgic connections but to show how the present linear market and the present crafting competences too often look at the problems through the lens of linear economy, putting forward replacement to repair, with obvious consequences of impoverishment of the available skills and lack of trust by the final consumer, beyond the loss of qualities and values of cultural heritage. It was merely our re-design skills to impose additional investigation, which would never happen in the case of normal citizens, especially from the newest generations.

Traditional and artistic workmanship struggle in the contemporary market, with no exception for the cultural heritage reuse sector. The skills are more and more left on the enthusiasts and hobbyists, who perform proper re-design and repair works, led by the self-pleasure and satisfaction mentioned by Sennet, as well as the excitement of the challenge, of the problem, that stands outside the market routine.

Traditional and artistic workmanship does not get the deserved recognition, both because it is seen as a nostalgic practice and for its presumed inability to be aligned with progress and, as such, as marginal and likeable. Marginal because it is by nature not likely to create the exports and job flows needed for sustainable development; likeable in that its small scale conjures up images of generally harmonious work relations; and because its resources often come from the local environment, which they consequently contribute to sustaining. It is likely, too, that they are recognised as being low-energy consumers while injecting creativity into their environment and keeping local activity alive. The very concept of "small-scale mass production" speaks volumes as to the little credibility of this sector: its creativity does not seem as much of a solution! However, the crafts sector today remains what it has always been: a place for observing and sizing up the changing needs of the surrounding world and the answers that can be given to them in the everyday. It remains, in this sense, as much the vehicle for as the memory of material culture, namely the intangible valuing ladder of the traditional "residuals". The quest for quality, which combines the "creative hand" with more innovative equipment, remains the characteristic of an economy that sees competition as necessarily based on both quality and cost, thus giving an outstanding role to human resources, as underlined by Sennet.

Lastly, far from relegating the crafts sector to merely a supporting role, globalisation also offers new prospects for development, which, though challenging, are real and could find in the circular economy a sudden ally. In fact, the two core characteristics of today's economy—being knowledge-based and globalised—make the importance of the immaterial heritage passed on and promoted by the crafts sector. As a source of continually replenished heritage, the crafts sector irrigates creativity. As a source of diversity, cultural heritage casts light on niches that, when added together, become a market of global scale. All the crafts sectors, are flexible enough to produce both significant varieties and significant sizes due to their potential for redesign.

Five main difficulties, also perceived in the CLIC project in mapping and networking actors in the cultural heritage, are recurrent in the craft sector, even if measures are set in place as technical arrangements, training, funding, legal status, etc. (Greffe 2014):

- A first difficulty is inherent in craft activity, in an economy where large distances between producers and users of goods or even services are more and more frequently the rule rather than the exception. As a result, the risks incurred by a craftsperson offering a new product are substantial, far more so than in the systems of yesteryear, when craftspeople were often in direct contact with their customers. As long as that contact was very close, they were able to address the need before their eyes, and the risk was minimal. Today, the crafts sector is creating new needs in a context that is less precise, in which it is forced to interpret, thus leaving room for uncertainty or even error. The risk is thus significant. The question of proximity of crafts as well as contextualisation, has been at the core of CLIC activities, so much to be modelled in the cultural corridor idea in Rijeka, meant to accommodate in a precisely bounded area the core, circular businesses in the craft and heritage sector (ref. Acri, Dobricic, Debevec, Sustainability, 2021). The reference to cultural district cannot be missed in this respect.
- A second difficulty arises from the fact that crafts sector enterprises have always been small, often individual. As a result, most of the cost ends up in the expense of marketing, research, funding, or even costs. Although, in general, the technical side, the knowing how, is fairly achieved, the commercial and financial aspects are far less satisfactory. This problem has been noticed heavily in the CLIC project. Most of the companies and crafts interviewed in the project have highlighted difficulties in the starting phase, often underscoring the need for public assistance for the start. Looking at the craft market today, especially through the old generation of craftsmen, they locate in the black market or act as hobbyists for their impossibility to align with the contemporary financial and organisational requirements. One of the common solutions in the CLIC project has seen incubators as a solution, although the involvement and commitment of past generations of craftsmen has been often forgotten, although in itself would constitute an exceptional circular practice.
- A third difficulty derives from the present forms of training. Regardless of country and based on a number of noteworthy examples, young people today are enjoying longer educations, a trend that shifts preference to general curricula, at the expense of applied skills. Moreover, such long curricula are characterised by a lack of experiential moments, where the knowing that can transform into the knowing how. If traditional crafts are a potential response to a lack of specialisation, education must reform soon to grasp the opportunity of competence transmission from former generations of craftsmen. In a recent set of interviews, famous fashion designers lamented the absolute lack of traditional skills, those holding tacit knowledge and taste, being obliged to rely on the few tailors remained in the market, often very old (Ferrero 2014).
- A fourth difficulty must be identified in the present growing competition between craftsmen and artists, who are today distinguished for concepts of intellectual property that benefit the latters and not the formers (given that artists produce unique concepts, while artisans and craftsmen reproduce through perfecting crafts). As a result, their only protection is that granted to the processes they use, in the form of patents. However, it can be said that craftspeople are, by tradition,

not always ready to take on this problem, which was not as prominent to them up to this point being the businesses radically different. Worthy to notice though that too often skilled crafts tend to define themselves as arts, generating new forms of information asymmetry for the users, with clear unbalances in the market. The Murano Glass production is an example of such frictions.

Framing correctly crafts and arts would highly benefit establishing the proper supporting measures.

- The fifth difficulty lies in the intwining between tradition and innovation, especially in addressing the dichotomy of craft-design. As we observed in this paper, the term 'design' is fully connected to the idea of repair and repurpose in cultural heritage, because it requires a cultural act of displacement, decontextualisation, but is also an intrinsic constituent of traditional skilled crafting. Underlying this semantic pairing is a full-fledged redeployment of crafts sector activity.

3.3 *Clustering: The Governance of Proximity*

The CLIC project, though, highlighted that scales for the application of circular economy are essential, with necessary tailoring in the case of cultural heritage adaptive reuse. Single initiatives are not sufficient to create a real loop of circularity. Given that the intangible is the engine of cultural production and preservation, the very meaning of tangible heritage, integrated governance is indispensable. To achieve good urban circularity, there must be many forms and layers of intervention, from design thinking, to financial tools, to a governance system allowing initiatives to be integrated fully. As we have seen, the circular process in cultural heritage reuse does not merely depend on the actors of reuse, but on a combined and coordinated mechanism with a common goal. As per our assumption, the present connection between cultural heritage and circular economy is an artificial construction that is meant to preserve cultural values and the environment, and as artificial, it needs proper mechanisms. Dealing with cultural heritage, especially looking at it from the intangible perspective, is not only between preservation and protection, but cornering to the need of associating different local resources that have been disconnected and belong to various sectors; then the circularity implies a theory of re-coordination that cannot be only going back to something already existent. According to some observers, in addition to encouraging eco-design, there would be a need to include the reciprocal behaviour of producers in an industrial and territorial ecology (ITE). By this definition, it is meant to promote the exchange of flows (materials and energy) and the pooling of needs among actors in a logic of proximity. To this we can link the debate on industrial clusters and districts, with a certain caution: one thing is the effectiveness of these forms of organisation among enterprises once they are created; another thing is to bound them to constraints of different economic dynamics and those of administrative dynamics. On the other hand, public support for companies creating networks can only go in the desired direction. Districts should be defined as a set of various activities and not as a set of one unique type of activity (Santagata 2002, 2007). It means that a district relies on the idea that this diversity is nurturing the sustainability of the district, even if some of these activities may be partly complementary and/or partly competitive. To make this fully effective and loop orientated, the circular economy process should inspire the management of cultural districts as an objective and a value that can be achieved by introducing specific, detailed, and well-designed forms of public tenders for public financial support that perfectly channel contestants according to precise rules of sustainability.

4 Discussions and Recommendations

The CLIC project highlighted once more the essential role of the intangible to generate the values of the tangible. There is no tangible without the intangible and vice versa. What transforms residuals from the past into heritage is their intangible component, their set of intangible qualities that per definition cannot be touched, modified, extracted, or alienable. Such dimensions of heritage found expressions in knowledge, taste, and consequently in their projecting phases done by design and crafting, with infinite opportunities for and from the circular economy. The experiences had led to the following pro-grammatic conclusions, which open recommendations.

Considering waste as an uncertain spatial and temporal category. As we have mentioned, cultural heritage represents an interesting example that demonstrates how waste is mostly a spatial and not temporal category, even though it is matter of time. In simple words, from the point of view of alternative narratives of waste, heritage might be seen as a fragmentary territory of residuals with strong symbolic assets, that embed obsolescence as a value and not as a lack. To become waste, as an obsolete thing, it has to be out of place, somehow de-contextualised. Hence, the status of waste depends on how we order, classify, categorise, and relate to our living context. As Douglas argues, it matters 'Where objects are' and not 'What they are', in other words, its status depends on our capacity to re-arrange and re-contextualise. In this sense, waste reveals our capacity for inhabiting. But, as heritage testifies, time matters: when we bring time into analysis, relieving the history, the lifecycle that obsolete or non performing embodies, we actually make a first step towards its recovery. By making use of time that is entrapped within, we actually relieve the space of its replacement.

Intangible as a driver. We assume that tangible is something physical having a concrete materiality, perceivable by senses, that does not need a mental construction to be known and recognisable; on the contrary, intangible as something that is recognisable by the individuals or community, because of the intellectual, spiritual, i.e. social constructions they have in mind, such as knowledge, practices, expression, and skills (UNESCO guidelines on ICH). However, if we consider tangible and intangible as inseparable, then we may acknowledge the rise of information asymmetry since physical tangibility does not per se guarantee that the user possesses its recognisable representation, because of the lack of experience with it (Laroche et al. 2001). In this circular loop, it is impossible to consider tangible cultural heritage as such without guaranteeing accessibility to its embedded intangible information, and vice versa, the performative aspect of intangible as tangible communicative competence (Bauman 1975). It is interesting to note that the Latin word *intactilis* as intangible meant "that remain intact, inviolable, sacred", which cannot be changed and remains constant. In this respect, considerations about reuse driven merely by development run often the risk to lose the intangible component of heritage, its capacity to move the knowing how, thus what makes of it unique.

"Proximity" as a model for reactivation of circular logic in urban landscape. If we consider that in landscape any outside, if "as perceived" (CoE 2000), or as experienced by people, is just another additional Inside, then there is no External as opposite of Inner. Actually, in Landscape there is no public space, which is not and has not, at the same time profoundly private dimension. The experiential dimension that (historic urban) Landscape imposes folds our inner private dimension from one context to another, in a way that we are obliged to perpetually re-design it; it is the contour, the context, that reactivates a new engineering of Inner and gives meaning to the co-existence of both. Then, the success of this coexistence depends on how fine their "proximity" is designed. We can even extend this contiguity of inner and external imaginary to the "proximity" design of new in or next to old, of innovation in tradition, leisure, and production, as the alchemical glue that sticks together all components of Landscape in a single story. As a matter of fact, Historic Urban Landscape can also be seen as a step towards the dramatic recognition that there is no outside (natural or cultural), no detachment between inner and outer urban conglomerates, there is no external, peripherical, and new whose purpose is to nourish an inner, central, and historical, and vice versa. In other words, we should give important emphasis to proximity, to what is perceived as in between, as a glue that sticks things together all components in a coherent ensemble; this exactly was the effort of co-designing a cultural corridor in Rijeka.

An alternative idea of action. As introduced in the chapters investigating craftmanship and design attitude, the idea of care, repair, and maintenance is standing right on the opposite of the building, constructing, demolishing, and innovating at any cost. Something much closer to the realms of cultivation and its capacity to contrast obsolescence. As per craftsman repair, gardener cultivation represents a model of a different culture of doing and producing where learning is not separated from doing and disturbance is a constitutional part of innovation forces. The idea of "innovation in cultivation" demands for dynamic experiential knowledge that does not stop where implementation problems begin and is inseparable from as well as driven by the action.

A new production grammar should be inspired by co-design with cultural heritage and nature might be introduced, beyond the mainstream learning from natural cyclical metabolism. In introductory chapters, we have investigated how the complexity of designing with residuals and the cyclical notion of time that tradition imposes might represent an occasion for an alternative circular design path. This unprecedent pattern of re-designing with residuals that affects circular economy relies at the same time on conservation (designing with heritage and its time dimension) and innovation (designing with nature and its metabolism), merging both under the same important "normative" question. The circular economy design imaginary embraces the complexity in a way that it cannot be exhausted exclusively in an empirical, quantitative, or doctrinal way. As a matter of fact, the estimations as true, false, correct, or wrong, result irrelevant within this design realm, because there are only things that are well or badly designed. This is exactly the normative ambiguity that affects decision makers, although it might represent its main driver: we consider the rise of ambiguity when we cannot embrace the complexity of multiple and different representations of the same thing under the single evaluation umbrella (Kovacic and Di Felice 2019).

Open character of circular economy design pattern. The CE approach should always leave some space to the unpredictable, an opening for a new interpretation, an additional re-design action by users, as buildings become dwellings only when they are inhabited and books are completed only when read, when the user's imaginary enters in between the written lines. The circular economy project should not be (and things, because behind every single object there is a project) intended as given once for all, but as a reactivator of interrelation and experience with the user, as an opening where change and vitalism drive continuous slippage towards successive adaptations. Only through this opening of the post-co-design through use does that ante-co-design participatory act as empowering practice.

Acknowledgments

Conceptualization: Acri, Dobricic, Greffe
Data curation: Acri, Dobricic, Greffe
Formal analysis: Acri
Funding acquisition: Acri, Dobricic
Investigation: Acri Dobricic Greffe
Methodology: Acri Greffe
Visualization: Acri Dobricic
Writing—original draft: Acri Dobricic Greffe
Writing—review & editing: Acri Dobricic Greffe

References

Acri M, Bonafede L (2000) The Serenissima woods and their utilisation in the city building trade. About Cadore. In: Suikkari R, University (eds) Historical European towns—identity and change. University of Oulu, p 7

Acri M, Dobričić S, Jokilehto J (2019) The circular economy in adaptive reuse: respecting authenticity and integrity. Il Patrimonio Culturale in Mutamento Le Sfide Dell'uso. Scienza e Bani Culturali 2019, Venezia. http://www.arcadiaricerche.it/editoria/2019.htm

Adorno TW (2009) Teoria estetica. Einaudi

Agamben G (2015) Gusto. Nottetempo

Appadurai A (ed) (1986) The social life of things: commodities in cultural perspective. Cambridge University Press. https://doi.org/10.1017/CBO9780511819582

Arendt H (1989) Vita Activa. La condizione umana (A dal Lago Ed). Bompiani

Barkley Rosser J Jr (2003) A Nobel Prize for asymmetric information: the economic contributions of George Akerlof, Michael Spence and Joseph Stiglitz. Rev Polit Econ 15(1):3–21. https://doi.org/10.1080/09538250308445

Bauman R (1975) Verbal art as performance. Am Anthropol 77(2):290–311. JSTOR

Benjamin W (1991) L'opera d'arte nell'epoca della sua riproducibilita' tecnica. Einaudi

Bond M (2017) The hidden ways that architecture affects how you feel. https://www.bbc.com/future/article/20170605-the-psychology-behind-your-citys-design

Brandi C (2000) Teoria del restauro (Rist). G. Einaudi

Clark A, Chalmers D (1998) The extended mind. Analysis 58(1):7–19. https://doi.org/10.1093/analys/58.1.7

Coccia E (2014) Il bene nelle cose: Pubblicità come discorso morale. Il mulino

Codello R, Torsello A, Scarpa C (2009) Architetture veneziane di Carlo Scarpa: Percorsi e rilievi di cinque opere, 1st edn. Marsilio

CoE (2000) European landscape convention. Council of Europe. https://www.coe.int/en/web/landscape/text-of-the-european-landscape-convention

Cornette J (1989) La révolution des objets. Le Paris des inventaires après-décès (XVIIe-XVIIIe siècles). Revue d'Histoire Moderne & Contemporaine 36(3):476–486. https://doi.org/10.3406/rhmc.1989.1507

Daalhuizen J, Badke-Schaub P, Batill S (2009) Dealing with uncertainty in design practice: issues for designer-centered methodology. In: ICED 09—the 17th international conference on engineering design, Vol 9: human behavior in design, pp 147–158. https://orbit.dtu.dk/en/publications/dealing-with-uncertainty-in-design-practice-issues-for-designer-c

Dal Co F, Mazzariol G (2006) Carlo Scarpa (1906–1978), 2. ed. paperback. Mondadori Electa

De Fusco R, Fiorino I (2004) Storia dell'arredamento: Dal '400 al '900. Francoangeli

Di Giacomo G (2013) La questione dell'aura tra Benjamin e Adorno. Rivista di estetica 52:235–256. https://doi.org/10.4000/estetica.1626

Doing P (2011) Review essay: Tacit knowledge: discovery by or topic for science studies?: Michael Polanyi, The Tacit Dimension (University of Chicago Press, 2009), 128 pp., £11.00/24.77/$16.00. ISBN 978-0-226-67298-4 (pbk). Harry Collins, Tacit and Explicit Knowledge (University of Chicago Press, 2010), xiii + 182 pp., £21.00/26.03/$32.50. ISBN 978-0-226-11380-7 (hbk). Soc Stud Sci 41(2):301–306. https://doi.org/10.1177/0306312710397690

Dorfles G (1972) Il Kitsch. Antologia del cattivo gusto. Mazzotta

Douglas M (2005) Purity and danger: an analysis of concept of pollution and taboo. Routledge

Ferrero, C. (2014, giovedì agosto 06:00:00). Moda: Troppi stilisti o aspiranti tali e pochi sarti, modellisti e figurinisti. Repubblicastagisti. http://www.repubblicadeglistagisti.it/article/moda-troppi-stilisti-o-aspiranti-tali-e-pochi-sarti-modellisti-e-figurinisti

Frey JM (2016) Spolia in fortifications and the common builder in late antiquity. Brill

Goodman N, Morizot J (1998) Langages de l'art: Une approche de la théorie des symboles. J Chambon

Grant J, Fox F (2007) Understanding the role of the designer in society. J Art Design Educ 11:77–87. https://doi.org/10.1111/j.1476-8070.1992.tb00689.x

Greenberg C (1939) Avant-garde and Kitsch. Partisan Review

Greffe X (2014) Innocraft final report

Greffe X (2017) The artist–enterprise in the digital age, vol 1. Springer. https://doi.org/10.1007/978-4-431-55969-6

Grlić R, Potokar B (2014) HiLoVv hidden live(r)s of Venice: three omnibuses of short student documentary films. School of Arts

Hannah D (1969) Architecture of alienation: the double bind and public space. IDEA J:43–58. https://doi.org/10.37113/ideaj.vi0.219

Hennion A, Latour B (1993) Objet d'art, objet de science. Note sur les limites de l'anti-fétichisme. Sociologie de l'art 11

Illich I (1984) Dwelling. (631ES)—Atlas of Places. https://www.atlasofplaces.com/essays/dwelling/
Jackson MO, Morelli M (2011) The reasons for wars: an updated survey. Edward Elgar. https://cadmus.eui.eu//handle/1814/17631
Jokilehto J (2018) A history of architectural conservation. Routhledge, Taylor and Francis
Jokilehto J (2019) Questions of authenticity. Conversaciones…Con Herb Stovel 8:18
Kakiuchi E, Greffe X (2015) Culture, creativity and City. SUIYO-SHA
Kant I (1993) Critica del giudizio. UTET
Kirchherr J, Reike D, Hekkert M (2017) Conceptualizing the circular economy: an analysis of 114 definitions. Resour Conserv Recycl 127:221–232. https://doi.org/10.1016/j.resconrec.2017.09.005
Klamer A (2017) Doing the right thing. Ubiquity Press/JSTOR. http://www.jstor.org/stable/j.ctv3t5r85
Kovacic Z, Di Felice LJ (2019) Complexity, uncertainty and ambiguity: implications for European Union energy governance. Energy Res Soc Sci 53:159–169. https://doi.org/10.1016/j.erss.2019.03.005
Kovacic Z, Strand R, Völker T (2019) The circular economy in Europe: critical perspectives on policies and imaginaries, 1st edn. Routledge. https://doi.org/10.4324/9780429061028
Laroche M, Bergeron J, Goutaland C (2001) A three-dimensional scale of intangibility. J Serv Res 4(1):26–38. https://doi.org/10.1177/109467050141003
Latour B (1991) Nous n'avons jamais été modernes: Essai d'anthropologie symétrique. Éditions La Découverte. http://spire.sciencespo.fr/hdl:/2441/5l6uh8ogmqildh09h2ppn89a8
Latour B (2007) Petites leçons de sociologie des sciences. La Découverte
Latour B (2008) A cautious Prometheus? A few steps toward a philosophy of design. In: Hackney F, Glynne J, Minton, Viv (eds) "Networks of design", annual international conference of the design history society. Universal Publishers, pp 2–10. https://hal-sciencespo.archives-ouvertes.fr/hal-00972919
Lazzarini L (2008) I pavimenti alla veneziana. Cierre
Leroi-Gourhan A (1943) L'homme et la matière. Évolution et techniques, I. Albin Michel
Liverani P (2011) Reading Spolia in Late Antiquity and contemporary perception. R. BRILLIANT—D. KINNEY (a Cura Di), Reuse Value. Spolia and Appropriation in Art and Architecture, from Constantine to Sherrie Levine, Farnham (Surrey) 2011, pp 33–51. https://www.academia.edu/21644887/Reading_Spolia_in_Late_Antiquity_and_Contemporary_Perception
Menary R (ed) (2010) The extended mind. MIT Press
Merleau-Ponty M (1945) Phénoménologie de la perception. Gallimard
Murray A, Skene K, Haynes K (2017) The circular economy: an interdisciplinary exploration of the concept and application in a global context. J Bus Ethics 140(3):369–380. https://doi.org/10.1007/s10551-015-2693-2
Nonaka I, Takeuchi H (1995) The knowledge-creating company: how Japanese companies create the dynamics of innovation. Oxford University Press
Norberg-Schulz C (1979) Genius loci: Paesaggio, ambiente, architettura. Electa. http://books.google.com/books?id=smJQAAAAMAAJ
Penzo G (2000) Navi veneziane: Catalogo illustrato dei piani di costruzione = Venetian ships: an illustrated catalogue of draughts. LINT
Pickles D (2016) Energy efficiency and historic buildings: insulating solid walls. Historic England. https://historicengland.org.uk/images-books/publications/eehb-insulating-solid-walls/heag081-solid-walls/
Pickles D, McCaig I, Wood C (2014) Traditional windows—their care, repair and upgrading. English Heritage. https://historicengland.org.uk/images-books/publications/traditional-windows-care-repair-upgrading/heag039-traditional-windows-revfeb17/
Polanyi M (1966) The tacit dimension. Routledge & Kegan Paul
Pratt M (1981) La filosofia dell'arredamento: I mutamenti nel gusto della decorazione interna attraverso i secoli dall'antica Roma ai nostri tempi. Longanesi

Rea R (1999) Il Colosseo. Destrutturazione e riuso tra IV e VIII secolo. Mélanges de l'école française de Rome 111(1):183–195. https://doi.org/10.3406/mefr.1999.3685
Reckwitz A (2002) The status of the 'material' in theories of culture: from 'social structure' to 'artefacts'. J Theory Soc Behav 32(2):195–217. https://doi.org/10.1111/1468-5914.00183
Rivallain J (2001) Cabinets de curiosité, aux origines des musées. Outre-Mers 88:17–35. https://doi.org/10.3406/outre.2001.3878
ROTOR (2021) About – Rotor Deconstruction. Rotor Deconstruction i. https://rotordc.com/about/
Ruskin J, Rosenberg JD (1987) Le pietre di Venezia, Rizzoli
Ryle G (1945) Knowing how and knowing that: the presidential address. Proc Aristot Soc 46:1–16
Ryle G (1949) The concept of mind. Barakaldo Books, Chicago
Santagata W (2002) Cultural districts, property rights and sustainable economic growth. Int J Urban Reg Res 26(1):9–23. https://doi.org/10.1111/1468-2427.00360
Santagata W (2007) La fabbrica della cultura: Ritrovare la creatività per aiutare lo sviluppo del paese. Il Mulino
Saqer Mustafa Sqour (2018) Attaining human aspects to avoid alienation in architecture. J Civil Eng Archit 12(2). https://doi.org/10.17265/1934-7359/2018.02.006
Schnapper A (1988) Collections et collectionneurs dans la France du XVIIe siècle. Flammarion
Sennett R (2013) L'uomo artigiano. Feltrinelli
Settis S (ed) (1996) Dalla tradizione all'archeologia, 2. rist. Einaudi
Simmel G (1900) A chapter in the philosophy of value. Am J Sociol 5(5):577–603. https://doi.org/10.1086/210919
Stiglitz JE (1975) The theory of 'screening,' education, and the distribution of income. Am Econ Rev 65(3):283–300. JSTOR
Stovel H (2007) Effective use of authenticity and integrity as world heritage qualifying conditions. City Time 2(3):16
Tanaka S (2011) The notion of embodied knowledge. In: Stenner P, Cromby J, Motzkau J, Yen J, Haosheng Y (eds) Theoretical psychology: global transformations and challenges. Captus Press, pp 149–157
Temesgen A, Storsletten V, Jakobsen O (2021) Circular economy – reducing symptoms or radical change? Philos Manag 20(1):37–56. https://doi.org/10.1007/s40926-019-00112-1
Tuan Y-F (2011) Space and place: the perspective of experience, 7. print. University of Minnesota Press
Webster K (2017) The circular economy: a wealth of flows, 2nd edn. Ellen MacArthur Foundation Publishing
Wittgenstein L (1966) Lectures on aesthetics, psycology and religious belief. Blackwell
Wolf JR, Myerscough MA (2007) Reputations in markets with asymmetric information: a classroom game. J Econ Educ 38(4):393–405. JSTOR
Zevi B (1948) Saper vedere l'architettura. Einaudi. https://www.einaudi.it/catalogo-libri/arte-e-musica/architettura/saper-vedere-larchitettura-bruno-zevi-9788806201067/

Chapter 4
The Intimate Circular Dimension of Heritage Conservation and Historic Urban Landscape

Saša Dobričić, Marco Acri, and Jukka Jokilehto

1 Introduction

Humanists, philosophers, archaeologists, planners, architects, anthropologists, sociologists, applied scientists, and experts from a variety of backgrounds have contributed to and debated the doctrine of conservation over the course of centuries. Their ideas have always been cautious, object- and context-related, and continuously updated on the accomplishments of the past. It has also been viewed from various angles by critics, who see it as either a modern movement that poses as anti-modern or as an anti-modern movement itself (Glendinning 2003). Experts are aware of the complexity of the conservation doctrine. Nevertheless, even when working with traditional and vernacular architectures, there are certain benchmarks that cannot be overlooked, misapplied, or avoided. This is especially true when it comes to the concept's financial sustainability. Using the lens of authenticity and integrity, the heritage preservation movement has always placed a strong emphasis on the various aspects of culture, such as cultural tangible and intangible heritage, and conservation, such as restoration, preservation, protection, maintenance, preventive maintenance, etc., (Cristinelli and Trabucco 2017; Jokilehto 2017; Tyler et al. 2018), stressing the significance of cultural heritage for the advancement of society towards more sustainable habitats. The movement has made numerous contributions that are rich in meanings and approaches. Because heritage is a byproduct of

Jukka Jokilehto died before publication of this work was completed.

S. Dobričić · M. Acri (✉)
Graduate School, Cultural Heritage Studies, University of Nova Gorica, Nova Gorica, Slovenia
e-mail: sasa.dobricic@ung.si

J. Jokilehto (Deceased)
Graduate School, Cultural Heritage Studies, University of Nova Gorica, Nova Gorica, Slovenia

L. Fusco Girard, A. Gravagnuolo (eds.), *Adaptive Reuse of Cultural Heritage*,
https://doi.org/10.1007/978-3-031-67628-4_4

society, it's a special kind of tangible witness of historic presence that disappeared (Leroi-Gourhan 1943). It is an inestimable capital. The preservation movement has been insisting on this for decades in contrast with development pressures that led to extensive losses of cultural and natural heritage worldwide. Such losses are not only enormous in terms of quantity, but mostly in terms of uniqueness and diversity, as many international documents report (Gracia Aldaz et al. 2016, among others). Heritage, cultural and natural, is diversity, and the global linear economy we have been supporting for decades is per definition an antagonist of diversity, because it proliferates thanks to the standardisation of production and the fast global dissemination within the use and replace logic. Diversity should be seen at different scales, as if we were looking at an ecosystem: macro, the forest—the city; meso, the tree—the building; micro, the insects—the techniques. We could even add a smaller scale, the material, and its composition. All these scales have the same equal importance in the authenticity and integrity of a heritage asset; thus, all perspectives should be enabled and respected. If "landscape is as perceived by people" (CoE 2000) all cultural layers contribute to the richness of heritage, living to the expert the humble role to highlight, whenever needed—namely through conservation or adaptive reuse—what ultimately needs to be given up in order to prolong the cultural heritage asset's life.

Said that, the concept of adaptive reuse was used by CLIC with a serious commitment: showing how is has been turned into the mere idea of "culture has to adapt", while the doctrine and the expert view is keen in conceiving adaptation as a one-to-one status, namely cultural assets vs humans and humans vs cultural asset.

Circular economy in this respect is a true ally for conservationists and humanists, due to its intrinsic mission to preserve the embodied energy, to reduce waste of resources and to get sustainable satisfaction from existing assets. The 7 Cs of circular economy, namely Recycling, Repurpose or upscaling, Remake or remanufacture, Refurbish and repair, Reuse, Reduce, Rethink and refuse (Girard and Gravagnuolo 2017; Potting et al. 2017) have been for centuries daily criteria of production, thus intrinsically part of the process of heritage making. The different that we face today in respect to the past, especially from the turning point in history of industrialization and the diffusion of a globalised economy, is given by the risk to erase heritage qualities through small, imperceptible changes, that in the long run may reach an irreversible status. If a parallel can be made, the climate change provoked by human activities, has already cancelled centuries, millennia of natural unique creations that will never return, not even with the most ambitious and virtuous project of humanity.

2 Conceptual Framework

2.1 What Is Heritage and Why It Is Important

But what are we trying to preserve and why this has relations with our economy? To understand the roots of the conservation theory and consequently why it has an intimate connection with our present definition of circularity, it is necessary to pick just few historic elaborations of the concepts of heritage and culture, to show how much the international concern on it has produced the achievements of today (Musgrave 1959).

Cultural heritage is defined by UNESCO as "the legacy of physical artefacts and intangible attributes of a group or society that are inherited from past generations, maintained in the present, and bestowed to future generations", although in practice it is a much more complex issue. Indeed, heritage is a fundamental aspect of humanity and of its development from generation to generation, an essential part of the learning processes of human beings within the different cultural and environmental contexts, and it is also the way to maintain a memory of past achievements. Today, a part of the heritage may be legally protected, while in other cases its conservation is left to the individuals or institutions to take care of. In a community, learning was partly based on verbal transmission of, and partly through participation in customs and rituals. In a family, the property was normally inherited by the children. In Latin, this was called *patrimonium* ('*pater*' father), a paternal estate, inheritance from a father or ancestors: this root is fully adopted in Italy today as *patrimonio*. The Hebrew Torah (the Old Testament of the Bible) accounts for several different forms of inheritance, such as giving a significant name to an individual or to a place, preserving an object as a testimony, or setting up a monument or temple (Jokilehto 2017).

Another form of inheritance is writing about past events. Writing history can actually be understood as a form of recognising something as heritage. The Bible itself is a collection of such books and memories, indeed, it is one of the earliest written histories in the world. Basically, history means inquiry or study of the past or knowledge acquired by investigation, as already indicated centuries ago by Ibn Khaldun (Khaldun 1978, p. 5):

> *History is one of the most widespread disciplines between nations (umam) and races (ajyal). The vulgar would like to know it. Kings and rulers seek it over and over again. Ignorant people can understand it as well as educated people. Indeed, history is, in appearance, only the account of political events, dynasties (duwal) and circumstances of the distant past, presented with elegance and raised by quotes. It distracts large audiences and gives us an idea of human affairs. It shows the effects of disturbances, it shows how such and such a dynasty came to conquer such a vast piece of land, until the day when the Call sounded, when its time was over.*

Vico criticised the expansion and development of modern rationalism and focused on the fundamentals of social science and of semiotics. Through *Verum esse ipsum factum* ("What is true is what is made") Vico legitimised the truthfulness of what still exist. The Neapolitan historian and jurist wrote that as God has created the world, only God can understand its intrinsic and extrinsic aspects (Vico and Cristofolini 2016; Vico and Lomonaco 2014), and only God can know it as a *factum*. Human beings can only learn about nature externally by observation. Considering that humanity has created cultural heritage, humans have the possibility to know its true character. Cultural heritage, however, is characterised by its diversification not only from region to region but also over time in the same region, by overlapping of apparently different cultures and civilisations (Lowenthal 2013).

Modern definitions of heritage have tended to focus separately on the different aspects of cultural heritage, such as movable and immovable heritage, material or tangible and intangible. Probably the most updated definition of cultural heritage comes from the so called Faro Convention (CoE 2005) stating that '*Cultural heritage is a group of resources inherited from the past that people identify, independently of ownership, as a reflection and expression of their constantly evolving values, beliefs, knowledge, and traditions. It includes all aspects of the environment resulting from the interaction between people and places through time.*' A definition that well refers to the contemporaneous UN document on Cultural Expressions (UNESCO 2005), notes that culture takes diverse forms across time and space and that human culture is characterised by its diversity. In Article 4, the Convention states that '*"Cultural diversity" refers to the manifold ways in which the cultures of groups and societies find expression. These expressions are passed on within and among groups and societies. Cultural diversity is made manifest not only through the varied ways in which the cultural heritage of humanity is expressed, augmented and transmitted through the variety of cultural expressions, but also through diverse modes of artistic creation, production, dissemination, distribution and enjoyment, whatever the means and technologies used.*'

2.2 Human Creativity, Genius Loci, Traditions, Sense of Place

We can see that there are fundamentally two aspects that are necessarily present in all products of humanity and thus in cultural heritage. One of these is the cultural concept or idea that represents the origin of a work. The other aspect is the material presence of the work that represents a record of its past, i.e., its history. The contribution of humanity has evolved gradually based on, human creative capacity and as a response to territorial constraints and emerging requirements of groups of people and communities. The whole world is generated by a life force (*élan vital*) and the results of past creativity then endure in matter, becoming a record of the history of human cultural achievements (Bergson 1998, p. 15). As a result of human creativity and duration over time, there are traditions, consisting of beliefs and customs passed down from generation to generation within a group or society, maintaining the symbolic meaning or special significance with the origins in the past (Geertz 1993, pp. 50–51). Creativity has been debated intensively worldwide, seen as the engine for humanity in problem solving, transforming obstacles and difficulties into solutions and opportunities. As such, creativity has been for long promoted as the leading sector for a new cultural preservation and progress movement (Santagata 2007). The creative aptitude has permitted the development of solutions to adapt to climate conditions, geographic constraints, abundant or scares availability of materials, isolation, multiculturalism, and many more site-specific conditions. It is mostly thanks to the creative human behaviour that every single spot in human history became unique, for the capability to generate ad-hoc, site related skills and forms of language. We refer to this with the locution *geius loci* (Markeviciene 2008; Norberg-Schulz 1979; Vecco 2020). The *genius loci* became so intense in terms of specialisation that in some areas they generated outstanding products that involved entire communities, for such being intrinsically permeating the living space. Economists called these phenomena cultural districts (Hinna 2015; Sacco et al. 2015; Santagata 2002; Valentino 2003) and there are plenty of examples like Caltagirone, Murano, Gragnano, Sevres, Nabeul, Piran are only a few in the long list. The *Genius Loci* is reflected in infinite manifestations. See Venice, how many inventions, specific to the site, could be introduced as tangible and intangible forms. Such a set of specificities, which is heritage by definition, is also contributing to the tradition, which is by many considered the engine of identity (Honko 1995; Jones 2000), although the link between conservation and identity (Piaget 1976) highlights both a convergence and a divergence, in the light of the diversity in time and space we have already mentioned. The etymology of 'tradition' refers to the Latin concept of '*tradere*', meaning: to hand over, give up, surrender. Indeed, traditions were normally confined to specific communities, where they were handed down from generation to generation. When continuing the tradition as a learning process, the customs and beliefs needed to be appropriated by the receiver, actuating the rhizomes (Deleuze et al. 2011) that are nourished by socio-cultural interactions. At the same time, the implementation of the past lessons would also imply creative action (but also empty the creative action) reflecting the changing needs and requirements. In fact, as already seen in the meaning of the word *traditio*, change is possible. Some (Hobsbawm and Ranger 2010, p. 4) argue that traditions differ from customs and the former may be invented by formalisation and ritualization, which has clear effects on the "identitarian" dimension we have taken some distance from. Change is possible, but in line with the intrinsic meaning of tradition that is related to the knowledge and savoir faire. This change is relevant, for example, in the work of an artisan, who maintains the tradition while being involved in the creation of new artefacts as part of an existing settlement or rebuilding decayed or damaged structures. Taking into account the infinite components that make a site unique, namely history and historical layers, cultural diversity in time and space, *genius loci*, traditions, and all the tangible manifestations at all scales of all this, which are the waste, the residual that is talking about the site, is what Leroi-Gourhan referred to from an archaeological and anthropological perspective (Leroi-Gourhan 1943). From the perspective of tradition, the focus on cultural heritage is mostly on daily-use objects and vernacular architecture or built environment. Such relationships were already recognised by the international community of experts in the following definition.

Vernacular building is the traditional and natural way by which communities house themselves. It is a continuing process including necessary changes and continuous adaptation as a response to social and environmental constraints. (ICOMOS 1999).

2.3 *Integrity, Authenticity and the Intrinsic Qualities of Heritage*

The recognition of heritage previously defined posed questions about if and how it must be preserved. This is an individual question that each of us faces in life, when, for example, a relative dies, leaving his heritage, *patrimoine*, *patrimonium*. Some criteria affect individual decision making, while in the case of the public domain, common criteria are necessary, which move from objective and subjective assessment according to the perspective. Among them we should like to stress the concepts of integrity, authenticity, and significance that are consolidated in the international debate (UNESCO World Heritage 2019) and, as such, adopted by most countries.

Integrity is defined as '*a measure of the wholeness and intactness of the natural and/or cultural heritage and its attributes*." The essence of integrity is that it includes all elements necessary to express the specificity of the asset, and it should not suffer from adverse effects of development and/or neglect. This can be applied to assets of different scales, not only architectures. The Galeb ship in the port of Rijeka, a national monument for its history as a strategic movable location for diplomatic events during the Tito presidency of former Yugoslavia, is presenting issues of integrity for the consistent loss of the original furniture and finishing, which actually make of it an outstanding asset. Integrity also affects the furniture and design market, including historic vehicles: if an historic car is missing parts or components, even though others are replaced by new, but original ones, its price decreases, sometimes even consistently. Integrity, though, should be seen also in traditional, vernacular architectures at different scales, including construcion details, such as tiles, windows, doors, handles. Replacements tout-court with subsequent changes in the wholeness of traditional structures and finishing, especially carried out in systematic ways in historic areas, determine a loss of integrity, consequently a loss of memory and uniqueness.

We refer to authenticity *when it comes to "our ability to understand … heritage, in part, on the degree to which information sources about … may be understood as credible or truthful*." (ICOMOS 1994), and it strongly affects decisions in the adaptive reuse of cultural heritage. The struggle on authenticity is in the daily life of preservationists, as architects, artisans, and antique dealers, with relevant impact on the truthfulness of the assets at stake. The actions of adaptive reuse may intake replacements with inappropriate materials, sometimes even reconstructions, offering an "almost like" situation. Such, almost like, multiplied by hundreds (and in the building sector, hundreds is a very easy number to achieve), determines fakes. If we refer to the previous definition of culture and heritage as based on the remains, the residuals of past creativity and genius loci, obtaining a fake through adaptive reuse is clearly a societal and economic failure.

Authenticity and integrity necessarily affect any conservation approach, placing at the core the interests and priorities of interpretation, whether of monumental, average, tangible, or immaterial nature. This critical dialogue with the artefact's past in the process of its recognition (of physical, aesthetic, and historic consistency) involves and affects all aspects of design practice and becomes its constitutive and methodological moment. The intervention, acting in its polarities of the freedom of creativity, evolves into a complex set of operations finalised at the preservation of an artefact's material integrity, at the preservation of the signs of time, and the transmission of such significance to the future. It is in this interpretative regime, where scrutiny, humbleness, and respect tend to prevail over inventiveness, that recognition and preservation of integrity and authenticity evolve and take form. However, it must be underlined here that creativity and inventiveness are not against conservation, but the opposite: there is much need for creativity in front of conservation constraints, and such creativity and inventiveness look really like the ones that has generated heritage for centuries. As an example, the building of Pakhaus de Zwijger in Amsterdam, hosting a CLIC partner, was creatively converted into a new space, paying attention to its authenticity and integrity through a creative effort to design a project enabling conservation and adaptation to new standards.

Applied firstly to World Heritage sites, the concept of Significance seems appropriate to be used when approaching members of a society that look at their environment. The significance of a resource recognised as heritage is defined in reference to material or immaterial sources, such as form and design, materials and substance, use and function, traditions, techniques, and management systems, location and setting, language, as well as spirit and feeling, and personal attachments or beliefs. Much of it contribute authenticity and integrity. Such a concept affects heavily the process of adaptive reuse because it may guide the decision-making process about what should be done with the most complete knowledge possible about the characteristic of an asset. As we have seen, the higher the number of sources in the significance, the smaller will be the freedom to change the asset, moving the adaptation from the asset side to the user side.

Looking at significance, authenticity, and integrity, the position of preservationists has always been orientated to the emphasis of the intrinsic qualities of an asset, independently from the external factors that judge its potential adaptation. Although culture economics positively entered the debate providing arguments to justify preservation actions, such contributions started from the value-based approach (Forsyth 2013). The value of something can be considered the quality that makes it useful or desirable in reference to other properties that have similar characteristics. An historic building or object gives us a sense of wonder and makes us ask more about the people and culture that produced it. It can be associated with different values, ranging from aesthetic to economic and even political (Feilden 1982). The preservationists have though always preferred to look at cultural heritage for their intrinsic qualities, that are totally independent from comparison with similar assets, giving so a different perspective, the one of cultural heritage and not the one of the developers. As an example, a traditional house in a historic centre is "valuable" because it may display original masonry, with typical plastering, with handmade window frames and shutters, or with vernacular roof structures made by locals in community based works. As such, it deserves conservation approaches that merge preservation through innovation, maybe also viceversa: any change determines a loss of qualities and, consequently, of economic value. This perspective, in itself, is fully aligned with the circular economy.

2.4 Restoration…and Adaptive Reuse …and the Urban Scale

The flash concepts that we have been proposing so far are simply a condensation of the reasons behind the conservation theory. Although maintenance (as a practice of care, of extension of the life of an object, as we constantly do with our cars and our leather shoes) required daily practices in heritage making, they became topics with the invention of history and the development of industrialisation that rapidly annihilated practices, skills and knowledge that had been perfecting for centuries. Destructions deriving from fast urbanism and wars sped the common realization about the cultural and social losses and preservation replaced maintenance in the daily imaginery. The word "preservation" is though addressing the large set of actions to safeguard cultural and natural heritage, including the preventive, legal and administrative structures. Restoration, as part of preservation (or conservation) addresses the design and implementaition of the physical intervention.

> *Restoration consists of the methodological moment in which the work of art is recognized in its physical being and in its dual aesthetic and historical nature, in view of its transmission to the future* (Brandi 2000).

Referring to the words of Brandi, though we are here speaking of a special type of heritage, i.e., the work of art, this definition can also be taken as a reference to better understand the significance of cultural expression by humanity, whether dealing with individual objects or larger territories that have resulted from the design or planning by human beings. The fundamental issue here is to distinguish between the artist's idea or architect's concept, on the one hand, and the material execution or construction, on the other. The artistic idea or concept represents the intangible aspect of the cultural expression, which does not age over time. The material (matter) of the work represents the tangible aspect that also carries its history.

As we have seen above, the terminologies concerning the recognition and conservation of the cultural heritage find different expressions in various languages. For example, while the word conservation is normally used in English texts by UNESCO as a general concept related to heritage, in the USA, preference is given to historic preservation, in Latin countries to restoration, and in Eastern Europe to reconstruction. There are other more specific terms that also need to be properly understood, such as: monument, historic monument, property, buffer zone, setting, environment, cultural landscape, historic urban landscape. In the relevant international conventions, recommendations, and charters, the 'international English' uses these various terms with specific meanings. So, while respecting the local traditions, it is also necessary that the translation or explanations are clear in order to avoid any misunderstandings.

Adaptive reuse can be defined as partly in opposition with the conservation theory, more a fascinating architectural project (Plevoets and Cleempoel 2012, 2013) that could react to the growing tendency of demolishing old buildings in favour of new architectures that had less and less to dialogue with the remaining historic fabrics and with the historic tissue. Interesting contributions have been made in this approach, including sorts of "manus" for the adaptation of specific buildings according to their typology or structures (Cantacuzino 1975). What differentiates the adaptive reuse from the traditional conservation approaches, namely reconstruction, critical restoration, and pure conservation, is the start: looking at the object, usually a building, as a box to be used for new, potentially of any kind, uses, while the traditional approaches were asset oriented, namely worthy to be preserved per se, for their intrinsic qualities, with possible new uses if compatible with such qualities. In this light, adaptive reuse has not been considered yet a fundamental reference theory, but a sort of drift of the architectural design applied to cultural heritage. Nevertheless, outstanding examples in this field can be listed, such as the works of Carlo Scarpa in Venice (Codello et al. 2009; Dal Co and Mazzariol 2006), Massimo Carmassi in Italy (Ingersoll and Ciampi 2002; Matteoni et al. 1992) and Andrea Bruno in Paris and Turin, if we refer to contexts with an "orthodox" conservation tradition, there are many more growing examples worldwide (Bullen and Love 2011a, b).

The ICOMOS 1993 Training Guidelines (ICOMOS 1993) recognise that conservation of the cultural heritage now rests within the general field of environmental and cultural development. It is a cross-disciplinary activity based on sustainable management strategies while guaranteeing continuity without the loss of authenticity and meaning and respecting the cultural context. Conservation can also refer to the treatment of specific elements of a property. In Latin countries, there is a tendency to use '*restauro*' or '*restauration*' as a general concept (Urbani and Zanardi 2000) corresponding to the English use of conservation. The Venice Charter refers to 'the process of restoration' as a highly specialised operation, aiming to preserve and reveal the aesthetic and historic value of the monument while respecting the original material and documented evidence. As an extreme approach, even though to our perspective probably the closest to the perspective of the circular economy, the restoration of pure conservation pursues the rigorous conservation of the artefact in the completeness of its stratifications and the conservation of the patina, as a sign of the transformation of the material over time (Casiello De Martino 1996; Dezzi Bardeschi and Locatelli 1991).

The question of reconstruction or rebuilding in heritage areas continues to be subject to diverse interpretations. The Venice Charter notes in Article 15, while to referring to ruins, that '*all reconstruction work should however be ruled out a priori*'. The Charter only accepts anastylosis, i.e., the re-assembling of existing but dismembered parts. In Italy, the reconstruction of historic buildings can be called '*ripristino*', which means rebuilding lost elements and removing parts not considered consistent with the original form (Marconi 1988, 1993, 1999). Any reconstruction should be legible and reversible. It can also be noted that there is a tendency to distinguish between the reconstruction of historically documented elements or features, on the one hand, and the modern rebuilding of a lost area, on the other. This distinction, however, is not always clear. While a reconstruction is expected to be based on fully documented evidence, the interpretation of such evidence is not always successful in practice.

The international debate does not explicitly favour one approach over another but requires that the qualities and specificities of cultural heritage are guaranteed as much as possible. As transversal, leading concepts in the practice of conservation and restoration, the *patina*, the layer of history on cultural heritage, and the reversibility, namely the characteristic of implementing solutions that could be easily removed or dissembled to return to the original state of the cultural asset, are often mentioned.

The issue of historic areas developed almost in parallel with the restoration/conservation theory. The first Italian modern master plan for an historic town was prepared in the 1950s. This plan not only recognised the historic city as urban heritage, but it also extended the recognition to the entire municipal area, including the natural areas and related religious sites, in light of the importance of understanding the physical, social, and economic context in order to be able to determine required action (Astengo 1958). The planner was involved in the organisation of the conference of Gubbio, which laid down the principles of urban conservation in '*La Carta di Gubbio*', still considered a reference document for the preservation of historic cities. This charter rejected all reconstruction and stylistic additions, demolition of even modest buildings with environmental character, any "thinning" and "isolation" of monumental buildings obtained by demolitions, or new additions in ancient fabric.

The questions of urban morphology and traditional typology were further studied by Muratori and Caniggia. The former recognised that building typology was intimately associated with the characteristics of buildings in each period, evidencing a genuine process of organic growth, a creative response to the needs of society. While being an integral part of the urban fabric, it also allowed for individual solutions, obtaining its specificity and identity as a unit in the continuous building process (Cristinelli 2013; Jokilehto 2017, pp. 195–198; Muratori 1967). The latter recognised that a community that resides in a particular location, conditions the related structures with specific individual and codified characteristics (Caniggia 1976). Within this context, typology is a system on the basis of which existing objects are matched in temporal-spatial succession. Such processes were strictly associated with the planning and design of historic urban areas. These included the experiences of Bologna and Ferrara, where instruments were developed for the analysis of the historic urban fabric. This included the identification of the historical periods that were expressed in (1) the evolution of the urban morphology, (2) the analysis of the typologies of buildings and spaces resulting from the private and public functions, (3) verification of the present condition of typological characteristics and function of the built fabric, and (4) the proposed treatment in respect of the traditional urban fabric. The scope was not only the preservation of the historic areas, but also their rehabilitation by bringing people back to such areas. This also required introducing appropriate services and infrastructures to were tailored in the spirit of the traditional context. The results of these analyses were formulated in planning norms specified for the requirements of the historic town within its context. The re-integration of urban lacunae was tested in case studies using an architectural language sympathetic to traditional fabric (Cervellati 1977).

These initiatives were taken over in parallel to other countries, especially the UK, even though the practice did not spread as much as in Italy. In particular, the contribution of Conzen, evidenced the evolution of the urban tissue considering three systematic aspects expressed by distinct form categories of town plan, building fabric, and land utilisation. A special room was given to the role of the community in historically shaping the urban morphology (Conzen and Whitehand 1981), stressing the importance of morphology and typology as witnesses of community behaviour in space.

Such movement found a significant scientific adhesion by the international community, confirmed by the issuance and ratification of planning documents (ICOMOS 1975; UNESCO 1976), recognising the need to preserve the historic tissue as a homogeneous entity, thus worthy to be preserved through an integral approach. Such movement ended recently with the Historic Urban Landscape approach (UNESCO 2011), at the basis of the CLIC project in the urban scale application of circular economy.

The planning for integrated urban conservation differs from normal urban design because the focus is to maintain and rehabilitate existing historic urban areas. While modern planning aims to respond to development with new design, integrated urban conservation works with the existing reality. The question is therefore to survey, research, and document the existing situation, understand the evolution of the urban morphology, and its component elements, the typologies of buildings and spaces, and their relationship with the social-cultural condition. The Integrated Urban Conservation Plan results from the critical assessment based on detailed surveys and analyses, and the establishment of a strategy for the improvement and rehabilitation. The aim is to bring back the lost population, and to provide the inhabitants with appropriate quality of living with appropriate required services and infrastructures, without compromising the historical character and qualities of the traditional neighbourhood.

3 The Circular Economy and the Conservation Approach

3.1 The Heritage Making and Maintenance as the Optimal Circular Practice

The apparently long, even though extremely shortened theoretical background of this paper on the concepts of heritage and conservation has the fundamental focus to re-establish the fundamentals of restoration and conservation approaches beyond the purely arbitrary practice of adaptive reuse, re-framing the conditions for the application of the principles of circular economy. In particular:

- showing that cultural heritage is the result of continuous interactions between humans and their environment (Dobričić and Acri 2018), made of constant daily creative efforts of adaptation and perfecting living techniques. As such, heritage, with an important emphasis on traditional built environment and assets, is per definition circular for it mostly humble generation;
- showing that the international debate on conservation of heritage has committed and engaged thousands of experts from different disciplines, backgrounds, traditions, and beliefs, taking to some achieved and shared conclusions;
- the conclusions being that heritage has intrinsic qualities, in some contexts defined as values, that need to be considered as a priority in any restoration project and that the project itself should be designed to their respect. The international movement on conservation has not been moved by the intention of stabilising the status quo, as too often stated, but to allow a proper management of the change, capable of grasping the creative acts of transitions in dialogue with the past;
- that the reference to cultural heritage adaptive reuse is not purely orientated to the built environment, thus buildings, monuments, and engineering construction of the past, but should be seen at other scales, as the object and the urban scale. In this perspective, there is not a heritage dimension that allows designers and developers to exclude details because they are fully contributing to the wholeness of heritage. As an example, the hand-made iron nail is a detail of the country chest, the chest is a detail of the traditional farmer house, and the traditional farmer house is a detail of the cultural productive landscape.

If we consider the urban and rural living environment through the lens of the circular economy, thus reconsidering the already mentioned rhizomes (Deleuze et al. 2011), the principles of destructive economy (Schumpeter and Forte 2001) and de-growth (Latouche 2012), beyond the 7Cs steps, it is impossible not to put in evidence how the present practices are no longer sustainable and that the traditional ones instead were. Maintenance is one of them.

The making of heritage in the past was dependent on low costs of labour work and high costs of materials. Before the industrial revolution, namely when the globalised economy started, material extraction, elaboration, and transport were extremely difficult, thus costly, with a consequent obligation to work on and with the local materials. Today, on the opposite, the cost of materials has become extremely affordable thanks to the infinite innovation in industry and chemistry that facilitate extraction and transport, as well as the application on site, while the labour cost, the manpower, increased enormously thanks to the improvement in the protection of human rights.

This means that, if circular economy was once the only option, today we need a new narrative for its application. We could say, at the extreme, the circular economy is a business model in itself for the fact that it represents the mission of a productive process. If we need a new narrative, we need to use the consolidated knowledge on cultural heritage conservation to our advantage, recognising the fundamental creative bond between heritage and sustainability.

In this respect, we must open a window inside the conservation doctrine and look if the past could suggest solutions. As we have seen, conservation and restoration had become necessary as theoretical approaches when the concept of history and identity became crucial. Until then, towns and rural settlements had developed according to the principles of reuse and maintenance. We all know that some of the most important cities in the world grew by reusing and reconstructing existing materials. Venice, for example, is a well-known example of *genius loci* for its capacity to transform, by cleverly managing climate and logistic constraints, marshlands into a prolific and wealthy civilization. The history of Venice talks about continuous reuse of the bricks of the Roman city of Altino first and of the settlement of Torcellum/Torceo after. This practice of reuse continued all along its history, including discarded wood from the Arsenal, remaining pieces of stones from former buildings, and infrastructural works, up to any form of debris to invent the terrazzo floor. A practice romantically represented by Ruskin in his books (Ruskin and Rosenberg 1987). A Venetian history of reuse and recycling that can be visible in many other forms of heritage, as well as in the socio-economic structure of the city. The *genius loci* in traditional architecture must be seen also for what concerns the comfort of the living and of the urban space. This is told by the architectural and urban solutions, such as the building mass to hold energy (fresh in summer and cold in winter), the plastering to prevent humidity and overheating, the water collection systems not to waste rainwater (Gianighian et al. 2010; Laureano 2001), etc.

As a consequence, traditional settlements have guaranteed their architectural preservation through the practices of maintenance, relying on both the capability of city users, including citizens, to act through basic maintenance works, and on the relatively limited number of solutions and materials for the maintenance works, but mostly on the reproducibility, replicability, and substitutability of parts.

Almost all the conservation documents mentioned in this paper, from the Athens Chart and the Venice Chart to the Amsterdam Chart and the UN Conventions and recommendations, as well as the different conservation approaches that have been briefly introduced, endorse maintenance as the preliminary, essential practice to avoid conservation. In fact, as we have seen, conservation is a traumatic moment for a cultural asset, because it comes in an emergency and with the need for selective decisions. The experiences in the CLIC project often refer to maintenance.

The CLIC database on best practices and adaptive reuse circular models, as well as numerous proposals in the case studies of the partnerships, namely Salerno, Rijeka, and Amsterdam, consider maintenance a node of circular economy. In Rijeka, for example, during the CLIC project, the municipality opened the REPERAI centre, a renovated space in the inner city meant to provide space for citizens to practice or learn maintenance of their daily use devices, such as bikes, furniture items, electronics, etc. The maintenance is here guaranteed by the availability of sufficient room and facilities, like tools, as well as by the presence of "experts" that may guide users of the space to operate through awareness and safety. This model is becoming common in Europe and represents one of the successful stories in terms of business creation within the circular economy. What is missing in the story, often highlighted by CLIC, is the maintenance of the built environment carried out at the same level, namely through the cooperation with citizens.

Maintenance in itself has a different geographical "distribution" in Europe. It is very well rooted in some Nordic European countries, such as the UK, Netherlands, Norway, and Finland, while it had difficulties spreading in the southern countries and the Balkans. There is no evidence about the reasons for this, considering that maintenance is a constantly word in citizens' daily life (from the car engines regular checks, to the heating pumps annual performance verifications, etc). However, the tradition in the UK is probably related to the learning by doing approach that has also been founding the conservation position of the different, numerous, preservation bodies of the country. If we move from institutional English Heritage and Historic England, towards independently organised organisations like SPAB and National Trust, including its distinct office in Scotland, relevant attention is dedicated to maintenance and capability building, recalling the famous quote by Morris "to stave off decay by daily care" (Morris and Webb 1877). Historic England and English Heritage devoted important amounts of funding in issuing strategic guidelines for the historic environment preservation and maintenance (Historic England 2017; Impey 2015; Lane and Menuge 2006; Pickles 2016; Pickles et al. 2014). The SPAB and the National Trust, are trying to accompany building and cultural asset users along in the awareness and maintenance process. SPAB has a special session devoted to maintenance, named "Maintenance Advice", where users may find interesting, easily comprehensible suggestions on daily maintenance of rainwater fittings, mortars, lime and cement renders, old brick walls, doors, metal frames, etc. Such English practice is favoured by a societal aptitude to voluntarism in commons care (Nijhuis 2011).

Attempts in guiding users and experts in the maintenance processes, to avoid expensive and unsustainable restoration works have been done elsewhere, with different results. In Spain, for example, a successful story has been recorded in Ademuz, where local citizens and building site workers have been supported by an easy-to-use guidebook issued by the Polytechnic University of Valencia (Vegas López-Manzanares and Mileto 2008, 2014). A similar initiative was taken in Gorizia, where outstanding traditional architectures from the Austro-Hungarian period, rich in solutions, details, and materials, are rapidly disappearing in their intrinsic qualities due to an adverse economic contingency in the city and a related lowering of the building and conservation skills (Acri and Biasi 2019). Such an idea of producing manuals for the daily care of cultural heritage assets is propitiously supported by the general public and the social media, which are constantly upgrading through tutorials the available knowledge on maintenance practice. This share of information in the Creative Commons sphere helps the circular economy in cultural heritage preservation, although the information offered is not always aligned with the doctrine and the highest standards in conservation.

Maintenance has been recently upgraded with the concept of preventive maintenance, introduced in the Netherlands, Belgium, and UK already in the 1980s, but spreading as a good practice for traditional and monumental heritage. The Monumentenwacht, Monumentwatch, is organised as an assisting program for cultural heritage owners that permit preventive actions to decay: the process, in general financially managed through an annual member fee, guarantees regular inspections to advise owners about urgent preventive actions; owners themselves are considered as stewards of their property, thus asked to monitor regularly on the most vulnerable parts of the heritage asset. Duplications of the initiative have been tested recently in other European contexts thanks to research and academic cooperation, involving Flanders, Netherlands, Spain, Italy, and Germany, but the idea seems far from developing on a large scale. One of the possible answers to this difficult question is rooted in the traditional scepticism on maintenance: whether for cars and small devices the costs for regular checks are affordable, the costs for buildings are higher and thus worthy to be postponed. Another reason may be positioned in the "ignorance" by the general public about housekeeping and monitoring, which was a common fundamental in the past, today removed for the increasing specialisation of knowledge and asymmetry of information (Barkley Rosser Jr 2003) generated by the globalised economy.

Maintenance permits a continuation of the use of cultural heritage, regardless of the activity to be accommodated in it. It preserves the intrinsic qualities of heritage as poses an additional question when planning potential adaptive reuse ideas. Per se, maintenance and preventive maintenance are the quintessence of circularity and conservation, *by reproducing de facto* a practice performed for centuries.

In this respect, few words could be spent to mention the practice of minimal intervention. This practice consists in intervening on decaying cultural heritage by performing the minimal intervention possible to set up basic use conditions, including safety, comfort, and health solutions. This practice often considers cultural heritage as a liveable "ruin" by adopting the pure conservation approach plus minimal installation provisions. The effect is guaranteeing a non-critical preservation of the intrinsic qualities, even the non-necessary or irrelevant ones, and opening longer room for appropriate design and destination options. The practice of minimal intervention is also characterised by a genuine reversibility of interventions, allowing the cultural object to return fully to its status if needed. Very good examples in Europe are the World Heritage Site of Suomenlinna, Finland, where the interiors of the fortification were made available to the large public also for big events hosting, making clear that the temporary adaptation had to be one-to-one, user-building (Lind 2014). Also, the CLIC project witnessed this approach in Rijeka in the case of the museum of contemporary art located in the Bencic complex, to permit better assessment of the most suitable conservation project to be implemented in a second phase.

3.2 The Matter of Norms and Regulation

Heritage conservation practice is often bogging down when dealing with norms and regulations, and, as practitioners beyond academics, we must admit this is due to several drawbacks that range from knowledge and education at all planning and management levels to the inconsistency of norms, the pressures of the globalised market, and the flattening of competences and skills. This paragraph will insist on the former, given that a specific chapter in this book is dedicated to the knowledge and the market (Acri, Dobricic, Greffe).

The legislation at EU level is still pretty incongruous, despite the ratification and adhesion by most countries to the international charters, conventions, and recommendations on cultural heritage preservation, previously described. If on one side there is enough knowledge, as well as plenty of successful stories on cultural heritage conservation and reuse, countries are struggling to be aligned, in terms of legislation and competences, at the highest standards. This is also why we are still debating about the positive or negative connotation of the term "adaptive reuse." As we have seen, the conservation theory has very stable principles, that put the intrinsic qualities of cultural heritage in the front, even dealing with traditional and vernacular architecture, and ensembles, postponing the users need by assigning to the creative act the role of dealing with preservation bonds. The international community of preservation, though, was not efficient enough in transmitting the positive relationship between the respect of the conservation bond and creative projects, which is to say, preservation and adaptation of intrinsic qualities of the heritage asset. This asymmetry and misalignment still admit different levels of monitoring of such intrinsic cultural qualities in the EU countries. If legislation is not positive and user-friendly, evident gaps in the application will occur; in addition, if the legislation is not clear and capillary, providing all possible solutions and advantages to the conservation approach, shortcuts will always be possible. Examples may be provided.

The Green Deal has opened numerous possibilities in terms of retrofitting and financing for the upgrading of the European built environment (Rodrigues and Freire 2017). All possible innovative solutions on the market are becoming affordable thanks to the investments in the fight against climate change provided by EU and European countries, even though with different intensity. Nevertheless, the effects of these opportunities were not properly evaluated with respect to the preservation principles. If European quality principles (ICOMOS 2020) were recently issued, with undoubted advantage in the field, they still lack in addressing the traditional architectures, those that determine the quality of the historic urban landscape. The green deal promoting tools, with the focus on buildings retrofitting, openly allow the application of any solution that permits better performances, releasing additional responsibilities to the cultural heritage monitoring public and private organisations, as well as on the professionals involved. If, for example, the replacement of traditional windows can be done, with similar performances by means of PVC or wood frames, heritage owners may opt most of the time for the cheapest solutions, namely PVC. Same applies to external wall insulation: if the upgrading requires, upon financing, an insulation, operators and heritage owners that do not have specific knowledge or enthusism about heritage would opt for its application, even if in opposition with the conservation principles. As professionals, we were asked to design in Gorizia the coating for historic buildings replicating the old Austro-Hungarian stone mouldings by applying lightened concrete ones, *de facto* achieving a fake façade. Similar examples could be made for new installations, which are often relatively improving the comfort condition of housing but imply very heavy substitutions, demolitions, and consequent inappropriate finishing. Floor heating is an example of this trend, inviting heritage owners to find almost impossible solutions to the detriment of historic techniques, such as the terrazzo floor. The terrazzo floor is a specificity of many architectures, from Venice to the entire eastern Europe. Its replacement determines a loss, especially if done with other finishing such as ceramic or stoneware tiles, parquet, or, even worse, laminate planks. As a matter of curiosity, the cost of a terrazzo floor ranges today between 200 and 400 euros per square meter, while the ceramic or stoneware tiles range between 30 and 150 and laminate between 20 and 60, a comparison that explicates clearly even the monetary loss.

1. The urban conservation approach is fundamental for the appropriate management and governance of the change because it is based on a deep understanding of the morphological and typological developments of the urban areas, including traditional materials and techniques adopted in history. Too often we assist in EU to a facadism approach that associate the value of the urban landscape merely to the façade, to what specialists would call instead urban panorama. The typological approach, on the other hand, favours the highest levels of preservation with obvious advantages for the circular economy. Moreover, the typological conservation holds the advantage to generate an administrative process that is releasing conservation architects and civil servants from liability on wrong decisions, imposing preservation rules, with connected application cards as recipes, with clear distinctions about the restrictions and the opening. This may seem too strict, but in reality, it has evident advantages for the local sustainable economy because it pushes for creative action to accommodate new compatible designs in historic structures as well as support local businesses that get accustomed to the specificities of the traditional fabrics. Venice has adopted the same typological approach as Bologna and Ferrara and other EU cities. At present, Venice can count on an almost detailed specification of possible actions for each building typology, closing since the very beginning possible bizarre requests by the investors. At the same time, building sector operators in Venice are aware of the specific rules and have specialised on specific maintenance and recurrent interventions that are aligned with the basic conservation principles, with enormous advantages in terms of waste generation.

The already mentioned EU-ICOMOS Quality Principles address the topic of public tenders, which is another important potential enemy to quality and circularity. The principle of low price has opened the market to an infinite set of interferences that have reduced the qualities in heritage conservation. This is true because there are not appropriate quality certification tools and there is no guarantee about the specific skills required in the specific context (the *genius loci*). Rijeka, tendering for the Rihub and the Bencic, highlighted the fact that the imposition of local materials and local skills was impossible due to the tendering principles of best offer. In this respect, a possible solution, beyond the revision of the tendering rules not only for public and monumental buildings, but for all heritage assets in heritage areas, would be to act as the Soprintendenza in Venice on the occasion of the restoration of the Accademia Galleries (Codello 2014). In that case, the tender for the best offer was bound to the respect of intervention specifications to be priced per unit: this meant for the Soprintendenza to draw up an exhaustive list of possible quality interventions for each potential conservation problem, imposing the participants to implicitly declare their responsibility and competence for the actions taken. The effort made by the Soprintendenza in reality demonstrated to be extremely efficient for the restoration of the Accademia Galleries, but also offered a new tool for future tenders on monumental building conservation.

3.3 The Conservation Building Site Process in the CE Loop

The intensive work of CLIC highlighted the need to work on different planning levels (Foster 2019, 2020) with respect to numerous circular criteria when dealing with the building sector and projects of adaptive reuse in urban contexts (Gravagnuolo et al. 2019). The steps identified by the research are the design phase, the building materials sourcing, the building site, the operation and use of the newly renovated asset, and eventually the repurpose and demolition. How does conservation deal with these phases, and how is this impacting the enhanement of the circular economy?

The Design Phase The Designer, namely architects, conservation architects, engineers and designers, has an essential responsibility in the overall impact of the process: on one side the responsibility is about the quality of conservation, the respect of the intrinsic qualities of the heritage asset, on the other side the responsibility is on the decision taken in terms of materials, structures, services, and energy consumption. The perfect design would be able to find the right compromise to preserve the cultural qualities, define a tailor-made set of solutions for the asset to improve existing or host new destinations, self-impose the concept of reversibility of intervention, and commit to address local crafts, skills, competences, and materials. But there are other persons responsible in this phase, namely the client, who is supposed to be aware of the intrinsic qualities of the asset, thus both accepting the design constraints of the designers and imposing quality design, but at the same time being conscious of the compromise to be made to guarantee authenticity and integrity. The third actor responsible of the design phase is the civil servant that, representing the authorities, gives the building permit; he/she is entitled to guide both the designer and the client towards the most correct solutions for the adaptive reuse of a common. In the list, another actor is responsible for the success of the process, namely the contractor, who is performing the conservation works; he/she is usually informed about the process in advance and thus has to provide evidence about the competence in dealing with adaptive reuse of cultural heritage appropriately. An important stress must be made about the use destination: the more a destination is multifunctional and operating, the more it is circular.

The Building Material Sourcing As stated in the previous paragraph, the designer should have the proper skills to indicate respectful materials for the implementation of an appropriate adaptive reuse intervention. Given that cultural heritage is mostly made of local materials, the decision should not be difficult. The bigger question is raised in front of installations and modern finishing. In this case, the decision to be taken in line with CE should select producers and providers as close as possible to the building site to limit the logistics and the footprint. This was a general rule also in the past. Another story is the governance of these logistics and how they are supported by the system. Providing skills and materials from the area is also under the responsibility of local authorities that need to promote specialisation, employment opportunities in the crafting sector, and education, especially referred to as the *genius loci*.

The Building Site The history of cultural heritage is played in the building site, and here there must be a lot to gain in the future. Applying the principles of circular economy in the building sites means organisation in the demolition works, namely by dismantling with care and savoir faire and storing in a safe and logical way. In CLIC, the experiences of the Belgian company ROTOR have demonstrated that this can be an excellent business opportunity involving specialisation in repurposing, including the selling of dismantled materials. A similar example comes from the Bank of Materials in Porto, where different local decorative elements, such as the ceramic tiles—the so-called azulejos—are stored by the municipality and sold for free to citizens in need during restoration works. Beyond careful demolition, the building site needs the skills to perform conservation properly, and this may include a set of "old style" craftsmen, such as carpenters for windows, roofs, doors, ceiling maintenance and restoration, stone workers for walls and plastering renovation, as well as plumbers and electricians capable of dealing with historic fabric, thus avoiding useless breaks in the structure. A good team composition is characteristic of local companies that are used to deal with the local genius loci and its requirements, optimising the use of resources and the reuse of the building site "waste".

The Cultural Heritage Asset Operation Once renovated, cultural heritage may be used in circular ways, applying the principles of circular economy for innovative sustainable business model. This may include a numerous set of options. From the conservation perspective, the use has to be aligned with the capacities of cultural heritage and with respect to its intrinsic qualities, authenticity and integrity. In Rijeka, for example, the renovation of Rihub served to create a multifunctional space acting as an incubator for new companies, but also as a social space in the evenings and nights; a bigger dimension has taken the Pakhuj de Zwijger in Amsterdam, which after a renovation through careful retrofitting has become a multifunctional space hosting community meetings and social events, thus maximising the embodied energy of the fabric. A similar final use was decided in Salerno with the Monumental complex of S. Sofia, that was converted into the so-called Palazzo Innovazione, where many of the principles of the circular economy have been implemented to host new start-ups.

The Urban Scale An additional dimension that came out in CLIC from the debates on Historic Urban landscape must be added for the urban governance of conservation. The link between circular economy and adaptive reuse at the urban scale rely on the capacity to put in place solutions that regenerate the urban space as place. An attempt was made in Rijeka through the model of the cultural corridor, inspired by the Ostrom theory of governing the commons (Ostrom 1990). The cultural corridor is an urban space where the praxises of conservation are experienced jointly with the principles of circular economy, considering users as members of an aesthetic heritage community (CoE 2005; Greffe 2010). The HUL approach in circular economy perspective must be based on the capitalisation of the human and social capital by offering exemplary spaces of creation and fruition of cultural heritage and circular economy in all its dynamic exchanges: somehow it is like recreating the human-centred city based on human natural speed, somehow seen as the *passages* (Benjamin et al. 2000), where a shortcut becomes a full place of self-cultural enhancement.

From a different perspective, the concepts of conservation, more than adaptive reuse, can also be crossed with the 7R of circular economy, revised by Green Peace into 9R.

Recovering or Downscaling The materials from the building site are partly considered waste. This is the present practice. Wood, iron, copper, brass, bricks, tiles, ceramics, plastics, etc., are accumulated without preliminary separations and sent to a local disposal centre, where they are eventually separated. This step was common also in the past, but the amounts of waste were radically different in composition and low in quantity. This may have occurred more in the case of patronal or monumental constructions, less in traditional buildings.

Recycling The materials from objects and architectural functional elements are selected and collected separately (onsite or out of the site). The most commons are beams, bricks, tiles, windows and door frames, glasses. The term better refers to the treatment of some materials that could have been reused in the building site in other functional elements, as wooden elements for interior walls, bricks debris for terrazzo floors or plasters and mortars, stones for walls filling or corner joints.

Repurposing or Upscaling Materials, mostly objects and architectural elements, are taken from the building site to be used or applied in other contexts (as an old window used in a pub to display liqueurs). Also, this may occur with elements that assume a different role in the building, as parts of previous beams becoming architraves, stairs structural elements, floor wooden boards, and even parts of the furniture if tailor-made, like benches, tables, shutters, etc. Historic architecture has plenty of such examples.

Remaking or Re-manufacturing Materials and objects or architectural elements are removed to be substituted with others or to be adapted/consolidated. The traditional building site was often avoiding full replacements, preferring adaptations and partial reconstruction, as in the case of wooden beam heads, wall openings, window and door frames, floors. This is still possible in most cases and suggested by manuals.

Refurbishing and Repairing Directly related to maintenance, this is when objects and architectural elements are preserved due to their present good state of conservation as well as their structural consistency (i.e., a terrazzo floor consolidation or reintegration, wooden beams cleaned and impregnated, etc.).

Reusing Materials, objects, and architectural elements that are removed from the site are reused somewhere else with the same function.

Reducing This is part of the design process, namely tailor-made solutions to reduce waste of materials and space. In particular, as done, for example, by Carlo Scarpa, the design of interior solutions was happening on site according to the existing recovered materials. As said, the adoption of the minimal intervention can favour the reduction of materials and the reduction of future efforts of repurposing.

Rethinking and refusing This is the essence of conservation in its pure conception. Materials, objects, and architectural elements and details are kept, favouring the "individual adaptation" instead of the "building adaptation." It may also include giving the building a role in the urban dynamics.

Remembering Green Peace refers to this as a consequence of refusing and adopting alternative solutions (do not accept plastic bags, thus remember to take yours from home), which has an important parallel also in the conservation project. In particular we refer to the design process, where new ideas in adaptation must be predictive of future modifications, thus considering reversibility as well as recyclability. Not to exclude in this step the tailor-made design, which gives value to the space and potentially prevents future demolitions.

Respecting This step is extremely valuable for the conservation approach because it, although to a lower scale in respect to the problems of climate change, asks for respect of cultural richness and diversity, thus for a consequent aptitude to the less, which can be translated into "do I really need such a drastic renovation?"

4 Recommendations/Discussions

The experience of CLIC is fundamental to adding another step forward in the conservation theory, considering new economic models that could enhance awareness of the embodied intrinsic qualities of the built environment. At the same time, the conservation movement and an increasing attention towards the cultural capital as contributing to the social well-being offer positive supporting hooking. Nevertheless, the combination of the topics in a successful governance framework needs to be further investigated and framed, with extensive dissemination and education actions at different scales. Such needs are confirmed by the experiences carried out through extensive academic research, as well as through the collaborative activities performed with the CLIC case studies, namely the cities of Salerno, Rijeka, the Pakhuijs de Zwieger organisation and the Vastra Gotaland Region. Four case studies coming from countries with different traditions on heritage preservation and on socio-economic orders, but all displaying enormous cultural and social capital willing to enter as soon as possible into a sustainable change. Considering all these components, the following brief recommendation could be made with a focus on cultural heritage preservation:

- There is still a need to insist on the achievements of the conservation doctrine. Cultural heritage preservation is a starting point of development movements because it represents the true uniqueness of Europe as a whole as well as of each local reality in Europe. The higher the number of cultural expressions we preserve, the more these can be turned into resources at the right moment. In this perspective, the locution "adaptive reuse", although with very clear meaning for the experts in the field, may be perceived as freedom of transformation, of change, somehow denying the main principles of circular economy, which is reducing. We see too often that with the flag of adaptive reuse, cultural heritage is losing its most important role: telling the story. Sociologists, anthropologists, and archaeologists realised already that heritage is telling us what past generations cannot anymore: is what remains from the past, in a sense, the waste of the past that we can upscale.
- Preservation constraints must be seen as positive challenges by the planners and designers. Inside the design of conservation there is an infinite space for imagination, creativity, and consequently job opportunities. Heritage must be taken as a capital that needs to be preserved at best to become resource and the cultural tourism theories have been telling this loud since the 80's.
- To approach cultural heritage correctly, we need a new use theory. The powerful vision of the circular economy wave seems intrinsically dependent on the local awareness of heritage significance. Cultural heritage preservation refers to local resilience in protecting and using. As we have seen though, circularity means also creating social benefit, generating circular flows beyond traditional ownership models. To be fully implemented, the circular economy must re-embracing

local synergies and mutual dependence agreements; namely, the object should be related to the context and the relationship micro, meso and macroscale always respected through a transitive relationship.

- Awareness raising and education are essential. The practice as professionals in cultural heritage conservation (Acri-Dobricic) shows that there is a decreasing competence at all levels in the building sites, which makes the overall conservation process increasingly difficult. Building companies are less and less specialised and mostly workers are less and less experienced in traditional techniques, reducing enormously the trust in the building site about the quality of interventions. In parallel, the general public is getting more and more distant from the inputs of the past due to normal generational changes that cannot guarantee the experiential dimension.
- The experiential dimension in heritage is fundamental. To perform appropriate maintenance works, users, cultural heritage owners, must get familiar with traditional materials and techniques and consequently with the practices that produced and maintained them. As a suggestion, manuals and tutorials could be promoted for each specific location with unique heritage assets. Moreover, new educational offers must be compulsory all over Europe to move new generations closer to their living environment, both natural and cultural, performing new forms of experience aimed at learning by doing. This lesson was also learnt by the conservation doctrine, which has suffered from the fact of being considered too academic and abstract. Conservation and maintenance are not abstract and not purely intellectual, but are in reality forms of life, as the heritage making has been for centuries.
- Actions should also be taken from the legislative perspective. The recent green deal applications at the national level have annihilated the conservation practice, opening huge financial benefits to barren conservation works purely orientated to energy efficiency. In this respect, such measures should be better tailored to cultural heritage adaptation, by imagining ad-hoc solutions less invasive and destructive than the ones proposed by the market. Moreover, such measures should be better distributed, including research in the field, for example, promoting start-up initiatives for the invention of sustainable, cultural heritage solutions for retrofitting.
- In terms of legislation, the EU should promote typological conservation as a general rule. Typological conservation holds numerous advantages in managing cultural heritage sites, by reducing designers and officers' responsibilities in case of cultural heritage losses; consequently reducing such losses. Typological conservation also promotes local crafts and skills upgrades, as well as the creative design for conservation. As a result, it maintains the intrinsic qualities of sites, guaranteeing the cultural diversities that have characterised the European territory for centuries.

5 Conclusions

As is evident, the circular economy was initially motivated by environmental concerns and has sought to safeguard ecosystems and natural resources, prioritising primary resources. As we have seen, a cultural resource (built environment, heritage, traditional buildings, etc.) involves a set of inherent qualities that form its significance and are inextricably linked to the complexity of the whole. As an illustration of the circular process, if we think of beams as just ordinary pieces of wood, we lose their connection to the layer to which they belong, we ignore the historical experience they represent, and we cancel the circular process that allowed them to be a part of the ceiling and building. This does not mean that replacement is not possible, but means that it cannot be done *tout-court,* as even single architectural elements are what remains of a knowledge, the residual that talks of a moment in time and, by extent, are contributing to the cultural landscape and/or the historic urban landscape in the relationship between typology and morphology, between private and common goods. Cultural heritage and the preservation theory may represent the (positive) bond of the Circular Economy model, as they do not allow its perverse drift in the productive chain based on the logic of substitution.

Although the research on this relationship between the circular economy and adaptive reuse of cultural assets is still in its early stages, it is essential to provide the conservation theory with a further dimension for accreditation. Though the principles of the circular economy reformulate the dynamics of traditional building sites from a modern perspective, they provide a framework for appropriate interventions on heritage, even if there is potential concern about the truly innovative nature of the adaptive reuse wave in Europe due to its potential to flatten the conservation theory. Nonetheless, there exist hazards associated with the possible misuse of the circularity concept for the benefit of a traditional economic linear process that involves waste and demolition or straightforward substitution. Such risk is related both to the potential creation of a parallel profitable market of reused objects, pushing the replacements, and to the misunderstanding of the importance of the part of a whole (the elements) and their constituting materials (the potential waste).

References

Acri M, Biasi A (2019) Piccola guida per la manutenzione dell'architettura austroungarica a Gorizia. Giavedoni

Astengo G (1958) Assisi: Piano generale e Piani particolareggiati di primo intervento. Urban Forum 24–25(24–25):1–124

Barkley Rosser J Jr (2003) A Nobel prize for asymmetric information: the economic contributions of George Akerlof, Michael Spence and Joseph Stiglitz. Rev Polit Econ 15(1):3–21. https://doi.org/10.1080/09538250308445

Benjamin W, Tiedemann R, Benjamin W (2000) I 'passages' di Parigi

Bergson H (1998) Creative evolution. Dover

Brandi C (2000) Teoria del restauro (Rist). G. Einaudi
Bullen PA, Love PED (2011a) Adaptive reuse of heritage buildings. Struct Surv 29(5):411–421. Scopus. https://doi.org/10.1108/02630801111182439
Bullen PA, Love PED (2011b) Adaptive reuse of heritage buildings: sustaining an icon or eyesore. Scopus, pp 1652–1662
Caniggia G (1976) Strutture dello spazio antropico. Studi e note, 1st edn. UNIEDIT
Cantacuzino S (1975) New uses for old buildings. Architectural Press
Casiello De Martino S (ed) (1996) La cultura del restauro: Teorie e fondatori. Marsilio
Cervellati P (1977) La nuova cultura delle citta': La salvaguardia dei centri storici, la riappropriazione sociale degli organismi urbani e l'analisi dello sviluppo territoriale nell'esperienza di Bologna. Mondadori
Codello R (2014) Architetture contemporanee a Venezia =: Contemporary architecture in Venice (Prima edizione). Marsilio, Fondazione di Venezia
Codello R, Torsello A, Scarpa C (2009) Architetture veneziane di Carlo Scarpa: Percorsi e rilievi di cinque opere, 1st edn. Marsilio
CoE (2000) European landscape convention. Council of Europe. https://www.coe.int/en/web/landscape/text-of-the-european-landscape-convention
CoE (2005) Conventino on the value of cultural heritage for society—Faro convention. CoE. https://www.coe.int/en/web/culture-and-heritage/faro-convention
Conzen MRG, Whitehand JWR (eds) (1981) The urban landscape: historical development and management. Academic
Cristinelli G (2013) Saverio Muratori e Egle Renata Trincanato: La nascita del restauro urbano in Italia. GB EditoriA
Carbonara G, Trabucco G (2017) Fondamenti per una dottrina del restauro architettonico. Ginevra Bentivoglio
Dal Co F, Mazzariol G (2006) Carlo Scarpa (1906–1978), 2. ed. paperback edn. Mondadori Electa
Deleuze G, Guattari F, Massumi B (2011) A thousand plateaus: capitalism and schizophrenia (Repr). Continuum
Dezzi Bardeschi M, Locatelli V (1991) Restauro, punto e da capo: Frammenti per una (impossibile) teoria. Franco Angeli
Dobričić S, Acri M (eds) (2018) Creative cities: which (historic) urban landscape. Mimesis
Feilden BM (1982) Conservation of historic buildings. Butterworth Scientific
Forsyth M (2013) Understanding historic building conservation. Wiley
Foster G (2019) Planning the circular city: focus on buildings' environmental impact. Bulletin of THE Calza Bini center, pp 117–123. https://doi.org/10.6092/2284-4732/7063
Foster G (2020) Circular economy strategies for adaptive reuse of cultural heritage buildings to reduce environmental impacts. Resour Conserv Recycl 152:104507. https://doi.org/10.1016/j.resconrec.2019.104507
Geertz C (1993) The interpretation of cultures: selected essays. Fontana Press
Gianighian G, Pavanini P, Del Pedros G (2010) Venezia come. Gambier & Keller editori
Girard LF, Gravagnuolo A (2017) Circular economy and cultural heritage/landscape regeneration. Circular business, financing and governance models for a competitive Europe. BDC Bollettino Del Centro Calza Bini 17(1):35–52
Glendinning M (2003) The conservation movement: a cult of the modern age. Trans R Hist Soc 13:359–376. JSTOR
Gracia Aldaz JM, Zhang H, Bokova I, Unesco (2016) Culture: urban future: global report on culture for sustainable urban development
Gravagnuolo A, De Angelis R, Iodice S (2019) Circular economy strategies in the historic built environment: cultural heritage adaptive reuse. https://doi.org/10.3217/978-3-85125-668-0-08
Greffe X (2010) Urban cultural landscapes: an economic approach. University of Turin
Hinna A (2015) Tipologie di distretti culturali a confronto: Politiche, governo e gestione. IL Capitale Culturale Stud Value Cult Herit 3:43–59. Paginazione. https://doi.org/10.13138/2039-2362/1237

Historic England (2017) The setting of heritage assets | historic England. Historic England. http://historicengland.org.uk/images-books/publications/gpa3-setting-of-heritage-assets/
Hobsbawm EJ, Ranger TO (eds) (2010) The invention of tradition (19th pr). Cambridge University Press
Honko L (1995) Traditions in the construction of cultural identity and strategies of ethnic survival. Eur Rev 3(2):131–146. https://doi.org/10.1017/S1062798700001435
ICOMOS (1975) European charter of the architectural heritage. ICOMOS International. https://www.icomos.org/en/resources/charters-and-texts/179-articles-en-francais/ressources/charters-and-standards/170-european-charter-of-the-architectural-heritage
ICOMOS (1993) Guidelines for education and training in the conservation of monuments, ensembles and sites. ICOMOS International. https://www.icomos.org/en/charters-and-texts/179-articles-en-francais/ressources/charters-and-standards/187-guidelines-for-education-and-training-in-the-conservation-of-monuments-ensembles-and-sites
ICOMOS (1994) The Nara document on authenticity. ICOMOS International. https://www.icomos.org/charters/nara-e.pdf
ICOMOS (1999) Charter on the built vernacular heritage. ICOMOS. https://www.icomos.org/images/DOCUMENTS/Charters/vernacular_e.pdf
ICOMOS (2020) European quality principles for EU-funded interventions with potential impact upon cultural heritage. ICOMOS International http://openarchive.icomos.org/id/eprint/2436/1/EUQS_revised-2020_EN_ebook.pdf
Impey E (2015) Conservation principles, policies and guidance—for the sustainable management of the historic environment 78
Ingersoll R, Ciampi M (eds) (2002) Massimo & Gabriella Carmassi. Senigallia Public Library, Verba Volant
Jokilehto J (2017) A history of architectural conservation, 2nd edn. Routledge
Jones MO (2000) 'Tradition' in identity discourses and an Individual's symbolic construction of self. West Folk 59(2):115. https://doi.org/10.2307/1500156
Khaldun I (1978) Ibn-Khaldūn: Discours sur l'histoire universelle; Al-Muqaddima. T. 3: … (V. Monteil, Ed.; 2., ed.revue). Sindbad
Lane R, Menuge A (2006) Understanding historic buildings: a guide to good recording practice. Historic England. https://historicengland.org.uk/images-books/publications/understanding-historic-buildings/heag099-understanding-historic-buildings/
Latouche S (2012) Vers une société d'abondance frugale: Contresens et controverses sur la décroissance
Laureano P (2001) Atlante d'acqua: Conoscenze tradizionali per la lotta alla desertificazione, 1st edn, Bollati Boringhieri
Leroi-Gourhan A (1943) L'homme et la matière. Évolution et techniques, I. Albin Michel
Lind T (2014) The tradition of the use of the monument: case study Suomenlinna. Essays on the Conservation of the Built Environment, Building Conservation in our time, 11
Lowenthal D (2013) The past is a foreign country—revisited. (Revised and updated edition). Cambridge University Press
Marconi P (1988) Dal piccolo al grande restauro: Colore, struttura, architettura (1a ed). Marsilio
Marconi P (1993) Il restauro e l'architetto: Teoria e pratica in due secoli di dibattito (1. ed) Marsilio
Marconi P (1999) Materia e significato: La questione del restauro architettonico (1. ed). GLF Editori Laterza
Markeviciene J (2008) Genius Loci and Homo Faber: a Heritage-making dilemma. Proceedings of 16th ICOMOS general assembly and international symposium: finding the Spirit of place between the tangible and the intangible, 12
Matteoni D, Carmassi M, Polano S, Mulazzani M (1992) Massimo Carmassi: Architettura della semplicità. Electa
Morris W, Webb P (1877) SPAB manifesto. https://www.spab.org.uk/about-us/spab-manifesto
Muratori S (1967) Civilta' e territorio (Vol. 1–2). Centro Studi di Storia Urbanistica

Musgrave RA (1959) The theory of public finance: a study in public economy. McGraw-Hill
Nijhuis DO (2011) Explaining British voluntarism. Labor Hist 52(4):373–398. https://doi.org/10.1080/0023656X.2011.632512
Norberg-Schulz C (1979) Genius loci: Paesaggio, ambiente, architettura. Electa. http://books.google.com/books?id=smJQAAAAMAAJ
Ostrom E (1990) Governing the commons: the evolution of institutions for collective action. Cambridge University Press. https://www.actu-environnement.com/media/pdf/ostrom_1990.pdf
Piaget J (1976) Identity and conservation. In: Inhelder B, Chipman HH, Zwingmann C (eds) Piaget and his school: a reader in Developmental Psychology. Springer, Berlin/Heidelberg, pp 89–99. https://doi.org/10.1007/978-3-642-46323-5_8
Pickles D (2016) Energy efficiency and historic buildings: insulating solid walls. Historic England. https://historicengland.org.uk/images-books/publications/eehb-insulating-solid-walls/heag081-solid-walls/
Pickles D, McCaig I, Wood C (2014) Traditional windows—their care, repair and upgrading. English Heritage. https://historicengland.org.uk/images-books/publications/traditional-windows-care-repair-upgrading/heag039-traditional-windows-revfeb17/
Plevoets B, Cleempoel KV (2012) Adaptive reuse as a strategy towards conservation of cultural heritage: a survey of 19th and 20th century theories. Reinventing architecture and interiors: the past, the present and the future, 6
Plevoets B, Cleempoel KV (2013) Adaptive reuse as an emerging discipline: an historic survey. In: Reinventing architecture and interiors: a socio-political view on building adaptation. Libri Publishers, p 11
Potting J, Hekkert M, Worrell E, Hanemaaijer A (2017) Circular economy: measuring innovation in the product chain. Netherlands Environmental Assessment Agency, p 47
Rodrigues C, Freire F (2017) Adaptive reuse of buildings: eco-efficiency assessment of retrofit strategies for alternative uses of an historic building. J Clean Prod 157:94–105. https://doi.org/10.1016/j.jclepro.2017.04.104
Ruskin J, Rosenberg JD (1987) Le pietre di Venezia. Rizzoli
Sacco PL, Ferilli G, Tavano Blessi G (eds) (2015) Cultura e sviluppo locale: verso il distretto culturale evoluto. Il mulino
Santagata W (2002) Cultural districts, property rights and sustainable economic growth. Int J Urban Reg Res 26(1):9–23. https://doi.org/10.1111/1468-2427.00360
Santagata W (2007) La fabbrica della cultura: Ritrovare la creatività per aiutare lo sviluppo del paese. Il Mulino
Schumpeter JA, Forte F (2001) Capitalismo, socialismo e democrazia. ETAS
Tyler N, Tyler IR, Ligibel T (2018) Historic preservation: an introduction to its history, principles, and practice, 3rd edn. W.W. Norton & Company
UNESCO (1976) Recommendatino concerning he safeguarding and contemporary role of historic Ares. UNESCO. http://portal.unesco.org/en/ev.php-URL_ID=13133&URL_DO=DO_TOPIC&URL_SECTION=201.html
UNESCO (2005) The COnvention on the protection and promotion of the diversity of cultural expressions. UNESCO. https://en.unesco.org/creativity/convention
UNESCO (2011) Recommendation on the historic urban landscape. https://whc.unesco.org/en/hul/
UNESCO World Heritage (2019) The operational guidelines for the implementation of the world heritage convention. UNESCO World Heritage Centre. https://whc.unesco.org/en/guidelines/
Urbani G, Zanardi B (2000) Intorno al restauro. Skira
Valentino PA (2003) Le trame del territorio. Sperling & Kupfer
Vecco M (2020) Genius loci as a meta-concept. J Cult Herit 41:225–231. https://doi.org/10.1016/j.culher.2019.07.001
Vegas López-Manzanares F, Mileto C (2008) Renovar conservando: manual para la restauración de la arquitectura rural del Rincón de Ademuz. Mancomunidad de Municipios Rincón de Ademuz

Vegas López-Manzanares F, Mileto C (2014) Aprendiendo a restaurar: un manual de restauración de la arquitectura tradicional de la Comunidad Valenciana
Vico G, Cristofolini P (2016) Principj di una Scienza nuova: 1725. Edizioni ETS
Vico G, Lomonaco F (2014) Princìpi di una scienza nuova: Intorno alla natura delle nazioni per la quale si ritruovano i princìpi di altro sistema del diritto naturale delle genti. Diogene

Part II
Practices, Experimentations and Tools for the Adaptive Reuse of Cultural Heritage in the Perspective of the Circular Economy Model: Evaluation Tools

Introduction

The adaptive reuse of cultural heritage represents a vital intersection between preservation and innovation, offering opportunities to breathe new life into historic buildings and sites while contributing to sustainable development. This section explores a range of practices, experimentations, and tools specifically designed to evaluate adaptive reuse projects. By integrating circular economy principles, these tools enable stakeholders to assess not only the economic and environmental impacts but also the social and cultural value of repurposed heritage assets.

Chapter 5
In Search of a Circular Model for Cultural Heritage Adaptive Reuse: Building Evidence-Base

Antonia Gravagnuolo, Gabriella Monteleone, and Luigi Fusco Girard

1 Introduction

The CLIC framework for circular adaptive reuse of cultural heritage (Fusco Girard this volume, Chap. 2) was tested through the assessment of European adaptive reuse cases, to identify the elements of success in diverse contexts according to specific and multidimensional circularity criteria. One of the objectives of the CLIC project was to strengthen efforts to harness data for heritage management in EU regions, developing participatory mapping, unified access to heritage-related databases, data visualization, citizen-generated data.

A data revolution is taking place in our world (UN-Habitat 2016a, b, 2017). More data are produced today than ever before, from a multitude of sources. Geo-spatial monitoring, citizen-generated and crowd-sourced data, and big data, are increasingly available in real time and complement official statistics. The data is no longer structured and in relational formats, but *unstructured and heterogeneous* and the nature of these heterogeneous data has no limits. Access to relevant data has become a non-trivial task for the user. The private sector, academia, and civil society are using this growing variety of data to make profits, inform adaptive reuse of cultural heritage, drive innovation, and support advocacy. In this scenario, an effective and precise search solution is required that ensures knowledge of what information is available, and provides a quick retrieval of that information. The solution should

A. Gravagnuolo (✉)
CNR ISPC Institute of Heritage Science, National Research Council, Naples, Italy
e-mail: antonia.gravagnuolo@cnr.it

G. Monteleone
Research and Innovation, FacilityLive, Politecnico di Milano, Pavia, Italy

L. Fusco Girard
Emeritus Professor at University of Naples Federico II; CNR IRISS, National Research Council, Institute for Research on Innovation and Services for Development, Naples, Italy

L. Fusco Girard, A. Gravagnuolo (eds.), *Adaptive Reuse of Cultural Heritage*,
https://doi.org/10.1007/978-3-031-67628-4_5

provide a quality user experience that is naturally intuitive and the entire process effortless.

Enhancing accessibility and capacity to use data for decision-making is strongly advocated in the New Urban Agenda for sustainable urban/territorial development (see Habitat III events on Data Revolution). Integrated and coordinated data generated through collaborative, participatory processes are essential to planning and monitoring cities in the twenty-first Century, and empowering citizens. Partnering across sectors to harness the explosion of available data, technologies, skills, and opportunities to connect multiple data sources is essential to unlocking data for evidence-based decision making (U.S. Department of State 2016).

However, data on cultural heritage in European countries are not easily available. Sectorial databases directly and indirectly related to cultural heritage resources have been developed by different bodies: EU Institutes for Statistics (e.g. Eurobarometer for extensive surveys), Central governments, National Institutes for Statistics, Religious bodies (e.g. the Catholic Church), Municipal institutions; Foundations, Universities and Adaptive reuse of cultural heritage centres, local Professionals associations and Civil Society Organizations (CSOs). Additional databases are provided by past and ongoing EU funded projects. Data gathering, access and use/management pose many difficulties to potential end-users/producers of data.

New technology for information management can empower local governments and citizens, enabling cooperation processes based on trust, transparency and engagement in decision-making. Citizens become "prosumers": heritage prosumers, data prosumers, decision prosumers (Ritzer and Jurgenson 2010). Technology can boost local heritage-led economies reducing costs of adaptive re-use processes and creating new markets for useful services that enhance the quality of life.

To collect, analyse and make available a large and dynamic set of heritage data for different users, a database of 126 cultural heritage adaptive reuse practices was built. The CLIC Survey on the adaptive reuse of cultural heritage was launched in 2018 engaging heritage adaptive reuse of cultural heritage, managers and authorities to collect relevant data useful for the assessment of performances in the perspective of the circular model. The selection of adaptive reuse practices was conducted according to basic criteria: projects already concluded or in the final phase of their implementation, for which information on their performances, results and impacts was available and reliable; located in European countries, covering almost all regions; clearly, projects that focused on heritage recovery and reuse, covering all typologies of heritage assets (religious, civic, military, residential, rural, commercial, etc.). The data collected for each adaptive reuse practice focused on environmental sustainability, social impact, economic and financial performances, cultural values protection and regeneration. Pictures and detailed descriptions of the heritage asset, its reuse process and the new current uses were sought, to generate a usable and useful tool for heritage professionals, adaptive reuse of cultural heritage and communities.

The CLIC Survey database was developed into the IT platform named "CLIC Knowledge and Information Hub" (KIH) using the FacilityLive © innovative technology for data organization, search and management.

This contribution aims to present the methodology and tools used in the CLIC adaptive reuse of cultural heritage to collect, organise, analyse and interpret relevant data on European cultural heritage adaptive reuse practices, to identify good practices, success factors and barriers towards the implementation of the circular model for heritage reuse and regeneration. The CLIC Survey rationale and structure is described, followed by the KIH platform features, data analysis and assessment. A number of case studies is provided as an overview of the relevant information collected and how it was used to advance knowledge on successful adaptive reuse cases.

2 Methodology: The CLIC Survey on the Adaptive Reuse of Cultural Heritage

The CLIC methodological approach was based on the analysis of empirical evidence to explore whether and how the experiences of cultural heritage adaptive reuse have been able to turn abandoned heritage/landscape assets into a resource for new jobs, wellbeing, health, social cohesion, regional competitiveness and environmental regeneration—as advocated by all international policy documents and scientific literature.[1] The theoretical debate and scientific literature on the adaptive reuse of cultural heritage highlighted specific knowledge gaps which needed to be investigated in order to identify viable and potentially transferable models and solutions to turn the underused and abandoned cultural heritage from a "cost" to an "investment" for society (Gravagnuolo et al. 2021; CHCfE 2015): effective ownership and management models; compatibility of new uses with heritage values; self-sustainability in terms of financial resources needed for its functioning and maintenance, as well as energy and natural resources self-sufficiency; capacity to activate local cultural, entrepreneurial and innovation ecosystems; capacity to generate net positive economic, environmental, social and cultural impacts in the territory.

The CLIC Survey on the adaptive reuse of cultural heritage was designed, developed, tested and implemented to collect useful information on the characteristics and impacts of adaptive reuse practices. The specific aim of the CLIC Survey was to collect a large and detailed set of information on cultural heritage adaptive reuse practices, identifying specific "typologies" of adaptive reuse interventions based on heritage main characteristics (e.g. use/function, localization, level of protection), as well as to explore the eventual relationships between the "performances" in terms

[1] Among the many reference documents, it is worth to recall here: Council of Europe 2000; FARO Convention, Council of Europe 2005; UNESCO Recommendation on the Historic Urban Landscape 2011; ICOMOS Burra Charter 2013; European Commission, Towards an integrated approach to cultural heritage for Europe 2014; European Commission, Getting cultural heritage to work for Europe 2015; Cultural Heritage Counts for Europe 2015; UN Agenda 2030 for Sustainable Development and Sustainable Development Goals 2015; UN New Urban Agenda 2016; European Parliament, Decision on a European Year of Cultural Heritage 2017; European Framework for Action for Cultural Heritage 2018.

of circularity and the specific management models, considering the overall costs/investments of the interventions and the cultural, economic, social and environmental impacts generated.

Thus, the CLIC Survey was designed to test through empirical evidence the actual implementation of the theoretical model of circular adaptive reuse of cultural heritage, re-interpreting the success (or unsuccess) of adaptive reuse practices under the point of view of their "autopoietic capacity", "generative capacity" and "symbiotic capacity" (See Chap. 2 of this volume).

Scholars in the heritage sector highlight that cultural heritage conservation/valorisation can enhance social cohesion, employment, quality of landscape, etc. However, empirical evidence is lacking. The CLIC survey aimed to map/assess innovative and successful business, financing and governance models. In particular, the survey contributed to explore in which cases the adaptive reuse of cultural heritage:

- enhanced social cohesion and heritage communities;
- generated economic wealth through adequate revenues/cash flows vs. investment and operating costs (economic/financial self-sustainability);
- enhanced the attractiveness of the area for new investments, in particular attracting cultural and creative industries, and enhancing the perceived visual quality of the landscape;
- was implemented through a specific role of the third sector (which is more and more considered as a key actor between public and private);
- demonstrated win-win-win solutions combining private short term interests with public longer term objectives;
- employed circular models through closed micro-metabolisms in water, energy and materials use.

The survey questions proposed had the objective to answer and articulate these key questions.

2.1 Survey Structure

According to the knowledge gaps identified, the CLIC Survey was structured into four main parts:

1. Introduction: aims and objectives of the survey;
2. Characteristics of the cultural heritage building/site/landscape and its ownership and management model;
3. New uses/functions activated through the adaptive reuse;
4. Assessment on Circularity—based on 31 questions to heritage managers to assess the impacts generated in a circular perspective through qualitative data; Quantitative data (optional) on heritage reuse impacts;

The specific sections of the CLIC Survey were designed to first introduce the respondent to the concept of circular economy in cultural heritage adaptive reuse, and guide him to correctly fill-in the information. “Closed” answers options were provided as far as possible in order to clearly categorize the answers in the analysis phase and build a usable database. The sampling and data collection campaign was developed to ensure homogeneous quote sampling with respect to heritage typologies and localisation in European regions.

1. The Survey introduction provided the explanation of aims and objectives of the survey, introducing the definition of adaptive reuse of cultural heritage and circular economy.
2. The section on the characteristics of the adaptive reuse practice included mandatory information such as the Project Name, brief description, city, country, localization, city size, and an open question on the motivation for including it into the adaptive reuse of cultural heritage database. A specific section on construction information included construction period, state of conservation, reuse period, cultural significance, typology or original function(s), square meters of adaptive reuse intervention. Further detailed information were sought about the governance and financing model of the adaptive reuse practice. This section referred specifically to the ownership of the location before and after the adaptive reuse intervention, vacancy before and after the intervention, managing body, funding sources, involvement of local community in decision making, barriers and bottlenecks, total investment for adaptive reuse.
3. The CLIC survey focused also on the uses/functions of the building or site after the intervention. The suggested list of uses was deduced from a frequency analysis of KEA Report of 71 case studies of cultural heritage regeneration (KEA 2017), and integrated with specific uses related to circular economy functions (e.g. repair café, materials bank, sharing economy functions). The new uses/functions were grouped into categories as follows. Moreover, it was requested to identify the uses/functions associated to revenue streams, in order to collect information on which functions are likely to contribute more to the economic-financial self-sustainability of the cultural heritage adaptive reuse practice (Table 5.1).

4. The last section was dedicated to the Assessment on Circularity. This part aimed to identify the most interesting practices of cultural heritage adaptive reuse in the perspective of circular economy, considered not only as a model for “wastes” reduction, reuse and recycling, but also introducing a larger perspective in line with the CLIC circular approach, based on the assessment of the self-regeneration capacity, the generation of impacts and the connection with the local community and entrepreneurial/innovation ecosystem..

A number of 31 questions were articulated according to the scientific article of Gravagnuolo et al. (2017). A short paragraph with definitions and explanations was provided for each group of answers to support the process of data collection using the online survey tool. In order to make the survey accessible to a wide and

Table 5.1 Suggested new uses/functions of the adaptive reuse practices

Category	Specific uses/functions
Residential	Residential units Co-housing units
Tourism accommodation	Hotel accommodation Non-hotel accommodation (e.g. B&b, hostel, house rental)
Commercial	Commercial units Health and/or wellness centre Restaurant Café
Public spaces and cultural functions	Public library Open air recreational spaces (e.g. garden, park) Education and aesthetic appreciation (e.g. touristic visits, educational visits with schools) Museum and exhibition centre Adaptive reuse of cultural heritage and education hub Cultural events, festivals and concert spaces Theatre and performing arts Conferences and fairs Sport facilities
Social services	Social services Community Centre/Hub Living Lab Fab Lab Creative Centres/Hub Artists' residencies
Entrepreneurial services	Enterprise and startup incubator/Hub Hub of cultural and creative enterprises Innovative startups and companies (e.g. digital, artificial intelligence, industry 4.0, internet of things, robotics, pharmaceutical adaptive reuse of cultural heritage) Circular economy enterprises Co-working and workspaces Workshop spaces
Circular economy specific activities	Bank of materials Repair café Bike-sharing places

non-expert public, qualitative answer options were chosen using a 5-values Likert scale (I don't know, No, Yes scarcely, Yes moderately, Yes highly).

Below the questions proposed to heritage managers to assess the level of circularity of the adaptive reuse practices (Table 5.2).

Moreover, quantitative data showing key impacts of the adaptive reuse practice were sought (optional):

- N. of enterprises/startups localised in the heritage building/site
- N. of permanent jobs created
- N. of volunteers
- N. of visitors/year
- Annual revenues from the new uses/functions

Table 5.2 Assessment on Circularity of cultural heritage adaptive reuse practices

Dimension	Questions (1–5 Likert scale values)
1) CULTURAL VALUES CONSERVATION / ENHANCEMENT It refers to the "enlargement of the lifetime of goods, assigning them new functions in a long time perspective", that is one of the objectives of the circular economy as well as the objective of the adaptive reuse of cultural heritage. How the reuse contributed to preserve, maintain and enhance the place identity and tangible and intangible values (heritage historic, architectural and artistic values; local skills, techniques and knowledge)? Good practice of cultural heritage adaptive reuse in terms of circularity should enhance the cultural values enabling their conservation in the long term.	In this section, it is assessed whether the reuse process has contributed to: Conservation/enhancement of tangible and intangible heritage values (historic, architectural and artistic values; local skills, techniques and knowledge)—Avoiding loss of authenticity and integrity; Awareness raise for cultural heritage; Awareness raise for circular economy;
2) CIRCULARITY OF CONSERVATION INTERVENTIONS (in terms of circular metabolism at the micro scale) This section highlights the most "visible" aspect of circularity, that is the conservation of natural resources and the avoidance of new resources extraction. It includes energy, waste, raw and processed materials as important resources in to be reused, recycled, re-generated a circular economy model. The technical choices of adaptive reuse are able to reduce resource consumption and negative environmental impacts.	Specify whether and how the reuse process contributed to implement circular metabolisms: Realization of low energy consumption systems; Implementation of renewable energy sources; Implementation of water storage and reuse systems; Use of local traditional materials, bio-materials, and/or reuse materials; Reduction of construction waste to landfill; Recovery/increase of green spaces and/or nature based solutions (avoiding loss of biodiversity).
3) CIRCULARITY OF OUTCOMES COMING FROM REUSE INITIATIVES This category introduces another level of circularity between the cultural heritage and its broader context. How the adaptive reuse project contributed to enhance the attractiveness of the surrounding area/ neighbourhood/city/region, and through which criteria this change in place attractiveness can be expressed? It is linked to the relevant impacts that can be observed, directly and indirectly related to the adaptive reuse project (as a consequence of the reuse).	In order to assess this second level of circularity, it was asked whether the reuse process contributed to economic spillover effects and impacts in the quality of public spaces: Enhance jobs creation (avoiding loss of local jobs); Attract innovative start-ups and companies (e.g. digital, artificial intelligence, industry 4.0, internet of things, robotics, pharmaceutical adaptive reuse of cultural heritage, innovative technologies, science and art adaptive reuse of cultural heritage); Attract cultural and creative industries (craft, design, architecture, film, music, fashion); Attract new commercial activities (accommodation facilities, bars and restaurants, shops, services); Attract cultural visitors (avoiding mass tourism, "AirB&b effect": Displacement of the residents in favour of short-term tourists); Attract new residents in the area;

(continued)

Table 5.2 (continued)

Dimension	Questions (1–5 Likert scale values)
	Enhancement of quality of public spaces; Increase real estate values in the area (avoiding gentrification in the area). Other important impacts linked to this level of circularity are those on social inclusion, wellbeing and health. For this purpose, participants are asked to specify whether the reuse process has contributed to: Enhance safety in the area; Enhance place attachment and local identity; Enhance social cohesion (social cohesion is defined as the willingness of members of a society to cooperate with each other in order to survive and prosper); Enhance the landscape visual quality of the area; Create/enhance a heritage community (that according to the FARO convention consists of people who value specific aspects of cultural heritage which they wish, within the framework of public action, to sustain and transmit to future generations); Increase/enhance cultural activities and events, increasing also citizens' participation; Enhance people's wellbeing; Enhance people's health (psychological and/or physical). Finally, a last category of questions explores the circularity of the business, financing and governance model applied. This circularity is linked to the economic-financial self-sustainability: Is the project able to generate the needed resources to be successful in the long term? We explored here how third sector actors are involved, as *mission-driven* actors able to create value in multiple dimensions (not only economic, but also social, environmental and cultural dimension), assessing whether the reuse management model has the following characteristics: It is financially self-sustainable; It generates diverse revenue flows; The third sector (e.g. NGOs, foundations, ethical banks, social enterprises, associations, civil society organizations) has been involved in partnerships/cooperation, stimulating new local investments; Different stakeholders have been involved in decision making processes (avoiding social conflict); Profits—if any—are reinvested to fund more local oriented initiatives and/or other heritage initiatives

The following section presents the results of the data collection and the innovative data management platform developed through the use of Facility Live Human-centric Technology to enable cooperation between data providers and data users, allowing data co-production, innovation and cross-disciplinary adaptive reuse of cultural heritage and communication.

3 CLIC Survey Results and Data Analysis

The CLIC Survey included 126 cultural heritage adaptive reuse practices from European countries. The analysis performed aimed to investigate the characteristics of cultural heritage adaptive reuse practices and their 'performance' in terms of circular economy, as well as the relationships between different typological, governance and management characteristics and the level of achievement of circular economy goals.

3.1 Geographical Distribution and Heritage Typologies

Regarding the geographical distribution of the projects, a coverage of more than 80% of EU Countries was reached. The distribution of projects between the geographical areas (Central-Northern Europe, Eastern Europe and Southern Europe) results sufficiently homogeneous. A sufficient number of items (projects) for each geographical area was ensured in the selection of cultural heritage adaptive reuse practices in the database, allowing to perform reliable statistical analyses (Fig. 5.1)

Regarding the typology of the building/site or landscape area, a homogenous distribution was also ensured. Data were recoded for more synthetic and robust analysis. The recoding generated seven categories: Religious, Civil/Residential, Military, Productive/Industrial/Commercial, Leisure, Mixed, Other—with a slight prevalence of productive typology (e.g. food production), industrial heritage and commercial typologies, and 'mixed' typologies which include larger complexes with a mix of buildings typologies (Fig. 5.2).

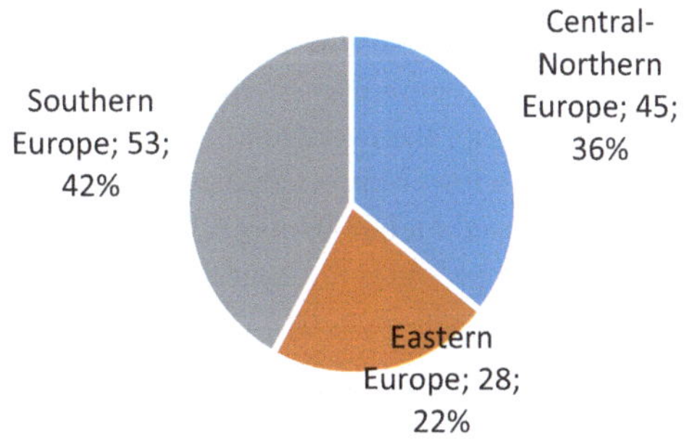

Fig. 5.1 Distribution of projects in the sample for EU geographical areas

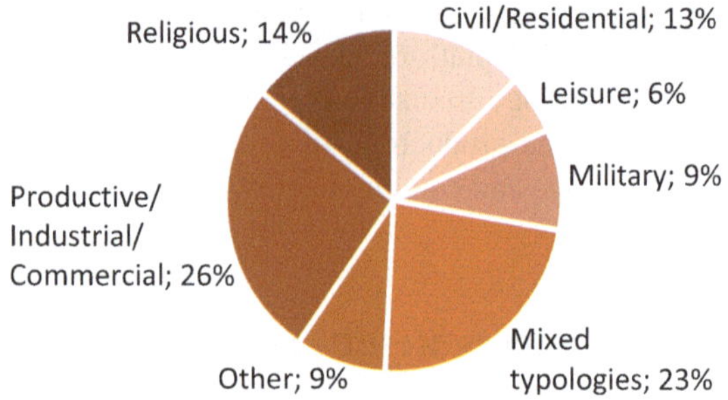

Fig. 5.2 Distribution of projects in the sample for typologies

Adaptive reuse practices in the three categories counted 28 units in Eastern Europe (22,22%), 45 units in Central-Northern Europe (35,71%) and 53 units in Southern Europe (42,06%).

3.2 Assessment on Circularity

Based on the 31 questions on circularity proposed in the survey questionnaire, the capacity of the adaptive reuse project to generate positive economic, social, environmental and cultural impacts was explored, as well as their capacity for generating economic-financial and environmental self-sustainability. It was possible to make some first considerations based on the distribution of answers provided by survey respondents.

Figure 5.3 provides the aggregated results according to this first level of circularity, which is related to a specific interpretation of cultural heritage values conservation and enhancement as a way for circular economy implementation. Cultural heritage has economic use values and independent-of-use values. This section of the circularity assessment is related to the "intrinsic circularity" generated by the conservation and enhancement of independent-of-use values, that generate in turn new use values, conserving and improving historical, aesthetic, symbolic values (as well as economic, social and environmental impacts in the context—generative capacity).

None of the respondents answered "I don't know" on the question on cultural heritage values conservation, demonstrating a very good knowledge and understanding of this aspect between respondents.

Figure 5.4 provides the aggregated results according to the second level of circularity, related to environmental externalities and closed metabolisms realization.

In this group, the average of "I don't know" answers is 34%, showing that there is generally a lower knowledge and scarcer awareness of the environmental aspects linked to closed metabolisms realization in the heritage conservation sector.

Figure 5.5 provides the aggregated results according to the third level of circularity on the impacts generated in the local area. For the statements related to the

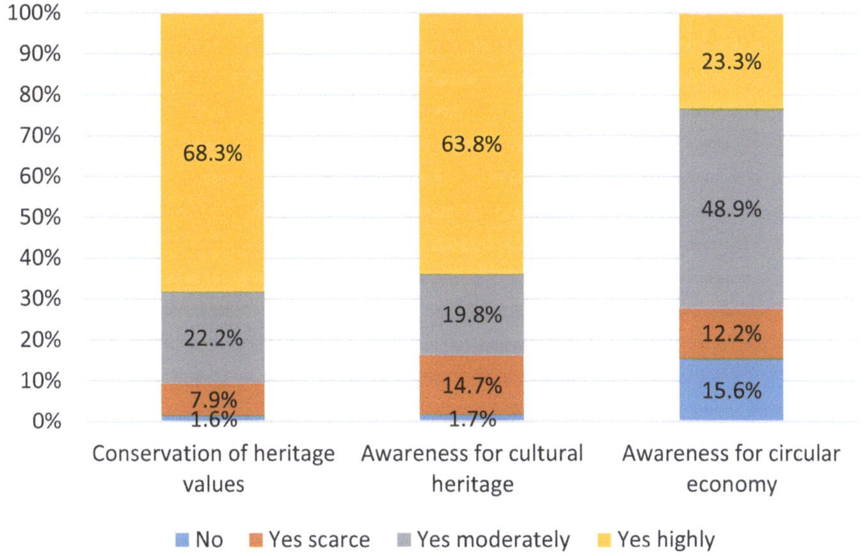

Fig. 5.3 Aggregated results for the first circularity level on conservation/enhancement of cultural values

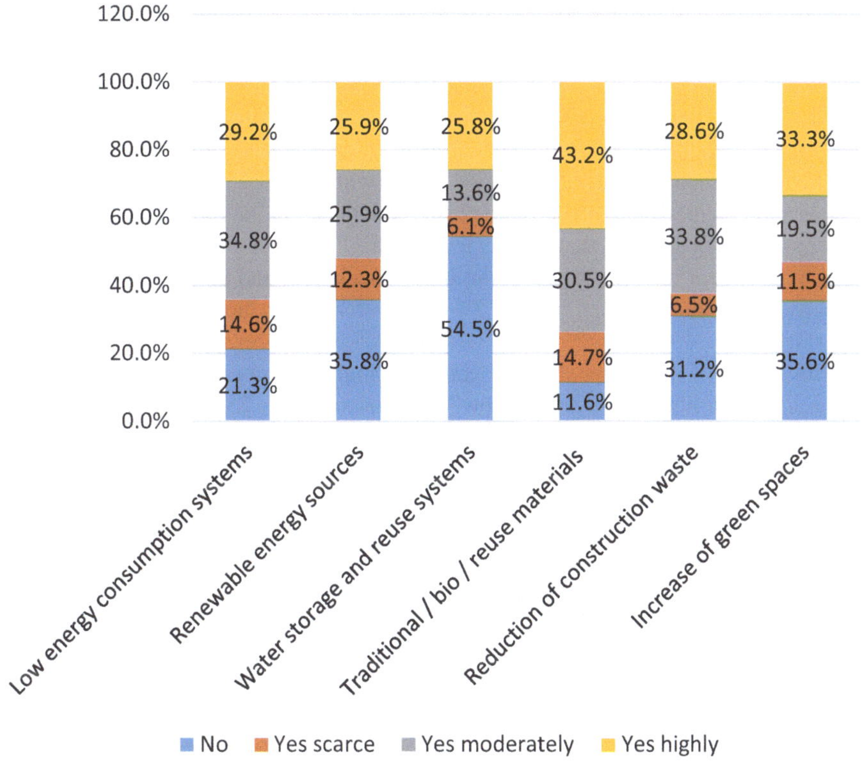

Fig. 5.4 Aggregated results for the second circularity level on closed metabolisms realization

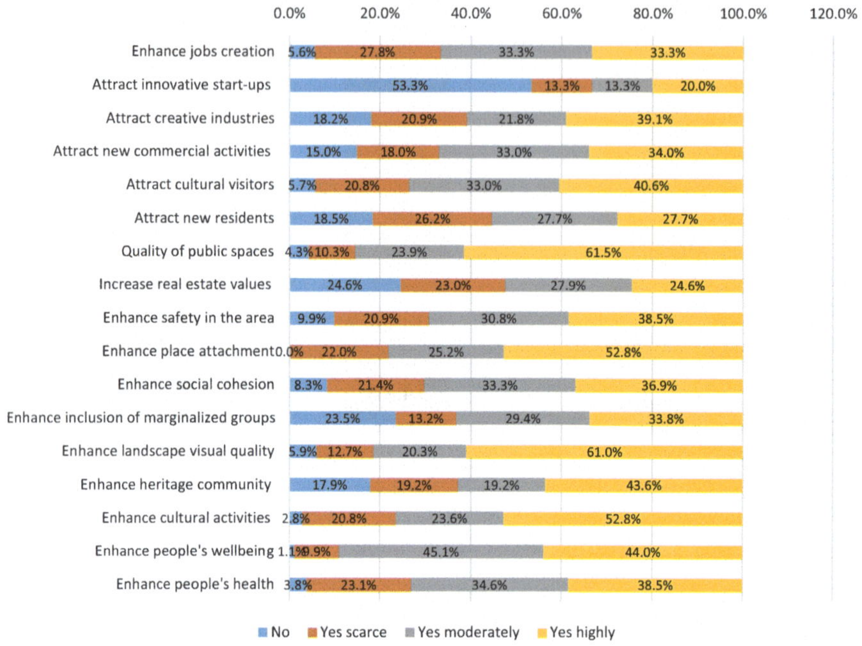

Fig. 5.5 Aggregated results for the third circularity level on generation of positive impacts in the local area

circularity of outcomes of conservation interventions, on the one hand it is possible to observe that the analysed case studies that show a high level of achievement of this kind of circularity are included between 24,6% (as far as the increase in real estate values is concerned) and approx. 61% (with reference to quality of public spaces and enhancement of landscape visual quality). On the other hand, the analysed case studies that do not show any level of this kind of circularity show very low percentages, near to zero (as far as the place attachment and enhancement of people's wellbeing is concerned) and 53,3% (with reference to the attraction of innovative start-ups).

This group of questions shows a generally higher number of positive answers and a lower number of "I don't know" answers (approx. 25%), demonstrating a generally acceptable level of knowledge and awareness about different impacts of cultural heritage adaptive reuse. As stated, the projects analysed have a quite long track-record of operation, since the majority of them was realized before 2015. Thus, the impacts reported can be considered a reliable information that represents the starting point for in-depth analysis of the impacts generated by cultural heritage adaptive reuse practices, through ex-post evaluation.

This section of the circularity assessment provides interesting information on the relationship between cultural heritage adaptive reuse projects and their context. Specific impacts of cultural heritage adaptive reuse on social cohesion and inclusion of marginalized groups were not always reported. These two questions showed a percentage of unavailable information between 33% (for social cohesion) and 46% (for inclusion). The available answers were generally positive, despite 23,5% of projects reporting no impact on the inclusion of marginalized groups.

The group of questions related to business, financing and governance models highlighted the internal structure of management, exploring the economic-financial self-sustainability of the adaptive reuse model, as well as the eventual involvement of different stakeholders. Also, there were explored the "aims" of the businesses and management models adopted through two indirect questions, exploring the involvement of third sector actors and the eventual reinvestment of the profits generated for new social activities, and thus focusing on the potential involvement of "social" businesses.

Figure 5.6 below shows the results of this group of questions. The average percentage of "I don't know" answers was 28% for this group, which was considered acceptable for the analysis of data. It also shows a quite good level of knowledge on the management model of the practices, which could be associated to the choice of target respondents, mostly adaptive reuse of cultural heritage with good knowledge of the reported cases, as well as directors or presidents of the managing organizations, CEOs, staff members and other active roles in the respective managing organizations.

The overall most positive result was given to the question related to the generation of diverse revenue flows, with a low rate of negative answers (9,4%) and low rate of unavailable information (around 15%), which shows that the large majority of the cultural heritage adaptive reuse practices collected are able to generate revenues through the uses/functions activated. However, observing the rates of answers

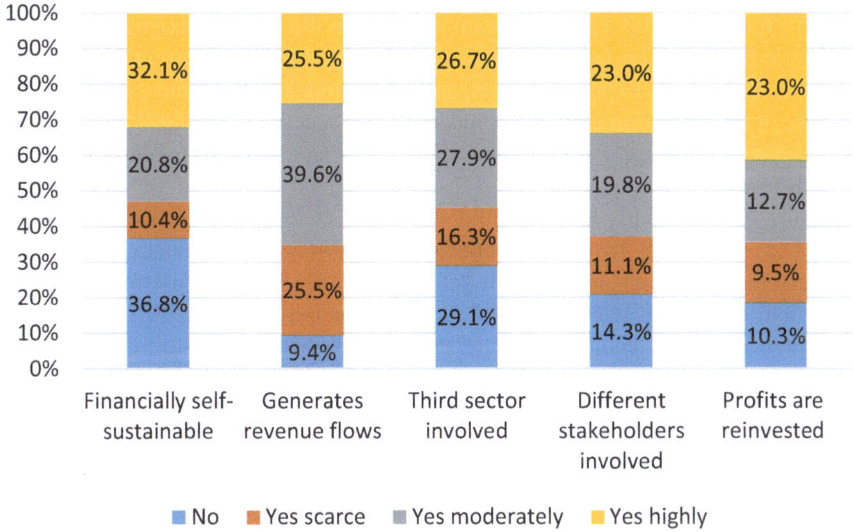

Fig. 5.6 Aggregated results for questions related to management characteristics and self-sustainability

for the financial self-sustainability, it can be noted that not in all cases the revenues generated are able to cover the operating and financial costs, since for this question the negative answers increase to 36,8% (at equal rates of unavailable information).

3.3 The CLIC Knowledge and Information Hub Platform

The CLIC Knowledge and Information Hub (KIH) was developed as a innovative prototype for the use of heritage adaptive reuse of cultural heritage in the CLIC project, however it was built in view of the extended use to heritage managers, professionals, owners and authorities. The platform was designed in an innovative way, testing and exploiting the FacilityLive semantic-based web technology, as work tool for adaptive reuse of cultural heritage to engage and cooperate with local stakeholders for co-producing information and knowledge on cultural heritage adaptive reuse practices. The use of the innovative FacilityLive data and information management technology had the aim to empower adaptive reuse of cultural heritage providing an open platform for cataloguing, presenting and sharing their adaptive reuse of cultural heritage findings, from best practice analysis to indicators to financing, business and governance models, literature and regulations enabling a rich and effective knowledge sharing/transfer and empowering the transdisciplinary adaptive reuse of cultural heritage process (Monteleone et al. 2019, 2021).

In the implementation of the Knowledge Information Hub, FacilityLive used a methodology based on co-creation, an approach that focuses on making stakeholders and users constructive and active participants in the definition and construction of the platform. The methodology developed and implemented by the FacilityLive adaptive reuse of cultural heritage team is based on the widely recognized innovation cycle process. The latter articulates itself around four concurrent phases: co-creation, exploration, experimentation, and evaluation. The fundamental concept at the base of the work is to gain direct and unfiltered access to end users' ideas, experiences, and knowledge, based on their daily needs and desire of feeling supported by a Knowledge Information Hub. Users are directly involved in co-creating, exploring, experimenting and evaluating new ideas, concepts and technological artifacts. The users are all involved from the early stage of the development process and throughout its duration. Insights are gathered directly from the users in order to define and implement realistic, useful, desirable and effective artifacts by using a number of tools and techniques (focus groups, interviews, brainstorming sessions and questionnaires).

The platform provides an overview of the reuse project (Fig. 5.7).

Here the user can find project details (project name, city and country in which the project is located, specific localization of the project, size of the city in which the project is located, approximate amount of the investment realized, year of realization of the adaptive reuse intervention, sq. meters of Reuse Intervention:, how the building/site/landscape is managed, type of managing body, how the adaptive reuse has been funded, barriers and constraints experienced in the realization of the

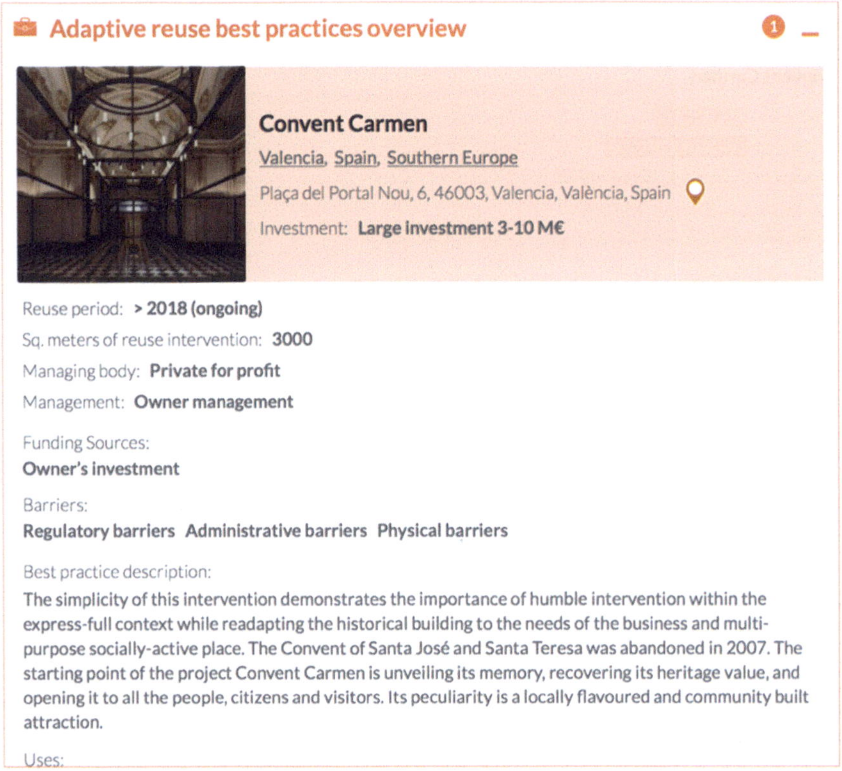

Fig. 5.7 Adaptive reuse practice overview

adaptive reuse intervention, a brief description of the economic, social and environmental impacts of the project, the final uses in the adaptive reuse intervention).

The platform provides also the characteristics of the involved cultural heritage building/site/landscapes (typology of the original building / site / landscape, level of protection of cultural heritage or recognition of its value by the local community, period of construction of the oldest part of the building / site / landscape elements, historic Period, state of conservation of the heritage building/site/landscape before the adaptive reuse intervention and a brief description of the heritage) (Fig. 5.8).

The specific localization of the project on Google Map is also provided.

The user can find information about the Impacts in the area due to increased attractiveness or impacts on social inclusion, wellbeing and health to which the reuse process has contributed and the type of owner and the state of use of the cultural heritage building/site/landscapes before and after the adaptive reuse intervention (Fig. 5.9).

Images of the cultural heritage building/site/landscapes and links to additional sources for more information are also provided.

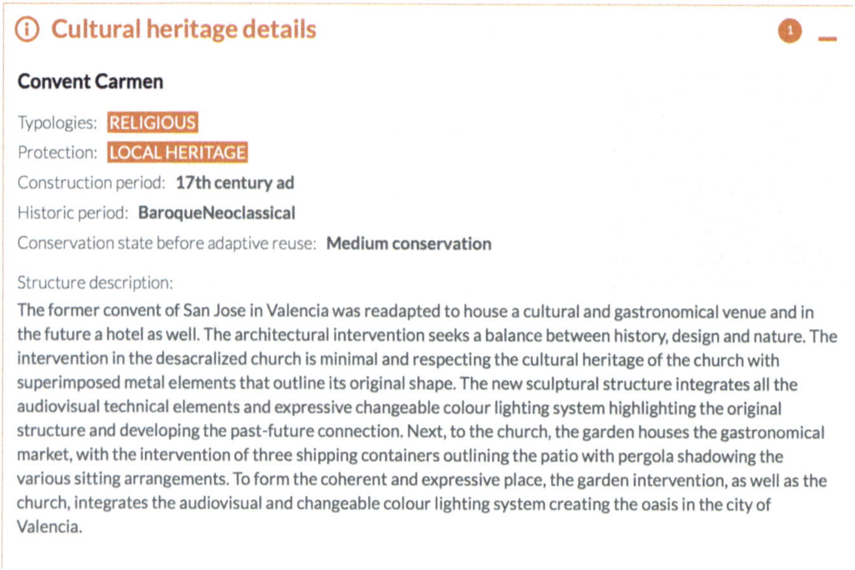

Fig. 5.8 Cultural heritage details

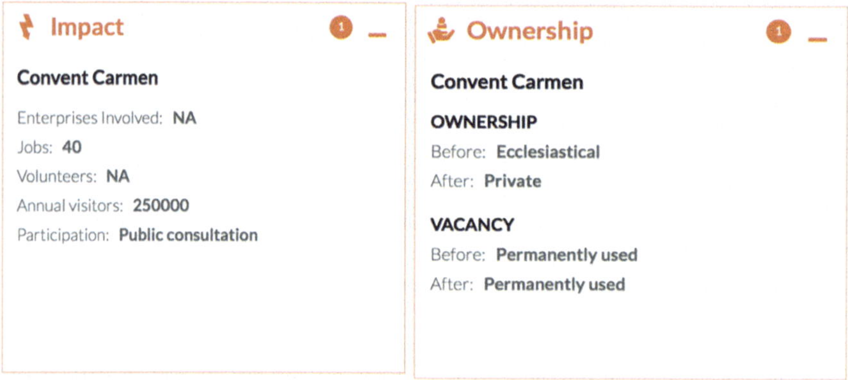

Fig. 5.9 Impact and ownership data

The Platform contains also specific data on different **aspects of circularity in the adaptive reuse of cultural heritage**. They include, for example, information on energy efficiency, water recovery and materials reuse, information on the conservation of the cultural value of the building/site/landscape through the adaptive reuse intervention, as well as information on the economic, social, cultural, and environmental impacts generated by the adaptive reuse project in the local context. Annex 1 provides more information on the features and functions of the KIH platform.

Box 5.1 Technological Innovation (Gabriella Monteleone)

The platform was based upon the proprietary technology of FacilityLive, a search driven next generation software platform for the organization and management of information, with patents granted in 46 countries worldwide. FacilityLive's disruptive technology finds exactly what you are looking for, thinking and treating information in the same way humans do. This allows people and organisations to have a compelling user experience in information search.

FacilityLive is an innovative software platform using a patented revolutionary semantic search engine. Abandoning the full text retrieval approach, including statistics and analytic algorithms, of the traditional search engines has allowed FacilityLive to have a higher level of precision compared to other existing search engines.

The platform is designed and built using a "human centric" approach to the management of information. The semantic search engine is extremely effective in linking information from disparate sources stored either in a structured or unstructured way. It is able to collect results and order them precisely in a contextualized and personalised way.

FacilityLive technology has the following advantages:

1. **Organisation of the information**—The platform organises digital *unstructured* and *heterogeneous* content (even of *poor quality*) through:
    - An accurate catalogue of all the information stored in any available information sources.
    - Recognition of different kinds of data and information, regardless of its origin, size, its format, and physical location.
    - The organisation of data and information from all sources based on the principle of its **relevance** with an indexing that is precise using a semantic approach which is typical to the human approach to organising information.
2. **Search**—The platform assists the user in an *intuitive way* when searching for information through the semantic interpretation based on the search text and returns the results with a high level of relevancy, identifying the most significant and precious information.
3. **Result presentation**—The platform returns the available information to the users in a simple and effective way that is highly intuitive, achieving maximum user experience. The information is aggregated, correlated and presented to the user in a workspace that represents a cogitative approach to information and its use.

When users access the platform on the web or on the mobile and begin to enter their search text in the search box, immediately they are presented with several suggestions for composing the rest of the search text. The search

(continued)

Box 5.1 (continued)

engine intelligently interprets what the user is writing and then guides the user further, showing the reasonable search paths for their needs based on the existing information sources it already knows about. The user at this point interacts with the search box to perfect and complete the search text.

Once the search text is composed and executed, the search engine is able to detect and associate the *semantic content* to different *concepts*. It can resolve complex search texts as it understands and associates the meaning of what the user is looking for even if it is written with different terms or languages. Therefore, the search text provided by the user should not be a list of words but a complete sentence from which then the search engine is able to extract the most relevant concepts, disambiguate them if necessary and return the most appropriate and accurate results quickly.

The FacilityLive technology uses a search model unlike the other search engines and similar to human methods. It is based upon **concepts**. The human mind is directed when searching information using a "knowledge domain". This is the set of all the terms that the cerebral system connects to the desired information. The memorised terms are elaborated in the mind by cogitative processes that create the ties between them. "Linguistic capability" is the instrument that allows these cerebral connections to be made.

The FacilityLive search engine simulates the capacity of the human brain to organise the connections between the terms being searched and other terms that a person creates. These connections are related using the concept of *semantic similarity* in the search engine by which sets of words are similar in meaning or content. The engine creates an ontological model that allows it to simulate the linguistic abilities of humans, thereby able to search for all documents present within an information repository by identifying analogies with the real meaning of the searched terms.

Once the user completes the input of the search text and submits it, based on the information sources available, the semantic engine searches for results simultaneously from all the different information sources available and returns only the most significant ones, correlating and integrating them in a logical manner.

The results are organised and presented to the user in an intuitive and personal workspace that facilitates human interaction. This innovative workspace provides the user with *conceptual* and *visual tools* that provide easy navigation through the results returned by the search engine and creates a *mental model* on which the user can work to perform further duties.

The search engine creates the user's workspace dynamically based upon the search performed. It organises the visual representations using methodologies that are close and similar to the human cognitive approach. The search engine divides the visualisation of the returned results in **conceptual areas** to

(continued)

Box 5.1 (continued)
promote detail in a *contemporary manner* instead of the usual sequential manner. The use of these presentation areas allows the returned results to be easily interpreted and satisfy the *associative logic* of the user more effectively than the traditional visualisation as lists. Finally, collections of results that are complimentary each other are aligned in various areas of the workspace to facilitate access to the information and the decisional process.

4 Discussion: Lessons Learned from the CLIC Survey

First considerations can be made regarding the results of the CLIC Survey, based on the answers received. In particular, regarding the assessment on circularity, it can be generally observed that almost all adaptive reuse projects, with very few exceptions, reported positive impact on the conservation/enhancement of cultural heritage values, as well as on the awareness raise for cultural heritage. On the other hand, while on low energy consumption systems and use of traditional /bio/reuse materials the level of not available information was lower and the answers were in general positive, the realization of water storage and reuse systems clearly emerged as a critical aspect. In fact, for more than 40% of the projects the information was unknown, while those reporting the information mostly answered as no water storage and reuse systems were employed. This poses a critical question on the awareness of heritage managers, institutions and professionals and in general of the heritage sector on the importance of these environmental aspects, showing a "loss of memory" of traditional but efficient construction systems adequate to local climate and landscape conditions, and particularly on the need of effective systems to conserve scarce resources such as water. These results could be alarming particularly considering that traditional buildings, sites and landscapes were realized with great attention to these aspects, employing in most cases the traditional/vernacular knowledge in the realization of energy-efficient constructions and rainwater recovery systems, through wise climate resilient construction techniques. An example is the Villa Campolieto in Italy, a historic villa of 1700 located in the Vesuvian area of Naples, were a complex system of cisterns was realized to recover and store rainwater for gardening use. This old system was restored through the adaptive reuse interventions, and it still serves its original scope today. Another example is the Giardino della Minerva (Minerva Garden) in Salerno, were ancient canals and cisterns realized on terraces and connected to mountain springs were recovered through the adaptive reuse interventions and are still in use to irrigate the various parts of the garden, but they also contribute to the atmosphere and micro-climate of the garden itself. Moreover, at the landscape scale it can be highlighted the ancient water channels and cisterns systems of traditional agricultural landscapes, which contributed to the stabilization of slopes in hill and mountain areas and ensured the conservation of water as a fundamental resource for local communities.

The general low level of awareness regarding the aspects related to energy efficiency, water recovery and storage systems, and generally climate-neutral construction solutions emerges as an open issue in the heritage sector. The adaptive reuse of cultural heritage is an opportunity to apply advanced retrofit techniques, recover ancient but still efficient "circular" technological systems, and enhance existing buildings /sites/landscapes through implementation of the most advanced technologies. This would generate multiple benefits, by reducing operation costs, avoiding negative environmental externalities and most importantly, generating positive impacts on human health.[2]

Regarding the realization of renewable energy technologies in heritage buildings /sites /landscapes, a relatively high percentage of "no" answers was reported. This could be related to the difficulties in finding a balance between the need of preserving the cultural values of the sites, maintaining their features and attributes in terms of materials used, volumes and "forms", and the need of enhancing the energy efficiency in the existing built environment. An example is the limitations in the use of photovoltaic panels in historic urban areas in Italy, which are not formally banned but, in the practice, retrofitting projects including roof photovoltaic systems which could have a negative impact on the integrity of the urban landscape are not easily allowed by heritage protection institutions. Technological alternatives are available and can still be developed to enhance the retrofitting and efficiency of historic buildings/areas. However, the high costs and general low level of knowledge remains a barrier, as demonstrated by these results to the CLIC Survey. Despite the existing barriers, this represents a relevant field for new jobs through circular economy applications in historic urban areas and cultural landscapes, not yet fully explored.

Almost all projects reported positive impacts in the enhancement of quality of public spaces and landscape visual quality of the area in which they are located. Also the safety of these areas was enhanced, probably due to the activation of new activities/functions in previously abandoned or degraded urban areas. Absolutely all projects reported positive answers to "place attachment". This confirms the great relevance of the adaptive reuse of cultural heritage as contribution to urban regeneration, also as key strategy to implement the Historic Urban Landscape recommendation.

Regarding the impacts on the liveability and general quality of life led by cultural heritage adaptive reuse, it can be noted that all projects reported an increase in cultural activities and positive impacts on wellbeing. Particularly, on this aspects, more than the 85% of available answers reported high or moderate contribution to wellbeing (only 1,1% of "no" answers). Moreover, almost all available answers were positive on health enhancement (only 3,8 of "no" answers), despite some difficulties in answering this question, with a rate of unavailable information around 38%.

Another critical aspect that is often evoked in theoretical literature and recommendations is the role of the "Heritage Community" as defined by the FARO

[2] See World Health Organization, Regional Office for Europe 2018. Circular Economy and Health. Risks and Opportunities.

Convention on the role of cultural heritage for society (Council of Europe 2005). The Heritage Community is represented by formal and informal association of citizens 'taking care' of the cultural heritage and spending efforts for its preservation and valorisation, ensuring that its cultural and societal value is conserved, enhanced and communicated. This aspect does not emerge clearly from the results of this Survey. Although available answers are relatively positive (with more than 80% reporting high or moderate impact), there is still a high number of "I don't know" answers reported (around 38%) for the related question. This poses a critical open question on how and 'how much' the Heritage Community is perceived and implemented as a "commons", as fundamental way to conserve cultural heritage and to ensure that it generates values for society.

With respect to spillover effects generated through the re-activation of abandoned and underused cultural heritage assets, results were encouraging for some aspects, while not fully satisfying for some specific questions. Positive answers with low rate of "I don't know" (between 10% and 20%) were given to questions related to jobs creation, attraction of creative industries, increase of commercial activities and increase of cultural tourism in the area. This confirms the positive role of cultural heritage adaptive reuse for the generation of economic spillover effects at local and regional level, and confirms previous studies on the localization choices of creative entrepreneurs in vibrant heritage areas (Smit 2011; Della Lucia and Trunfio 2018; Esmaeilpoorarabia et al. 2018).

Particularly, regarding the number of jobs created, it is interesting to analyse the quantitative data reported for 55 projects. This question, not "mandatory" for the respondent, was aimed at collecting reliable information when possible, without compromising the larger data collection through too much detailed information requested in the first level of Survey. Analysing the results, a number of 17 projects reported jobs creation between 1–10, and 26 projects reported jobs creation between 10–50. Moreover, 8 projects created a number of 50–100 jobs, and 6 projects reported the creation up to 500 jobs, with a relevant impact also at regional level.

Although they represent diverse typologies of cultural heritage and very different management experiences, these projects deserve further exploration to understand the factors that led to these performances in terms of jobs creation, as well as potential positive but also negative social impacts that could be generated along with economic spillovers.

Regarding attraction of innovative start-ups, the answers to this question show controversial results. On one side, on this aspect it was reported the highest number of "no" answers (53,3% of available answers). Moreover, almost 28% of the project do not report this information at all, probably for lack of knowledge and understanding of respondents.

However, it is interesting to note that 20% of the available answers reported a high impact in the attraction of innovative start-ups through the adaptive reuse of cultural heritage. More detailed exploration of this data showed 30 projects with high or moderate attraction of innovative start-ups, which include 20 projects which reported also a specific number of jobs created, the half of them with a high number of jobs between 30 and 350. Thus, it can be observed a relationship between cultural heritage adaptive reuse, localization of innovative start-ups and jobs creation.

More controversial results emerged for the role of "social businesses", since the rates of unavailable information was higher than 30%. However, the available answers given to these questions further confirm an emerging role of third sector organizations and 'mission-oriented' businesses in the reuse and management of cultural heritage assets, and the perception that a more effective combination between public "sustained" functions and self-sustainable functions should be identified. Another aspect that can be highlighted is related to the involvement of different stakeholders, which can have a role of service providers, marketing support and/or services users. The distribution of yes and no answers is highly favourable for the involvement of stakeholders as relevant element for the success of adaptive reuse projects. As expected, the capacity of heritage managers to engage a large number of stakeholders can be considered critical to support adaptive reuse initiatives, both in the start-up and in the longer-term operation phases.

The survey results showed that adaptive reuse is an approach able to face the continuous obsolescence (technological, functional, positional, etc.) of a site/asset during the time: to reduce the speed of the entropy through maintenance, repair, recycle, etc., thus producing an extension of the life-cycle of the asset/resource.

Adaptive reuse of cultural heritage is related to an increase in cultural activities and positive impacts on wellbeing, with interesting results highlighting also a potential impact on health enhancement, opening up to a new adaptive reuse of cultural heritage field in heritage studies; the question on "how" these impacts should be evaluated, particularly on wellbeing aspects, was investigated later in the CLIC adaptive reuse of cultural heritage.

Cultural heritage adaptive reuse generates economic spillover effects at local and regional level, through jobs creation, attraction of creative industries, increase of commercial activities and increase of cultural tourism in the area, also confirming previous studies on the localization preferences of creative entrepreneurs in heritage places.

The most critical barriers are related to costs (economic barriers) and regulations; careful analysis of costs, and particularly how these costs can be reduced through circular economy (e.g. new technologies and collaborative/synergistic approaches) should be conducted; moreover, to overcome the difficulties due to "standardized" heritage regulations to be applied in enormously different heritage sites and contexts, "free" space for experimentations, supported by suitable evaluation tools, could be enabled, in order to develop the needed innovations in heritage management.

While the cultural value in heritage adaptive reuse is clearly perceived, conserved and in some cases enhanced through contemporary new uses and transformations/additions, the same does not happen for the awareness and implementation of environmental friendly and retrofitting technologies for "closed metabolisms realization", which would be able to reduce operating costs and avoid waste of key environmental resources such as water, energy, raw materials for constructions; moreover, "circular" building construction models for adaptive reuse would be able to enhance the health of people through employment of safe materials and avoidance of pollution and emissions.

Finally, increases in real estate values consequently to the adaptive reuse of heritage assets could be more often and more clearly assessed, also in relation with financing models based on land value capture mechanisms.

5 Conclusions

The CLIC Survey represented the initial, but relevant step of the CLIC adaptive reuse of cultural heritage, informing all subsequent phases of in-depth best practices assessment, circular adaptive reuse modelling, circular business model definition and implementation, innovative financial models identification, participatory evaluation and decision-support for the development of circular adaptive reuse projects in specific case studies. The CLIC Survey allowed to identify success factors and barriers to circular adaptive reuse of cultural heritage based on evidence-base, compared with the theoretical model. The CLIC circular model was informed by the results of the Survey, identifying the key "building blocks" of circularity in cultural heritage and implementing them concretely through a set of specific criteria and indicators used in the ex-ante evaluation, enabling transparent participatory decision-making processes. Through the Survey, specific knowledge gaps were addressed related to adaptive reuse management and financing models at European level, impacts generated and level of financial and energy/environmental self-sustainability, allowing to better understand the "value for money" of diverse interventions in the perspective of the adaptive reuse as an "investment" instead than a cost for society. The sample of adaptive reuse projects collected through the Survey cannot be considered as complete nor exhaustive of the different contexts and situations, however it represents a large knowledge base sufficiently diversified and covering almost all countries and homogeneous heritage typologies. The IT heritage data platform of the "Knowledge and Information Hub" developed by the innovative technology startup FacilityLive had the merit of providing a structured, usable and human-centric tool for organising heritage data at European level. The flexible, human-based intelligent features of the FacilityLive data search and organisation technology can represent a future benchmark for heritage data platforms, enabling advanced, quick and easy adaptive reuse of cultural heritage towards different users categories. It represents a "hub" of heritage data with unique and specific features designed for heritage managers, adaptive reuse of cultural heritage, professionals, and public authorities, where the contents and results of each adaptive reuse practice are clearly presented and easily visualised.

This contribution specifically focused on the results of the Assessment on Circularity section of the CLIC Survey as it was particularly relevant for understanding the overall "success" of the adaptive reuse practices from the circular perspective. However, the full report of data analysis including information on reuse practices characteristics, previous and new uses, financing aspects, and other data characterizing the heritage assets is available in the CLIC repository (Gravagnuolo et al. 2021; Buglione et al. 2021). According to the Open Data approach, the database was made available publicly as inter-operable tool for further exploration.

Based on the results of the CLIC Survey and the observed potential for evidence-based knowledge production, it seems desirable that the process of data collection, organisation, analysis and interpretation of cultural heritage adaptive reuse practices continues through further efforts of heritage managers and adaptive reuse of cultural heritage, able to provide relevant and reliable data that can be used to enhance knowledge and processes (projects, interventions) on cultural heritage towards the circular reuse model as proposed by the CLIC adaptive reuse of cultural heritage.

Annex—The CLIC Knowledge and Information Hub Platform (*Gabriella Monteleone*)

The CLIC Knowledge and Information Hub platform provides a common space for adaptive reuse of cultural heritage and practitioners to share and exchange knowledge about cultural heritage adaptive reuse, in terms of "best practice". Furthermore, it develops a pilot function based on requirements from the cities of Salerno, Rijeka, Amsterdam and Västra Götaland Region to showcase local reuse practices and "present" the abandoned heritage as assets/opportunities for adaptive reuse investments towards the implementation of a "circular city" model. The application in the four cities/territories, representing different geographic, historic, cultural and political contexts is replicable in any other European (and not only European) city. The platform provides also a knowledge base of all existing literature on Adaptive Reuse and Circular Economy topics (Fig. 5.10).

The user can find the information he/she needs by using the search bar at the top of the page. Some suggestions of search are presented below the search bar, divided by topics.

The User can find information regarding:

- projects of **cultural heritage adaptive reuse** from European countries, provided by respondents to the CLIC Survey ("Best Practices" button);
- **opportunities of cultural heritage adaptive reuse** from European **cities and regions** ("Cities and Regions" and "Opportunities buttons");
- all existing literature on Adaptive Reuse and Circular Economy topics ("Literature and Regulations").

There is also an **About us** section, a **Help** page and the possibility to contribute to a **survey** and/or add a **city/region** to the platform.

Due to the immediate visibility of the most relevant results and their relationships, the user has a simple and evident visual representation at their disposal that provides access to results including their physical and logical locations that are most relevant to their search. This is done through instruments that enable the

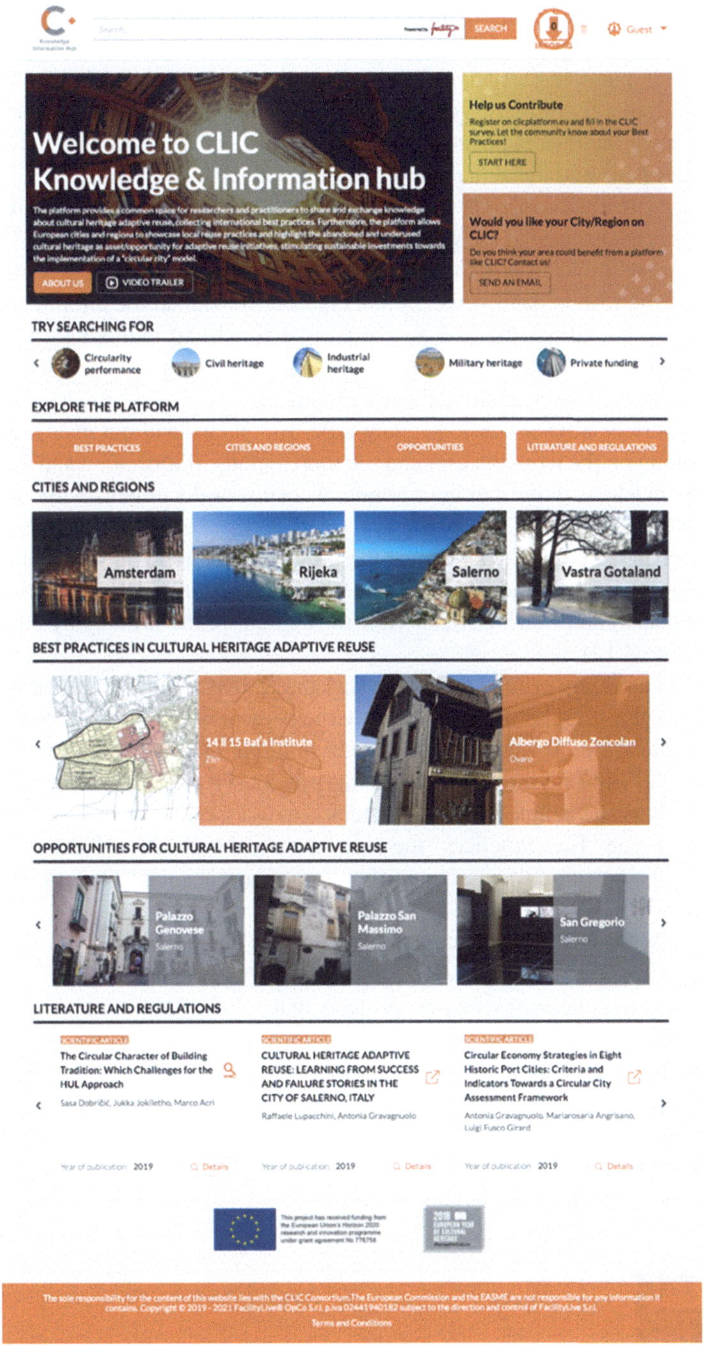

Fig. 5.10 Home page

simultaneous vision of multiple levels of information. Clicking on a result provides accessibility to that result without needing to leave the workspace: The Hyperlens® component implements a patented feature of FacilityLive that allows navigation locally and remotely directly from the result on the results page. The user is never forced to leave the results page and therefore does not have to toggle between the search result page and the actual source of the user activated result.

The user can easily interact with the results, choose and save them by simply dragging the results around the workspace. The Infobag® component is designed to collect and save a selection of the results chosen by the user creating an information pack. From the results page you can then email or download the entire contents of the Infobag. This allows the user to create a personal and individual information package containing all the selected artefacts from the search. The user can then review these results later or share them with others.

The Hyperlens allows you to see any type of content, without ever leaving the page. Clicking on the icon, the user can see more details, information and data about the content he/she is interested in or remote websites or documents without leaving the page he/she is on (Fig. 5.11).

The infobag is a smart folder. Inside the Infobag® the user can save and organize his/her search results. To add items to the Infobag the user can drag-and-drop them in the designated area. The user can view the content of the Infobag by clicking on this icon (Fig. 5.12).

The user can freely browse the contents of the Knowledge Information Hub platform as a guest user. If he/she wants to save the content of the Infobag, he/she needs to register and log in to create an account. By logging into his/her account on other devices he/she will find the items he/she has previously saved in the Infobag.

Cultural heritage adaptive reuse practices

By clicking on "Best Practices" the user can access to the knowledge of the platform about the best circular practices of cultural heritage adaptive reuse in Europe and beyond (Fig. 5.13).

It's possible to search, collect and compare the characteristics and impacts of more than a hundred cultural heritage and landscape reuse projects facilitating the enrichment of the platform with new projects in a collaborative way.

The platform not only empowers stakeholders and practitioners providing an open platform for cataloguing, presenting and sharing reuse projects but it is also of

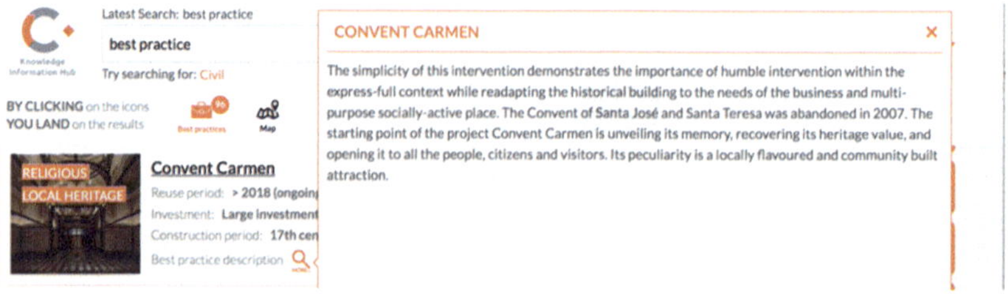

Fig. 5.11 The Hyperlens

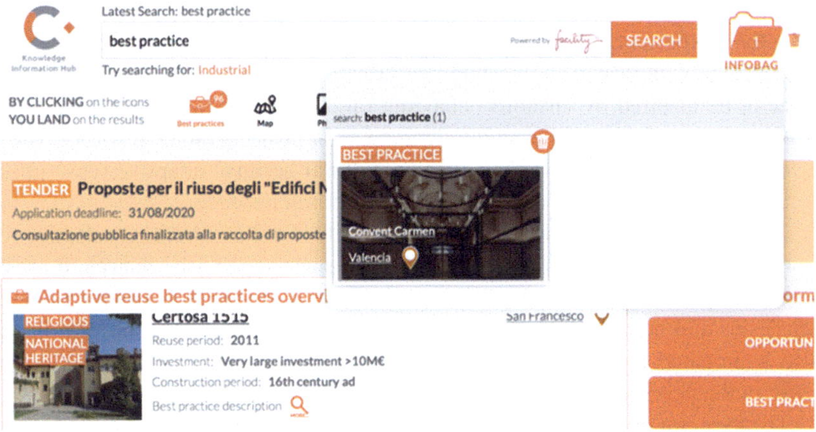

Fig. 5.12 The Infobag

Fig. 5.13 Best practices

interest to anyone who wishes to look into the adaptive reuse practices and learn from their experiences such as citizens interested in seeing the projects of their city, local officials interested in possible new uses of buildings in their area of competence or be inspired by virtuous behaviour on other territories, investors interested in being able to evaluate the economic impacts of projects on the territory. The aim is to make accessible the information on cultural heritage adaptive reuse projects collected within the CLIC adaptive reuse of cultural heritage.

The platform contains information on projects of cultural heritage adaptive reuse from European countries, provided by respondents to the CLIC Survey. It presents general information on the adaptive reuse project and its location, its structural characteristics and management aspects, and its uses/functions. The platform includes also aggregated data on different potential impacts of cultural heritage adaptive reuse in the perspective of the circular economy.

The user can find the information he/she needs by using the search bar at the top of the page. Some suggestions of search are presented below the search bar, divided by topics.

Through the platform, the user can consult information on single project and he/she can search specific groups of projects according to their characteristics for instance typology of the heritage building/site/landscape, type of *owner* before the adaptive reuse intervention (public, private, ecclesiastical, other), type of *managing body* (public, private for profit, private non-profit, public-private-partnership, mixed), and type of *funding* (public, private foundation, private donations, owner's investment, manager's investment, crowdfunding campaign, other).

When the user starts typing, the search assistant, containing further search suggestions, appears (Fig. 5.14).

If none of the suggestions satisfies his/her search, the user can add more details in the search bar.

Fig. 5.14 Search assistant

By selecting "religious", the refreshed page displays the widgets containing search related results, all projects related to religious heritages.

The interface of the Knowledge Information Hub platform is composed of Widgets. Widgets are containers that organize results based on search pertinence. The page displays a number of Widgets divided by information type. They contain all the information regarding the search.

Each new search displays new results and new widgets. For example, the user can access the information of a specific project clicking on the name of a single project.

The user can modify the visualization of the page by minimizing widgets he/she is not interested in. To do so, he/she has to click on the top right "_" button on the individual widget he/she wants to hide. The widget is now hidden.

To visualise it again, the user can simply click on its icon in the widgetbar. The widgetbar is a tool displayed beneath the search bar. It shows the icons representing the widgets. It shows the number of items in each widget. It allows the user to see which widgets are visible and which are hidden. By clicking on the icon of a widget in the widgetbar, the page scrolls to that widget's position (Fig. 5.15).

The aim of the CLIC Knowledge and Information Hub is to integrate the dispersed available information, facilitate the retrieval of information and making the information available for decision-making.

The platform has been implemented to manage and visualize in an innovative way the information related to heritage assets to inform investors, policy-makers, entrepreneurs and civil society organizations on the opportunities and costs related to available heritage assets to start heritage regeneration actions.

The platform allows a city/region to showcase local reuse practices and present the abandoned heritage as assets/opportunities for adaptive reuse investments towards the implementation of a "circular city" model. Detailed data and information for the reuse of a building/site are provided as well as the communication of eventual calls for tender.

The application, based on requirements from four cities/territories, the CLIC pilot cities of Salerno, Rijeka, Amsterdam and Västra Götaland Region, representing different geographic, historic, cultural and political contexts is replicable in any other European (and not only European) city.

The platform:

- enables institutional communication of adaptive reuse objectives and opportunities in a city/region
- facilitates stakeholders' cooperation and access to information
- allows the launch of adaptive re-use strategic projects testing the proposed circular models

The platform not only empowers city and regions providing an open platform for cataloguing, presenting and sharing reuse opportunities but it is also of interest to anyone who wishes to look into the adaptive reuse practices of European cities and regions and learn from their experiences as well as citizens interested in seeing the

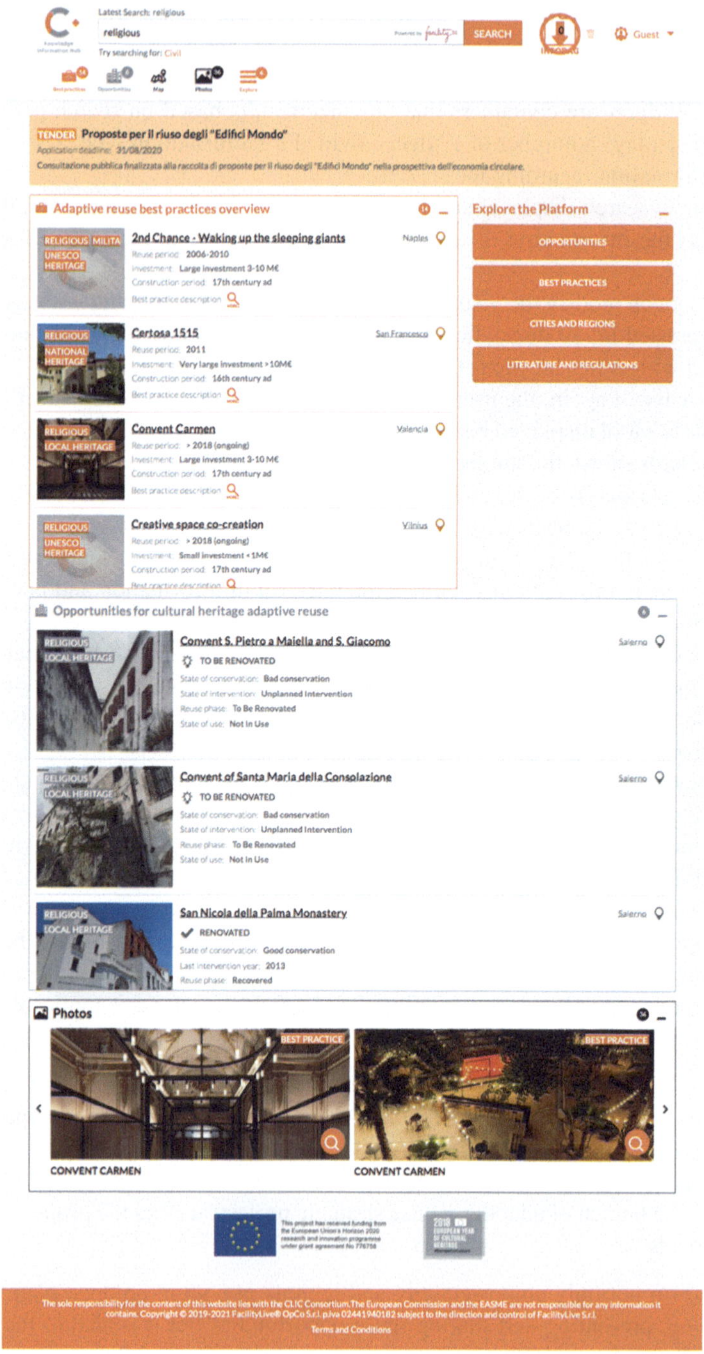

Fig. 5.15 Search "Religious"

projects of their city, investors interested in being able to find opportunities on a territory.

The ambition is to use and further develop highly innovative data and information technology to enable cooperation between multiple stakeholders and inform their investment and management choices.

The platform organizes, manages and provides innovative access to all the information on opportunities of cultural heritage adaptive reuse from European cities and regions. It collects and presents specific information on the adaptive reuse opportunity and its location.

Through the platform, the user can consult information on all opportunities across Europe, search specific groups of opportunities (available spaces, to build and reuse, to design and reuse) or according to their characteristics. For instance, typology of the heritage building/site/landscape (Civil, Residential, Religious, Military, Industrial, Commercial, Leisure, Agricultural), state of conservation, state of intervention, etc (Fig. 5.16).

The user can consult information about local reuse practices and opportunities for adaptive reuse investments in a specific city/region as well find call and tenders published by the local administrator, news, documents and contacts (Fig. 5.17).

The platform presents a detailed description of an opportunity (Fig. 5.18).

The user can find specific data related to the involved heritage: its typology, protection, the typology of construction, the indoor and outdoor sq. meters, number of floors, the owner, the urban planning zone, the energy performance as well as specific documents as planimetries (Fig. 5.19).

Images, contacts, news, links to websites or social where find further information, opening times to visit the building/site are also provided.

Furthermore, the platform provides all the details related to the adaptive reuse process (Fig. 5.20):

The user can find details about the reuse status of the heritage building/site (to be renovated or renovated) and the current reuse phase, its state of use, the allowed uses, the last intervention year and the intervention needed, the estimated or actual investment range and costs for the reuse project, its intervention timeframe timeline and managing aspects as the governance model, management type and manager.

FacilityLive's technology is available on Cloud as "software as a service". Making and upkeeping Web and Mobile solutions is made easy and intuitive thanks to its smart components-based platform. FacilityLive Cloud platform generates a digital Platform, that gives the end user access to information within the complete and intuitive experience. Specifically, for this purpose FacilityLive studies, designs and develops new and innovative technological components.

The FacilityLive Cloud Platform allows the creation of platforms that offer an innovative user experience when it comes to the fruition of information. This experience is enabled by an innovative search engine that provides precise answers to complex questions. That's how the platforms created with FacilityLive's technology allow humans and machines to run searches and get a dynamic composition of data and information, within a specific domain.

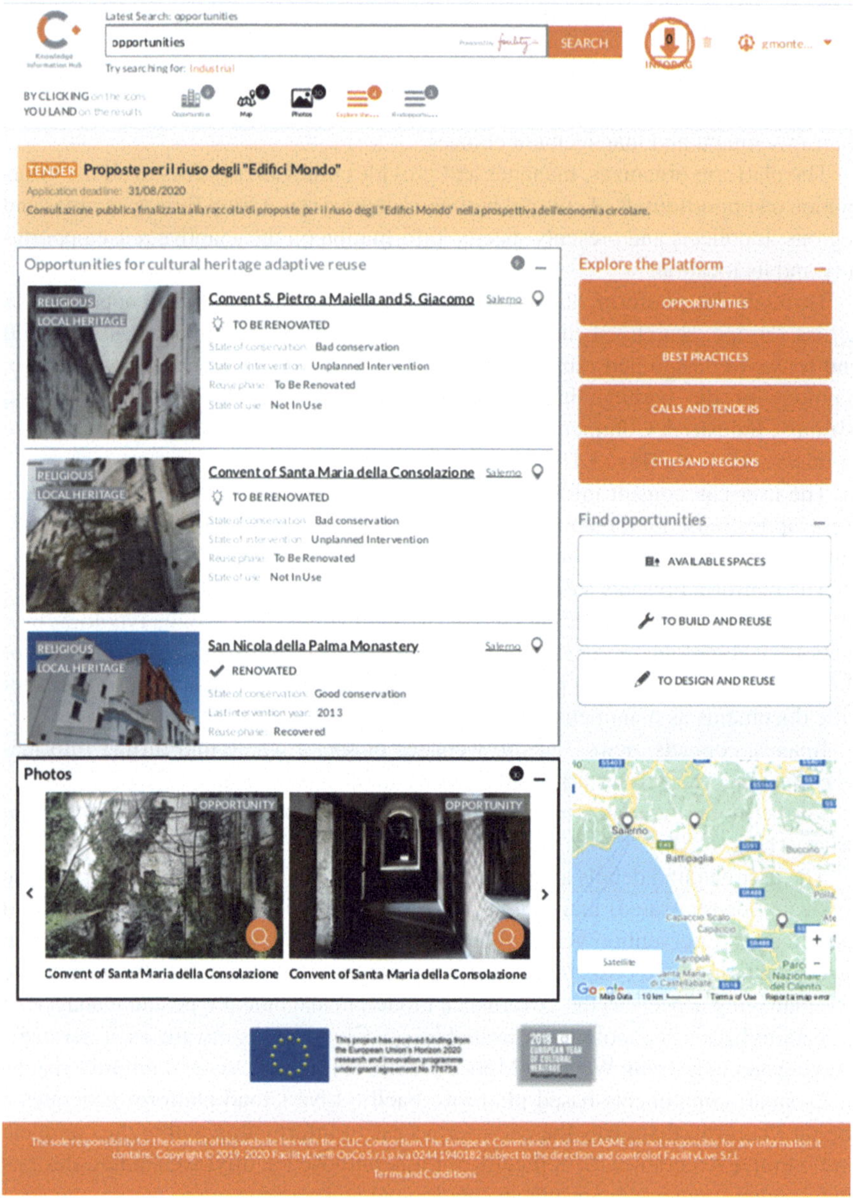

Fig. 5.16 Opportunities

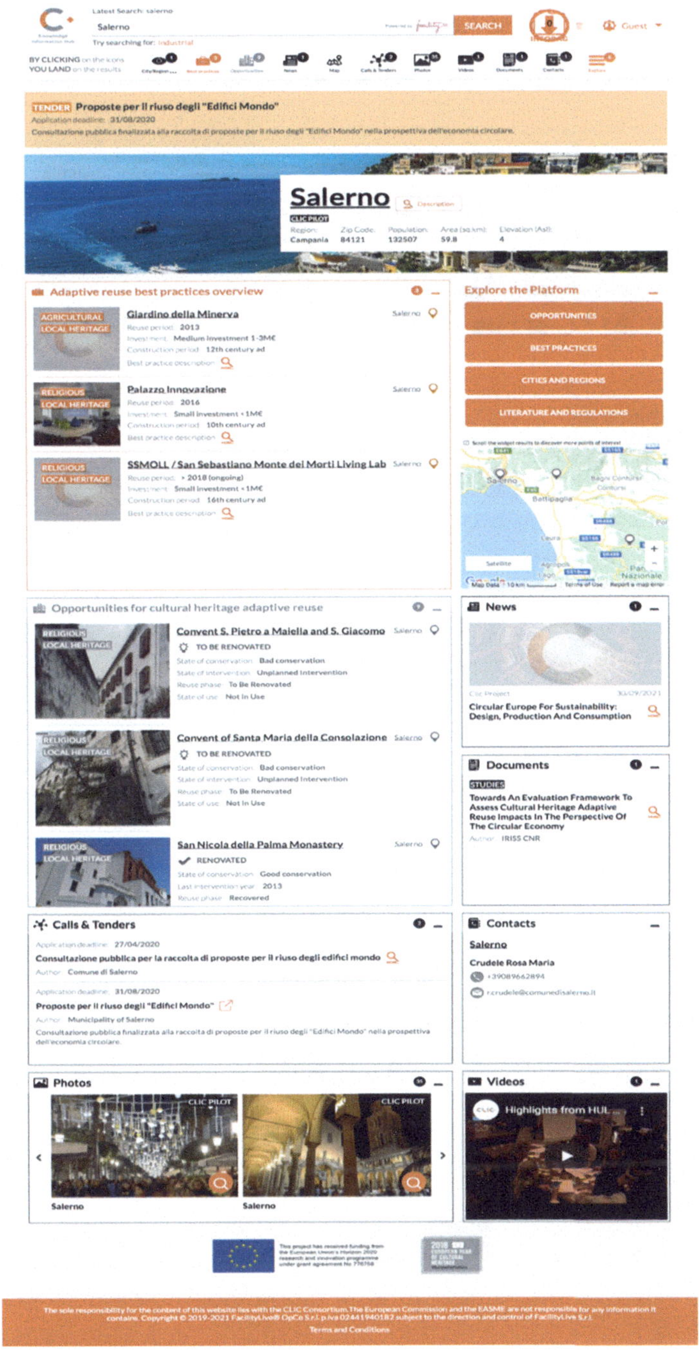

Fig. 5.17 Overview of city/region

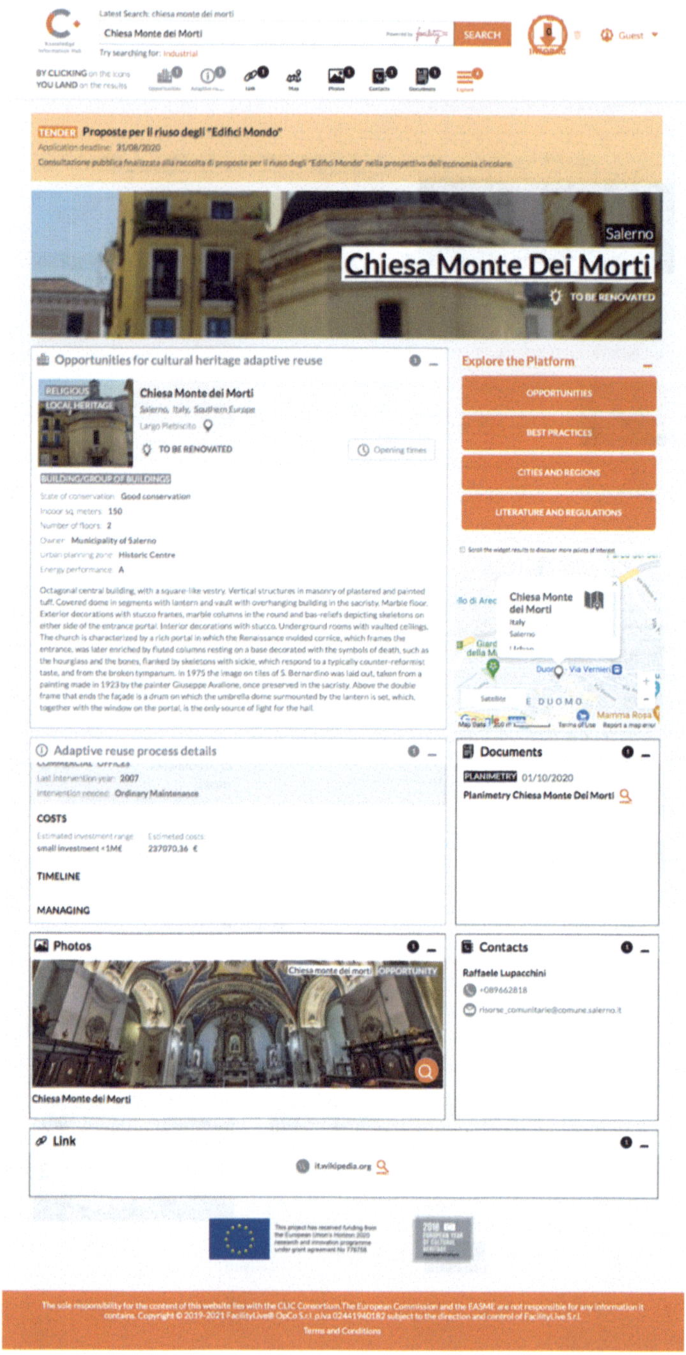

Fig. 5.18 Opportunity

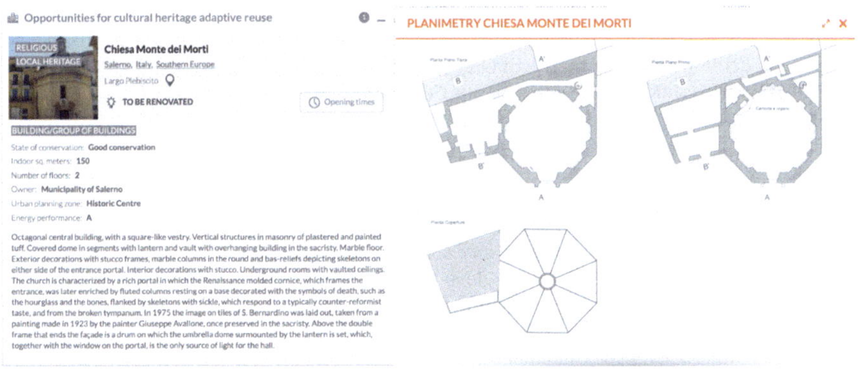

Fig. 5.19 Heritage building information

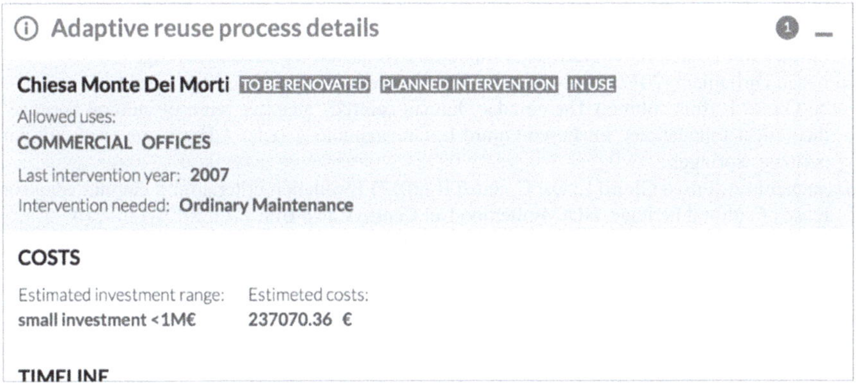

Fig. 5.20 Adaptive reuse main data

The architecture of the platform is a microservices model, that proved effective over the past years, because it allows a keen evolution when it comes to *Scalability*, *Fault-tolerance* and *Resilience*.

FacilityLive Cloud Platform is developed according to the pillars of a Cloud technology *Data security*, *Data management*:, *Performance*, *Dependability and availability of the service* and *Legal aspects*.

References

Australia ICOMOS (2013) The Burra Charter: the Australia ICOMOS Charter for places of cultural significance. Available at: https://australia.icomos.org/wp-content/uploads/The-Burra-Charter-2013-Adopted-31.10.2013.pdf

Buglione F, Bosone M, Angrisano M, Iodice S, Gravagnuolo A, Fusco Girard L (2021) Database of best practices (Horizon 2020 CLIC deliverable D1.4)

CHCfE Consortium (2015) Cultural heritage counts for Europe. International Cultural Centre, Krakow. http://blogs.encatc.org/culturalheritagecountsforeurope//wp-content/uploads/2015/06/CHCfE_FULL-REPORT_v2.pdf
Council of Europe (2000) European Landscape Convention. Florence 20.10.2000. European Treaty Series. No. 176. 7 p. Available at: https://rm.coe.int/CoERMPublicCommonSearchServices/DisplayDCTMContent?documentId=0900001680080621
Council of Europe (2005) Convention on the value of cultural heritage for society (Faro Convention, 2005). Available at: https://www.coe.int/en/web/culture-and-heritage/faro-convention
Della Lucia M, Trunfio M (2018) The role of the private actor in cultural regeneration: hybridizing cultural heritage with creativity in the city. Cities 82:35–44
Esmaeilpoorarabia N, Yigitcanlar T, Guaraldab M (2018) Place quality in innovation clusters: an empirical analysis of global best practices from Singapore, Helsinki, New York, and Sydney. Cities 74:156–168
European Commission (2014) Communication from the commission to the European Parliament, the Council, the European Economic and Social Committee and the Committee of the Regions - towards an integrated approach to cultural heritage for Europe, COM(2014) 477 final, Brussels, 22.7.2014
European Commission (2015) Getting cultural heritage to work for Europe. Report of the Horizon 2020 expert group on cultural heritage
European Commission (2018) European framework for action for cultural heritage
European Parliament (2017) Decision on a European year of cultural heritage
Fusco Girard L (this volume) The circular "human-centred" adaptive reuse of cultural heritage: theoretical foundations. In: Fusco Girard L, Gravagnuolo A (eds) Adaptive reuse of cultural heritage. Springer
Gravagnuolo A, Fusco Girard L, Ost C, Saleh R (2017) Evaluation criteria for a circular adaptive reuse of cultural heritage. BDC Bollettino Del Centro Calza Bini 17(2/2017):185–216
Gravagnuolo A, Vellecco I, Fusco Girard L, Lauro NC, Ikiz Kaya D, Pintossi N, Acri M, Dobricic S, Debevec M, Jokilehto J, Lupu A, Allegro I (2021) Survey on best practices of cultural heritage adaptive reuse (horizon 2020 CLIC deliverable D1.3)
KEA (2017) Future creative cities. Why culture is a smart investment in cities. Report of "Culture for cities and regions" project. Available at: https://keanet.eu/projects/culture-for-cities-and-regions/
Monteleone G, Milani G, Ring C (2021) CLIC platform for pilot cities (Horizon 2020 CLIC deliverable D6.22)
Monteleone G, Milani G, Spiaggi M, Gravagnuolo A (2019) CLIC knowledge information hub web platform (Horizon 2020 CLIC deliverable D3.5)
Ritzer G, Jurgenson N (2010) Production, consumption, Prosumption: the nature of capitalism in the age of the digital "prosumer". J Consum Cult 10:13–36
Smit AJ (2011) The influence of district visual quality on location decisions of creative entrepreneurs. J Am Plan Assoc 77(2):167–184
UNESCO (2011) The UNESCO recommendation on the historic urban landscape. Available at: https://whc.unesco.org/uploads/activities/documents/activity-638-98.pdf
United Nations (2015) UN agenda 2030 for sustainable development and sustainable development goals
United Nations (2016) New Urban Agenda 2030
U.S. Department of State (2016) Harnessing the data revolution for sustainable development: the U.S. government and the global partnership for sustainable development data. Available at: https://2009-2017.state.gov/r/pa/prs/ps/2016/09/262294.htm
UN-Habitat (2016a) "Smart city strategies and data revolution for sustainable development", Habitat III Events on Data Revolution. http://habitat3.org/the-conference/programme/all/smart-city-strategies-and-data-revolution-for-sustainable-development/
UN-Habitat (2016b) Habitat III 2016. https://unhabitat.org/habitat-iii/
UN-Habitat (2017) Habitat III "New Urban Agenda". http://habitat3.org/the-new-urban-agenda/
World Health Organization, Regional Office for Europe (2018) Circular economy and health. Risks and opportunities

Chapter 6
A Statistical Model Representation and Analysis of Cultural Heritage Adaptive Reuse Practices Based on Latent Variables for Circularity Assessment

Natale Carlo Lauro, Antonia Gravagnuolo, Luigi Fusco Girard, Immacolata Vellecco, and Maurizio Lauro

1 Introduction

Cultural heritage is a non-reproducible cultural resource, contributing to communities' identity and wellbeing, to be preserved for present and future generations ensuring cultural identity and diversity as human right.[1] However, cultural heritage buildings, sites and often entire historic urban areas or villages can be subject to abandonment and degradation, if not used and maintained over time. Between diverse approaches to preserve cultural heritage (e.g. restoration, recovery, maintenance), adaptive reuse is a process of re-use of buildings and sites that have lost their original function, identifying new functions/uses, compatible with heritage values. Adaptive reuse can be a valid solution to keep cultural heritage in a proper conserva-

[1] UNESCO Universal Declaration on Cultural Diversity.

N. C. Lauro
Associazione Professori Emeriti Federico ll, APEF, Naples, Italy
e-mail: carlo.lauro@unina.it

A. Gravagnuolo (✉)
CNR ISPC Institute of Heritage Science, National Research Council, Naples, Italy
e-mail: antonia.gravagnuolo@cnr.it

L. Fusco Girard
Emeritus Professor at University of Naples Federico II; CNR IRISS, National Research Council, Institute for Research on Innovation and Services for Development, Naples, Italy

I. Vellecco
IRISS Institute for Innovation and Services for Development,
CNR National Research Council, Naples, Italy
e-mail: i.vellecco@iriss.cnr.it

M. Lauro
Istitituto di Ricerca e Cerificazione Statistica e Informatica Applicata, IRCSIA, Naples, Italy
e-mail: mauriziolauro@ircsia.it

L. Fusco Girard, A. Gravagnuolo (eds.), *Adaptive Reuse of Cultural Heritage*,
https://doi.org/10.1007/978-3-031-67628-4_6

tion state, preserving cultural values at risk of irreversible loss due to abandonment, and ensuring access and enjoyment of heritage assets (Misirlisoy and Günçe 2016). Re-use is also considered the key process through which implementing the circular economy, not only in manufacturing industry but also in the construction industry and urban/regional development, thus the adaptive reuse of cultural heritage can represent an effective approach to realise circular cities and regions, bringing abandoned and underused historic buildings and sites to new life. However, from the point of view of conservation, adaptive reuse implies possible changes to adapt heritage to new uses, ranging from structural modifications, to technological improvements or lighter, reversible interventions. The degree of change of cultural heritage needed for implementing new uses can vary, and should be carefully assessed in a multidimensional perspective, to avoid the risk of threatening or reducing cultural value through the reuse process. Thus, appropriate evaluation methods and tools should be identified to support the decision-making process, particularly when multiple stakeholders are involved and different solutions can be proposed.

To effectively implement the circular economy and circular city/region model, the adaptive reuse of cultural heritage should align with circularity principles and criteria, which have been described in literature (Fusco Girard 2021; Luigi Fusco Girard this volume, Chap. 2; Gravagnuolo et al. 2017, 2018). The circular economy model foresees, for example, the reduction of energy use and greenhouse gas emissions, reduction of raw materials extraction, wastes minimisation, green areas regeneration and use of nature-based solutions, contribution to biodiversity regeneration, cultural and social values enhancement, place attractiveness enhancement, as well as economic and financial sustainability—to ensure maintenance and continuous use and improvement over time -, generation of new jobs, and also increase of social cohesion, higher care of cultural heritage by local communities. The multiple and diversified objectives of a circular adaptive reuse of cultural heritage require multi-criteria evaluation tools, allowing to handle priorities and values in a multidimensional framework, in particular in contexts where multiple stakeholders and groups are involved in decision making or subject to impacts of reuse actions (e.g. residents in a reused historic urban area, end-users).

The assessment on circularity of adaptive reuse alternatives for different cultural heritage typologies and contexts can be supported by the knowledge of good practices that can represent valid examples to transfer and replicate. Moreover, good practices demonstrating positive impacts in the medium and long term can support the identification of specific criteria and indicators of circularity, based on the results obtained through the adaptive reuse. In the first stage of the Horizon 2020 CLIC research, the question on the specific characteristics of the "ideal" circular adaptive reuse of cultural heritage was addressed under several perspectives and disciplines: heritage conservation, architecture, environmental studies, economics, urban studies, social science. Thus, several potential impacts of circular adaptive reuse were identified, also based on previous studies such as "Cultural Heritage counts for Europe" (2015), which identified diverse dimensions of heritage conservation impacts for cities and communities. The CLIC research aimed to go beyond the state-of-the-art, identifying guiding principles and criteria for a circular adaptive reuse of cultural heritage, to generate positive results in the cultural, environmental, social and economic dimensions of sustainability. However, circular adaptive reuse of cultural heritage was considered an emerging research topic, which required to build knowledge through extensive literature review and case studies assessment, to identify the "best" practices that could guide future interventions.

The aim of this chapter is to propose an innovative methodology to assess the circularity of cultural heritage adaptive reuse best practice, and testing, at the same time, the validity of the theoretical framework discussed in the previous chapters. The evaluation of several practices on the base of suitable indicators collected through a European survey allowed to validate specific circularity criteria and definitions that can support decision-makers.

According to the CLIC framework (Luigi Fusco Girard this volume, Chap. 2), the specific research questions on the circular adaptive reuse of cultural heritage were focused on whether and how, in which cases, and at which conditions, the adaptive reuse of cultural heritage can be able to generate self-sustainability of diverse resources, as well as external positive economic, social, environmental and cultural impacts, exploring which elements can inform its synergistic relationship with the context.

This stage of the research operationalizes the CLIC theoretical framework through the analysis of a suitable set of practices, and different typologies of circularity through a suitable model approach based on the above mentioned concept treated as latent variables.

2 Methodology

Based on the review of previous studies on cultural heritage adaptive reuse and circular economy, main principles and criteria for circular adaptive reuse of cultural heritage were identified—see Fusco Girard (2021) and Gravagnuolo et al. (2017). The CLIC methodological approach is based on the analysis of empirical evidence to explore whether and how the experiences of cultural heritage adaptive reuse have been able to turn abandoned heritage/landscape assets into a resource for new jobs, wellbeing, health, social cohesion, regional competitiveness and environmental regeneration—as advocated by all international policy documents and scientific literature.[2]

The methodology adopted for the identification of best practices of circular cultural heritage adaptive reuse was thus based on three main phases:

[2] Among the many reference documents, it is worth to recall here: European Landscape Convention (2000), FARO Convention, Council of Europe (2005), UNESCO Recommendation on the Historic Urban Landscape (2011), ICOMOS Burra Charter (2013), European Commission, Towards an integrated approach to cultural heritage for Europe (2014), European Commission, Getting cultural heritage to work for Europe (2015), Cultural Heritage Counts for Europe (2015), UN Agenda 2030 for Sustainable Development and Sustainable Development Goals (2015), UN New Urban Agenda (2016), European Parliament, Decision on a European Year of Cultural Heritage (2017), and European Framework for Action for Cultural Heritage (2018).

1. Structuring of a survey to collect data on cultural heritage adaptive reuse practices (Gravagnuolo et al. this volume);
2. Data collection targeting 126 examples/practices of cultural heritage adaptive reuse;
3. Data analysis through statistical methods to test the validity of the theoretical framework and assess the level of circularity of each practice.

The CLIC Survey was designed to test—through empirical evidence—the validity of the theoretical framework of circular adaptive reuse of cultural heritage, re-interpreting the success (or insuccess) of adaptive reuse practices under the point of view of their "autopoietic capacity", "generative capacity" and "symbiotic capacity".

The aim of the CLIC Survey was thus to assess the level of circularity of cultural heritage adaptive reuse practices through its determinants and components, providing useful insights to support decision making. This chapter analyses the results of the CLIC survey using the statistical method of the Structural Equation Model (SEM) to test the validity of the theoretical framework based on empirical evidence (see methodological notes on SEM method below). The SEM statistical method (Carroll 1968; Wold 1983; Chin 1998; Amato et al. 2004; Tenenhaus et al. 2005; Vinzi et al. 2010; Lauro et al. 2016; Cataldo et al. 2016; Addinsoft 2019) was chosen as it allows to explore the relationships between complex, non explicit (latent) concepts through the related explicit (manifest) variables, enabling an exploratory analysis to statistically test the validity of a theoretical model.

Methodological Notes on Structural Equation Model (SEM)

The statistical method of PLS-PM to build the Structural Equation Model. The component based structural equation modeling has been used in particular with Partial Least Squares Path Modeling (PLS-PM) method, employing XLSTAT software (Addinson, 2019). This is an innovative method for representing complex relationships between observed variables and Latent Variables.

Partial Least Squares Path Modeling (PLS-PM) is a statistical approach for modeling complex multivariable relationships (structural equation models) among Manifest (observed) and Latent Variables. Since a few years, this approach has been enjoying increasing popularity in several sciences (Esposito Vinzi et al. 2007). Structural Equation Models include a number of statistical methodologies allowing the estimation of a causal theoretical network of relationships linking latent complex concepts, each measured by means of a number of observable indicators.

From the standpoint of structural equation modeling, PLS-PM is a component- based approach where the concept of causality is formulated in terms of linear conditional expectation. PLS-PM seeks for optimal linear predictive relationships rather than for causal mechanisms thus privileging a prediction-relevance oriented discovery process to the statistical testing of causal hypotheses. Two very important review papers on PLS approach to Structural Equation Modeling are Chin (1998, more application oriented) and Tenenhaus et al. (2005, more theory oriented).

(continued)

Furthermore, PLS Path Modeling can be used for analyzing multiple tables and it is directly related to more classical data analysis methods used in this field. In fact, PLS-PM may be also viewed as a very flexible approach to multi-block (or multiple table) analysis by means of both the hierarchical PLS path model and the confirmatory PLS path model (Tenenhaus and Hanafi 2007). This approach clearly shows how the "data-driven" tradition of multiple table analysis can be somehow merged in the "theory-driven" tradition of structural equation modeling so as to allow running the analysis of multi-block data in light of current knowledge on conceptual relationships between tables.

The PLS Path Modeling algorithm. A PLS Path model is described by two models: (1) a measurement model relating the Manifest Variables to their own Latent Variable and (2) a structural model relating some endogenous Latent Variables to other Latent Variables. The measurement model is also called the outer model and the structural model the inner model.

The measurement model. A Latent Variable (LV) is an unobservable variable (or construct) indirectly based on its relationship to a block of observable variables which are called Manifest Variables (MV) or indicators. According the nature of such relations of Manifest Variables to their Latent Variables, they are respectively called reflective or formative. In the first case, the MVs values reflect the characteristics of the latent concepts that can be identified using the MVs through simple regressions. In the second case, the LV is caused (formed) by its MVs, identifying the corresponding LVs through multiple regressions.

The reflective way has been adopted in our model, which implies that the variables are highly correlated between themselves, to express the latent concept univocally and in a robust way (unidimensionality condition). This condition is evaluated according different criteria such as Dillon Goldstein Rho index (>0,7), or highest PCA eigenvalue (>1).

PLS Path Modeling is a mixture of a priori knowledge and data analysis. In the reflective way, the a priori knowledge concerns the unidimensionality of the block and the signs of the loadings. The data have to fit this model. If they do not, they can be modified by removing some Manifest Variables that are far from the model. Another solution is to change the model and use the formative way.

The operation of the model. The PLS (Partial Least Squares) operates by means of an iterative algorithm consisting of three fundamental phases. The first phase consists of an estimate of the so-called external weights. These weights are associated with the Manifest Variables (indicators) and refer to the interrelations between each Manifest Variable and the corresponding Latent Variable (representing the latent multidimensional "concept", not directly measurable, described by the group of Manifest Variables associated). For each block of variables, based on the type of relations, the weights are calculated as simple regression coefficients (reflexive indicators): each Manifest Variable is considered as a dependent variable of a model in which

(continued)

the role of explanatory variable is assumed by the corresponding Latent Variable. It is possible to make a regression with respect to a Latent Variable not directly observed since the latter is estimated, in the first step of the iterative algorithm, as a linear combination of its Manifest which coefficients are arbitrarily chosen.

In a second phase, the values derived in the first phase for the latent, exogenous or endogenous explanatory variables in the structural model, are summed algebraically on the basis of the sign of the correlation coefficient between adjacent variables, i.e. linked by a causal relationship. The results of this calculation introduce a new weight estimate for the Latent Variables. If the results of the estimates of the weights obtained in the two phases converge, the iterative estimation process ends and the coefficients of the structural model are then estimated through the classic least squares regression between the newly estimated Latent Variables. The PLS estimation process is able to handle a complex model without problems and requires a reduced number of observations since not all the model is estimated simultaneously.

The algorithm used during the estimation, aimed at explaining the values of latent and Manifest Variables, leads to satisfying model prediction properties, since it is aimed at maximizing the variability explained for Latent Variables and, possibly, for Manifest Variables. Furthermore, the PLS estimate is non-parametric as it does not assume a particular probability distribution for the Manifest Variables or a particular measurement scale for them. The coefficients for causal relations and the weights for the relations between latent and Manifest Variables have the important statistical property called "weak consistency", in the sense that, as both the sample size and the number of items detected by means of the questionnaire increase, they better approximate the true values of the statistical population from which the sample of analyzed cases was extracted.

Model Validation. A path model can be validated at three levels: (1) the quality of the measurement model, (2) the quality of the structural model, and (3) each structural regression equation. The communality index measures the quality of the measurement model for each block. The redundancy index measures the quality of the structural model for each endogenous block. The average redundancy for all endogenous blocks can also be computed. A global criterion of goodness-of-fit (GoF) can be proposed (Amato et al. 2004).

The survey collected data on the name, type and localisation of the adaptive reuse practice, as well as information on the heritage construction period, state of conservation, adaptive reuse period, cultural significance, typology or original function(s), and square meters of adaptive reuse intervention. Also, information about ownership, management and investment was collected. The new uses/functions activated through the adaptive reuse was explored, including a set of possible new uses, from residential to commercial, cultural, educational and various social uses. Thus, data on the specific characteristics of the heritage building/site, the process of adaptive reuse, and the new uses introduced was collected. Finally, a specific section on the "assessment on circularity" included a set of questions to assess the circularity performance of the adaptive reuse practice.

According to the CLIC theoretical framework of circular adaptive reuse of cultural heritage, specific *circularity criteria* were identified, that reflect the theoretical model. First of all, two main Latent Variables were considered, as determinants of the diverse expressions of circularity:

- *Cultural value enhancement*: the capacity to preserve, conserve and regenerate cultural values;
- *Management characteristics towards self-sustainability*: the capacity to become financially self-sustainable;

In turn, these two variables determine the four fundamental building blocks of circularity (outcomes):

- *Closed metabolism realization*: the capacity to realise circular metabolism flows of energy, water, materials, becoming environmentally self-sufficient;
- *Landscape quality enhancement*: the contribution of the adaptive reuse intervention to generate positive impacts in the local context such as visual quality, safety, green areas and public spaces;
- *Social impact*: the social and cultural impact of the adaptive reuse intervention, including raising the level of residents' and visitors' awareness for cultural heritage, increase cultural activities, place attachment, social cohesion, inclusion, wellbeing and health;
- *Economic spillover effects*: the generation of economic spillovers in the city/region through the adaptive reuse intervention.

Thus, to explain these complex elements/criteria of circularity ("Latent Variables", LV), a set of related explicit variables ("Manifest Variables", MV) were identified, as showed below (See Annex for detailed report of SEM model reliability factors).

List of Latent Variables and related Manifest Variables:

LV1—Enhancement of cultural value	MV4.2 –Quality of public spaces
MV1.1 –Conservation of heritage values	MV4.3 –Enhance safety in the area
MV1.2 –Awareness raise for circular economy	MV4.4 –Enhance landscape visual quality
LV2—Management characteristics and self-sustainability	**LV5—Social impact**
MV2.1 –Economic and financial self-sustainable	MV5.1 –Awareness raise for cultural heritage
MV2.2 –Generates revenue flows	MV5.2 –Enhance place attachment
MV2.3 –Third sector involved	MV5.3 –Enhance social cohesion
MV2.4 –Different stakeholders involved	MV5.4 –Enhance inclusion of marginalized groups
MV2.5 –Profits are reinvested	MV5.5 –Enhance heritage community

(continued)

MV2.6 –Total number of uses (classified in 5 groups)	MV5.6 –Enhance cultural activities
	MV5.7 –Enhance people's wellbeing
	MV5.8 –Enhance people's health
LV3—Closed metabolism realization	**LV6—Economic spillover effects**
MV3.1 –Low energy consumption systems	MV6.1 –Enhance jobs creation
MV3.2 –Renewable energy sources	MV6.2 –Attract innovative start-ups
MV3.3 –Water storage and reuse systems	MV6.3 –Attract creative industries
MV3.4 –Traditional / bio / reuse materials	MV6.4 –Attract new commercial activities
MV3.5 –Reduction of construction waste	MV6.5 –Attract cultural visitors
LV4 – Landscape quality enhancement	MV6.6 –Attract new residents
MV4.1 –Increase of green spaces	MV6.7 –Increase real estate values

The six "Latent Variables" (LV) represented specific criteria or "building blocks" of circularity for cultural heritage adaptive reuse projects. A seventh LV was identified, describing the "*Global Circularity Performance Index*" of each adaptive reuse practice, derived from the direct contribution of four LVs (Closed metabolism realization, Landscape quality enhancement, Social impact, Economic spillover effects), and the indirect contribution of two LVs (Cultural value enhancement, Management characteristics towards self-sustainability).[3]

The survey proposed a set of specific questions to assess the performance of the adaptive reuse intervention with respect to each aspect expressed by the MVs. The level of performance was assessed by asking to respondents "whether and to which extent" the adaptive reuse intervention contributed to each of the aspects identified through the specific MVs. To make the survey accessible to a wide and non-expert public, "simple" answer options were chosen, on a qualitative self-explaining Likert scale: No, Yes scarcely, Yes moderately, Yes highly. The option "I don't know" was included (see Chap. 5 of this book for more details). The Manifest Variables were recoded, starting from the questions of the CLIC Survey questionnaire. The answers were recoded into numerical (ordinal) values considering the different answers given: NA—I don't know;[4] 1—No; 2—Yes scarce; 3—Yes moderately; 4—Yes highly.

Adaptive reuse practices were collected through a online survey tool (see Chap. 5) involving heritage experts and organisations throughout Europe, maintaining a geographical balance with respect to European macro-areas (Central-Northern,

[3]At the "Global Circularity Performance Index" are assigned 24 Manifest Variables (MVs), in a hierarchical SEM perspective, which correspond to the 4 LVs directly impacting on it: Closed metabolism realization, Landscape quality enhancement, Social impact, Economic spillover effects (outcomes). This allows to estimate the relevance of their impact on the Global Circularity Performance Index according to the usual practice of hierarchical SEM.

[4]*Missing Data Treatment.* Specific treatment for missing data (NA—I don't know) was applied to estimate missing values as the "nearest neighbour". This method allows to have the most reliable estimation based on the analysis of the answers given to similar practices included in the database which present valid data.

Eastern and Southern Europe), and with respect to heritage typologies (religious heritage, civil/residential buildings or sites, industrial heritage, military structures, mixed use including more than one typology, and others). The subjects involved in the data collection selected a set of 126 good practices that were assessed in the perspective of circularity. The sample of projects included in the database was not meant to be statistically representative of the population of all adaptive reuse interventions in Europe as it is not known, nevertheless they represent a large, geographically balanced and relevant examples of adaptive reuse practices that achieved general objectives of heritage conservation, new uses in previously abandoned or underused heritage sites, and that were still in operations after at least 3–5 years, thus it was possible to observe the impacts generated by a sufficient variety of diverse projects typologies (Figs. 6.1 and 6.2).

The following sections describe the results of the data collection and statistical analysis conducted to assess the circularity performance of cultural heritage adaptive reuse practices.

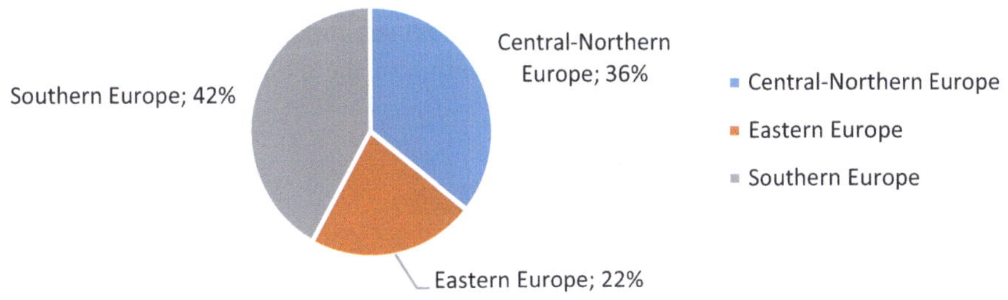

Fig. 6.1 Distribution of analysed projects for EU geographical areas

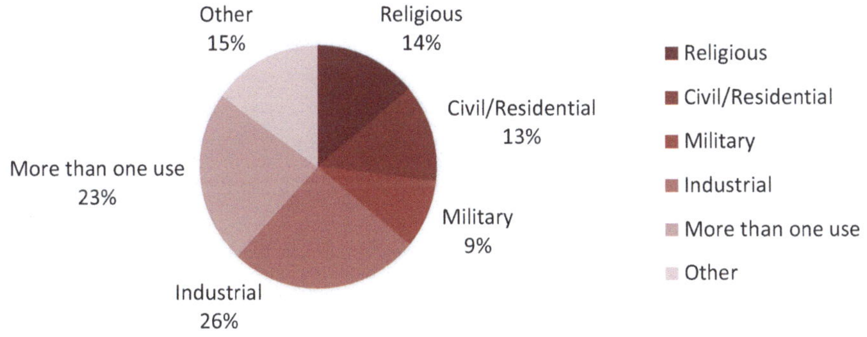

Fig. 6.2 Distribution of analysed projects for typologies

3 Identification of Good and Best Practices Through a Modelling-Based Approach

3.1 The Circularity SEM Model: Definition and Results

For the evaluation of adaptive reuse practices, the answers provided by respondents on the CLIC survey assessment on circularity section was analysed. The SEM model was designed for the statistical analysis, describing the relationships between the Latent Variables (key circularity criteria or "building blocks").

The structural model expresses in statistical terms both the relationships between specific circularity "determinants" and circularity outcomes, and the relationships between outcomes and circularity results. Figure 6.3 shows how the SEM was built, and the relationships between the LV on which a Global Performance is based.

Figure 6.4 shows the results of the SEM, which identify the relationships between the Latent Variables.

The results of the estimation of the structural model parameters reported in Fig. 6.4 show the average score for each latent variable (m) impact (Reg) and contribution (cR2) of each explanatory LV on its own dependent LV, as well as their average scores (m). In the SEM for Circularity, all the impacts are significantly different from zero,[5] confirming the good definition of the model in terms of cause-effect relations between its latent variables.

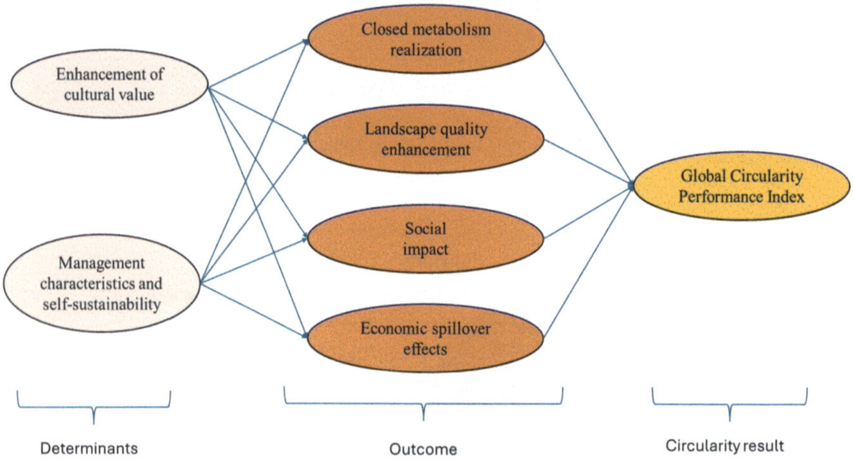

Fig. 6.3 Circularity Structural Equation Model used to assess the level of performance of circularity

[5] The model passed all verifications of reliability (showed by alfa values), complete details can be found at www.clicproject.eu. The significance of the relations between LVs in terms of impact, evaluated in terms of the probability of a t-test obtained according to a bootstrap approach, che ne nostro caso assume valori di $p < 0{,}01$, helps to affirm that the impact of a LV on another target LV is significantly different from zero.

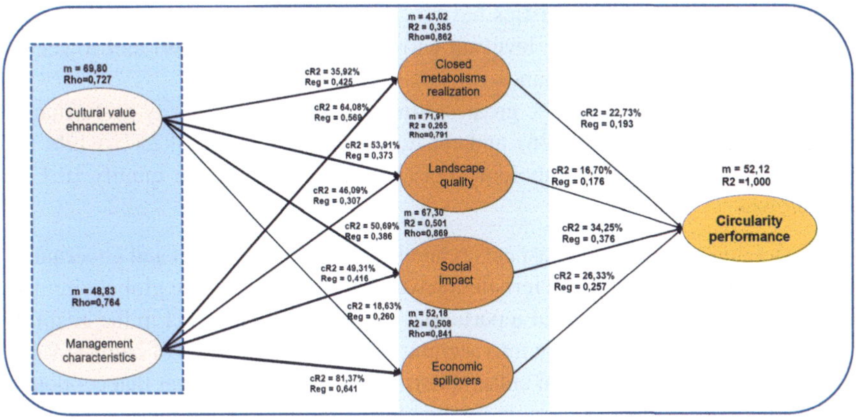

Fig. 6.4 Results of the Structural Equation Model on Circularity performance

The model allows an initial consideration of the average values of the latent variables, which are expressed on a 0–100 scale. The highest visitor evaluations concerned the landscape quality (71,91), social impact (67,30) and cultural value enhancement (69,80). Other latent variables reported average score values around 50, such as economic spillovers (52,18) and management characteristics (48,83). Around this level is also the assessment of the global circularity score.

Looking at the R^2 values varying between 0 and 1, it can be seen that the latent variables of management characteristics and cultural value enhancement explain well the variations in the social and economic components of circularity, presenting values around 0,50, while they explain less well the environmental and landscape aspects, as it was obvious to expect. The value $R^2 = 1,00$ for the 4 components of circularity with respect to the overall circularity score is not a judgement of the quality of the model, but simply an identity since we have assumed that the manifest variables of the latter are the same as its components.

In terms of contribution to the R^2 the model shows the following results:

- contributes more to the variability of the Landscape quality than the Management characteristics (53,9% against 46,1%);
- The Cultural value contributes slightly more than the Management characteristics on the Social Impact (50,7% against 49,3%);
- Management characteristics contribu mentre te significantly more than the Cultural value to Economic spillovers (81,4% against 18,6%);
- The two Latent Variables that contribute most to the variability of Circularity Performance are the Social Impact (34,2%) and the Economic spillovers (26,3%).

In sintesi il management .... contribuisce maggiormente a determinare gli aspetti ambientali ed. economici mentre siaqivale al valore culturale nel determinare gli aspetti sociali e paesaggistici della circolarità.

Analysing the direct impact (reg)s, the model showed the following results:

- The Cultural value has its greatest impact on the Closed metabolism (0,569) and then on the Social impact (0,386) and finally on the Landscape quality (0,373) while impacting less on the Economic spillovers (0,260);

- The Management characteristics have a greater impact on Economic spillovers (0,641), on an intermediate level on Closed metabolism (0,569) and lower on Social impact (0,416) and Landscape quality (0,307);
- The Latent Variables with the greatest direct impact on Circularity performance are the Social impact (0,376) and the Economic spillovers (0,257) while the impact of the Closed metabolism (0,193) and the Landscape quality (0,176) are lower.

The model thus shows that the aspects related to the business model and governance of the site (management characteristics) have a large impact on the global circularity score, suggesting the need of a particular attention to this aspect in the adaptive reuse of cultural heritage, taking into account that the enhancement of cultural value should be a primary objective in cultural heritage sites. It is also clear that economic and social components impact more on the global circularity performance, suggesting that a circular adaptive reuse of cultural heritage should particularly focus on the economic and social impacts.

The resulting scores for each LV were normalized in a 0–100 scale to provide a finer measurement and allow comparisons. A total score for each LV was calculated, representing different aspects influencing the circularity performance, as well as the Global Circularity Performance Index. Once the general model was estimated, the scores of the LVs could be computed in relation to the geographical areas of location of the case studies, and to the heritage typology. Analysing the scores on the basis of the geographical area to which they belong, we can immediately see how Circularity Performance is higher in interventions carried out in Central-Northern Europe (57,08) than in Southern Europe (50,75) and even more than in the East Europe (41,41) (Table 6.1).

For the Cultural value, both Central-Northern and Southern Europe express high scores (over 72) while Eastern Europe reaches the score of 60. It could be thus observed a possible higher attention of Central-Northern and Southern countries to cultural heritage values conservation.

The scores for Management Characteristics are also higher in Southern Europe and in Central-Northern Europe, although with a lower score (around 50 on a scale of 0–100), while the value for Eastern Europe is decidedly low (37,6). Management

Table 6.1 Latent Variables scores for EU geographical regions

EU Region	Cultural value	Management characteristics	Closed metabolisms	Landscape quality	Social impact	Economic spillovers	Circularity performance
Central-Northern Europe	72,86	49,62	53,48	74,09	67,55	59,88	57,08
Eastern Europe	60,16	37,63	30,36	70,46	61,15	41,03	41,41
Southern Europe	72,15	50,90	39,39	70,92	69,69	51,09	50,75
ALL	69,80	48,83	43,02	71,91	67,30	52,18	52,12

Characteristics are related to the capacity to generate diverse revenue flows, reaching economic and financial self-sustainability over time. Also, this variable is related to third sector involvement and stakeholders engagement, towards multi-level and participatory governance models. The low values reached by Eastern Europe countries in this variable could be related to more public-led investments in cultural heritage, with less reliance on private entrepreneurship, including third sector, for the adaptive reuse of cultural heritage. In fact, Eastern Europe showed generally a more "top-down" approach relying on public investments more than private or community-based interventions.

With regard to the Closed Metabolism, the Centre-Northern Europe reaches the score of 53,4 while the results of Southern Europe (39,4) and Eastern Europe (30,3) are quite distant. This variable is related to the adoption of low energy consumption systems and renewable energy sources, as well as water storage and reuse systems, use of traditional / bio / reuse materials and reduction of construction wastes. The results reflect the general higher attention in Central-Northern Europe towards the use of renewable energy sources, for energy retrofit also in cultural heritage buildings and sites. On the other hand, the low scores in Southern Europe could reflect a difficulty in using new technology for renewable energy generation in heritage sites, showing the necessity of establishing a dialogue with heritage authorities and professionals to enhance capacity and awareness with regard to environmental issues in the heritage field. Moreover, Eastern Europe coutries could lack capacity and resources for the identification and implementation of new compatible technologies for energy retrofit of historic buildings and sites.

Landscape quality scores were more balanced, all above 70 with a maximum in Central-Northern Europe (74,1) and at similar levels in the East (70,5) and Southern Europe (70,9), showing a general attention towards specific aspects such as increase of green spaces, enhancement of the quality of public spaces and safety, as well as landscape visual quality.

The Social impact shows fairly high scores with a peak in Southern Europe (69,7) and followed by Centre-Northern Europe (67,5) and Eastern Europe (61,1), related to how the adaptive reuse of cultural heritage contributed to raise awareness for cultural heritage, enhance place attachment and social cohesion, enhance inclusion of marginalized groups, stimulate cohesive heritage communities taking care of cultural heritage, and thus increase cultural activities in the target areas. Moreover, this variable shows the contribution of the adaptive reuse of cultural heritage to the enhancement of people's wellbeing and health.

Finally, the Economic spillovers scores reach a maximum in Central-Northern Europe (59,9) while less brilliant scores are reached in Southern Europe (51,1) and in the Eastern Europe (41). This result could be related to a higher capacity in Central-Northern Europe to generate new jobs in the heritage sector through the adaptive reuse, establish new uses able to make the heritage site attractive for innovative start-ups, creative industries and commercial activities, while attracting also new residents in the heritage site area and visitors for tourism and educational purposes. Also, the economic spillovers related to the enhancement of real estate values are here considered (Table 6.2).

Table 6.2 Latent Variables scores for heritage typologies

Heritage typology	Cultural value	Management characteristics	Closed metabolisms	Landscape quality	Social impact	Economic spillovers	Circularity performance
Religious	71,94	38,45	34,59	59,37	61,68	38,11	38,83
Civil/residential	67,40	37,30	32,89	72,79	73,80	53,68	51,53
Military	79,19	46,56	51,55	81,03	73,51	58,33	59,79
Industrial	72,42	53,01	46,82	74,01	67,48	59,77	55,38
More than one use	65,94	55,53	44,33	71,22	63,16	49,90	49,41
Other	64,56	43,00	41,36	75,11	67,36	49,20	50,86
ALL	69,80	48,83	43,02	71,91	67,30	52,18	52,12

Observing the results based on the typology of case studied, the main evidences are the following:

- For the global circularity performance the highest scores are realized for the Military heritage typology (59,8) and the Industrial (55,4) while the lowest ones concern Religious structures (38,8), which could be related to higher difficulties in religious heritage settings to activate a diversity of new uses and implement effective energy retrofit technologies, also observing the lowest scores in this category for the variables related to management characteristics, closed metabolisms and economic spillovers.
- The Cultural value is higher in the case-studies of Military typology (79,2), Industrial and Religious one (approximately 72), expressing a higher attention in the conservation of heritage values for these categories, probably linked to larger buildings and sites that highly impact the urban landscape, compared with civil and residential buildings categories and other mixed typologies.
- The highest scores about the Management characteristics are found in heritage typologies such as the Industrial (53,01) and in the mixed (More than one use) typologies (55,5) while they are lower for Civil / Residential (37,3), showing that these categories could be more adaptable to a diversity of new uses, allowing diverse revenue flows with less barriers also for participatory governance models and involvement of the third sector.
- The Closed Metabolism receives higher scores in typologies like the Military (51,55) and Industrial ones (46,8) and lower for the typology Civil / Residential (32,9) and Religious (34,6), here also it could be related to the higher flexibility of those heritage typologies to implement effective energy retrofit strategies.
- The Landscape quality is higher in the Military typologies (81) while it is lower in the Religious ones (59,4), expressing the relevant impact that forts and other large historic military buildings have for the urban and coastal landscapes. Also, it could be observed that the adaptive reuse of large military heritage could be more easily linked to increase of green areas and public spaces quality, while the reuse of small religious buildings could have less impact on the change of surroundings and related public spaces.
- The Social impact is greater for the Civil / Residential (73,8) and Military (73,5) and lower for the Religious typologies (61,7), however the scores obtained by all categories are relatively high, showing the importance of adaptive reuse of cultural heritage for generating positive social impacts, including the establishment of heritage communities to care for heritage, and the enhancement of wellbeing.
- Finally, the Economic spillovers variable score is higher for the typologies like the Industrial (59,8) and the Military ones (58,3) while it has lower scores for the Religious typologies (38,1), showing the existence of higher barriers in religious heritage sites to generate economic impacts in the urban/rural areas, as it is clearly more difficult for religious heritage owners, managers and communities to accept in a religious (or ex-religious) place a wide range of diverse profitable uses related to commercial activities, startups and companies.

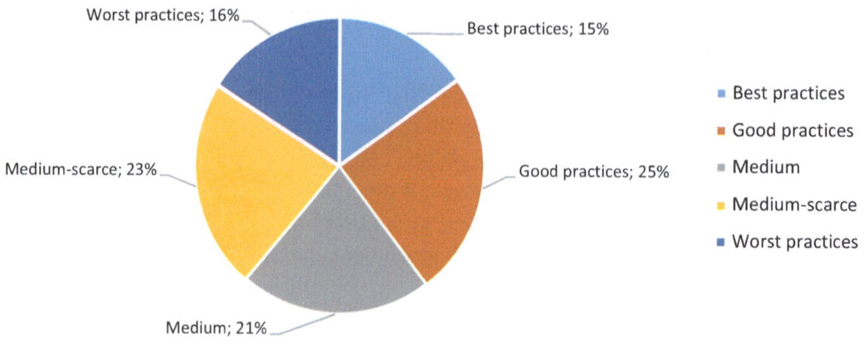

Fig. 6.5 Distribution of practices in each class of circularity

Finally, based on the scores reached in the Global Circularity Performance Index, a classification into five classes with equal intervals was defined to identify groups of projects as follows:

- Scores 80–100—"Best practices"
- Scores 60–80—"Good practices"
- Scores 40–60—"Medium performance"
- Scores 20–40—"Medium-scarce performance"
- Scores 0–20—"Scarce performances"

The percentage of practices included in each category is showed in the Fig. 6.5.

It should be noted that 40% of the analysed sites fall into the good and best practice classes with a score of over 60. The cases belonging to the first class are all interesting for a total circularity score of at least 80 in a 0–100 scale (Table 6.3). Analysing the best practices in terms of Circularity Performance, it is possible to underline the "Top-5" projects with the highest global scores: they are represented by C-Mine, Grand Hornu, De Hoorn, Tour à Plomb, Pianofabriek.

3.2 *Characterization of Circularity Classes: Best Practices Common Features*

This section analyses the characteristics of best practices emerging from an additional data mining analysis. The characteristics of each circularity class were identified through the technique of "variables characterization", considering the Classes of circularity as the variables to be characterized, and the general characteristics of the site and the project (based on the survey answers) as the characterizing variables. Below the characteristics of best practices are described.

The procedure called characterization of a categorical variable was used to describe the main characteristics of the best and worst classes. The statistical software used

Table 6.3 "Best" and "Good" practices of cultural heritage adaptive reuse for global circularity performance

	Best practices site name	Circularity performance score
1	C-Mine	100,000
2	Grand Hornu	94,511
3	De Hoorn	94,145
4	Tour à Plomb	92,851
5	Pianofabriek	90,752
6	Alden Biesen	90,086
7	Mercato Centrale San Lorenzo	89,129
8	Fort Monostor	88,953
9	Van Nelle Fabriek	88,378
10	Botanical Garden of National Palace of Queluz	86,839
11	Zitadelle Spandau	86,773
12	Hotel Katajanokka	86,628
13	Mercado Da Ribeira	86,616
14	Fort Vechten	86,300
15	Le Brass Centre Culturel de Forest	85,463
16	Kinsterna Hotel	85,162
17	Albergo Diffuso Zoncolan	83,696
18	Meelfabriek Leiden	83,332
19	Catacombe di Napoli	82,258
20	Hal 5	80,615
	Good practices site name	**Circularity performance score**
21	Zeche Zollverein	79,896
22	Villa Campolieto	79,419
23	H-Farm	78,684
24	Gasometer City	78,485
25	ReDock La Junquera	78,067
26	Omeriye Ottoman Baths	77,999
27	De Lakfabriek	77,052
28	Geofort	76,735
29	Fort K'IJK—Fort bij Krommeniedijk	75,583
30	Molino Stucky	75,565
31	Škratelj Homestead—Slovene Cinemateque museum	75,038
32	Caballero Fabriek	74,766
33	Convent Carmen	74,696
34	Urbact 2nd Chance	72,351
35	Pakhuis de Zwijger	72,344
36	Hotel Cais De Santarem	72,180
37	Lesczynski Manor: Elderly Healthcare & Residencies	71,666
38	2nd Chance—Waking up the sleeping giants	71,211

(continued)

Table 6.3 (continued)

	Good practices site name	Circularity performance score
39	Atlas building, Eindhoven University of Technology	70,055
40	Fort Resort Beemster	69,587
41	Riot studio	69,489
42	Ssmoll / San Sebastiano Monte dei Morti Living Lab	68,445
43	Battersea Arts Centre—performing arts centre	67,806
44	Spirito Santo Palazzo Storico	67,755
45	Lanthieri Manor	65,141
46	Train World Museum	64,030
47	L'Asilo	63,986
48	Bauska Fortress	62,981
49	Pfefferwerk	60,358

is SPAD, a well-known statistical software of the French school (e.g. Lebart, Morineau), which uses this procedure to describe for example sub-populations of a sample (Lebart et al. 2016). Through the comparison between a subpopulation (for example the case-studies that have achieved a score between 80 and 100 as Circularity performance level) and the total sample (which is composed of the case-studies that have achieved any score between 0 and 100 as Circularity performance) this procedure through the Test-value procedure makes it possible to respond to questions of the type: are the case-studies located in Southern Europe over-represented or under-represented in the class of cases with a Circularity performance score between 80 and 100 compared to the total sample? Is public ownership over-represented or under-represented compared to the total sample?

Methodological Notes on Characterization of Variables (DEMOD)

Characterisation of variables. The analysis of the characteristics of a nominal variable (or DEMOD procedure) is a statistical procedure used to describe the characteristic variables and the modalities of the response of specific categories of people/businesses/statistical units in a sample (e.g. survey questions with two or more modes of response) widely used in data mining. By comparing a sub-population (e.g. those who choose to visit a location for cultural purposes and the total sample), this procedure through the Test Value allows answering statistical hypotheses such as: are males over-represented or under-represented in the sample of those who choose a location for cultural reasons? Is the age group over-represented or under represented compared to the total sample? The Test Value allows to identify the response patterns (consumer characteristics) most associated with specific responses to the survey. Values of 1.4–1.5 and higher are considered as particularly interesting, while lower values can be relevant in particular cases or small samples. The results are described through three main elements (columns): % of category in group (referred to the selected sub-sample); % of category in set (referred to the total sample); Test Value (response pattern as described above).

Variable label	Characteristic categories	Test-value	% category in group	% category in sample
Participation	Public consultation	3,42	55,00	21,42
EUgeographicalregions	Central-Northern Eur	2,67	65,00	35,71
Heritageprotection	National	2,27	50,00	26,19
SELFSUSTAINABILITY	yes_high	2,17	50,00	26,98
Typology	Productive/Industria	1,76	45,00	26,19
InvestmentSize	Large inv. (3-10 M€)	1,47	45,00	28,57
Managementandoperation	Owner management	1,04	60,00	46,82
Stateofconservation	Bad conservation	1,02	50,00	37,30
Adaptivereuseperiod	2005-2010	0,94	30,00	19,84
Citysize	Metropolis	0,94	30,00	19,84
InvestmentSize	Medium inv. (1-3M€)	0,89	25,00	15,87
Ownershipbefore	Public own_bef	-0,87	45,00	56,34
EUgeographicalregions	Southern Europe	-0,94	30,00	42,06
Vacancybefore	Permanently used bef	-1,12	15,00	27,77
Constructionperiod	2-Renaissance	-1,21	0,00	9,52
Heritageprotection	Local	-1,53	25,00	42,85
Typology	Civil/Residential	-1,62	0,00	12,69
EUgeographicalregions	Eastern Europe	-1,84	5,00	22,22
InvestmentSize	Small inv. (<1M€)	-3,72	0,00	34,12

Fig. 6.6 Characterization of the circularity class of "best practices"

The group of "best practices" included 15% of the total number of projects (Fig. 6.6). It described the best practices in terms of circularity, shows the characteristics that differentiated the cases identified as best from the rest of the sample of the projects in the sample.

The cases identified as best practices were distinguished by a greater use of public consultation as a form of participation (55% in this group compared to 21,4% in the overall sample), for a greater presence in the Centre-Northern Europe area (65%) and minor in the South and Eastern Europe. These are cases that have a type of protection that was more often national (as much as 50%), with a greater frequency of self-sustainable projects (50% against 27% in the overall sample) and there is a higher presence of large investments or average (the first at 45% of the total, between 3 and 10 million euro). The prevalent projects' typology in this class were the ones in the production, industrial or commercial field in 45% of cases. These are projects that are most frequently developed in Metropolis areas (30%), with a period of re-use between 2005 and 2010 (30% of the cases) and in a state of poor conservation (50%). It is also worth noting that the typology of small investments under 1 million euro is absolutely not present in this group, while it represents the 34% of the entire group of projects, showing a lower capacity of small investments to generate high returns in terms of financial sustainability and positive impacts in the target area. Also, in this group of "best" practices are not present the categories of civil and residential heritage, as well as renaissance buildings or sites—probably mostly civil/residential, showing that other categories of larger buildings and larger investments such as industrial heritage sites and investments above 1 million euro tend to generate more visible impacts.

Observing the results, some considerations emerge from the analysis of "best practices". First, the best "circular" practices are those where stakeholders' involvement in decision making has been realized (mostly through public consultation with stakeholders, but also through other forms of citizens' engagement); this confirms the positive role of collaboration and cooperation between public institutions, private organizations able to provide new services and uses, and the local community as final "user" of the adaptive reuse projects; however, a specific role of the Heritage

Community has not emerged in the practices observed, thus more clear investigation will be needed to explore the real implementation and potential of the Heritage Community for the adaptive reuse of cultural heritage.

Then, it could be observed that the investment and management model influences the global circularity performance, generating economic-financial self-sustainability and social impact through the new uses. The best practices are characterised by a high level of financial self-sustainability. These are the projects in which revenue flows are higher, and are able to cover investment and management costs. The most common management model observed within the best practices sample is direct management from the owner, however "use concessions" for free or "renting" models emerge as common practice in the entire sample of adaptive reuse projects. Through these models, the abandoned heritage for which the legal owner has no resources for recovery and reuse can be still maintained and conserved for future generations, avoiding to "sell" properties also from public bodies due to lack of financial resources, in a win-win strategy that guarantees benefits in the short time for private subjects, and net benefits in the long time for public body.

It is observed also that large and medium investments (over 1 million euro) characterize the best practices, while small investments are not present (0,00% in the group, compared to 34,12% in the whole sample). This could lead to explore the possibility of mixed private and public investment, including also contribution from local communities (public-private-people partnerships), to stimulate higher performance of the adaptive reuse intervention. Within the best practices group, it emerges also the presence of industrial heritage reuses, as well as interventions in large cities. Thus, it is possible to hypothesize that interventions in rural or peripheral areas have currently less probability of success, as they are located in less accessible areas with lower numbers of users and lower investments, thus a specific support could be planned for this kind of interventions.

The group of "good practices" characterised by a score in the global circularity between 80 and 60, showed specific characteristics with a higher presence of small investments (less than 1 million euro), private funding and financing, a low but possible financial self-sustainability, and non profit organisations as managers of the heritage sites. This shows a particular group which also includes a higher percentage of religious heritage linked to ecclesiastical bodies as owners. The link between religious heritage and third sector actors as possible managers, with smaller investments from mostly private organisations can be a interesting model, showing the compatibility of social uses and socially oriented manager organisations with the particular category of religious heritage.

After analysing the emerging characteristics in the groups of best and good practices, the average scores obtained for each Latent Variable ("building block" of circularity) were assessed within a Impact-Performance matrix, providing practical suggestions for future adaptive reuse projects, based on the empirical evidence built through the statistical analysis results.

3.3 *An Operational Use of the Latent Variables to Support Decision the Impact-Performance Matrix (IPM)*

The results of the circularity structural model can be used to identify critical variables of circularity for decisional aim by comparing the score reached by each project with the value of its impact on the variable of the global circularity performance. Accordingly, 4 intervention areas are identified based on the level (low, high) teached by these criteria. For instance the most critical area concern latent variable that present low performance score and high impact (Fig. 6.7).

A scatter diagram (Impact, Performance Map) based on the representation of the 6 explanatory latent variables affecting indirectly (Management characteristics, Cultural Values) or directly (Closed metabolism, Landscape quality, Social impact, Economic spillover) the Global circularity defined in the model is then built.

The diagram is divided into four quadrants separated by a vertical line that represents the threshold that distinguishes the acceptable average scores from the unacceptable ones, and from a horizontal line that represents the threshold that separate low impacts from high impacts.

In the specific case, the thresholds are respectively based on the average of the 6 explanatory variable mean score and the average of their impact on the Global Circularity. The total impact considered for the determinant Vs include both direct and indirect effects whereas as for the 4 circularity outcomes only their direct effect on the Global circularity LV is considered.

The upper left quadrant is characterized by the variables that have a worst performance and have a higher impact on Circularity performance (or another target variable); therefore, it represents an area of intervention or priority for improvements to be designed/foreseen. The lower left quadrant describes variables that have a worse performance but have a lower impact on the Circularity performance; therefore, they represent a monitoring area. The score can be improved but the impact is not high. The upper right quadrant in which are positioned the variables that have a better performance and that have a greater impact on the Circularity performance, represents an area that must be maintained. It will be increased easily if the scores are not so high. The lower right quadrant shows the variables with good performance and a lower impact on Circularity performance and represent an area to be enhanced.

Fig. 6.7 Impact performance matrix

		Latent variable average score	
		Low	*High*
Total Impact	*High*	**Priority intervention Area**	**Maintenance Area**
	Low	**Monitoring Area**	**Enhancement Area**

Observing the overall scores of Latent Variables in the sample as a whole, and their associated weights (importance), some considerations can be made. Analysing the intervention matrix for the Latent Variable "Circularity performance", in the priority intervention area (upper left quadrant) the only Latent Variable is the one measuring the Management Characteristics: in this area, the impact on the outcome index represented by the Circularity is high, but the scores for the case-studies are low. It is therefore a priority area to be improved to bring reuse interventions closer to a higher level of circularity. The Maintenance Area at the top right is also interesting. These are Latent Variables with high but also high in terms of impact. It is a level of performance absolutely to be maintained because the absence of these elements would hinder the ability of an intervention to achieve a good level of circularity (Cultural Value and Social impact). The bottom left quadrant is the Monitoring Area, characterized by variables with low scores for the case-studies and low impacts too. This is an area to be monitored, where the impact is not high but the scores can be easily improved (Closed metabolisms and Economic spillovers). At the bottom right there is an area to be enhanced. The Landscape quality has high scores but low impact on the circularity and the challenge is to favour a greater capacity of this variable to bring returns to the circularity score.

To identify critical intervention items in order to improve the circularity approach results in the area under study, these maps can be replicated for each of the six LVs of the SEM model, with respect to the MVs that contribute to them, using the weights and average values of each MV (see Annex).

In the following, due to its position in the Priority intervention area (Fig. 6.8), we report the impact-performance map realized for the Management Characteristics (Fig. 6.9).

Critical indicators (in the upper left part of the map) are not present, however some lower scores indicators are present (e.g. Financial self-sustainability, Third sector) which require some intervention to move in the right part of the map. At the same time the map (upper right quadrant) highlight indicators which performance should be maintained such as revenue flows and reinvestment of profits, which are localized in the upper right part of the graph with due to their high impact and high performance as well.

4 Conclusions

The contribution of the present chapter concerns both an innovative representation for the Circular adaptive reuse of cultural heritage by a suitable model based on latent variables, and the possibility to operationalize the theoretical framework on the subject matter established by the CLIC project (Luigi Fusco Girard this volume, Chap. 2) by analysing the data issued from the CLIC Survey (Chap. 5).

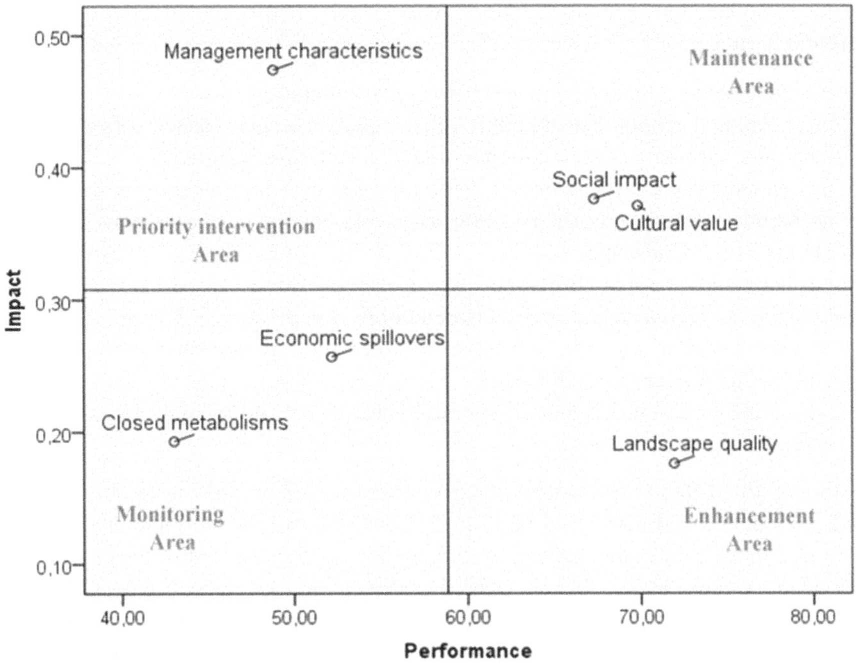

Fig. 6.8 Impact-performance map based on Latent Variables of circularity

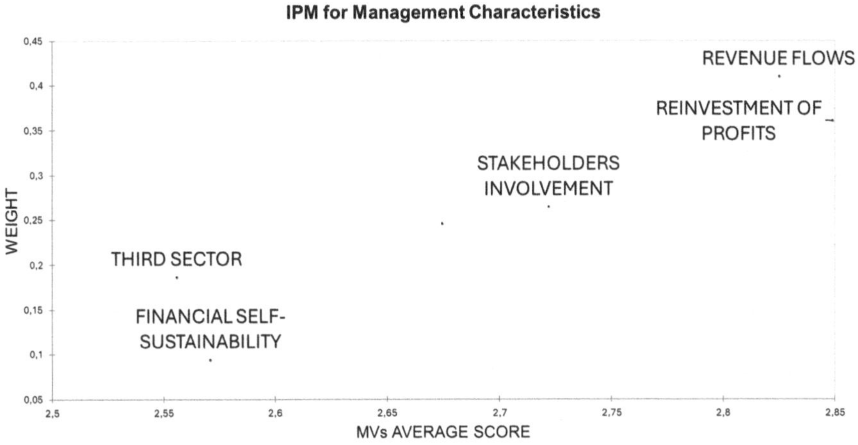

Fig. 6.9 Impact-performance map for the manifest variables on management characteristics

The Survey on Circular adaptive reuse of cultural heritage allowed to identify some best and good practices, as well as less satisfying reuse practices, and also to test the validity of the circular theoretical model for cultural heritage adaptive reuse through empirical evidence.

In particular, through this Survey it was asked whether the adaptive reuse of cultural heritage:

- has enhanced social cohesion and heritage communities;
- has generated economic wealth through adequate revenues/cash flows vs. investment and operating costs (economic/financial self-sustainability);
- has enhanced the attractiveness of the area for new investments, in particular attracting cultural and creative industries, and enhancing the perceived visual quality of the landscape;
- has been implemented with a specific role of the third sector (which is more and more considered as a key actor between public and private);
- has demonstrated win-win-win solutions combining private short term interests with public longer term objectives;
- has employed circular models through closed micro-metabolisms in water, energy and materials use.

Although the different levels of circularity observed, all projects were able to provide abandoned and underused buildings / sites / landscapes with new use values, avoiding the loss of cultural memory and enhancing their cultural values, confirming adaptive reuse as a viable and effective strategy to give "new life to old buildings" (but also, as said, sites and landscapes, as well as objects), thus contributing to the implementation of the circular city model; almost all projects reported positive impacts in the enhancement of quality of public spaces, safety and landscape visual quality of the area in which they are located. All projects reported positive impacts for "place attachment", confirming the relevance of the adaptive reuse of cultural heritage as contribution to urban regeneration.

The aim of the statistical analysis performed was to assess the level of circularity of the analysed case studies, specifying whether the practices regenerated resources (human, environmental, economic, cultural, etc.). Data on each project was collected through the CLIC Survey and analysed through the statistical method of Structural Equation Model identifying the relevant 'building blocks' underlying the Global Circularity Performance Index for 126 cultural heritage adaptive reuse projects. The SEM allowed to validate key concepts of circularity in cultural heritage (latent variables) and to explore the relationships between them, "overcoming" the traditional approach of "pillars" of sustainability and making operational the theoretical "circularity model". The "weights" of Latent Variables emerged from the SEM showed that the social impact variable is stronger: this represents on the one hand a signal of the importance of the generative capacity in the "social" dimension. The best practices were thus analysed more in-depth through a data mining technique (characterisation of variables—DEMOD), identifying the emerging characteristics in the specific group of the "best" adaptive reuse projects in terms of circularity. Finally, an Impact-Performance Matrix was used to explore the possible uses of the indicators of circularity in the adaptive reuse of cultural heritage, for further implementation and test towards planning and design decisions aid / support.

Beyond the cognitive and decision-supporting aspects that derive from estimating the assumed circularity model, it should be noted that its mathematical formalisation also allows it to be used for predictive purposes, making it possible to simulate the effects of intervention policies or the calculation of the circularity score for new sites.

To conclude, the statistical analyses performed in this section allow to confirm the validity of the proposed model for circularity assessment and understand the nature and relationships of data collected through the CLIC Survey on cultural heritage adaptive reuse practices. This study represents the result of a specific stage of the CLIC adaptive reuse of cultural heritage project aiming at developing multicriteria methodology and multidimensional evaluation tools to support the implementation of innovative "circular" governance, financing and business models for cultural heritage adaptive reuse. The complexity of the data analysed in this experimental analysis could not lead yet to definitive conclusions, nevertheless the statistical analysis approach pursued enabled the validation of the theoretical model, identifying "building blocks" and "best practices" of circular adaptive reuse of cultural heritage, and allowing to glimpse some interesting hypotheses that could be explored further through qualitative and quantitative analyses (in-depth case studies).

Annex—Reliability of the SEM Latent Variable Model, Scores and Weights of Manifest Variables

Latent variable	Manifest variables	Mean	Outer weight	Outer weight (normalized)
Cultural value (Rho = 0,727)	HERITAGE VALUE	3571	0,921	**0,571**
	CIRCULAR ECONOMY (CE) AWARENESS	2762	0,693	0,429
Management characteristics (Rho = 0,764)	FINANCIAL SELF-SUSTAINABILITY	2571	0,094	0,064
	REVENUE FLOWS	2825	0,409	**0,278**
	THIRD SECTOR	2556	0,186	0,126
	STAKEHOLDERS INVOLVEMENT	2722	0,264	0,179
	REINVESTMENT OF PROFITS	2849	0,360	**0,245**
	TOT N. OF USES	2524	0,160	0,109
Closed metabolisms (Rho = 0,862)	LOW ENERGY SYSTEMS	2516	0,208	0,179
	RENEWABLE ENERGY	2111	0,212	0,183
	WATER RECOVERY	1833	0,276	**0,237**
	MATERIALS REUSED	2802	0,236	**0,203**
	WASTES REDUCTION	2278	0,230	0,198

(continued)

Latent variable	Manifest variables	Mean	Outer weight	Outer weight (normalized)
Landscape quality (Rho = 0,791)	GREEN SPACES	2619	0,230	0,160
	PUBLIC SPACE QUALITY	3389	0,384	0,267
	SAFETY	3095	0,423	**0,294**
	LANDSCAPE QUALITY	3310	0,402	**0,280**
Social impact (Rho = 0,869)	CULTURAL HERITAGE (CH) AWARENESS	3397	0,229	0,137
	PLACE ATTACHMENT	3310	0,256	**0,153**
	SOCIAL COHESION	3175	0,214	0,128
	INCLUSION	3000	0,151	0,090
	HERITAGE COMMUNITY	2730	0,151	0,090
	CULTURAL ACTIVITIES	3302	0,196	0,117
	WELLBEING	3310	0,274	**0,164**
	HEALTH	3135	0,202	0,121
Economic spillovers (Rho = 0,841)	JOBS CREATION	3008	0,238	0,171
	STARTUP ATTRACTION	2151	0,185	0,133
	CREATIVE INDUSTRIES	2841	0,159	0,114
	COMMERCIAL ACTIVITIES	2794	0,194	0,139
	CULTURAL TOURISM	3024	0,212	**0,152**
	RESIDENTS ATTRACTION	2595	0,216	**0,155**
	REAL ESTATE INCREASE	2540	0,190	0,137

References

Addinsoft (2019) XLSTAT statistical and data analysis solution. Boston

Amato S, Esposito Vinzi V, Tenenhaus M (2004) A global goodness- of-fit index for PLS structural equation modeling. In: Proceedings of the XLII SIS scientific meeting, vol. contributed papers, CLEUP, Padova, pp 739–742

Australia ICOMOS (2013) The Burra Charter: The Australia ICOMOS Charter for places of cultural significance

Carroll JD (1968) A generalization of Canonical Correlation Analysis to three or more sets of variables. In: Proceedings of 76th convention of the American Psychological Association, pp 227–228

Cataldo R, Grassia MG, Lauro NC, Marino M (2016) Developments in higher-order PLS-PM for the building of a system of composite indicators. Qual Quant. https://doi.org/10.1007/s11135-016-0431-1

Chin WW (1998) The partial least squares approach for structural equation modeling. In: Marcoulides GA (ed) Modern methods for business adaptive reuse of cultural heritage. Lawrence Erlbaum Associates, pp 295–336

CHCfE Consortium (2015) Cultural heritage counts for Europe. International Cultural Centre, Krakow. Available at: http://blogs.encatc.org/culturalheritagecountsforeurope//wp-content/uploads/2015/06/CHCfE_FULL-REPORT_v2.pdf

Council of Europe (2000) European landscape convention. Florence 20. 10. 2000. European Treaty Series. No. 176. 7 p. Available at: https://rm.coe.int/CoERMPublicCommonSearchServices/DisplayDCTMContent?documentId=0900001680080621
Council of Europe (2005) Faro convention. The framework convention on the value of cultural heritage for society. Faro, Portugal
Esposito VV et al (2007) Handbook of partial least squares: concepts, methods and applications. Springer
European Commission (2014) Communication from the Commission to the European Parliament, the Council, the European Economic and Social Committee and the Committee of the Regions – towards an integrated approach to cultural heritage for Europe, COM(2014) 477 final, Brussels, 22.7.2014
European Commission (2015) Getting cultural heritage to work for Europe. Report of the Horizon 2020 expert group on cultural heritage
European Commission (2018) European framework for action for cultural heritage
European Parliament (2017) Decision on a European year of cultural heritage
Fusco Girard L (2021) The evolutiorary circular and human centered city: towards an ecological and humanistic "re-generation" of the current city governance. Hum Syst Manag 40(7):753–775. https://doi.org/10.3233/HSM21121
Fusco Girard L (this volume) The circular "human-centred" adaptive reuse of cultural heritage: theoretical foundations. In: Fusco Girard L, Gravagnuolo A (eds) Adaptive reuse of cultural heritage. Springer
Gravagnuolo A, Fusco Girard L, Ost C, Saleh R (2017) Evaluation criteria for a circular adaptive reuse of cultural heritage. BDC Bollettino Del Centro Calza Bini 17(2/2017):185–216
Gravagnuolo A, Saleh R, Ost C, Fusco Girard L (2018) Towards an evaluation framework to assess cultural heritage adaptive reuse impacts in the perspective of the circular economy. Urbanistica Informazioni, pp 28–31
Gravagnuolo A, Monteleone G, Fusco Girard L (this volume) In search of a circular model for cultural heritage adaptive reuse: building evidence-base. In: Fusco Girard L, Gravagnuolo A (eds) Adaptive reuse of cultural heritage. Springer
Lauro NC, Grassia MG, Cataldo R (2016) Model based composite indicators: new developments in partial least squares-path modeling for the building of different types of composite indicators. Social indicators adaptive reuse of cultural heritage. https://doi.org/10.1007/s11205-016-1516-x
Lebart L, Morineau A, Piron M (2016) Statistique exploratoire multidimensionnelle. Dunod, Paris
Misirlisoy D, Günçe K (2016) Adaptive reuse strategies for heritage buildings: a holistic approach. Sustain Cities Soc 26:91–98
Tenenhaus M, Hanafi M (2007) A bridge between PLS path modeling and multi-block data analysis. In: Esposito VV et al (eds) Handbook of partial least squares: concepts, methods and applications. Springer
Tenenhaus M, Esposito VV, Chatelin Y-M, Lauro C (2005) PLS path modeling. Comput Stat Data Anal 48(1):159–205
UNESCO (2011) Recommendation on the historic urban landscape. In: Records of the general conference – 31st session
United Nations (2015) UN agenda 2030 for sustainable development and sustainable development goals
United Nations (2016) New urban agenda 2030
Vinzi VE, Trinchera L, Amato S (2010) PLS path modeling: from foundations to recent developments and open issues for model assessment and improvement. In: Handbook of partial least squares. https://doi.org/10.1007/978-3-540-32827-8_3
Wold H (1983) Quantitative systems analysis: the pedigree and broad scope of PLS (partial least squares) soft modeling. In: Martens H, Russwurm H Jr (eds) Food adaptive reuse of cultural heritage and data analysis. Applied Science Publisher Ltd

Chapter 7
Understanding Best Practices of Cultural Heritage Adaptive Reuse in the Perspective of the Circular Economy: In-Depth Assessment of Case Studies

Francesca Buglione, Antonia Gravagnuolo, Mariarosaria Angrisano, Silvia Iodice, Martina Bosone, Pasquale De Toro, and Luigi Fusco Girard

1 Introduction

The CLIC adaptive reuse of cultural heritage argued that cultural heritage adaptive reuse can represent a viable strategy for the implementation of circular cities and regions (CLIC project – Circular models Leveraging Investments in Cultural heritage adaptive reuse 2017), contributing to the reuse of natural and cultural resources, the re-generation of values of heritage sites and buildings, and in many cases also to social and community bonds regeneration thanks to new social uses of heritage sites (Fusco Girard 2018; Fusco Girard and Gravagnuolo 2017). In the first phase of the adaptive reuse of cultural heritage, the theoretical circular model of cultural heritage adaptive reuse was developed applying inter-disciplinary and trans-disciplinary approaches involving adaptive reuse of cultural heritage and practitioners from

F. Buglione · M. Bosone · P. De Toro
Department of Architecture, University of Naples Federico II, Naples, Italy
e-mail: martina.bosone@unina.it; detoro@unina.it

A. Gravagnuolo (✉)
CNR ISPC Institute of Heritage Science, National Research Council, Naples, Italy
e-mail: antonia.gravagnuolo@cnr.it

M. Angrisano
Engineering Department, Pegaso Telematic University, Naples, Italy
e-mail: mariarosaria.angrisano@unipegaso.it

S. Iodice
Territorial Development Unit, European Commission, Joint Research Centre, Ispra, Italy
e-mail: silvia.iodice@ec.europa.eu

L. Fusco Girard
Emeritus Professor at University of Naples Federico II; CNR IRISS, National Research Council, Institute for Research on Innovation and Services for Development, Naples, Italy

L. Fusco Girard, A. Gravagnuolo (eds.), *Adaptive Reuse of Cultural Heritage*,
https://doi.org/10.1007/978-3-031-67628-4_7

diverse fields in a common reflection. The circular model was discussed, and further explored through the analysis of cultural heritage adaptive reuse practices, applying statistical analysis tools to identify the "best practices" in the perspective of circularity. The experiences analysed were not meant initially to reach specific circularity goals, and therefore they do not represent yet "ideal" solutions for circular adaptive reuse of cultural heritage. However, relevant lessons can be learnt from these practices, that can be applied in the design and implementation of future circular solutions for abandoned and underused cultural heritage.

In this chapter, three best practices of cultural heritage adaptive reuse are explored in-depth to understand in particular their business, financial and governance model and the impacts they generated in their territory: De Hoorn (Belgium), C-Mine (Belgium), the Catacombs of San Gennaro (Italy).

The analysis of best practices informed the CLIC adaptive reuse of cultural heritage as they represented useful examples of how circularity goals are reached through diverse interventions. The "ideal" circular model for cultural heritage adaptive reuse should include the achievement of all goals (cultural, environmental, social, economic), however these best practices were important to contribute to structuring specific circularity criteria for evaluation of future practices, in line with real setting experiences and representing a benchmark and inspiration for the future interventions.

The following sections present the methodology implemented to select and analyse the case studies, and the results obtained through direct interviews conducted with heritage sites/buildings managers. The discussion and conclusions section summarises the lessons learned and the key success factor in the case studies, paving the way for the identification of circularity criteria and indicators for ex-ante evaluation in the subsequent adaptive reuse of cultural heritage phases.

2 Methodology

The best practices were selected according to the statistical analysis conducted in the previous phase of the CLIC adaptive reuse of cultural heritage (Gravagnuolo et al. 2023). In particular, three cases were identified in which specific aspects of circularity are particularly expressed: environmental goals (C-Mine), economic goals (De Hoorn), and social-cultural goals (Catacombs of San Gennaro). According to the circularity performance analysis conducted through statistical tools, the three practices were all considered exemplary for the circular economy model (Gravagnuolo and De Lucia 2019). However, their overall performance is the result of diverse actions, models and impacts (Gravagnuolo et al. 2021a). The aim of the in-depth analysis is to understand which factors can explain the high performance of those best practices, supporting the definition of effective circularity criteria and indicators (Bosone et al. 2021; Gravagnuolo and Fusco Girard 2021), and the business, financial and governance models behind their success that could be transferred in other contexts (Allegro et al. 2021; Pickerill 2021; Ruba et al. 2020).

To explore the details of the best practices experiences, a semi-structured interview was conducted with the heritage sites/buildings managers, retrieving the needed information to define their adaptive reuse circular model. According to the CLIC framework of circular adaptive reuse of cultural heritage (Fusco Girard 2023), three main domains were considered to structure the interviews and evaluate the practices: the auto-poietic capacity, the generative capacity and the symbiotic capacity (Fusco Girard 2021b).

1. Auto-poietic capacity

   The "auto-poietic" capacity is here expressed as the capacity of economic-financial self-sustainability, the capacity of self-regenerating the resources needed for its "life" in the long term (e.g. energy, water, materials...), and the capacity of self-regenerating its cultural value over time through, for example, restoration works, appropriate maintenance, new interpretation.

   - Cultural value enhancement
   - Management characteristics and self-sustainability

2. Generative capacity

   The "generative" capacity is expressed by the endogenous variables: closed metabolism realization, landscape quality enhancement, social (and cultural) impacts generated, and economic spillovers generated. These variables represent the capacity of the cultural heritage adaptive reuse model to generate resources for the local context, also 'closing' the metabolism and thus avoiding excessive environmental resources consumption. In this perspective, the closed metabolism realization could be conceptually included also within the "auto-poietic capacity", as a way to avoid negative environmental externalities.

   - Closed metabolism realization
   - Landscape quality enhancement
   - Social impact
   - Economic spillover effects

3. Symbiotic capacity

   The "symbiotic" capacity (also in terms of symbiotic exchanges with the context) expresses the capacity to establishing relationships with the "context" in which the cultural heritage adaptive reuse project is realized. It is linked to the cooperation and collaboration approaches that enable a more efficient use of resources (such as those realized in "industrial symbioses"), as well as clustering processes in the territory (implementing an "economy of relationships").

   - Industrial symbioses
   - Economy of relationships

The variables identified helped to define the structure through which the case studies were assessed, considering the conservation of cultural values, circular metabolisms activated at micro level through materials, energy and water management in a closed life-cycle assessment perspective, the self-sustainability of the economic model adopted in the long run, as well as social and economic impacts at local and regional level. The detailed investigation of the "best" practices (as well as 'worst practices' in terms of barriers & bottlenecks occurred) (Ikiz Kaya et al. 2019) included the analysis of costs and investments needed for the adaptive reuse of cultural heritage and how these costs can be reduced through synergistic and cooperative models, through innovative technologies, as well as the actual role of third sector actors in terms of enhanced management capacity to generate positive social impacts, and engage in public-private-social partnerships (social / economic productivity).

The best practices show how the adaptive reuse of cultural heritage achieve or satisfy the above-mentioned capacities (Gravagnuolo and De Lucia 2019). The interviews were structured in two main sections: the first included the 4 forms of capital (manufactured, natural, social and human) involved in the adaptive reuse of cultural heritage, the second included information on the circular business, financing and governance models. More in depth:

- The **Manufactured Capital** describes the cultural heritage (type, age of construction, state of preservation and abandonment, central or peripheral location, owners, current use).
- The **Natural Capital** focuses on Circular Metabolism. It explores the measures / technologies used to make the building/site "circular" (e.g. energy efficiency, renewable energy sources, natural ventilation and lighting, water recovery and recycling systems, waste-to-energy systems, green roofs and/or facades and/or other nature-based solutions, local, eco-sustainable and/or recycled/demolition materials, application of digital technologies, natural transition, decarbonization, CO_2 sequestration, particulate matter, microclimate, regeneration of natural capital, climate change adaptation measures, etc.).
- The **Social Capital** efforts on Community responding to this questions: what role does the local community play in reuse? How is the inclusion of disadvantaged people improved? To what extent is the community involved in choices? How does management promote forms of cooperation and social interaction, and thus trust community networks?
- The **Human Capital** focuses on the way in which functions contribute to actively engaging people in educational pathways, increasing their knowledge, skills, and competencies through heritage reuse (e.g., rehabber, traditional skills, self-construction, repair, hospitality, etc.).
- The **Circular Business Model** explains what activities / functions are included; who is the managing entity; which activities / functions generate revenues, and which activities / functions are sustained; what are the annual costs and revenues and if the operating revenues cover the annual costs; how many jobs are created as a result of reuse.
- The **Funding Model** outlines which entity(s) co-founded the reuse; who initiated the idea of reuse (leadership); how were the works funded (e.g., public funds? Private funds? Crowdfunding? Investment funds? Foundations/third sector, etc.); how much was spent in the recovery / restoration phase (costs).

- The **Circular Governance Model** describes who is the owner, who manages the site, what is the relationship between public, private and third sector institutions and finally what is the role (if any) of the Heritage Community.

The best practices address different types of heritage, from the industrial cultural heritage to archaeological sites (Monteleone et al. 2021). They are an example of achieving multidimensional productivity in the adaptive reuse of cultural heritage. They are capable of generating economic, social, environmental and cultural benefits. For example, they have all contributed to job creation by improving the quality of life of the people involved, regenerating degraded or unproductive urban contexts, and promoting inclusion.

In some cases, they do not need external funding because their activities build on existing local resources and thus become economically productive (Catacombs of San Gennaro). They are able to be self-sustainable with the internal revenues and also to reinvest the profits in other programmes and activities increasing the circularity vision.

Reuse practices are attentive to ecological aspects with the realization of low energy consumption systems, the implementation of renewable energy sources, the use of local traditional materials, bio-materials and/or reuse materials, the implementation of green spaces and/or Nature Based Solutions, recovery of traditional systems avoiding loss of biodiversity. Furthermore, some of them represent the transposition of the logic of industrial symbiosis in the field of cultural and creative enterprises (C-Mine, De Hoorn). Similarly to cases of industrial symbiosis, developed within the framework of industrial ecology, where the interaction between different industrial plants is aimed at maximising the re-use of resources (material, energy, water, services and skills) normally considered as waste, the activities of localised cultural and creative industries generate a system of synergistic relationships and exchanges. Table 7.1 shows the main characteristics of the three adaptive reuse cases analysed.

The following section presents the results of the interviews, with detailed information on the characteristics and impacts of the best practices analysed.

Table 7.1 Best practices analysed in-depth

Name of the asset	City	Typology	Reuse year	Use	
				Before	After
De Hoorn	Leuven, Belgium	Industrial heritage	2007	Former brewery	Innovation and community hub
C-Mine	Genk, Belgium	Industrial heritage	2005	Mine site	Creative hub
Catacombs of San Gennaro	Naples, Italy	Archaeological/ Religious heritage	2009	Closed archaeological site	Open archaeological site

3 Results of Best Practices Analysis

3.1 De Hoorn, Leuven—Belgium

De Hoorn[1] is a former brewery located in Leuven, specifically in the renewed Vaartkom district, where Stella Artois was invented in 1926 and produced over the years. The building has subsequently been adaptively reused and transformed into a vibrant place hosting different kind of activities.

De Hoorn became a protected monument and cradle of Stella Artois. At the end of 2007, after 25 years of vacancy, 7 entrepreneurs decided to jointly transform the building into an innovative workplace for small and large companies working in the creative economy as well as into a meeting place. Since 2012, De Hoorn, with its mixed combination of creative workplaces, event halls and catering establishments, has become an urban catalyst on the Vaartkom area. This complex adaptive reuse project received the Leuven Architecture Prize 2015 and the Europa Nostra / EU prize for cultural heritage 2016.

A complex extension (5600 sqm) next to and over the current building was planned, through which additional offices for the creative economy, an upgrade a conference location and the addition of a short-term residence were established. For the operation, a completely new exploitation/membership formula was developed that enables more solutions for flexible work and extra cross-fertilization. The project started in the spring of 2018 and became operational in the end of 2019.

3.1.1 Manufactured Capital

In the fifteenth century the company 'Brouwerij Den Hoorn' became one of the largest ones in the city, also because drinking water was contaminated and drinking beer was considered healthier than drinking water. In 1537 Den Hoorn became the most important enterprise in the city.

The building, characterized by a significant architectural and structural detail, was designed by architect L. Monnoyer and it dates back to 1922. It is characterized by an exceptional structural design, with a 5 meter high Vierendeel structure creating a 20 meters wide free span in the brewing all. In 1992 the building was abandoned because Stella Artois brewery moved to the other side of the Leuven Canal. The abandoned buildings of the former Brewery Den Hoorn were bought in 2007 and completely renovated while retaining the old brewing installation. The building was given a new use as a business complex from October 2012.[2]

[1] https://www.dehoorn.eu/nl

[2] https://www.urbex.nl/brouwerij-den-hoorn

3.1.2 Natural Capital

A soft approach was chosen for the restoration; the original "Garret" roof structure was damaged during the bombardments of the Second Word War and was replaced by a more industrial structure. Different procedures were experimented to steam wash the existing brick facades, the existing floors and walls. For the machineries, a new technique of ice blasting was used to remove all oil, dirt, without even removing the original paint on the cast iron engines or machinery. Materials were reused and eco-solutions were implemented throughout the building.

The restaurant's cuisine is in line with the principles of waste recovery. It is a seasonal cuisine, nature in its purest form. Organic meat is used in preparing dishes.

3.1.3 Human Capital

The way in which the space is organized and the functions chosen contribute to actively involve the users who, being in an open space, can easily interact and facilitate the development of new networks and links. Furthermore, creativity, cross-breeding and entrepreneurship are key elements of this open meeting place.

3.1.4 Social Capital

De Hoorn is a workspace and an active community space (8000 visitors/year). This site encourages a sense of community and interactions between the users. The co-working spaces bring more people into contact with the businesses on site and to learn more about sustainability thinking.

The renovated brewery is now home to the owners' organizations and represents their own ideals. They aim to provide an innovative, open meeting place where social economy is the key. The completed result is an exemplary restoration of a beautiful industrial monument, with clever additions to accommodate its new occupants.

This social and private initiative has contributed greatly to preserving what is an important example of industrial heritage in Leuven and has brought an abandoned building back to life with a truly imaginative use of the spaces within the well conserved, historic fabric. The new additions are well integrated and are respectful of the integrity of this prime building.

3.1.5 Values Co-creation

The maintenance of the industrial character linked to the production of beer allows to implicitly tell the history of the place, its cultural value and the meaning it represented in the past for the inhabitants, guaranteeing historical continuity.

Aside from the office spaces, the building has event areas which are mainly in the historic brewing rooms and a Grand Café which is open to the public. As well as the excellent conservation work carried out by the owners and the architects, the accessibility of the building to the public via the heritage trail and the restaurant, which has been based on the principles of social economy, is exemplary. Industrial heritage is very widespread in Europe, yet it is often overlooked and forgotten. It is hoped that the potential of other such buildings to foster creativity and to enrich Europe's culture and heritage will also be recognized.

3.1.6 Circular Business and Financing Model

Core Business

The business model of De Hoorn is focused on renting the spaces of the renovated building. Through venue rentals the organization is able to maintain its financial self-sustainability. The spaces are the following:

- *Workspace*: De Hoorn offers the possibility to rent a office space, provides with desk in a co-working room, small and flexible private offices suitable for organizing workshops, booking meeting rooms, launching brainstorming sessions. De Hoorn is Leuven's prime landing base for creative professionals and ambitious digi-wizards. Here can works freelance web designer or digital native.
- *Housing:* De Hoorn has 9 lofts in its private housing area on the Vaartkom. Rooms with pure design and the beautiful view over the city. The furnished lofts can be rented per week or month, or for an even longer period. Tested and approved by expats, travelers, artists, employees and digital nomads. Two-bedroom lofts are also available for groups of up to 4 people.
- *Meeting rooms*: small and large meeting rooms for brainstorming sessions, (board) meetings, workshops and seminars. There is also an inspiring meeting room with a terrace and catering options. The meeting rooms are for 5–150 people.
- *Parties & Events*: De Hoorn is an historical event location in Leuven. An event team will customize the event.
- *Bar & Restaurant*: the bar is the heart of all social life in De Hoorn. Customers, co-workers, friend, family, locals and lost travelers, they all end up in De Hoorn bar, a great place to meet people.

In the following table the rental prices for each revenue streams are shown (Table 7.2).

De Hoorn has become a dynamic entrepreneurial site with a strong focus on sustainable jobs and wealth creation. About 300 jobs were created in the reuse process.

Table 7.2 Analysis of revenue streams in De Hoorn, Belgium

Venue		Rental cost
Workspace	**Hot Desk**: co-working space	€15/person/day
	Private Desk: 3 up to 10 people	€300/person/month
	Private Office: 3 up to 10 people	€300/person/month
	Big Office: from 11 people	€270/person/month
	Virtual Office: postal address	€150/person/month
Housing	**Loft**: 9 bedrooms	€350/midweek €440/week €1.590/month
Meeting rooms	**Brainstorm**: up to 10 people (40sqm). Artistic meeting rooms, suitable for brainstorming and creative sessions.	From €50
	Focus: up to 6 people (13sqm). This small meeting room is suitable for meetings and customer appointments.	From €50
	Boardroom Circle: up to 10 people (36sqm). Original meeting room with round table, suitable for board meetings and customer appointments.	From €50
	Boardroom Square: up to 12 people (32sqm). Artistically furnished conference room with square table.	From €50
	Multifunk: up to 60 people (114sqm). Combinable meeting room, suitable for seminars, parties.	From €620 < 4 h From €890 > 4 h
Parties/events	**Skybox king**: 50–80 persons	From €775
	Skybox Queen: 23–50 persons	From €575
	Atrium: 50–160 persons	From €650
	Brewing room: 22–120 persons	From €1150
	Machine room: 26–300 persons	From €995
	Machine room + Hangar: 180–400 persons	From €1350
	Restaurant: 50–85 persons	From €575

Funding Model

It is a very large investment, more then €10,000,000 was spent for the restoration, initiated by a group of young and creative entrepreneurs who bought and enabled the development of De Hoorn. Miss Miyagi,[3] a society which coordinates real estate projects developed for and by its end users, coordinated the project management since 2009 through the establishment of a phased development.

[3] https://missmiyagi.eu/

Circular Governance

In 2006, seven young and local creative and cultural entrepreneurs decide to buy and develop De Hoorn as a place that represents what their organizations stand for: an innovative, open meeting place where creativity, cross-breeding and entrepreneurship are key. A mixed program of offices spaces (solely for the creative economy), events spaces (mainly in the historic brewing rooms) and a Grand Café (bar/restaurant). The choice for events spaces and a restaurant restores the public importance of the building and offers interesting additions to the office spaces for creative minds.

The managing body is a private for profit body and the management is carried on by the owner. Miss Miyagi is responsible for the project management (concept development, financial feasibility, preparation of operating model, coordination of design, tendering and site). The end users were involved in the redevelopment project.

3.2 C-Mine, Genk—Belgium

In 2005, the city of Genk, in Belgium, launched a vast project to regenerate a disused industrial area with the aim of converting it into creative hubs.

In the past the C-mine[4] site was part of former Winterslag coal mine. Nowadays the entire area has been converted in a creative hub and cultural center, where the focus is on education, creative economy, recreation, and art. The site is developed in a green park with different buildings: a cinema, a theatre, a faculty of the Luca School of Arts and space for events. It is considered a creative incubator center for start-ups. Furthermore, it attracts tourists with its unique virtual tour of the former mining site.

C-mine forging strong partnerships and entering into dialogue with actors from various artistic fields; by involving the wider community in a versatile environment, focused on study, adaptive reuse of cultural heritage, design, innovation and reflection.

In the C-mine site four main functions take place:

- Artistic creation: space for contemporary culture and artistic productions;
- Recreation: C-Tour is a tour in the leads you from the underground passages up to the mine buildings;
- Higher education: establishment of the Media and Design Academy (product and graphic design, animation and video as well as continued professional training);
- Creative economy: creating links between knowledge, innovation and entrepreneurship.

[4] Web site: https://www.c-mine.be/en

3.2.1 Manufactured Capital

In 1900 the city of Genk (a quiet village of about 3000 inhabitants) had three mines: Zwartberg, Waterschei and Winterslag. The mining site at Winterslag was particularly vast, in fact, in addition to the buildings above ground, it consisted of a fully-developed underground concession stretching many kilometers from the pithead. Nevertheless, in 1988, the mine closed permanently. The complex of mining buildings at Winterslag has been an officially protected site since 1993.

Following the mine closure, the city of Genk was on the lookout for a new role for the immense mining complex at this location. The idea to locate a hub for the creative industries at the site of the former Winterslag coal mine was born in 2000. In 2001, the city of Genk acquired the site from LRM (Limburgse Reconversie Maatschappij) and in 2005, 'C-mine' was inaugurated.

In 2013, C-mine was awarded the Vlaamse Monumentenprijs (Flemish Monument Prize).

The surviving buildings are located on area of 5000 sqm. All the protected monuments on the mining site have been assigned new roles. The coal washing plant, brickworks, workhouses and cooling towers have not been preserved. Still present on site are the Lamp Room ('lampisterie'), the Office Buildings, the machine building with power plant and the collection machines (in the Energiegebouw, or Energy Building), the Warehouses, the Horse Stables and the two Pit Towers.

- *Energy building:* is C-mine's main building. Today there are the visitor reception, a culture and design centre and catering facilities. The old machine rooms have been restored and two new theatre rooms have been added to the existing buildings. The interior of the machine rooms has been preserved and has been carefully restored. The Barenzaal, the mine's former power plant, serves as a space for corporate functions.
- *Office Building*: at its peak, the mine at Winterslag employed 6350 people. The main offices of the mine management and the offices of the administration were built in neo-Flemish Renaissance style. The old mining offices now house C-mine Crib, the incubator for young and creative entrepreneurs.
- *Lamp Room*: miners used to pick up their number and lamp at the start of their shift before going underground, now it houses a cinema complex, a number of companies and some catering establishments.
- *Warehouse*: turned into a studio and living space by and for Pieter Stockmans, the porcelain artist. In past to operate a mining company, it was necessary an enormous amount of supplies. All the materials required were stored in large warehouses like these. Smaller warehouses were also built underground on various levels.
- *Horse Stables*: at a later stage, the horse stables were used as garages. Today they serve as business space.
- *Pit Towers*: they are two towers, originally both were equally tall, at 40 meters. The Large Pit Tower in 1963 was replaced by a new one, over 72 meters high. The new tower was placed on top of the old one. The Small Pit Tower (1915–16)

is the oldest one in the Kempen basin. Miners used to call these pit towers the 'belles fleurs' ('beautiful flowers'). The towers have been preserved ensuring that the site retains a high visibility in the area. In 2015 a sculpture has been created under the towers: the Labyrinth Sculptural maze on the C-mine plaza, an impressive steel construction designed by Gijs Van Vaerenbergh on the occasion of C-mine's 10th anniversary. It has a wander through 1 km of steel corridors.
- *Luca School of Arts*: it's a campus that offers students specializations in Product Design, Photography, Animation Film, Communication & Media Design (Interaction Design & Game Design) and Film for Television. The school collaborates with C-mine Crib, FabLab, IDE (Innovation & Design Euregio), FLACC and other creative organizations and companies. This creates an interesting cross-fertilization between the worlds of school and work.

3.2.2 Natural Capital

The lighting of the structure has been carefully designed, each facade has its own particular lighting, as do the squares and green areas. This creates a very special atmosphere, especially in the evening hours. The maintenance costs of the lighting are exceptionally low. There is a mix of energy-efficient, standard and customised LED solutions.

By 2008, the Limburg region had already declared its intention to become carbon neutral, and the city of Genk was involved in the Accelerating and Rescaling Transitions to Sustainability (ARTS) project, which identified 90 actions to be developed to support sustainability in the city. These initiatives have become a huge driver for sustainability and have also deepened cooperation between policy makers and stakeholders outside the project development consortium.[5] The Acceleration Roadmap, developed in the ARTS Project, supports the City of Genk to govern its activities towards accelerating the sustainability transition. Genk's combination of multicultural urbanity and green surroundings makes the city a unique place. Starting from C-mine, it's possible to explore the area on foot or by bike. A bicycle rental system is available at C-mine. Also, a project to produce syngas and hydrogen out of the energetic fraction of municipal waste started in 2020.

3.2.3 Human Capital

C-mine works with all campus residents towards developing a community, offering a high-quality artistic visitor programme in the visual and performing arts, design and film, with innovative and international programme in the areas of development, production and participation. At C-mine the event programming is vast. Presentations

[5] https://ec.europa.eu/energy/topics/oil-gas-and-coal/EU-coal-regions/resources/genk%E2%80%99s-ongoing-transition_it

and expo have been organised here about several issues. Musical, theatre, family events, lectures and workshops, dance, design and circus events.

Types of educational or recreational activities are organized in C-mine site in the form of visits and guided tours within the buildings. The visits are for a fee and reservations are required. One of the tours organised is the C-mine Expeditie, which offers a different perspective on the reality of the mine, a museum-expedition that talks about the ex function of the site through sensory experiences. C-mine expedition starts in the Energy Building and descend to a depth of six metres, where are surprising installations. In the story tubes, life-like stories are combined with animations, holograms, stage elements and special effects, like interactive sounds produced by operating levers, buttons and wheels. The expedition ends with the ascent of the Pit tower in the plaza. 700,000 visitors a year are counted.

3.2.4 Social Capital

C-mine is a workspace and an active community space. This site fosters a sense of community and encourages unique interactions between the users who are encouraged to manage it as a cultural and community hub to come together, enjoy shared experiences and get involved in creative projects. Every day, entrepreneurs at C-mine work to create and build such creative products as games, apps, websites, television decors, drones, light shows, design objects, stage productions.

C-mine tries to elevate artists, performers, entrepreneurs, students above their own individual capabilities and ambitions. The aim of C-mine regeneration is to inspire the culturally interested, amateurs of heritage sites, students, children, visitors, tourists and locals alike, with new insights and experiences and also to invigorate the urban environment and the wider region by offering artistic and design-oriented innovation and creation.

One of the partner companies using the C-mine spaces (Erfgoedcel Mijn-Erfgoed) wants to map and help to preserve cultural heritage, make it as accessible as possible to a wide public and collaborate with heritage actors in the region. Much of this cultural heritage is linked to the rich mining past of these municipalities, but the field of heritage in the Mining Region is much broader: one need only think of religious heritage, numerous museums and private collections, local history clubs dealing with dialects. and street names, old customs and traditions, festivals, war memories and commemorations.

3.2.5 Values Co-creation

The redevelopment of this disused mine benefits an entire community. Citizens of the Belgian mining town of Genk and visitors now enjoy the advantages of a cultural arts centre that is self-financing with various commercial initiatives on the site. Genk has become a dynamic entrepreneurial city with a strong focus on sustainable jobs and wealth creation. In order to secure and strengthen future economic competitiveness in this transitional phase to a more sustainable way of life.

The old coal-mining complex of Winterslag is converted into a cultural and leisure centre with a set of installations focused on education and the creative industry. In the assessment on circularity (D1.3) C-mine was assessed with the higher score related to circularity performance as it's circular functioning based on symbiosis. Indeed, the C-mine complex, after being renovated, was designed as a cultural centre in which several creative industry make this place a vibrant place characterized by the complementarity and the interconnection between their activities.

In its everyday use it accommodates several leisure and educational spaces being a hub of all the activity associated with a university campus and the different cultural and productive spaces.

Similarly to cases of industrial symbiosis, developed within the framework of industrial ecology, where the interaction between different industrial plants is aimed at maximising the re-use of resources (material, energy, water, services and skills) normally considered as waste, in the C-mine complex the activities of localised cultural and creative industries generate a system of synergistic relationships and exchanges. Therefore, this case represents the transposition of the logic of industrial symbiosis in the field of cultural and creative enterprises.

3.2.6 Circular Business and Financing Model

Core Business

C-mine is a hybrid, creative site founded on the Winterslag coal mine in Genk. The mine's industrial heritage has been carefully renovated: more than 30 venues on offer, from small to large, indoor and outdoor; from theatre halls and meeting rooms to open spaces for catering and exhibitions. The buildings are located around the centrally located C-mine square. All spaces are very well equipped.

C-mine is able to guarantee economic self-sustainability thanks to the involvement of different stakeholders and a mix of functions including commercial units, restaurant, café garden, education museum, exhibition adaptive reuse of cultural heritage, cultural events, theatre, area for conference, social uses, community hub, innovative start up hub, co-working spaces, workshop spaces, design centre, music hall, theatre, Design School.

The day-to-day business of C-mine is focused on renting the spaces, summarized in the following table. Through venue rentals, C-mine is able to maintain its financial self-sufficiency (Table 7.3).

Table 7.3 Analysis of revenue streams in C-Mine, Belgium

Venue			N. of participants	Rental cost
1	Outdoor/ Conference space	**C-mine Cribe:** the cozy courtyard at C-mine Crib is a location or outdoor breakout space for meetings. Play a game of ping-pong, there is an area with picnic tables.	50 persons	Free, but only in combination with the rental of a meeting room or the Creativity Loft at C-mine Crib.
2	Outdoor/ Conference space	**Main hall terrace:** it is a terrace, suitable for meetings, workshops, training sessions, receptions.	15–30 seats	€30 for associations, schools and residents from Genk.
3	Outdoor/ Conference space	**Ruin on C-mine square:** The 'ruin' is what remains of the smaller Pit Tower's collection building. This outdoor space is suitable for meetings, receptions, networking events, small concerts.	100 persons standing	
4	Multipurpose space/theatre space	**Kleine Zaal:** is located in the Energiegebouw (Energy Building). This venue is suitable as a theatre space (with or without a grandstand) or as a multipurpose room.	206 persons seated/ 850 persons standing	Daily rate: €200 for associations, schools and persons located in Genk. €400 for associations, schools and persons located outside Genk €650 for commercial organizations.
5	Multipurpose/ Conference space	**De Veranda:** It comes with a fully-equipped kitchen, a bistro and a spacious terrace.	120 seats or 200 standing places	€425 per half day (4 hours) €625 per weekday €2200 per weekend
6	Banquet space/ Multipurpose space/ Conference space	**Studio Pieter Stockmans: t**his is the location of the atelier, shop and exhibition space for Studio Pieter Stockmans. Presentation area: 300 persons (seated) Standing reception/dinner: 300 persons (standing) Seated reception/dinner: 125 persons (seated)	300 persons	Available upon request
7	Conference space	**Creativity Loft:** an inspiring space on the second floor of C-mine Crib, is divided into three parts: a large meeting room, an open space and a cozy sitting area for conference space, seminars, meetings, brainstorming sessions, workshops, training sessions.	24 persons in U-shape, 65 persons in theatre style	€300 per half day (4 hours)

(continued)

Table 7.3 (continued)

Venue			N. of participants	Rental cost
8	Conference space	**Meeting room C-mine Crib**: two perfectly equipped meeting rooms in a unique historical setting, with all amenities	14 to 16 seats	€50 per hour €300 per day
9	Theatre space	**Euroscoop:** in addition to regular movie programming, Euroscoop also offers the possibility to organise events in its 10 rooms.	100 to 340 seats	Depending on a number of factors, including the nature of the activity, whether or not a film screening is included, the choice of venue and timing.
10	Outdoor	**C-mine plaza**: an impressive location for larger events.	5000 persons standing	Variable: depending on several parameters like as the consumption of utilities, ticket price.
11	Multipurpose space	**Collection room:** is an historic space suitable for birthday parties, receptions, ceremonies.	50 persons standing	€250 per day
12	Banquet space/ Party space/ Multipurpose space	**Barenzaal**: is in the Energiegebouw (Energy Building). The perfect location for conferences or parties in an environment that has an authentic and industrial look.	400 persons seated 800 persons standing	€1500 per day
13	Exhibition space	**C-mine design centrum:** is alocation for ground-breaking design projects.	100–170 persons standing	Per day: €150 – €500 Per weekend: €250 – €1000 Per week: €450 – €1500
14	Conference space	**Meeting rooms in the Energiegebouw:** plenty of space and quiet for holding meetings, workshops, etc	25 seats (3 rooms) 15 seats (1 room)	Half-day (4-hour) rate: €30 for associations, schools and persons located in Genk. €60 for associations, schools and persons located outside Genk. €120 for commercial organizations.

(continued)

Table 7.3 (continued)

Venue			N. of participants	Rental cost
15	Theatre space	**Main Hall:** this theatre space is located in the impressive Energiegebouw (Energy Building) and comes equipped with all modern technology required for conferences and theatre plays.	487 persons	€400 for associations, schools and persons resident Genk. €900 for associations, schools and persons located outside Genk. €1600 for commercial organisations.
16	Multipurpose space	**Foyer:** is the pivotal connection between all areas of the Energiegebouw (Energy Building).	400 persons standing	€200 for associations, schools and persons located in Genk. €400 for associations, schools and persons located outside Genk. €650 for commercial organisations.
17	Multipurpose space/exhibition space	**Compressorenhal:** is located on the first floor of the Energiegebouw.	720 persons standing	€400 for associations, schools and persons located in Genk. €800 for associations, schools and persons located outside Genk. €1300 for commercial organizations.

Funding Model

The redevelopment of the former mining sites in Genk was financed in 2014 by the Flemish government with an investment of €217,000,000. Then, an integrated territorial development plan (ITI) was drawn up, which allowed European funds (ESF, ERDF and Cohesion Fund) to be used. In order to strengthen transformative capacities, the government spent 8.5% of the city's budget in 2015 on investing in knowledge building, with the aim to increase their efforts in citizen empowerment.

The Winterslag site was acquired in 2001 by LRM that changing its name to C-mine. 500 square meters of floor space was regenerated in collaboration with the University of Genk. The Brussels-based architecture firm 51N4E supervised the entire conversion of the buildings and outdoor areas. The existing brick buildings were converted into exhibition areas, through light renovation. In addition, two new buildings were constructed to house two theatres.

The project was financed under the Limburg Objective 2 Programme 2000–2006 and the Flanders RCE Programme 2007–2013. The project started in June 2005 and was completed in November 2010. The total cost of the interventions was about €8,917,442 includes ERDF founds of €3,178,197.

Summarizing, all the investment amounted about to around €10 million, started with public funding from the EU and national public funding through a public consultation process.

Partnerships

Many different organizations were involved in a wide partnership. Finance came from ERDF Limburg and the city of Genk (European Union Regional Policy. Workshop on the Re-using Brownfield Sites and Buildings).

Over time, 330 jobs were created in 42 companies and organizations, of which about 200 were in the creative sector in 33 creative companies.

Organizations work in different buildings of the complex. In the Office Building works C-mine Crib, an incubator where young companies and start-ups are established. In the Lampisterie there are film companies, digital commerce experts, providers of cloud, security and software, cultural heritage preservation and valorization companies (Euroscoop, Vaimo, EASI, Erfgoedcel Mijn-Erfgoed and Speelmijntje). In the Horse Stable there are: Altrio Group (home care service), Painting with Light (lighting experts), NASCOM (digital products). In the Warehouse is established the Studio Pieter Stockmans, a porcelain artist. Finally, in the new C-mine business zone there are several organizations (White Light, AED Partner Deusjevoo, UPspace, FabLab, MiMsoftware, Yungo).

Circular Governance

Today the site is in public ownership, whereas before it was privately owned. The managing body is characterized by several start-ups. For example C-mine Crib, an incubator supportting young companies/start-ups in their entrepreneurial growth by offering creative and team-oriented workplaces. There are 37 companies in the C-mine Crib business family.

The redevelopment of the former brownfield site was carried out through a lengthy public consultation process to understand how to turn the brownfield sites into new opportunities to boost the city's economy. Several stakeholders were involved to understand how to give these areas a new life and access public funding: politicians, citizens, local entrepreneurs.

3.3 *Catacombs of San Gennaro, Naples, Italy*

The Catacombs of San Gennaro are an archaeological site/religious heritage located in one of the most densely populated and characteristic neighbourhoods of Naples, the "Rione Sanità", and are an important part of the city's history strongly intertwined with that of its patron saint, San Gennaro. Access to the Catacombs today is possible either from the forecourt of the Basilica dell'Incoronata del Buon Consiglio in Capodimonte, or from the Basilica di San Gennaro extramoenia. The site is managed by the social cooperative 'La Paranza', formed by young people from the Sanità District, after being recovered thanks to the collaboration of "l'Altra Napoli" Onlus, Don Antonio Loffredo, and the cooperatives and citizens' associations already present in the neighbourhood. It is a cultural, social and economic revolution promoted by local young people with an intervention of social inclusion and enhancement of a common good for the development of a community in a particularly difficult area. Since 2006, La Paranza has been helping to revitalise the city's image by rediscovering the artistic and cultural heritage of a working-class neighbourhood steeped in history and charm.

The beauty of the area attracts tourists and helps to integrate the district into the city's productivity circuit. A social economy has developed, giving rise to a network of small cooperatives and artisans. The site makes use of fair trade and small businesses. Local products are used from social cooperatives and citizens. The Catacombs of San Gennaro have revived an entire district. The Foundation has put together all the realities operating in the community to consolidate and support even more the activities that are not able to support themselves economically. In 2009 the Catacombs were opened to the public and guided tours and events are organized.

3.3.1 Manufactured Capital

The Catacombs of San Gennaro are ancient cemetery areas that are located underground in Naples and date back to the second and third centuries AD. They represent one of the oldest monuments of Christianity in Naples. When the activities at the Catacombs began, the upper level was open to a select few: it was not open every day, it did not have employees but only the Inspector of the Pontifical Commission who was also responsible for accompanying visitors to the Catacombs. There was no lighting system, and the lower vestibule could not be visited.

The archaeological site includes the Catacombs of San Gennaro[6] and the Catacombs of San Gaudioso and two other Basilicas annexed to them:

- Basilica of San Gennaro extra moenia which is owned by the Local Health Authority. This basilica, which is located in the Rione Sanità, was the gateway to the Catacombs.
- The church of San Severo has a small hypogeum which cannot be visited at the moment, but La Paranza (the social cooperative managing the archaeological site) is having a confrontation with the Pontifical Commission for managing it.

There is a project for the Cristallini Church to use the Basilica as a space for cinematographic activities. Next to the Basilica there is a space of the Municipality

[6] Web site: https://www.catacombedinapoli.it/it/luoghi/catacombe-di-san-gennaro-napoli

which has been entrusted to the parish, for a project to carry out sports activities, which are currently held in the sacristy of the church of S. Maria della Sanità.

In addition to the archaeological heritage of the Catacombs, other spaces have also been recovered for a total of 12,000 square meters:

- Rectory House: Casa del Monacone for religious hospitality (6 rooms and a living room with a kitchen made available to tourists).
- "L'Altra Casa" (canonical house near the Basilica of San Severo where take place the activities of Sanità Ensemble).

Indoor areas are made up of 6000 sqm of Catacombs of San Gennaro and 12,000 sqm of canonical houses, disused churches and other outdoor spaces. Outdoor areas (garden) are 1500 sqm of overall outdoor garden (1000 sqm of gardens near the Catacombs of San Gennaro and 500 sqm Garden of Oranges near the Church of San Severo).

From 2006 to 2008 there has been the training and experimentation of guided tours with voluntary work at the Catacombs of San Gaudioso. In 2009 the Catacombs were opened to the public.

3.3.2 Natural Capital

The Catacombs of San Gennaro is aligned with the circularity principles adopting Nature-Based Solutions in the heritage site. There are 500 sqm of outdoor gardens and there is no particular level of pollution, on the contrary, the Hill of Capodimonte, where the Catacombs of San Gennaro are located, is characterized by healthy air.

The site makes use of fair trade and small businesses. Local products are used in the bar from local social cooperatives, produced by NCO (New Organized Kitchen), artisan coffee produced by a Neapolitan family.

A large investment was made in the latest generation LED electrical system (in 2009 it was mentioned by a famous magazine and compared to the Camp Nou system in Barcelona). All the guides are provided with an app on their mobile phones that allows them to manage the plant during the visits in order not to waste energy and to safeguard the frescoes. After 10 years it has been realized that some of the frescos have bacteriological attacks in some places, so they are experimenting with neon lights that have a compound inside that should avoid this fermentation of bacteria that damage the frescos. So at the moment there is a need for extraordinary maintenance and updating with respect to new technologies.

3.3.3 Human Capital

The choice to use the social cooperative as a legal form is also a result of the structural criteria of the strategic plan: in fact, this legal form tends more towards the growth of human capital than of invested capital. To date, ten cooperatives have been created and are engaged in various sectors, from social to construction, to improve the quality of life in the neighborhood.

Before establishing the cooperatives, young people were trained by giving them the opportunity to attend school and to travel in order to acquire a critical eye that would allow them to see beyond the ghettoizing logic imposed by social discomfort. The investment on human capital has contributed to the formation of skills in young people who have begun to interact with the religious cultural heritage of the neighborhood through actions of enhancement that have in turn triggered processes of redevelopment of many outdoor public spaces such as squares, open spaces and gardens, which have become triggers for social and economic activities. The action of recovery of associations and cooperatives born in Sanità on religious cultural heritage has gone beyond the boundaries of the neighborhood and the good practices that inspired it have been used to recover the Church's real estate assets also in other degraded contexts of the city of Naples (Giammetti 2019).

For ordinary work (plasters, external flooring, lighting, etc.) are employed the young people at Officina dei Talenti who have expertise in the field. Educational workshops for school children, painting, restoration and excavation workshops are here organized. The children first make the visit and then move to the hypogeum of the Basilica del Buon Consiglio where they do the workshops.

3.3.4 Social Capital

The recovering process of the Catacombs had to face cultural barriers that affected the entire Sanità Ward. Within the Ward there was maximum support because the community of the neighborhood has seen this project as an opportunity, a positive alternative for their future. It was difficult to work on the bad "reputation" of the Sanità District and the negative image that the Ward had and that blocked tourism.

Residents have increased their awareness of the value of their heritage, of the favorable strategic position of Sanità District with respect to the center of Naples (re-establishing a connection rather than feeling isolated). The Catacombs have entered into a symbiotic relationship with the neighborhood, entering into dialogue and supporting initiatives and encouraging the creation of other activities and, at the same time, the neighborhood has reacted well by connecting with the city. So there is a double circular process between the archaeological site and the neighborhood and between the neighborhood and the city.

The community and the neighborhood have entered into an osmotic relationship with each other, generating a virtuous process that has allowed the neighborhood to open up to the city. In this perspective, the strategic project is translating into action on the physical space and on the community one of the principles of circular economy: transforming waste into resource, generating value from waste (Giammetti 2019).

The area is composed of a very large condominium: in the same area there are nuns, a clergy house, the Basilica del Buon Consiglio, who use the outdoor spaces (co-fruition). This is the unique Catacomb accessible to the disabled: dedicated access directly from the Sanità Ward (behind the Basilica of San Gennaro extra moenia), slabs for the visually impaired and blind made by Iron Angels reproducing the frescoes.

3.3.5 Values Co-creation

The management plan is based on a systemic logic that generates mutuality between cooperatives associations and foundations, in the name of a circularity according to which the activities that are able to revenues support activities that produce less revenue or no revenue at all, such as social services, but are no less necessary for implementing human capital.

The Catacombs are the economic engine of the strategic plan and with their induced are capable of sustaining the functional reuse of many abandoned churches and the activities of associations involved in social work. In addition to the Catacombs, there is also funding coming from calls for proposals in which cooperatives, associations and foundations participate and contributions from private individuals who support the actions of recovery of the material and human heritage of the neighborhood. That same circularity that has set up a system of space and community, translates into an in an economy of gift, in an action of mutual support useful to "maintain over time the value of common goods" that the strategic project has produced and/or supports (Giammetti 2019).

Artistic Works are created by the cooperative "Iron Angels" and are made with waste materials. It is not possible to estimate a quantity of reused material, but these works have a symbolic value.

3.3.6 Circular Business and Financing Model

Core Business

The Catacombs can guarantee economic self-sustainability thanks to the contribution of different activities managed by the cooperative "La Paranza". The activities that generate revenues are:

- *Guided tours*: the Catacombs are always open and, in addition to ordinary tours, there are also guided tours in the entire Sanità District (the visit of the "Miglio Sacro" lasts 3.5 h and also includes the Fontanelle Cemetery in the center of Naples). The "story" of the places and the experience of reuse itself is an added value and focuses not only on the beauty of the places but on the whole process of growth of the local community. Visitors often leave a donation in addition to paying the price of the ticket. The full ticket is 9.00€ (taking as an example the standard of the Catacombs of Rome) and allows you to visit both the Catacombs of San Gennaro and those of San Gaudioso also in two different moments because the ticket can be used in one year. This action has been made according to a twofold perspective: the first one is to entice the visitor to cross the Sanità district to go from one archaeological site to the other, the second one is to encourage people to return to the neighborhood at zero cost (because the ticket is paid only once and is valid for the two sites even if the visit takes place at two different times) but at the same time they are carriers of other economy and know the district in all its potential. Ticket revenues in 2018 amounted to €675,000 (130,000 visitors). In 2019 there were 160,000 visitors.

- *Bookbar* close to the ticket office (open only to the public visiting the Catacombs). In 2018 generated €106,000 in revenues considering the activities of the bar, the sale of souvenirs, the sale of the book (With the birth of the local community foundation was born Edizioni San Gennaro (e.g. "Vico Esclamativo" in which are told stories of the Sanità District).
- *Events* during the year.

Activities that do not generate revenues and not related to the cooperative "La Paranza" are:

- Casa del Monacone: it is an activity in charge directly of the church of Santa Maria della Sanità, so the owner is Antonio Loffredo who is the parish priest. It was born as a religious hospitality activity and performs a double function: on the one hand it mainly supports the expenses of the parish, on the other hand for the children of "La Paranza" it works as a 'training school' to start work to welcome the tourist/visitor, an opportunity to improve the knowledge of foreign languages.
- Theatre, theatrical visits
- Family support, parish activities
- Parish expenses, utilities
- Educational, formative, scholastic activities
- Recreational leisure activities
- Sporting activities

The profits are reinvested mainly in restorations and, through the Local Community Foundation, in other socially useful projects (children's trips, school activities, etc.). Hundred percent of the profits are reinvested in social projects (approximately €160,000 of profits by subtracting the operating expenses (€700,000) from the annual receipts (€862,000) unless maintenance work is required.

The archaeological site is the engine to which the other "poles" that are also economically and financially supported by it are functionally connected. The tourists who frequent the site bring "cash economy" so, after deducting the management costs, also considering the legal instrument chosen for the management of the catacombs site, it has been possible to activate a virtuous mechanism that reinvests the profits on the asset to finance activities of social utility (theatre, music, film school, etc.).

Funding Model

La Paranza was founded in 2006 by a group of 5 young people led by Don Antonio Loffredo, the parish priest of the ward. The initial years were of analysis of the territory to understand what the problems were. The group was already active since 2002, with voluntary work to organize guided tours in the Catacombs of San Gaudioso. The first visits began in 2006. These were also the years in which the young people involved began to make experiences abroad in order to increase awareness of the value of their heritage and to develop ideas on how to enhance it. The spark was Father Antonio's ability to put local human capital at the service of the process of valorisation of archaeological heritage and local development. In 2011 parish priest Antonio met Ernesto Albanese, president of the "L'Altra Napoli" Association. This Roman entrepreneur experiences a traumatic event in Naples: during a robbery his father dies, but instead of leaving Napoli, he decides to found this association by putting together his entrepreneur friends and starting a fundraising campaign to activate some economy and calls it "L'Altra Napoli" to give a different vision of the city. The meeting with father Antonio Loffredo is fundamental because "L'Altra Napoli" is the first association that starts to financially support the activities of "La Paranza". In 2008, "L'Altra Napoli" became a partner of "La Paranza" on the occasion of the call for proposals launched by the Fondazione con il Sud for the recovery of archaeological sites. The cooperative in partnership with "L'Altra Napoli" took part in this call for proposals, proposing a project for the enhancement of the archaeological site of the Catacombs of San Gennaro. The call is won and Fondazione con il Sud gives an initial capital of €400,000 which has doubled with the increase in external sponsors. "L'Altra Napoli" has invested €100,000. In 2009 there is the reopening of the Catacombs to the public.

In 2008 the cooperative "Officina dei Talenti" was also founded, where all the young people with the greatest manual skills were brought together in order to take care of all the ordinary and extraordinary maintenance activities of the spaces. "Officina dei Talenti" was born as a rib of "La Paranza" and currently, in addition to taking care of the spaces, it also has private contracts for the recovery of apartments, so it is configured as a real construction company. La Paranza has 40 employees (direct employment). Officina dei Talenti has 15 young people (indirect occupation).

From an experience with Riccardo Dalisi was born another cooperative "Iron Angels" that currently has 3 people. Dalisi has established a craft workshop that reuses waste materials for the creation of works of art. These cooperatives with their activities are able to support themselves economically. There are many other cooperatives in the area that are involved in the recovery of school, theatre, but these are activities that must be supported and that the cooperative "La Paranza" is not able to support itself. For this reason, in 2014 the "Fondazione di Comunità San Gennaro Onlus" was created, composed of both the non-profit and profit sectors, in which all the social cooperatives (together they founded the "Cooperazione San Gennaro" group) and the network of merchants in the district who have also joined this Foundation. The Catacombs of San Gennaro have revived an entire district. The Foundation has put together all the realities operating in the community to consolidate and support even more the activities that are not able to support themselves economically. The projects supported are both social and urban (redevelopment of the area, such as squares, street furniture, urban green areas). Members of the foundation are different entrepreneurial realities (Feudi San Gregorio, Grimaldi Foundation, Pasquale di Costanzo Foundation, Buonafede family, etc.) that have formed the social capital of the Community Foundation. The Foundation was born with a project launched by Fondazione con il Sud, which doubles the initial capital.

This network of cooperatives has the commitment to produce in 10 years a capital of about €100,000: since there are 10 cooperatives, every year each cooperative tries to give the Foundation 1000€/year for a total of about 10,000€/year in order to increase the social capital which, together with the profits, can finance other social utility projects in the territory. Therefore, the same cooperatives and traders' network that are part of the "Fondazione di Comunità San Gennaro Onlus" also contribute economically to the increase of the social capital. The Foundation also participates in other projects and calls for proposals to recover funding, such as other Foundation calls with the South, regional and municipal projects.

A crowdfunding campaign "Teniamo in vita il passato" was launched for the maintenance of the frescoes which produced €30,000 all reinvested for the recovery of the frescoes in the Catacombs. The Young Industrialists Group of Confindustria, which donated €19,000, and 10 other private donors took part in the campaign.

The Foundation has invested about €4,500,000 on the whole Sanità district coming from private funds collected in 10 years from fundraising activities and donations (recovery of the squares, recovery of the basilicas, recovery of other spaces for activities, etc.). "L'Altra Napoli" has invested €1500,000 in the Sanità district.

Initially in 2009 for the reopening of the site of the Catacombs an investment of €800,000 was made: subsequently more money was invested for other interventions in outdoor spaces. Some investments cannot be quantified because many private sponsors did not contribute with financial support but through "in kind" contributions (e.g. Moccia sponsored the cement, there were many collaborations with professional associations, universities, etc.). In the experience of the Catacombs, gratuitousness has contributed a lot and has gone out physically in the city and has contributed to a physical-spatial regeneration that has also meant a social and economic regeneration. For example, some recovered spaces are Largo Vita, Slargo where a statue has been installed to celebrate the 50th anniversary of Totò's death and section of road at the beginning of the Virgins at the height of Poppella's activity. The Administration, not being able to financially support these activities, has granted the authorization to carry them out.

No public funding was received while private donations were made by the network of traders and cooperatives through the Community Foundation. The remaining donations came from visitors. It is always tried to divert all donations to the Community Foundation because Fondazione con il Sud doubles any amount of the social capital that Fondazione San Gennaro manages to raise. The San Gennaro Community Foundation in 10 years has to recover a maximum of €2,500,000 which, thanks to the agreement provided for by the Foundation with the South, is doubled to €5000,000 (the 10 years are counted from 2015 which is the year of establishment of the Foundation). "La Paranza" is able to support itself but is not able to support other projects as well. Therefore, the Foundation was created with the aim of financing other socially useful projects born in the Ward and "La Paranza" only acts as a support as a partner of the Foundation together with other cooperatives.

Partnership

The "La Paranza" social cooperative manages the site ensuring an economic sufficiency but works with other organizations as partners like "L'Altra Napoli" foundation, the "Officina dei Talenti" cooperative, the "Iron Angels" cooperative, the "Fondazione di San Gennaro Onlus Community", citizens making donations. This creates a community network regenerating the territory in economic terms and creating social and cultural values. In total the number of organizations currently collaborating are 12 traders and 10 social cooperatives. There is the network between cooperatives and local actors, especially after the establishment of the Foundation, but it is not formalized (50–60 people in the community linked to the parish). Sanità Ensemble started with a first module of 35 boys, the second module had 42 boys. So in total the Catacombs "magnet" attracted 80 people. The Catacombs from attracting force to aggregating force.

For extraordinary work specific skills in restoration are required and the protection is entrusted to the Pontifical Commission. It is therefore the latter that is responsible for identifying the company/professionals competent for the specific problem (especially with regard to the frescoes). For ordinary work (plasters, external flooring, lighting, etc.) are employed the young people at Officina dei Talenti who have expertise in the field.

In total 40 number of jobs are created (37 full time + 3 trainees). Volunteers do not work directly on the site but are mainly engaged in socially useful activities. These volunteers (10 participants) are sometimes also available to support stable workers (e.g. in case there are too many visiting groups). They work 3–4 days per month considering that they are also involved in the monthly staff meetings.

Circular Governance

Decisions are taken within the "La Paranza" cooperative founding members group, maintaining a continuous dialogue with Antonio Loffredo and also with Carlo Borgomeo (Fondazione con il Sud) for the company's choices.

The Paranza has no direct contact with the Pontifical Commission because the latter has entrusted the Catacombs to the Diocese / Curia of Naples. There is an agreement between the Pontifical Commission and the Curia of Naples which, in turn, has an agreement with La Paranza for the management of the sites. The agreement strongly desired by the Pontifical Commission provides that "La Paranza" should give 50% of its vitality to the Diocese. This has never been done because otherwise the cooperative's activity could not be supported and the Cardinal has never requested these amounts either. In 2018 this issue was discussed in order to put in place a new convention that could protect the activity by also making a contribution to the Pontifical Commission. The first agreement was provided for 5 years, renewable tacitly for another 5 years. In 2019 it was renewed, but a new form of agreement should be found.

4 Discussion and Conclusions

The three case studies analysed showed that different models can be developed to reach a high level of circularity. In De Hoorn, the circular economy was implemented in the recovery of building materials and adoption of high environmental standards for the reused building. Also, a high level of financial self-sustainability was observed, also thanks to the favourable localization of the building that could allow sufficient revenue streams from renting the spaces. Similar conditions were observed in C-Mine, where venues renting represents the main revenue stream. However, this aspect should not be confused with the "value proposition", which in both cases focused on the heritage values and the possibility to access hybrid spaces for creative workers, companies, freelancers, startuppers, and other people and organisations, generating a vibrant environment that makes the heritage sites attractive to potential users—which are considered not only as "clients", but as a "community" that identifies with the place also thanks to its regenerated cultural value. Also, in both cases the capacity to self-generate the natural resources needed for their functioning is an added value, as energy generation from renewable sources (compatibly with heritage values) represents a highly important aspect in view of the implementation of the European Green Deal, and particularly the Renovation Wave policy, as well as the New European Bauhaus and circular city initiatives (Nocca 2021).

A slightly different condition is showed in the Catacombs of San Gennaro, where the core business and revenue model is represented by guided visits to the archaeological heritage site, while the value proposition is focused on the "storytelling" of the re-discovered heritage place interlinked with the regeneration of an entire community in one of the most socially deprived neighbourhoods in the city of Naples. The capacity of young people to thrive and to rediscover their common culture and identity through the heritage site, is the main focus of the experience of each visitor to the Catacombs. The way in which the heritage site is presented, visited, and discovered through the lenses of "real" people living in the neighbourhood and taking care of the places, is undoubtedly an emotional value that is able to attract more and more visitors over time. The social cooperatives and foundations created for the management and maintenance of the site generated around 40 jobs in the area, which can be seen as a extraordinary result compared with the average levels of unemployment. Moreover, the choice of third sector actors as managers of the heritage place contributed to generate additional benefits for the neighbourhood, with revenues re-invested in new regeneration projects for local families, youths and the entire city. In the case of the Catacombs, the adaptive reuse of cultural heritage became a symbol of the redemption of an entire community and urban area, usually depicted as "dangerous" in newspapers and social media, turning it into a place of "regeneration" and optimism.

Another interesting common aspect observed in the best practices is the capacity to regenerate and "beautify" the surrounding urban/rural areas. In line with the New European Bauhaus approach, "beauty" represents an important value strictly linked with human wellbeing and ecosystems health (Fusco Girard 2019). Here, beauty is intended not only as the particular and very personal perception of aesthetic values—even if neurosciences and behavioural studies are still exploring common features of "beautiful" places as perceived by human beings—but mostly as a sense of cleanliness, welcome, safety, integration with nature and care for the place. The adaptive reuse of cultural heritage, in the most interesting cases observed during the CLIC adaptive reuse of cultural heritage, often has led to the regeneration of their urban/rural settings, contributing to the wellbeing and health of local populations, as well as visitors. It is clear, in fact, that abandoned and degraded historic places communicate unease and unsafety, and can be even impacting human and ecosystems health. Thus, circular adaptive reuse of cultural heritage should aim at regenerating natural ecosystems—also in urban areas through nature-based solutions integrated in the heritage site—and achieving "beauty" and sustainability. Beauty and sustainability are indeed fundamental elements of a "human-centred" circular economy, intended not only in the technological sense, but also as a economic model inspired by nature that can be able to put human and ecosystems health and wellbeing at the centre of development goals. The New European Bauhaus interprets this ambition highlighting that sustainability goals are strictly interlinked with wellbeing, expressed through three main keywords: beauty, sustainability and participation.

Thus, last but not least, participation can be seen as one of the most important aspects of a circular and "human-centred" model of cultural heritage adaptive reuse. The capacity of the analysed experiences to structure a participatory, open circular governance model and maintain it over time, is one of the success factors that can be identified (Garzillo et al. 2019). The more the community is involved since the first stages of the adaptive reuse process, the more stable and long-term success could be expected for the reuse initiative (Gravagnuolo et al. 2021b). Participatory decision-making processes can be however challenging, both for lack of resources and lack of interest by owners, managers and diverse community groups (Bosone et al. 2019). Here, the CLIC adaptive reuse of cultural heritage focused on participatory evaluation methods to support and facilitate choices in cultural heritage adaptive reuse (Izulain et al. 2021), which become very relevant in contexts of conflicting interests and scarce resources. These evaluation models are further explored in the following chapters of this volume, supported and integrated through the results of best practices assessment (ex-post evaluation informing ex-ante evaluations).

To conclude, the "ideal" **circular symbiotic heritage ecosystem** (Fusco Girard 2023) should include all elements found in the "best practices", from energy generation from renewable sources to financial self-sustainability, jobs generation, heritage communities stimulation, cultural and social regeneration, enhancement of landscape beauty, and circular governance based on active participation of local communities in decision-making processes.

Best practices show that it is possible, and desirable, to reach circularity goals in cultural heritage adaptive reuse. Thus, they suggest specific criteria and indicators of circularity that contribute to operationalise the CLIC framework of circular adaptive reuse of cultural heritage (Fusco Girard 2021a).

These three good practices can be interpreted as a demonstration that the circular symbiotic heritage ecosystem model can be successfully implemented. This is due to the ability to identify specific innovative uses/functions that are demanded by the market and are not provided at other sites. Circular adaptive reuse thus expresses creativity in finding unique solutions. These unique uses/functions are able to produce a flow of multidimensional benefits that justify the investment in the adaptive reuse.

References

Allegro I, Lupu A, Pickerill T (2021) CLIC Deliverable D4.2—Circular financing models for cultural heritage adaptive reuse. https://www.clicproject.eu/wp-content/uploads/2022/01/D4.2-Circular-financing-models-for-cultural-heritage-adaptive-reuse.pdf

Bosone M, De Toro P, Fusco Girard L, Gravagnuolo A, Iodice S (2021) Indicators for ex-post evaluation of cultural heritage adaptive reuse impacts in the perspective of the circular economy. Sustain For 13(9):4759. https://doi.org/10.3390/su13094759

Bosone M, Micheletti S, Gravagnuolo A, Garzillo C, Wildman A (2019) Towards a circular governance for the adaptive reuse of cultural heritage. BDC Bollettino Del Centro Calza Bini 19(2):279–305. https://doi.org/10.6092/2284-4732/7270

CLIC project—Circular models Leveraging Investments in Cultural heritage adaptive reuse (2017) CLIC project—circular models Leveraging Investments in Cultural heritage adaptive reuse. Horizon 2020

Fusco Girard L (2018) Discourse of Professor Luigi Fusco Girard at the European Parliament high-level European parliament conference "Cultural heritage in Europe: linking past and future." European parliament conference "Cultural heritage in Europe: linking past and future." https://www.clicproject.eu/wp-content/uploads/2018/09/CLIC-Luigi-Fusco-Girard-Discourse-at-the-European-Parliament.pdf

Fusco Girard L (2019) The human-centred city development and the circular regeneration. In: Fusco Girard L, Trillo C, Bosone M (eds) Matera, città del sistema ecologico uomo/società/natura il ruolo della cultura per la rigenerazione del sistema urbano/territoriale. Giannini Publisher

Fusco Girard L (2021a) CLIC Deliverable D2.7—CLIC framework of circular human-centred adaptive reuse of cultural heritage. https://www.clicproject.eu/wp-content/uploads/2022/01/D2.7-CLIC-Framework-of-Circular-Human-centred-Adaptive-Reuse-of-Cultural-Heritage.pdf

Fusco Girard L (2021b) The evolutionary circular and human centered city: towards an ecological and humanistic "re-generation" of the current city governance. Hum Syst Manag 40(7)

Fusco Girard L (2023) The circular "human-centred" adaptive reuse of cultural heritage: theoretical foundations. In: Fusco Girard L, Gravagnuolo A (eds) CLIC circular models for cultural heritage adaptive reuse. Springer

Fusco Girard L, Gravagnuolo A (2017) Circular economy and cultural heritage/landscape regeneration. Circular business, financing and governance models for a competitive Europe. BDC. Bollettino Del Centro Calza Bini 1(1):35–52

Garzillo C, Balenciaga I, Izulain A, Rangil-Escribano T, Wildman A (2019) Circular governance models for adaptive reuse of cultural heritage, Horizon 2020 CLIC Deliverable D3.4

Giammetti M (2019) Dismissione e riuso degli spazi del sacro. BDC—Bollettino Del Centro Calza Bini 19(2):395–416

Gravagnuolo A, De Lucia M (2019) CLIC Deliverable D1.4—database of best practices. https://www.clicproject.eu/wp-content/uploads/2022/01/CLIC_WP1_D1.4_Database-of-best-practices.pdf

Gravagnuolo A, Fusco Girard L (2021) CLIC Deliverable D2.4—database of indicators and data in pilot cities. https://www.clicproject.eu/wp-content/uploads/2022/01/D2.4_Database-of-indicators-and-data-in-pilot-cities.pdf

Gravagnuolo A, Bosone M, Fusco Girard L (2021a) CLIC Deliverable D2.5—methodologies for impact assessment of cultural heritage adaptive reuse. https://www.clicproject.eu/wp-content/uploads/2022/01/D2.5-Methodologies-for-impact-assessment-of-cultural-heritage-adaptive-reuse.pdf

Gravagnuolo A, Micheletti S, Bosone M (2021b) A participatory approach for "circular" adaptive reuse of cultural heritage. Building a heritage community in Salerno, Italy. Sustain For 13(9):4812. https://doi.org/10.3390/su13094812

Gravagnuolo A, Vellecco I, Fusco Girard L, Lauro NC (2023) Circularity assessment through ex-post evaluation of "best practices": a modelling-based approach. In: Fusco Girard L, Gravagnuolo A (eds) CLIC circular models for cultural heritage adaptive reuse. Springer

Ikiz Kaya D, Pintossi N, Koot CAM, Colenbrander BF (2019) CLIC Deliverable D1.5—report on barriers and bottlenecks. https://www.clicproject.eu/files/D1-5.pdf

Izulain A, Garzillo C, Wildman A (2021) CLIC Deliverable D5.2—local action guide: collaborative approaches to adaptive reuse of cultural heritage. https://www.clicproject.eu/wp-content/uploads/2021/04/CLIC-D5.2-Local-Action-Guide_ICLEI_Final.pdf

Monteleone G, Milani G, Ring C (2021) CLIC Deliverable D6.22—CLIC platform for pilot cities. https://www.clicproject.eu/wp-content/uploads/2021/12/D6.22-CLIC-Platform-for-pilot-cities.pdf

Nocca F (2021) CLIC Deliverable D2.6—the circular city model: conceptualization and tools. https://www.clicproject.eu/wp-content/uploads/2022/01/D2.6-EU-policy-oriented-publication-defining-a-European-circular-city-territory-model-of-sustainable-development-through-adaptive-reuse-of-cultural-heritage.pdf

Pickerill T (2021) CLIC Deliverable D4.1—overview of hybrid financial instruments and investment leverage enablers for cultural heritage adaptive reuse. https://www.clicproject.eu/wp-content/uploads/2021/06/D4.1-Overview-of-Hybrid-Financial-Instruments-and-Investment-Leverage-Enablers-for-Cultural-Heritage-Adaptive-Reuse.pdf

Ruba S, Drouillon P, Ost C (2020) CLIC Deliverable D4.5—circular business model workshops for cultural heritage adaptive reuse. https://www.clicproject.eu/files/D4-5.pdf

Chapter 8
The CLIC Multidimensional Impacts Assessment Framework: Criteria and Indicators for Circular "Human-Centred" Adaptive Reuse of Cultural Heritage

Antonia Gravagnuolo, Martina Bosone, and Luigi Fusco Girard

1 Introduction

Heritage is present in everyday society and it is a resource linked to social capital, economic growth, and environmental sustainability (Bandarin and van Oers 2014). Adaptive reuse is defined as "any building work and intervention aimed at changing its capacity, function or performance to adjust, reuse or upgrade a building to suit new conditions or requirements" (Douglas 2006). In Historic Preservation: Curatorial Management of the Built World (1982), James Marston Fitch points out that the adaptive reuse of historic buildings "is more economic" not only in terms of the "conservation of the energy represented by the built environment," but also for the "relative costs of old and new built space". Adaptive reuse can be an effective conservation strategy allowing present and future use of abandoned heritage buildings, groups of buildings, landscapes or sites, changing and enhancing their functions and adapting the existing features to new needs (Bullen and Love 2010). However, the adaptive reuse intervention should not compromise heritage values, thus the threshold of transformation versus conservation should be carefully identified. Cultural heritage adaptive reuse is a complex activity, where multiple and often conflicting values need to be considered (CHCfE consortium 2015). Therefore, adaptive reuse should be supported by adequate multidimensional and multicriteria

A. Gravagnuolo (✉)
CNR ISPC Institute of Heritage Science, National Research Council, Naples, Italy
e-mail: antonia.gravagnuolo@cnr.it

M. Bosone
Department of Architecture, University of Naples Federico II, Naples, Italy
e-mail: martina.bosone@unina.it

L. Fusco Girard
Emeritus Professor at University of Naples Federico II; CNR IRISS, National Research Council, Institute for Research on Innovation and Services for Development, Naples, Italy

L. Fusco Girard, A. Gravagnuolo (eds.), *Adaptive Reuse of Cultural Heritage*,
https://doi.org/10.1007/978-3-031-67628-4_8

evaluation tools which enable to deal with multiple values and needs, also considering the diverse range of stakeholders, users and beneficiaries of the interventions.

Cultural heritage can be considered an economic good (Nijkamp 2012; Ost 2016; Getty Conservation Institute 1998). Cultural heritage, as a non-renewable cultural capital (Bourdieu 1986), is "linked" to the economy (Ost 2009, 2016) since economics refers to the management of scarce and non-renewable resources (Ost 2009). This means first of all the risk of loss of the socio-cultural memory. But also, an economic cost.

Adaptive reuse of cultural heritage is considered as one of the most effective and environmentally friendly tools of modern urban development towards sustainability (Yung and Chan 2012), allowing to address "urban waste" of territorial abandoned resources/capital. However, adaptive reuse should take into careful consideration the heritage conservation theories, which have been shaped by the contribution of outstanding thinkers (Boito, Ruskin, Benjamin, Riegl, Brandi, Jokilehto, etc.) highlighting the importance of the concepts of authenticity and integrity, which are at the base of a real transmission of values, avoiding to take choices based only on economic viability criteria, or at the opposite based on a-critical preservation and relying only on public funding. In fact, the sites recognized as cultural heritage are increasing, but the costs for functional maintenance/reuse are growing, while public resources available are becoming scarcer, and private actors are increasingly focused on the short time for payback. The consequence is that there is a growing risk that the decay of heritage increases year by year, because lack of funding support.

For these reasons, heritage conservation is also an economic choice (Vecco 2007). As cultural capital, cultural heritage has an *intrinsic* value but also some *instrumental* ones: it is necessary to adopt a holistic approach to its evaluation, characterized by the integration of use and non-use values. Costs and values of cultural heritage systemic/adaptive reuse should be compared in a multidimensional space. The relationship between costs and "complex values" influences the willingness to invest in the functional recovery of cultural heritage. The higher the perceived value for potential actors, the higher the willingness to take the risk of investment. This "complex value" of cultural heritage depends on its intrinsic characteristics, but also on extrinsic (context) features (Fusco Girard 2013, 2014).

The investment gap in cultural heritage and landscape regeneration can be addressed through careful evaluation of all costs, of "complex values" and impacts of adaptive reuse, selecting function(s) not only linked to tourism attractiveness, but also for the well-being improvement, providing critical evidence of the wealth, of jobs, social, cultural, environmental and economic returns on the investment.

Specific evaluation tools were developed within the CLIC project to support decisions towards the adaptive reuse of cultural heritage, adopting the circular economy perspective. This chapter develops an indicator framework able to support the ex-post and thus ex-ante assessment of circular "human-centred" adaptive reuse of cultural heritage, that has been test applied as a first test in the CLIC pilot case studies. The multidimensional indicators on the impacts of cultural heritage are both quantitative and qualitative indicators and they refer to different scales from the single building to the urban scale.

The CLIC evaluation framework developed aims to make explicit the multidimensional productivity that arises from the adaptive reuse of cultural heritage in the circular economy perspective, highlighting the complex notion of value embedded in cultural heritage (Fusco Girard 1987; Fusco Girard and Nijkamp 1997; Fusco Girard and Vecco 2021). This study proposes the multi-criteria impact assessment framework based on the concept of multidimensional productivity of cultural heritage (Hosagrahar et al. 2016), which has been grounded into the theory and practice of the circular economy as way to achieve economic growth and wellbeing "decoupling growth from resources consumption" (Ellen MacArthur Foundation 2012; Le Moigne 2014; Wijkman and Skånberg 2015; Ghisellini et al. 2016; Geissdoerfer et al. 2017).

This study presents the structured framework for the ex-post and ex-ante evaluation of the impacts of cultural heritage adaptive reuse practices in the perspective of the circular economy. It builds on previous analysis of more than 120 case studies of cultural heritage adaptive reuse (see Chap. 6 of this volume: also, CLIC Deliverable D1.3 "Survey on best practices of cultural heritage adaptive reuse"). The CLIC evaluation framework is theoretically based on Multi-Attribute Utility Theory (MAUT) (Keeney and Raiffa 1976; Zeleny 2006), identifying evaluation goals, objectives and criteria in multiple interrelated dimensions (Gravagnuolo et al. 2017).

Adaptive reuse interventions of buildings and sites contribute to sustainable urban regeneration (Fusco Girard 2019a). The reuse of abandoned and underused assets, which represent urban "wastes", supports the implementation of the circular economy model in the spatial dimension (Gravagnuolo et al. 2017; Angrisano et al. 2019; European Commission 2019). However, circular economy indicators are rarely applied to cultural heritage (Fusco Girard et al. 2019; Gravagnuolo et al. 2019a), even if circular economy started to be implemented in the built environment and building construction sectors in recent years (ARUP 2016; Ellen MacArthur Foundation and ARUP 2019). The economic evaluation of social cost-benefit financial analysis is a part of the multicriteria evaluation approach.

The concept of cultural heritage adaptive reuse as an instrument to achieve circular economy goals at the territorial scale emerged through the CLIC research. An integrated and multidimensional approach should be adopted within the systemic circular approach (Gravagnuolo et al. 2017; Fusco Girard 2019b). However, recent reviews of the literature on circular economy indicators showed that the most used indicators are currently related to waste management, raw materials, recycling rates, economic performance of circular businesses, energy, toxicity and clean materials cycles (Moraga et al. 2019; Parchomenko et al. 2019). Few studies are currently available on the relationships between cultural heritage adaptive reuse and circular economy. Indirect approaches can be identified in the ecosystems services assessment frameworks (MEA 2003; TEEB 2010; Costanza et al. 2014). For example, Stanik et al. (2018) analysed cultural heritage from the perspective of cultural ecosystem services, with the aim of identifying and developing an indicator-based framework formed by indicators related to historic land uses and historic elements. Still, Gravagnuolo et al. proposed an evaluation framework for circular economy implementation in the adaptive reuse of cultural heritage (Gravagnuolo et al. 2017), while Foster (2020) associates the concepts of adaptive reuse, cultural heritage and circular economy, focusing on the environmental benefits of heritage conservation. More in depth, the study of Foster (2020) performs a literature review to demonstrate the alignment between circular economy goals and adaptive reuse of heritage assets in a life-cycle perspective. Foster and Kreinin (2020) also realised an in-depth review of environmental indicators for the adaptive reuse of cultural heritage in the perspective of the circular economy, able to demonstrate the environmental savings of adaptive reuse. Then, Foster et al. (2020) developed a Circular Environmental Impact Indicator Framework for cultural heritage adaptive reuse, in order to integrate macro European Union-level indicators with environmental indicators at the micro scale. A different study focused on the objectives of using circular economy indicators, classifying them based on a taxonomy of ten categories: levels, loops, performance, perspective, usage, transversality, dimension, units, formats, sources (Saidani et al. 2019). In particular, the social and cultural dimension of the circular economy seem to be less explored, with fewer consideration on the impacts on people and local communities (Kirchherr et al. 2017; Lemille 2017; Moreau et al. 2017), while first studies indicate that implementing circular economy (World Health Organization Regional Office for Europe 2018) determines a potential positive impact on human health and more generally between health and climate change issues (Watts et al. 2018a, b, 2021; Gupta et al. 2019).

The existing literature body on the linkages between cultural heritage adaptive reuse and circular economy shows a potential in building a comprehensive framework of quantitative and qualitative indicators to assess performances and impacts of adaptive reuse interventions in a circular economy perspective. The article of Bosone et al. (2021) analysed and classified existing heritage indicators, comparing them with circularity criteria. Through the analysis of 76 literature sources on cultural heritage impacts, the study explored how indicators are currently used in heritage research and practice as impacts assessment tools. More than 3500 indicators were retrieved and classified. The study explored the concept of circular economy in cultural heritage adaptive reuse, identifying specific impact criteria and highlighting the knowledge gaps for further research. Despite many indicators are already in use in the cultural heritage sector and attempts to systematize heritage indicators have been developed (Labadi 2011; Fusco Girard et al. 2015; Nocca 2017), the circular economy perspective still needed to be fully implemented in this field.

The aim of this chapter is to identify multidimensional evaluation tools, as criteria and indicators, according to the CLIC framework of circular adaptive reuse of cultural heritage, starting from the analysis of previous studies and ex-post evaluation of adaptive reuse practices, to structure a comprehensive operational framework for ex-ante evaluation and participatory decision-support in the perspective of circularity. The following sections present the methodology and results of the identification of specific, multidimensional criteria, as well as quantitative and qualitative indicators of circularity for cultural heritage adaptive reuse. Conclusions highlight the novelty of the CLIC approach in the evaluation and the perspectives for the implementation of the proposed framework.

2 Methodology

To make operational the theoretical CLIC framework for circular adaptive reuse of cultural heritage, a set of multidimensional criteria and quantitative and qualitative indicators was identified, to be used in ex-ante evaluations for the aim of enhancing decisions in the diverse phases of the adaptive reuse process. The methodological process adopted in the CLIC research followed diverse phases (Fig. 8.1):

- analysis of previous studies and scientific literature on cultural heritage and circular economy;
- definition of the CLIC theoretical framework (auto-poietic capacity, symbiotic capacity, generative capacity)—see Chap. 2 of this volume;
- ex-post evaluation of best practices and previous experiences of cultural heritage adaptive reuse, including "worst" practices;
- ex-ante evaluation framework for decision support: identification of main dimensions, criteria and indicators of circularity in cultural heritage adaptive reuse.

The methodological process adopted a complex dynamic in which the process was not linear, as each phase influenced the previous and following one following a "circular" research process. In fact, the analysis of previous studies was not only a preliminary phase to define the CLIC theoretical framework, but also a base for the definition of ex-post evaluation criteria and identification of suitable indicators for ex-ante evaluations. Moreover, the CLIC framework definition represented the starting point for the in-depth analysis of previous studies and evaluation tools, as well as for the definition of circularity criteria in ex-post and ex-ante evaluation, however it was also influenced by research results in each phase, integrating

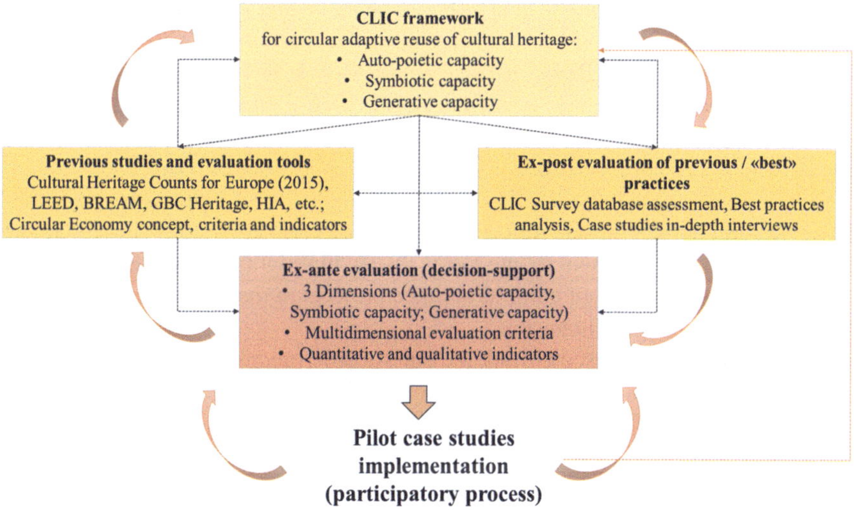

Fig. 8.1 Methodological process to identify multidimensional circularity criteria and indicators

relevant outputs and feedbacks and adapting through several refinements and assessments. In a dynamic and evolutionary perspective, which is characteristic of the circular processes, the final phase of pilot case studies experimentation provided relevant feedbacks for the further refinement and development of the CLIC framework, including criteria and indicators. This complex and dynamic research was addressed as a "back and forth" process, which provided a operational framework that can be adopted in "real" settings, as it was tested and validated also through practical experience, implementation and experimentation.

This reflection started from the theoretical definition of circular adaptive reuse of cultural heritage provided by Fusco Girard in the previous Chap. 2, and with particular reference also to Gravagnuolo et al. (2017), and further explored through the previous chapters of this volume. This chapter, in particular, moves forward towards operationalising the theoretical framework by focusing on previous studies (referring mainly to Bosone et al. 2021) and existing evaluation frameworks, and the ex-post evaluation of previous and best practices, to finally identify the proposal for ex-ante evaluation based on a multidimensional operational framework of circularity dimensions, criteria and indicators.

2.1 *Previous Studies on Evaluation Tools for Circular Adaptive Reuse of Cultural Heritage*

We define the circular economy in cultural heritage adaptive reuse as '*a sustainable economy that enables a continuous positive development cycles that preserve and enhance the created values, in an indefinite time, of cultural and natural capital, optimises resource yields and minimises system risks by managing finite stocks and renewable flows*' (Gravagnuolo et al. 2017). The main characteristics of the circular economy model for cultural heritage were identified, including diverse frameworks such as the 9 Rs approach (Reuse, Reduce, Reuse, Repair, Refurbish, Remanufacture, Repurpose, Recycle, Recover) (van Buren et al. 2016; Potting et al. 2017), and the ReSOLVE framework proposed by Ellen MacArthur Foundation (Regenerate, Share, Optimize, Loop, Virtualize, Exchange) (Ellen MacArthur Foundation 2015a, b, c). A first set of evaluation criteria for adaptive reuse of cultural heritage was developed and applied in scientific articles (Gravagnuolo et al. 2017; Fusco Girard and Gravagnuolo 2018; Bottero and Lerda 2019), highlighting potential and actual impacts of adaptive reuse of cultural heritage projects in the economic, social, cultural and environmental dimension.

According to Luigi Fusco Girard, the auto-poietic model of nature regeneration is embedded in the conceptual framework of circular adaptive reuse of cultural heritage (Fusco Girard 2020; Fusco Girard this volume, Chap. 2). As natural systems, the heritage site can be interpreted as a "lively regenerative/auto-poietic system", able to self-generate the resources needed for its functioning and to use all wastes as resources for new productive cycles. Through the circular economy approach applied, it is possible to interpret and evaluate the adaptive reuse process identifying diverse forms of capital that are re-generated: man-made capital, natural capital, social capital, human capital.

Based on this conceptual evaluation model of "Circular adaptive reuse of cultural heritage", and its initial implementation through case studies analysis, a structured set of evaluation criteria for circular adaptive reuse of cultural heritage can be identified. The main 'circularity dimensions' were considered:

- the 'regenerative capacity',
- the 'symbiotic capacity', and
- the 'generative capacity'.

The evaluation framework is based on a set of criteria and indicators that enable performance assessment of existing projects with respect to circularity objectives, and that can be used to orient choices towards circular «human-centred» adaptive reuse of cultural heritage.

The evaluation criteria proposed in the literature were synthesized and discussed during a series of Focus group sessions conducted with experts in heritage conservation, heritage economics, evaluation methods, circular economy, sustainable finance, economic spillovers of development projects and social impacts. General evaluation criteria resulted from different rounds of discussion, while indicators were synthesized based on existing literature and previous studies.

Indicators are synthetic tools to interpret reality: sound data collection, data analysis and data interpretation is needed to assess the indicators. The CLIC framework is based on three main types of indicators:

- Statistical indicators which are normally expressed as ratios or as percentages, allowing them to be assessed in relation to a baseline.
- Trends, whereby 'raw' numbers are monitored over time (e.g., number of visitors from one year to the next).
- Checklists which are not statistical (i.e., non-parametric), but enable some assessment of topics which cannot be captured through quantitative assessment (e.g., asking residents whether a certain cultural heritage site represents a factor of local identity). Even a checklist requires supporting evidence to permit validation of the responses.

Indicators can be diverse according to the evaluation phases: ex-ante (decision-making) and ex-post (impacts assessment) (Fig. 8.2).

The indicators have thus a double use: in the ex-ante phase, they express the goals and expected results that the circular adaptive reuse will achieve, and in the ex-post phase they become a tool for monitoring and assessing the results, providing evidence-base of actual results to stakeholders.

As showed in Fig. 8.2, the indicators set was developed also considering the diverse phases of the adaptive reuse intervention, mainly:

- Planning and design
- Building construction works
- Operations/management

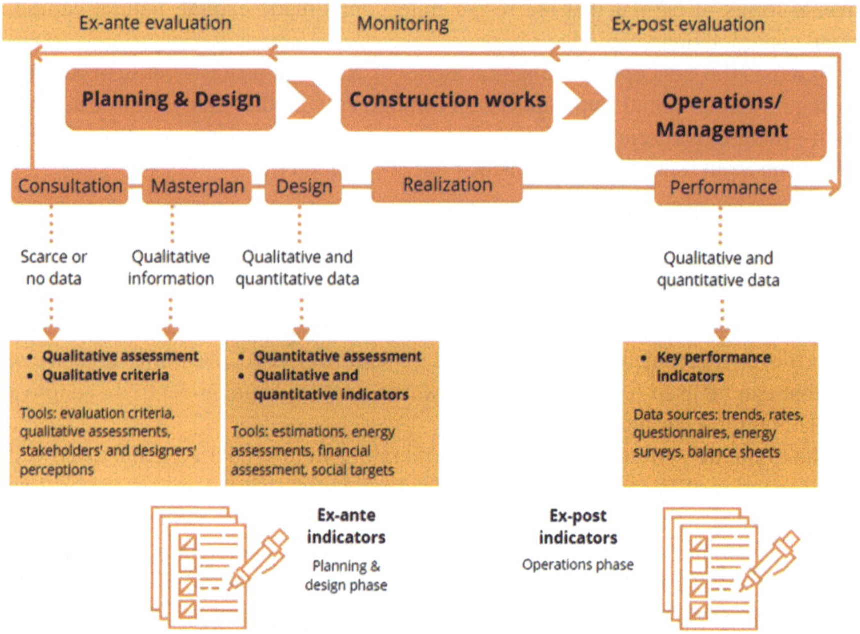

Fig. 8.2 Indicators set structuring according to the adaptive reuse phase and data/information availability. (Source: Gravagnuolo and Fusco Girard 2021, CLIC Deliverable D2.4 Database of indicators and data)

In each phase, diverse types of data and indicators may be available, while the circularity dimensions and criteria to be considered should be coherent in all phases. Also, the scale of the adaptive reuse intervention should be considered when adopting an indicators set. In fact, small interventions cannot be assessed based on the same indicators used in large-scale adaptive reuse projects, as some criteria may be more relevant than others in diverse typologies of adaptive reuse.

The relevance of criteria, and thus the weight assigned to each of them, can be highly variable, also according to the needs and preferences of stakeholders and local communities. The weight of criteria, and even criteria and indicators, could be variable in relation to different phases of the evaluation process itself, as some "hidden" information or data may arise during the process, supported by evidence and stakeholders involvement, and can thus lead to a change in preferences and thus evaluation results. This "adaptive" and evolutive process of evaluation and co-evaluation can take some time and effort for decision makers, however the benefits of reaching a more desirable solution for more stakeholders, and a more effective use of the financial resources available, should encourage public and private actors to adopt it.

Criteria and indicators were identified and synthesized, adapting them to the specific case study. Below are presented the criteria and indicators identified in line with the CLIC framework, to be applied in ex-ante and ex-post evaluations.

Previously published scientific articles analysing evaluation tools for cultural heritage adaptive reuse represented a base for the development of the CLIC evaluation framework. Particularly, Fusco Girard (2020), Gravagnuolo et al. (2017) and Bosone et al. (2021) summarised the interrelationships between circular economy and cultural heritage adaptive reuse, proposing a set of multidimensional criteria based on the CLIC theoretical approach. Particularly, the paper of Bosone et al. (2021) identified a set of criteria according to an extensive literature review, including research articles and grey literature (Fig. 8.3).

In the present work, an overview of existing protocols and tools for the evaluation of cultural heritage is presented (Box 8.1).

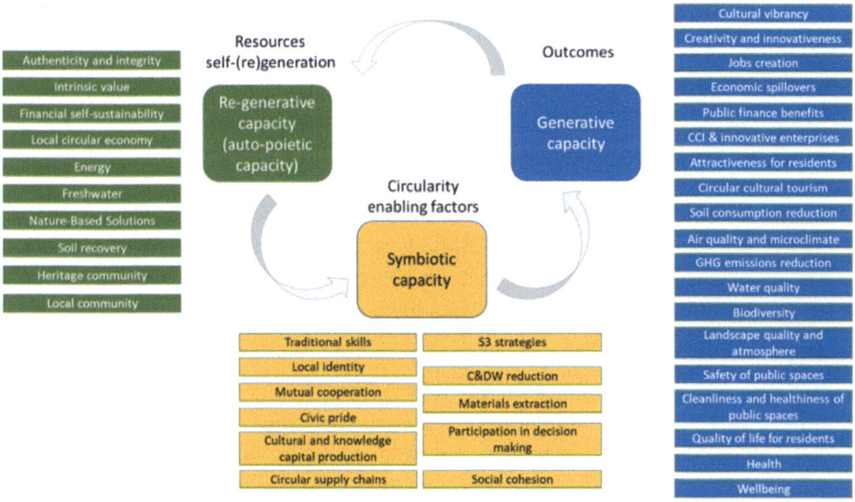

Fig. 8.3 Evaluation criteria of circular adaptive reuse of cultural heritage. (Bosone et al. 2021)

Box 8.1: Sustainability Protocols for Buildings

The sustainability protocols can be viewed as helpful instruments to explore initiatives for promoting the sustainable building culture (Gaballo et al. 2021).

LEED[1]

The LEED protocol was developed in 1993 in the United States, by the USGBC (United States Green Building Council). LEED certification is adapted by different countries according to their own regulations.

Working on the entire process, from design to actual construction, LEED requires a holistic approach.

(continued)

[1] https://gbcitalia.org/certificazione/leed/; https://greenwichsrl.it/certificazione-leed/; https://www.webuildvalue.com/it/curiosita-infrastrutture/certificazione-leed.html

Box 8.1 (continued)

The aim of the LEED protocol is to increase the level of energy and water savings in buildings while decreasing carbon dioxide emissions. That's not all: LEED protocols also take into account the ecological quality of indoor environments, site selection, and so on.

The LEED certification system is based on the awarding of a number of LEED® credits for each of the requirements characterising the sustainability of a building, of which there are eight (according to version 4.0 of the US protocol): (1) Transport and Location (LT), (2) Site Sustainability (SS), (3) Water Efficiency (WE), (4) Energy and Atmosphere (EA), (5) Materials and Resources (MR), (6) Indoor Environmental Quality (IEQ), (7) Innovation (I), (8) Regional Priority (PR).

In each area, there are several items, each of which can receive a different score. The sum of the credits gives an overall score: above a certain point threshold, certification is awarded. The four levels of LEED certification are: "Certified", "Silver", "Old", "Platinum".

The weight of the different areas is not homogeneous: the proportions change according to the type of buildings to be certified. However, the most important area is the energy aspect, which can weigh up to 30% of the total score, while the regional priorities area has a weight of less than 4%.

BREEAM®[2]

BREEAM certification is based on a comprehensive and integrated sustainability assessment methodology. It applies to various types of buildings or parts of buildings—whether new, redeveloped or existing—and to interior spaces. BREEAM assessment criteria include: Energy and water use, Indoor environment (health and well-being), Pollution, Transport, Materials, Waste, Ecology, Management processes.

One of the features that make BREEAM a versatile tool is that each assessment area has a different weighting depending on the application context and a national or local reference standard can be used to demonstrate the required performance. This results in great flexibility, allowing the protocol to be adapted to all countries. In Italy, protocols and technical manuals defined as 'International' apply, since Italy does not have country-specific protocols: NC International, In Use, IRFO, Community.

In the certification report prepared by the BREEAM Assessor and finally submitted to the BRE (Building Research Establishment Environmental

(continued)

[2] https://bregroup.com/; https://greenwichsrl.it/en/breeam-certification/; https://www.tuvsud.com/it-it/settori/real-estate/costruzioni/sistemi-di-valutazione-della-sostenibilita/schema-breeam

Box 8.1 (continued)

Assessment Method), the weighted score of BREEAM credits for each of the BREEAM categories is indicated.

The sum of the weighted credits determines the building's environmental performance (rating) and the corresponding certification level: "Pass," "Good," "Very good," "Excellent," and "Outstanding".

The aspects that characterise buildings certified with BREEAM are:

- Low environmental impact;
- The use of best practices and sustainable building products;
- The use of innovative solutions for building sustainability;
- Quality that exceeds legislative standards;
- Solutions to reduce costs

ITACA[3]

The Itaca Protocol is a tool for assessing the level of energy and environmental sustainability of buildings. It represents a guide for orienting design choices towards a coordinated system of virtuous solutions aimed at saving natural resources and quality living.

The Itaca Protocol is derived from the SBTool international assessment model, developed as part of the Green Building Challenge research process, and contextualised to the Italian territory in relation to the reference legislation and its environmental characteristics. The Itaca Protocol was born a few years ago out of the need for Italian regions to equip themselves with valid tools to support territorial policies to promote environmental sustainability in the building sector.

The protocol envisages that the performance of a building is evaluated on the basis of its consumption, energy efficiency, impact on the environment and impact on human health.

The aim is to encourage the construction of increasingly innovative, zero-energy buildings with low water consumption, as well as materials that in their production involve low energy consumption and at the same time guarantee high comfort.

The Protocol also guarantees the objectivity of the evaluation through the use of indicators and verification methods that comply with the technical standards and national reference laws.

The principles on which the tool is based are:

(continued)

[3] https://www.proitaca.org/guida-al-protocollo-itaca.php;https://biblus.acca.it/focus/protocollo-itaca-cosa-serve-e-come-si-usa/;

Box 8.1 (continued)

- identification of criteria to measure the various environmental performances of the building under examination
- definition of the reference performance (benchmark) against which the building's performance can be compared in order to assign a score corresponding to the relationship between the performance and the benchmark
- weighting of the criteria determining their greater and lesser importance
- final synthetic score defining the degree of improvement of the performance as a whole considering the standard level

For the assessment of environmental sustainability, the protocol adopts a multi-criteria analysis system, structured according to four hierarchical levels: Tools, Areas, Categories and Criteria.

The building performance score is calculated through an evaluation procedure that consists of 3 consecutive steps

1. characterisation: the building performance for each criterion is quantified through appropriate indicators
2. normalisation: the value of each indicator is made dimensionless and is 'rescaled' into a normalisation interval
3. aggregation: the normalised scores are combined together to produce the final score

For each criterion is defined:

- code and criterion name
- assessment area and category
- requirement (i.e. the quality objective to be pursued)
- performance indicator and its unit of measurement (if quantitative in nature)
- reference performance scale to be used for the normalisation of the indicator in the range from −1 to +5
- method and verification tools to calculate the value of the performance indicator

Through the building specific parameters, elaborated according to the procedures foreseen by ITACA, the Building Performance Indicators are calculated for each Criterion of the ITACA Protocol. The Performance Indicators may be quantitative or qualitative, depending on the type of Criterion and the relevant legislative and regulatory references. The Performance Indicator, related to the relative Performance Scale (benchmark), determines the Criterion Score between −1 and 5 (Rating Scale). The Score represents the normalisation of the value of the Performance Indicator in the ITACA Protocol and allows both the processing of all the scores of the building in aggregate form, and the comparison of performance between the Criteria. The Scores are determined through the Rating Scale (Table 8.1), where performance levels are defined.

(continued)

Box 8.1 (continued)

Table 8.1 The ITACA protocol rating scale

−1	Performance below standard and current building practice
0	Minimum acceptable performance defined by current laws or regulations. Represents current construction practice
1	Slight improvement in performance compared to current regulations and current construction practice
2	Moderate improvement of performance compared to current regulations and current building practice
3	Significant improvement in performance compared to current regulations and current construction practice
4	Moderate improvement over current best construction practice
5	Significantly improved performance compared to current best construction practice Experimental

The Global Score represents the environmental energy sustainability performance of the entire building according to the Evaluation Scale adopted by ITACA. It also represents the threshold set, from time to time, by regional regulations that provide for the granting of volumetric, fiscal and in some cases economic incentives.

Some considerations

In fact, the Ellen MacArthur Foundation (EMF) has identified ten policy levers (Ellen MacArthur Foundation 2013), including such protocols, as viable economic incentive levers with a view to designing, evaluating, and certifying the built environment's sustainability through a set of performance standards (Abastante et al. 2020).

All of these protocols are based on standardised evaluation processes but, in their most recent versions, criteria typical of the circular economy model are present: e.g. the use of energy from renewable sources, water reuse, waste recycling, material reuse. As previously illustrated. These protocols are based on the weighted summation evaluation system which is very simple, operational and understandable by all involved people. Really, it can be noted that there is the risk of leading of some undesirable results, following purely mathematical steps rather than taking into account the actual and natural use of weights by human beings (Zeleny 1991).

Multicriteria evaluation can be particularly suitable for cultural heritage adaptive reuse, as it is able to combine diverse evaluation dimensions and criteria, addressing complex decision problems (Fusco Girard 1987; Fusco Girard and Nijkamp 1997; Nijkamp 1977). According to Cultural Heritage Counts for Europe research (2015), multicriteria evaluation methods are currently not used in participatory processes, thus further research should be implemented to apply them in cultural heritage adaptive reuse processes that intend to engage stakeholders and local communities for shared decision-making. However, recent literature presents examples of application of multicriteria evaluation methods for participatory processes of cultural heritage adaptive reuse (e.g. Gravagnuolo et al. 2023 forthcoming). The CLIC research explored multicriteria evaluation as a possible tool for enhancing decision making engaging relevant stakeholders and communities.

2.2 Ex-Post Evaluation of Adaptive Reuse Practices

The ex-post evaluation phase is characterized by the use of indicators able to provide evidence base of the results achieved, to compare the expectations with the actual results and eventually adapt strategies and management models to achieve the goals. The matrix of quantitative and qualitative indicators proposed for ex-post evaluation is based on the CLIC evaluation framework, refined through the assessment of best practices and in-depth case studies of cultural heritage adaptive reuse described in previous sections. The matrix of indicators for ex-post evaluation was built through the collaboration of CLIC researchers and experts in diverse sectors, from heritage sectors to economics and finance, social science and ecological economics/environmental science, covering multiple dimensions, tools and approaches.

The set of multidimensional and multicriteria indicators is meant as a useful orientation for heritage managers to make sense of the diverse impacts of cultural heritage adaptive reuse, supporting circular management and business models which are based on the recognition of the positive and negative externalities of human activities, and their "internalization" in a systemic social-ecological model that takes into account the business as integral/interconnected part of the larger societal and ecological system, and thus responsible for its functioning and overall sustainability.

The high variability of heritage typologies, scales, historic-cultural value, level of conservation, does not allow to identify a large set of universally applicable indicators. The additional information elaborated for each indicator is aimed at facilitating the use of the tool in a flexible way, allowing "adaptation" in the selection of indicators which can be selected by site managers based on the scale and adaptive reuse phase.

Ex-post evaluation regarded a more in-depth investigation through the collection of detailed data on a smaller set of practices, shifting from a qualitative-based assessment to a quantitative and qualitative assessment. In particular, some of the aspects considered more relevant to be explored through ex-post evaluation are related to social and cultural impacts, environmental performance (e.g. energy, water, materials), as well as economic-financial results and management models.

The results of the CLIC Survey were explored further through 10 case studies, for which more detailed information was sought through in-depth interviews with site managers. The interview for managers was structured in different sections related to:

- Revenues and costs
- List of activities generating revenue or supported
- Overall yearly income
- Type and mission of the organization
- Detail of operating costs
- N. of jobs/permanent jobs/volunteers
- Funding information: which type of funding, purpose of funding (renovation, management, other)
- Fundraising
- Functions/uses of the place (tourism/recreation, education, cultural, production, residential/accommodation, community hub)
- Audience (how many clients or people in the community)
- Synergies and collaborations with other organizations

The list of case studies investigated is reported in Table 8.2.

A synthetic information is provided below for each case study investigated.

Villa Campolieto is a Vesuvian villa located along the "Miglio d'Oro" historic road, so defined for the historical and landscape beauty and the presence of Vesuvian villas, in the municipality of Ercolano, nearby Naples in Italy. The Villa and its garden dates back to XVIII Century. The site was restored since the 70s through diverse interventions funded by the Italian Ministry for Culture, Cultural Heritage and Tourism. Today the Villa is included in a network of four historic sites in the

Table 8.2 Cultural heritage adaptive reuse case studies investigated for ex-post evaluation

N.	Adaptive reuse project	City	Country	Organization	Role of interviewee
1	Villa Campolieto	Ercolano	Italy	Fondazione Ente Ville Vesuviane	Director
2	H-Farm	Treviso	Italy	H-Farm	EU projects responsible
3	Palazzo Innovazione	Salerno	Italy	Palazzo Innovazione	CEO
4	Not Quite	Fengersfors	Sweden	Enderlien & CO	Owner of the mill in Fengersfors
5	Pfefferberg	Berlin	Germany	Pfefferberg Foundation	Director
6	Catacombe di San Gennaro	Naples	Italy	Catacombe di Napoli	President
7	Pakhuis de Zwijger	Amsterdam	Netherlands	Tertium	Owner director
8	Musil	Brescia	Italy	MUSIL	Director
9	Toolbox	Torino	Italy	Toolbox	CEO
10	Edit	Torino	Italy	EDIT	CEO

area of Ercolano and it hosts cultural and educational events, as well as private events and training courses. The last interventions included the implementation of a special photovoltaic system ensuring very low impact on the cultural value and aesthetic of the building, as well as low energy consumption lighting system. The ancient water recovery system was also recovered and provides irrigation water for the garden and green areas. The management is fully public, exploiting the opportunities offered by access to additional funding for cultural events and other revenues from private events. The adaptive reuse of the historic Villa Campolieto led to the attraction of almost 10.000 visitors per year, and contributed to the urban area regeneration attracting foreign and domestic cultural visitors and new commercial activities.

H-Farm was established in 2005 on a former agricultural farm as an incubator of innovative SME in the field of ICT, developing into a diversified complex, including education activities. H-Farm is at present a innovation hub where Innovation, Entrepreneurship and Education are combined together, in the historic rural environment. The main premises of H-Farm were built in the agrarian land surrounding Venice and Treviso, around a typical rural building in need of conservation. The implemented restoration was conducted with careful attention to respect the historic-cultural values of the site, linking the conservation with contemporary design and environmental sustainability through the adaptive reuse. The current extension of the complex, including newly designed and built offices, is fully integrated with the natural assets. H-Farm is originally a ICT consultancy company, but the new localization in the historic rural area of Treviso in Italy promoted the diversification of the company's activities, the creation of a large community of innovators and a hub of entrepreneurial education for all ages, and the revitalisation of the rural area. The rural buildings in the area of H-Farm include co-working spaces, offices, guest rooms, bar and restaurants, and open spaces for relax in nature and gatherings, providing opportunities to the local community to take part in free innovation events.

Palazzo Innovazione is a co-working space and headquarter of Healthware company, which located in the historic building of the ex-Convent "Santa Sofia" in Salerno, Italy. The Convent was built at the end of the tenth century as the first monastery of the Benedictine Order. The Convent hosted many functions over time, up to the latest adaptive reuse as digital health enterprise headquarter, business incubator and co-working space. The private company managing the site invested about 800.000€ for the renovation of the building, which is used partially as its own headquarter—this allows to pay a rent of 6.000€ per month to the municipality who own the historic building. In the other parts, the building hosts a co-working space and incubator for startups in digital health sector, with a bar/kitchen and facilities for the enterprises. The place offers free educational events on entrepreneurship and digital innovation each week, opening-up its doors for interested citizens. This activity is part of its marketing strategy, linked to the consulting services offered for the enterprises of the territory. The success of these initiatives lies in the balance of private and public interest. The uses/functions are in line with the city strategic policies, and an additional benefit for the municipality is the private investment that the companies allowed to repair and maintain the buildings (Lupacchini and Gravagnuolo

2019). The adaptive reuse of the ex-Convent as "Palazzo Innovazione" contributed to the urban regeneration of the historic city centre, providing new opportunities for the local community and strengthening the local innovation ecosystem.

Not Quite is an artists' community located in Västra Götaland region, in a small village (350 inhabitants) surrounded by forests in the Swedish countryside. Not Quite is at the forefront in the development of an industrial heritage site into an art and culture driven community, exhibition and workshop spaces. The adaptive reuse of the industrial building conducted by Not Quite aimed at maintaining the buildings in their historic shape, avoiding rebuilding but valorising the industrial atmosphere of the place. The industrial site includes diverse buildings renovated, in a post-industrial area which needs extensive remediation. The private owner rents the buildings to the collective of artists, that manages the membership fees, workspaces, the bar and the shop. Many artists involved live in the village area. Educational projects are developed with local schools. Initial investment risk was lowered thanks to regional rural development funds through which the region provided guaranteed access to finance, fully repaid over the last years. A crowdfunding campaign was started to collect the investment capital to buy the old mill[4]. Guided tours and recreation activities are active in the site, with around 30,000 people participating each year. Educational and training courses and conferences are organized, where around 2000 cultural workers come and spend a weekend at the site. A informal heritage community is active in the place, specifically there is a working group on local identity aiming at collecting and disseminating the stories from the ones who worked at the mill in the past. The old paper machines are conserved in the site, providing a special setting for the artworks along with the industrial remains.

Pfefferberg is a old brewery with restaurants founded in 1841 in Berlin, Germany. The site has a long story of adaptive reuses over time. It turned into a enterprise for chocolate production and after that into a bakery, with offices, storerooms and restaurants. After the World War II, it became a printery with storerooms, garages, workshops and offices. In the late 80s the idea of "Factory of Culture" was launched and the Pfefferwerk association and Pfefferwerk GmbH (Ltd.) were founded aiming to carry out social and cultural activities on Pfefferberg area. In the 90s the Pfefferwerk Ltd. Bought the site with subsidies from Berlin Senate and private contribution, later the Pfefferwerk Foundation was established to manage the site under the "heritage building right" regulation. A leasehold contract was agreed with the construction company for the renovation of the site and the rental services to be started, maintaining the intended use as established by Berlin Senate: culture, art, social services, small businesses. During the 2000s the site was renovated and the Foundation started its activities to manage the diverse buildings with various uses. A number of 24 buildings is managed in Pfefferberg, with around 40 enterprises running the cafés, hostel, workspaces, exhibition spaces, offices and other cultural activities, sharing social criteria for running the activities, as

[4] https://bitforbit.notquite.se/

improving the chances of disadvantaged people to enter the labour market, thus generating a high social impact. Also, the Foundation invests in social projects external to the site of Pfefferberg, generating additional positive impacts in the urban area. Thanks to the success of the site with almost 180,000 users each year, Pfefferberg can rely on a sustainable business model able to support the social activities conducted and the renovation and maintenance of the heritage site.

The **Catacombe di San Gennaro** are a archaeological heritage site which dates back to the second and third centuries AD, recovered and reused as a cultural tourism and community place in a disadvantaged Sanità district in the city of Naples, Italy. The archaeological site includes the Catacombs of San Gennaro and the Catacombs of San Gaudioso and two other Basilicas annexed to them. They represent one of the oldest monuments of Christianity in Naples. In addition to the archaeological heritage of the Catacombs, other spaces were recovered for a total of 12,000 square meters. The site hosts around 160,000 visitors a year with a large economic impact in the urban area. The archaeological site represents also an engine to which other "poles" are economically and financially linked, such as the guest house and the laboratories for social, cultural ed educational activities. The Catacombs generated 40 new jobs for disadvantaged youths in this area of Naples over the last 10 years, stimulating also cultural and social development in the area. The initiative was taken by the local Church leader, starting with the training of a first group of youths as local guides and the renovation of the site promoted thanks to a first funding from Foundations. From 2006 to 2008 the training and experimentation of guided tours with volunteers was started at the Catacombs of San Gaudioso. In 2009 the Catacombs of San Gennaro were opened to the public. In 2006, "La Paranza" cooperative was founded by a group of 5 young people and the local Church leader, which has now 40 employees. In 2008, the "Officina dei Talenti" cooperative was also founded, where the young people with the greatest manual skills were brought together in order to take care of the maintenance activities of the spaces. "Officina dei Talenti" has now 15 employees and in addition to taking care of the Catacombs spaces, it has also private contracts for the recovery of residential units in the area, turning into a well-established construction company specialized in recovery works. From the experience of the artist Riccardo Dalisi, another cooperative called "Iron Angels" was founded, a craft workshop that reuses waste materials for the creation of works of art, employing currently other 3 local youths. The great generativity of this experience can be observed also in the visual enhancement of the neighbourhood, with the collaboration of local owners of commercial units and the entire community working to make the historic urban area clean, safe, green and more and more beautiful. In 2014, the Community Foundation "San Gennaro Onlus" was created, composed of both the non-profit sector and the local owners of commercial units. The Foundation has invested over 4,5 M€ on the whole Sanità district coming from fundraising activities and donations for the recovery of the squares, recovery of the heritage sites, and the recovery of other spaces for cultural and social activities. The large social and economic impact of this experience, conducted in a disadvantaged urban area with high criminality and poor education levels, makes it one of the best examples of "circular" adaptive reuse of cultural heritage.

Pakhuis de Zwijger is a cultural organization which opened its doors in 2006. It is located in a industrial heritage building listed as national heritage, a former storage warehouse in an area which was previously an abandoned port area called the Eastern Docklands. Over the past decades Pakhuis de Zwijger has grown to become a fundamental cultural organizations within the city of Amsterdam. At Pakhuis de Zwijger there are over 600 public, freely accessible events per year, that all aim at analysing, rethinking and redesigning solving the biggest urban challenges of our time. Sustainable solutions and true innovation is promoted by including all urban stakeholders in critical, constructive and action-oriented dialogue. These include the traditional parties -the (local) government, (big) businesses and academia- but also (social) entrepreneurs, societal organisations and above all: the citizens. Throughout the year, Pakhuis de Zwijger organizes various event series, city expeditions and festivals that aim to inspire, activate, inform, connect and empower citizens and stakeholders to arrive at "better cities for all". The creative and innovative approach to these matters has become the trademark of Pakhuis de Zwijger in Amsterdam, participating as key stakeholder in the Circular Amsterdam initiative. Stadsherstel, a public limited non-for profit company, manages the building implementing also renewable energy and retrofit interventions. A large investment of about 13 M€ was realized to renovate the building, combining private investment, grants from local government and subsidies to make sure that some areas in the building have a social price for rental. Pakhuis de Zwijger programming generates social value, produces knowledge, insight, creates a larger network, contributes to enhance the local community, to enhance knowledge in the local community, and provides a platform to initiatives, voices, and narratives, which otherwise would not have a large platform. The platform and expertise of Pakhuis de Zwijger is offered to diverse initiatives so they can enhance their impact, bringing people together to come to creative and innovative solutions for urgent and present-day challenges. Pakhuis de Zwijger builds *communities of practice,* and works together with partners to create programming on urgent matters concerning urban development.

The **MUSIL** (Museum of industry and work in Brescia, Italy) is a little system of 4 industrial museums. The project began in the 1980s and continued with the first collections at the beginning of the 1990s.The musil is the first Italian museum dedicated to industrialization as a phenomenon that involves the whole of society, represented through a great variety of materials, conceptually organized in a unitary way. The Foundation, established in 2005, has the task of jointly managing the 3 existing museums and coordinating the activities for the construction of the headquarters in Brescia city. Specifically, the Foundation is responsible for setting up the Museum Headquarters. The museographical path is linked to water, from its formation in the atmosphere to its fall on the earth—its condensation on ice, its gathering in the alpine lakes, the harnessing in artificial dams until the arrival in the hydroelectric power station, where it is turned into electricity. The communication of scientific contents is offered takes through a dialogue with technological artifacts full of history. The voice if workers, technicians and ordinary people provide the social frame of this epic history. In 2015, the museum was recognized as an anchor point of the European Industrial Heritage Route, the European "road" that collects the main sites of the continental industrial heritage. The project is fully funded and managed by public bodies and hosts 4000 visitors per year, educational and cultural activities. The public oriented governance model guarantees the functioning of the site, even if the generativity of the reuse intervention remains limited.

Toolbox industrial building was built starting in 1915 and is part of a much larger complex of about 51,000 square meters hosting industrial and manufacturing activities in Torino, Italy. The company marked an important activity for the city on a national and international level and collaborated in the expansion of Turin linked to the presence and success of the FIAT group's automotive industry. In addition to the phase of maximum building development of the area, 1962 represented the beginning of the deindustrialization process that went through several steps and changed the ownership of the site, until reaching the abandonment and degradation of the entire area at the end of 2007. The Toolbox proposal of coworking managed to take shape by proposing, in a complex period of economic crisis, a valid alternative to the work difficulties of those years. Toolbox represented the first structured coworking created in Italy, capable of offering new working areas able to attract and welcome freelancers in a dynamic space, in which it was possible to meet similar people, oriented towards the future, with which to exchange ideas, dreams and practical advice to in the face of a limited expense for the workstation. Toolbox was in fact built through step-by-step investments, and its size has expanded with the increase in demand, thus allowing to limit investment risks. The various adaptive reuse interventions have made it possible to recover the large spaces of the former Carlo Garrone foundry, which are the new areas of the coworking stations surrounded by "service boxes", volumes of limited dimensions designed to encourage socialization. Inaugurated in 2010, with the first intervention of a thousand square meters, Toolbox today covers an area of 10,000 square meters with a total investment of 3 M€ (Bottero and Lerda 2019). The philosophy of Toolbox is centred on the "economy of relationships", on values that enable the shifting from the sharing economy to the "caring economy". Users co-create the meaning and atmosphere of the place, co-generating added social value and a self-sustainable business model led by a socially-oriented private entrepreneur.

EDIT is a former industrial site in the city of Torino, Italy, built in 1888 in a block that is now known as the former INCET area, a factory that for years produced and exported internationally the technological innovation products of the period, such as wires for transmission of electricity, cables for telephone and telegraph transmission and ropes for transportation. INCET was an industrial symbol of the city of Turin and it contributed to the urban expansion of the Barriera di Milano industrial district. The industrial production lasted until 1968 when it was then moved elsewhere and the buildings were abandoned, starting a period of decay of the entire urban area for over 40 years. Today the Barriera district of Milan is a symbol of urban rebirth and EDIT represents the latest reconversion and redevelopment project that has allowed the old beauty of the industrial complex to be brought back to life, offering an innovative concept in the food and beverage sector. The adaptive reuse works were completed between 2015 and 2017 and involved an area of 5000 square meters, with a total investment of almost 12 M€. Before the intervention, the building consisted of an industrial volume with a rectangular plan in a severe state of neglect; this has been recovered in its entirety while maintaining its original volume. Internally, it was possible to exploit the height to insert a new slab, re-proposed in an industrial nature, in order to double the walkable surface. The original perimeter walls were covered with insulating panels, to improve the energy efficiency of the building, and walls and volumes were inserted from scratch to ensure the development of different areas but with a visual continuum, respecting the industrial heritage structure in a dynamic and flexible reality. The main functions of EDIT include areas intended for bakery cafes, restaurants and pubs with an adjoining brewery also made available to external users who can use them to produce their own beers with the help of master brewers. The offer also expands with the presence of four kitchens for workshops and show cooking events. The EDIT project has made it possible to bring an old dilapidated building back to life, giving it the opportunity to host a new innovative function that can increase the attractiveness of the neighbourhood, together with the other adaptive reuse interventions in the same area (Bottero and Lerda 2019). The place promotes a local economy based on local food and beverage products, establishing relationships with high quality product enterprises.

Table 8.3 presents the key data summarizing the in-depth adaptive reuse case studies analysed and the impacts generated.

Some case studies analysed demonstrated that the adaptive reuse and valorisation of heritage resources was a driver for the entire neighbourhood revitalisation, stimulating residents and owners of commercial activities to invest in the refurbishment/renewal of surrounding buildings, streets, green areas and public spaces. This was the case in particular for the "Catacombs of San Gennaro" in Naples, Italy, where a socially and physically degraded neighbourhood was substantially regenerated, enhancing the cleanliness, safety and beauty of the area and thus attracting more commercial activities and visitors nearby the heritage site. Thus, the indicators set proposed and the data collected through interviews with heritage site managers aim to capture the multidimensional impacts of cultural heritage adaptive reuse.

The relatively high number of "n.d." (no data), especially in the section regarding environmental data and circular technologies, makes clear how difficult it is, even for the managers of heritage sites, to realize the information needed for their sustainable management. Clearly, the sample of ten case studies is not representative of adaptive reuse projects, however a general lack of quantitative data was present also in similar studies, such as Cultural heritage Counts for Europe (2015), thus the difficulty in identifying quantitative data on several aspects related to the environmental dimension seems coherent with the CLIC results. The lack of skills and competences in environmental management could be probably a motivation for the lack of environmental data, even if common information such as the Building Energy Performance Level should be available for buildings in European countries. During the interviews conducted, it was observed a scarce attention of some site managers to the environmental circularity aspects of the adaptive reuse intervention,

Table 8.3 Indicators and data used for analysing in-depth the circularity of ten cultural heritage adaptive reuse projects

	DATA	Villa Campolieto	H-Farm	Palazzo Innovazione	Not Quite	Pfefferberg	Catacombs of San Gennaro	Pakhuis de Zwijger	Musil	Toolbox	Edit
GENERAL INFORMATION	**Country**	Italy	Italy	Italy	Sweden	Germany	Italy	Netherlands	Italy	Italy	Italy
	Region	Campania	Veneto	Campania	Vastra Gotaland	Berlin	Campania	North Holland	Lombardia	Piemonte	Piemonte
	City	Ercolano	Treviso	Salerno	Fengersfors	Berlin	Naples	Amsterdam	Brescia	Torino	Torino
	Urban context	Medium-sized city	Small city, Rural	Medium-sized city	Small city, Rural	Large city	Large city	Large city	Medium-sized city	Large city	Large city
	Heritage type	Residential Palace	Rural heritage	Monastery	Rural heritage	Industrial heritage	Archeological heritage	Industrial heritage	Industrial heritage	Industrial heritage	Industrial heritage
	Adaptive reuse (uses)	Cultural events, Education, Visits	Headquarter, Digital hub, Co-working, Entrepreneurial hub and school	Headquarter, Digital health hub, Co-working, Events	Co-working, Creative hub, Workshop space, Exhibition	Cultural events, Workshop spaces, Art galleries, Hostel, Bar & restaurant	Cultural tourism, Education hub, Hospitality	Cultural events, Dialogue platform, Bar & restaurant	Museum	Co-working, Events, Leisure, Bar & restaurant	Beer production, Bar & restaurant
	Adaptive reuse period	1970–2010 (various phases)	2000–2005	2017	2003	2000	2005–2010	2000s	2000	2006–2010	2015

MANAGEMENT	**Ownership**	Public	Private	Public	Private	Foundation	Church	Private	Public	Private	Private
	Management model	Public supported	Rent	Rent	Rent	Rent	Use concession	Rent	Public supported	Rent	Rent
	Manager organisation	Foundation	Private for profit	Private for profit	Private for profit	Foundation	Private non profit	Private non profit	Public	Private for profit	Private for profit
	Governance model	Public custodian	Private custodian	Private custodian	Private custodian	Community custodian	Community custodian	Community custodian	Public custodian	Private custodian	Private custodian
	Funding sources	Public	Private	Private	Private	Public	Community	Private	Public	Private	Private
	People involvement in decision making	No	No	No	No	No	No	Yes	No	No	No
	Total investment	7 M€	1–3 M€	800,000€	1,5 M€	100 M€	1–3 M€	>10 M€	1–3 M€	<1 M€	>10 M€
	Sqm adaptive reuse	2500	4800	4000	15,000	22,000	6000	6200	3200	10,000	5000
BUSINESS MODEL	**Leadership**	Foundation	Entrepreneur	Entrepreneur	Entrepreneur	Civil society	Church	Activist	Municipality	Entrepreneur	Entrepreneur
	Activities non generating revenue	Yes	Yes	Yes	Yes	Yes	Yes	Yes	Yes	Yes	Yes
	Yearly overall income	300,000€	60 M€	300,000€	275,000€	350,000€	860.000€	2.000.000€	100.000€	n.d.	3.000.000€
	Operating costs	300,000€	n.d.	250,000€	270,000€	100,000€	700.000€	1.600.000€	90.000€	n.d.	5.000.000€
	Reinvestment in the urban area	Yes	Yes	No	Yes	No	Yes	No	No	No	No
	Jobs (direct+indirect)	8	25 + 650	3	1	2 + 500	40	100	3	n.d.	50
	Volunteers	No	No	No	30	5	10	No	2	No	No
	Fundraising	Yes	No	No	Yes	Yes	Yes	Yes	Yes	No	No

(continued)

Table 8.3 (continued)

	DATA	Villa Campolieto	H-Farm	Palazzo Innovazione	Not Quite	Pfefferberg	Catacombs of San Gennaro	Pakhuis de Zwijger	Musil	Toolbox	Edit
ACTIVITIES AND OUTCOMES	**N. enterprises localised**	No	100	10	n.d.	40	3	3	No	200	n.d.
	Cultural tourism	Yes	No	No	Yes	Yes	Yes	No	Yes	No	No
	N. of visitors	10.000	200	800	30.000	180.000	160.000	No	10.000	No	No
	Education activities	Yes	Yes	No	Yes	Yes	Yes	Yes	Yes	Yes	Yes
	N. of people receiving education	5000	500	No	20	300	n.d.	1.000	400	400	100
	Cultural activities	Yes	Yes	No	Yes	Yes	Yes	Yes	Yes	Yes	Yes
	N. of participants in cultural activities	n.d.	4000	No	2000	n.d.	n.d.	77.000	300	8.000	500
	Productive activities	No	Yes	Yes	Yes	Yes	No	Yes	No	Yes	Yes
	Residential/ hospitality function	No	Yes	No	No	Yes	Yes	No	No	No	Yes
	Community activities	No	No	No	Yes	Yes	Yes	Yes	Yes	No	No
	N. organisations collaborating	n.d.	n.d.	n.d.	40	100	22	460	5	200	n.d.
	N. people in the community	n.d.	n.d.	n.d.	100	n.d.	60	80.000	50	600	n.d.
	Heritage Community	No	No	No	Yes	No	Yes	No	Yes	No	No

ENVIRONMENTAL DATA AND CIRCULAR TECHNOLOGIES	**Energy need**	60 KW	n.d.	n.d.	200 KWh	n.d.	n.d.	1.025.000 KWh	6.000 KWh/ sqm	n.d.	n.d.
	Renewable energy sources on site	Yes	Yes	No	Yes	Yes	No	Yes	No	n.d.	No
	Share of renewable energy in gross final energy consumption	80%	85%	0%	40%	n.d.	0%	68%	0%	n.d.	0%
	Building Energy Performance Certificate Rating	n.d.	n.d.	n.d.	n.d.	n.d.	n.d.	n.d.	G	n.d.	n.d.
	Energy Efficiency Label of major heating and cooling systems	n.d.	n.d.	n.d.	n.d.	n.d.	n.d.	n.d.	n.d.	n.d.	n.d.
	Fresh water consumption: Liters/person/ year	n.d.	n.d.	n.d.	1000 m^3/ year	n.d.	n.d.	2.640 m^3/ year	n.d.	n.d.	n.d.
	Water recovery systems	Yes	Yes	No	No	Yes	n.d.	No	No	n.d.	No
	Nature-based solutions	n.d.	Yes	No	Yes	No	Yes (green areas)	No	No	n.d.	No
	Recycled materials	n.d.	Yes	Yes	Yes	n.d.	Yes	No	No	n.d.	No
	Traditional and local materials	Yes	Yes	Yes	Yes	Yes	Yes	No	No	n.d.	Yes
	Remediation	No	Yes	No	No	n.d.	No	No	No	n.d.	Yes
	Biodiversity preservation	Yes	Yes	No	No	n.d.	Yes	No	No	n.d.	No

related to energy consumption, water consumption, nature-based solutions, reduction of raw materials extraction, biodiversity and remediation. Northern and Central Europe case studies demonstrated higher awareness, however quantitative data remained difficult to collect.

Another reason for the scarce attention on environmental aspects can be identified in the conflict often arising between heritage preservation regulations and environmental needs in view of higher building performances. In fact, renewable energy sources such as common photovoltaic panels, external walls insulation or change of windows and doors could not be applicable to heritage buildings, or determine higher costs to find alternative and more compatible solutions. It should be also noted that, since this in-depth assessment was conducted on case studies with a history of 5 or more years of activity, the environmental measures taken 5 or 10 years ago could be not updated. With the launch of the European Green Deal, it is likely that also heritage buildings will be object of careful environmental assessments, from basic energy assessment to more complex Life-Cycle Assessment of each intervention, however the issue of skills would need more attention.

As per the economic-financial performance, many sites investigated were sufficiently self-sustainable, not receiving subsidies from the public sector in the operational phase, except from Villa Campolieto and Musil managers who declared to receive public support for the functioning of the heritage sites. Large public investments were present in three sites (Villa Campolieto, Musil and Pfefferberg), however public co-investment was also included in other cases, showing the central role of the public sector for heritage conservation, especially in the case of low or null short-term returns. Circular business models could support heritage owners and managers to develop more sustainable business models for the adaptive reuse of cultural heritage, enhancing the attractiveness of investments in cultural heritage also for private and third sector actors, engaging the local community and stakeholders in co-investment, and ensuring larger participation and collaborations in the operation phase.

The in-depth assessment highlighted also rare attempts to assess the cultural and social impacts of the adaptive reuse project. In many cases, the number of visitors per year can be seen as a relevant proxy of the appreciation of the heritage site. However, visitors' numbers are influenced by the size of the heritage site, and by the management model, considering that not all places are open every day for the general public and many do not provide guided visits as they are mostly working places. Therefore, it could be interesting to conduct more careful assessments of the heritage appreciation after the adaptive reuse, investigating how people perceive the atmosphere of the place, its openness, creativity and vibrancy, as well as people's perceived wellbeing, mutual trust, and general attachment to the place in relation to the collective memory, introducing the "intrinsic value" as a key aspect of heritage cultural diversity and uniqueness.

The following section presents the proposal of evaluation criteria and indicators selected from previous studies and integrated with the results and reflections of case studies research.

3 Results: Circularity Criteria and Indicators for Ex-Ante Evaluation of Cultural Heritage Adaptive Reuse

According to the evidence-base built through the previous phases of literature review and case studies analysis, a structured set of operational criteria and indicators of circularity for cultural heritage adaptive reuse was developed. These criteria and indicators represent a direction/orientation for heritage sites managers and decision-makers, as well as for investors, to take more effective choices towards heritage reuse circularity in the ex-ante evaluation phase.

The "lessons learned" and the observations deduced from previous studies and ex-post evaluation led to the identification of 11 criteria that can support a multidimensional analysis:

- economic-financial indicators, necessary to assess the financial viability and self-sustainability of the proposed adaptive reuse intervention on abandoned and/or underused cultural heritage;
- environmental indicators, used to promote ecological self-sustainability, through closed cycles of energy, materials, water in material cultural heritage, and avoid other environmental costs such as soil consumption, biodiversity loss, pollution, greenhouse gases emission;
- social indicators, necessary to set social goals and targets and estimate the social impacts of alternative interventions, such as citizens and people inclusion, new opportunities for cultural participation, access to social services, increased well-being and quality of life; and last, but not least,
- cultural indicators, which are related to the conservation, regeneration and transmission of cultural heritage values, both tangible and intangible, instrumental and intrinsic, in line with the "complex" notion of value of cultural heritage proposed in the CLIC project.

A set of 67 indicators was selected in coherence with the three general principles of adaptive reuse in Chap. 2 (Fusco Girard this volume) to be used in the ex-ante evaluation stage to support decision-making processes towards circularity implementation in cultural heritage adaptive reuse (Table 8.4).

This set of indicators can be used by public, private and social investors, including Investment Funds and Revolving Funds, to (1) take financing decisions in line with the EU Taxonomy and the European Commission guidelines for sustainable finance, and (2) to monitor the achievement of goals and targets. The indicators identified represent goals to be monitored during the development of the project, ensuring the achievement of the proposed objectives, in particular in case of implementation of financing mechanisms based on "pay for result" and "pay for success", including revolving funds.

Table 8.4 Circularity dimensions and indicators groups

Circularity dimensions	Criteria
Regenerative/autopoietic capacity	1. CULTURAL CAPITAL REGENERATION 2. FINANCIAL AND ECONOMIC CAPITAL REGENERATION 3. NATURAL CAPITAL REGENERATION 4. HUMAN CAPITAL REGENERATION 5. SOCIAL CAPITAL REGENERATION
Symbiotic capacity[a]	6. HISTORIC URBAN LANDSCAPE QUALITY REGENERATION 7. ACCESSIBILITY OF THE URBAN AREA 8. PARTNERSHIPS AND SYNERGIES
Generative capacity	9. EMPLOYMENT GENERATION 10. LOCAL ENTREPRENEURSHIP AND INNOVATION 11. QUALITY OF LIFE, WELLBEING AND HEALTH 12. ECONOMIC IMPACTS

[a]Systemic relationships in the context where the heritage asset is localized

3.1 Regenerative Capacity Indicators

The regenerative capacity includes 33 indicators divided into five groups: regeneration of cultural capital (tangible and intangible), financial capital, natural capital, human capital, social capital.

This regenerative/"auto-poietic" capacity is here expressed as the capacity of economic-financial self-sustainability of the cultural heritage site intended in its management model, the capacity of self-regenerating the resources needed for its "life" in the long term, such as energy, materials, water resources, and the capacity of self-regenerating its cultural value over time. The "human-centred" perspective introduces also the capacity of regenerating the "human capital", including people's skills, knowledge, entrepreneurial attitude, and "social capital" as the capacity to support each other as a cohesive community and build the "heritage community" as defined in the FARO Convention (Council of Europe 2005) (Table 8.5).

3.2 Symbiotic Capacity Indicators

The symbiotic capacity (also in terms of symbiotic exchanges with the context) should be expressed through the relationships of the cultural heritage adaptive reuse project with the "context" in which it is realized. The concept of circular metabolism (i.e. circular and symbiotic exchanges with the context) can be applied at the "meso" level of the heritage area/neighbourhood, or even city/region.

This concept can be better understood through a simple example: the same adaptive reuse project with the same characteristics and same management model (e.g. the reuse of an abandoned church as location for community hub and cultural events) could have different performances in terms of overall circularity if placed in

Table 8.5 Regenerative capacity indicators

Criteria	Indicator	Assessment procedure
CULTURAL CAPITAL REGENERATION	Authenticity and integrity conservation of heritage values	Expert evaluation, based on a Likert scale 1–5 and/or list of measures adopted/not adopted
	Intangible heritage values regeneration[a]	Expert evaluation, based on a Likert scale 1–5, and/or assessment of community's perceptions
	Intrinsic values regeneration[b]	Linguistic evaluation including expert and non-expert assessment
	Accessibility of cultural heritage site	Expert evaluation, based on a Likert scale 1–5 and/or list of measures adopted/not adopted
FINANCIAL AND ECONOMIC CAPITAL REGENERATION	Net Present Value of investment (NPV)[c]	NPV is the difference between the present value of cash inflows and the present value of cash outflows over a period of time.
	Internal Rate of Return (IRR)[d]	IRR is the discount rate that makes net present value (NPV) of all cash flows equal to zero.
	Return on Investment (ROI)[e]	ROI is a performance measure used to evaluate the efficiency or profitability of an investment or compare the efficiency of a number of different investments.
	Payback period[f]	The payback period refers to the amount of time it takes to recover the cost of an investment or how long it takes for an investor to reach breakeven. Shorter payback periods mean more attractive investments, while longer payback periods are less desirable.
	Debt Service Coverage Ratio (DSCR)[g]	DSCR is a measure of the cash flow available to pay current debt obligations.
	Loan Life Coverage Ratio (LLCR)[h]	LLCR is a financial ration used to estimate the ability of a borrowing organisation to repay an outstanding loan.
	Project Life Cover Ratio (PLCR)	PLCR is the ratio of the NPV of the cashflow over the remaining full life of the project to the outstanding debt balance in the period.
	Public Sector Comparator (PSC) and Value for Money (VfM)[i]	PSC and VfM calculation aims at verifying the opportunity and convenience of realizing a project under a project financing scheme rather than a traditional tender.
	Share of public and private contribution	% equity, % debt; % public contribution (for construction and/or operations and maintenance)
	Local co-financing	Percentage of local co-financing on total investment, including crowdfunding, local co-investors, financial participation in community foundations and other local co-investment forms

(continued)

Table 8.5 (continued)

Criteria	Indicator	Assessment procedure
NATURAL CAPITAL REGENERATION (Circular solutions)	Energy generated on site through renewable sources	Percentage of KWh generated on total estimated energy need (Energy assessment)
	Energy performance level upgrade	Number of levels upgrades in Energy performance levels (e.g. from level G to level A = 7 levels)
	De-impermeabilization of soils	Percentage of permeable outdoor surfaces
	Heat island effect reduction	Percentage of surfaces interested by interventions of heat island reductions
	Reduction of freshwater consumption	Quantity of freshwater consumption avoided
	Rainwater recovered	Quantity of rainwater recovered through water filtering and recovery systems
	Reduction of raw materials consumption	Percentage of materials from reused and recycled products (volume)
	Green surfaces	Percentage of green surfaces including buildings (green roofs, green façade)
	Nature-Based Solutions	Use of Nature-Based Solutions in the adaptive reuse of the heritage building or site
	Environmental remediation	Percentage of soils remediated, and/or elimination of pollution causes and unhealthy materials
HUMAN CAPITAL REGENERATION	Traditional skills and construction techniques	Number of people employing traditional skills and construction techniques involved in the adaptive reuse intervention
	Entrepreneurship enhancement	Number of activities stimulating entrepreneurship and self-entrepreneurship, such as incubators, accelerators, co-working spaces)
	Education and training activities	Number of people educated and trained
SOCIAL CAPITAL REGENERATION	Social capital	Number of associations for 10,000 inhabitants
	Support to weak and marginalised social groups	Number of people from weak and marginalised social groups target of activities and services included in the heritage site reused

[a]As humanistic foundation of European culture, of the acknowledgment of human scale, the sense of beauty that is regenerated through adaptive reuse

[b]Fusco Girard and Vecco (2021)

[c]To calculate NPV, it is necessary to estimate future cash flows for each period and determine the correct discount rate. If the NPV of a project is positive, it means that the discounted present value of all future cash flows related to the project will be positive, and therefore attractive.

[d]When comparing investment options with other similar characteristics, the investment with the highest IRR would be considered more desirable to undertake.

[e]ROI is expressed as a percentage and is calculated by dividing an investment's net profit (or loss) by its initial costs. ROI does not take into account the holding period or passage of time, and so it can miss opportunity costs of investing elsewhere.

[f]The payback period is calculated by dividing the amount of the investment by the annual cash flow.

(continued)

Table 8.5 (continued)

[g]The formula for the DSCR requires net operating income (EBIT) and the total debt servicing for the project (interest and principal payments that are due in the coming year). A DSCR less than 1 means negative cash flow, which mean a borrower will be unable to cover or pay current debt obligations.

[h]LLCR is calculated by dividing the net present value (NPV) of the money available for debt repayment by the amount of outstanding debt. LLCR is similar to the DSCR: the DSCR captures a single point in time, whereas the LLCR addresses the entire span of the loan.

[i]PSC may be defined as the risk-adjusted cost for the Public Administration for an infrastructural project to be realized in a project finance scheme. PSC is especially useful to assess the convenience of the project financing considering the correct allocation of risks among the Public and the private partners. The PSC is defined as the sum of the following variables:

Present value of construction costs;

Present value of O&M costs;

Present value of the risks transferred to the private subject in case of project finance.

It is necessary to calculate the PSC for the realization of the project in a project finance scheme on one side, and with a traditional tender on the other side. The difference between the two indicators (tender vs project finance) is the Value for Money (VfM): if VFM is positive, then the activation of a project finance is convenient for the Public Administration

different context (e.g. rural vs. urban, metropolis vs. village, high-income livelihoods neighbourhood vs. marginalized neighbourhood, young 'hipster' neighbourhood vs. elderly residential area, etc.).

The symbiotic capacity includes 13 indicators divided into two groups: the capacity of stimulating partnerships and synergies between diverse territorial actors and multi-level policies, and the contribution to the reconnection of fragmented landscape through enhanced accessibility, public and green space integrity, quality and safety (Table 8.6).

3.3 Generative Capacity Indicators

The generative capacity is expressed by the capacity of the cultural heritage adaptive reuse intervention to generate resources for the local context, which in turn provide new financial, cultural, social and environmental resources for cultural heritage regeneration, in a circular perspective.

The generative capacity includes 17 indicators divided into three groups: employment generation, attractiveness for new businesses and entrepreneurs, enhancement of quality of life, wellbeing and health. It includes the "spillover" effects at regional level, the revitalisation of the local economy in the urban/rural area, the overall attractiveness for new businesses, commercial activities, residents, innovators, entrepreneurs, creative workers, thus in general the generation of a desirable environment turning a "dead" site into a vibrant "place" (Table 8.7).

Table 8.6 Symbiotic capacity indicators

Criteria	Indicator	Assessment procedure
HISTORIC URBAN LANDSCAPE QUALITY REGENERATION	Historic Urban Landscape quality regeneration	Expert evaluation, based on a Likert scale 1–5, and assessment of community's perceptions also through visualizations and simulations
ACCESSIBILITY OF THE URBAN AREA	Accessibility enhancement	Distance on feet from the nearest public transport node
	Public and green space accessibility	Percentage of public and green space recovered/regenerated or made more accessible
	Pedestrian mobility	Percentage pedestrian areas on total intervention surfaces
	Sustainable mobility	Percentage of surfaces dedicated to sustainable mobility such as bicycle routes
PARTNERSHIPS AND SYNERGIES WITH THE CONTEXT	Stakeholders involvement in decision-making	Number of organisations involved in the decision-making phase of the adaptive reuse process
	Stakeholders involvement in the management phase	Number of organisations involved in the management phase
	Third sector actors involved	Number of third sector actors involved in all phases of the adaptive reuse process
	People involvement	Involvement of people in the decision-making process, such as through consultation and co-creation
	Marginalised social groups involvement	Involvement of minor and marginalised social groups in the decision-making process, such as through consultation and co-creation
	People collaboration	Number of collaboration agreements and pacts signed for the collective care of cultural heritage
	Businesses collaboration and symbioses	Number of collaboration and symbioses contracts, such as through circular supply chains models and localisation of complementary businesses in the area
	Synergies with higher level policies	Assessment of the contribution to the realisation of regional, national and international policies, and/or local ecosystems policies
	Trust levels	Trust levels (e.g. Edelman trust survey)

Table 8.7 Generative capacity indicators

Criteria	Indicator adopted	Assessment procedure
EMPLOYMENT GENERATION	Jobs creation	Number of direct and indirect full-time equivalent jobs generated in sectors such as: professional and consulting activities, construction works, circular economy, research and development, creative and cultural activities, sustainable tourism
	Regional economy spillovers	Number of indirect jobs created through the adaptive reuse intervention (calculation of spillover effects)
	Local economy	Number of new businesses localised in the urban area, such as commercial activities, cultural and creative activities, circular economy activities as repairing and recycling, sustainable cultural tourism, and other sectors
	Jobs/Investments ratio	Number of jobs generated per 100.000 euro of investment
LOCAL ENTREPRENEURSHIP AND INNOVATION	Enterprises localisation	Number of enterprises localised in the heritage site reused
	Entrepreneurs and self-entrepreneurs localisation	Number of entrepreneurs and self-entrepreneurs localised in the heritage site reused
QUALITY OF LIFE, WELLBEING AND HEALTH	Beneficiaries of adaptive reuse intervention	Number of final beneficiaries target of the adaptive reuse intervention who are likely to enhance their quality of life, wellbeing and health
	Proximity activities	Number of new proximity activities localised in the area, such as commercial activities, neighbourhood services
	Cultural activities	Number of cultural activities likely to be activated in the area
	Cultural participation	Number of people estimated to participate in cultural activities per year
	Arts, craft, making and repairing activities	Number of arts, traditional craft, making activities (such as fab labs), and repairing activities localised in the area
	Creative and innovative spaces	Percentage of surfaces dedicated to arts, craft, innovation, culture
	Urban art	Percentage of surfaces equipped with urban art
	Public space and socialization	Percentage of surfaces dedicated to socialisation, such as squares, parks, community hubs, bars and restaurants, and other types of activities promoting gathering and socialisation
	Green space	Percentage of publicly accessible green spaces, including roads and streets equipped with green surfaces

(continued)

Table 8.7 (continued)

Criteria	Indicator adopted	Assessment procedure
	Heritage Community	Number of people involved in the care of cultural heritage as common good
	Perceived aesthetic quality of landscape	Percentage of people perceiving a enhancement of landscape aesthetic quality
	Use of regional resources	Percentage of materials from reduced distance (<50 km) (volume)
	Recovery of natural resources potential linked with the capacity to enhance wellbeing perception	Capacity of adaptive reuse solution to recover the potential benefit of local natural resources linked with the capacity to enhance wellbeing perception (e.g. thermal resources)
	Carbon emissions per sqm indoor area	Life Cycle Assessment, CO_2 eq/sqm of the adaptive reuse intervention
	Social sustainability: Diversity Capacity for self-organization Common meaning Capacity for learning Trust	Expected contribution of the adaptive reuse project to Social Sustainability, including: Diversity, Capacity for self-organization, Common meaning, Capacity for learning, Trust (see: Domaradzka, Chaps. 10 and 11 in this volume)
ECONOMIC IMPACTS	Real estate market values	% of increase of average market values of residential units in the heritage area (€/sqm)
	Increase of fiscal revenues for public administrations	% of increase of fiscal revenues for public administrations
	Attractive capacity of commercial activities	Number of new commercial activities in the heritage area

4 Discussion and Conclusions

The multidimensional indicators about impacts of cultural heritage conservation/regeneration were initially classified on the base of the 4 pillars of sustainability: cultural, economic, social, environmental. The development of the theoretical framework lead to the identification of three principles of circularity: the regenerative/autopoietic capacity, the symbiotic capacity, and the generative capacity. These three principles include indicators expressing sustainability dimensions, integrated in a systemic perspective, overcoming the "pillars" approach. The related criteria and indicators are both quantitative and qualitative and they are referred to different scales (micro, meso and macro scales). Indicators in each dimension were deduced from the analysis of best practices and from national and international sources. They represent a grid able to ensure that the assessment reflects all values and dimensions to be considered. They are a basis of information and, at the same time, allow developing a common language about impacts and benefits of cultural heritage adaptive reuse.

Indicators in the ex-ante evaluation phase represent specific "objectives" and targets of circular "human-centred" adaptive reuse of cultural heritage, to be assessed and monitored before, during and after the realisation of the intervention. They support specifically "result-based" financing instruments and governance models, building a framework of useful data which can inform choices of owners, managers and the local communities.

Moreover, criteria and indicators are an important tool to be used in participatory ex-ante co-evaluation processes to trigger evidence-based analysis and informed decisions, taking into account the point of view, needs and desires, costs and benefits of diverse stakeholders. Through prioritization techniques, weights can be assigned to criteria and indicators to build preferences matrices and analyse synergies and conflicts between stakeholders, in a evolutionary process in which preferences are not static but can change through discussion, reflection, understanding of impacts and simulations.

The matrix of criteria and indicators, adapted and "weighted" according to the specific contexts and preferences/needs, can be integrated with context-specific additional indicators, which can be linked to the specific heritage typology and conditions, as well as to the specific activities, stakeholders, beneficiaries and final users involved.

The proposed CLIC evaluation framework was developed to provide a flexible, adaptable but still usable set of indicators for ex-post and thus, also, ex-ante evaluation reliable for diverse heritage typologies, contexts, scales, and adaptive reuse phases. It should be used and tested extensively by diverse stakeholders to ensure its wide usability, adapting and refining it according to further findings achieved through extensive testing. However, the reflections developed and the space for experimentation was an opportunity to develop a coherent, first-tested and validated framework which can potentially work to assess the real contribution of cultural heritage adaptive reuse to sustainable development, in Europe and beyond.

The assessment of impacts is more and more necessary especially in view of the implementation of sustainability policies such as the European Green Deal, the New European Bauhaus, and the Sustainable finance initiative with the EU Taxonomy. Sustainable finance and sustainable policies should be grounded in careful assessment of impacts/results obtained, both in the public and in the private sector. The Third sector, including social enterprises, foundations, philanthropy, civic associations and other organisations with a "social" mission, already started the process of better accounting for the activities carried out. Third sector actors are more used to carry out impacts assessments, based on diverse methods and tools and mostly using ESG (Environmental, Social, Governance) criteria and indicators (Zamagni et al. 2015; Venturi and Perra 2018). Circular Economy policies are supported as well by a set of clear indicators which provide evidence base of the impacts of investments, projects, programmes and initiatives (Ellen MacArthur Foundation 2015a; Gravagnuolo et al. 2019b; Circular economy network 2021; EUROSTAT 2021; OECD 2021).

The cultural heritage sector, and especially the initiatives related to cultural heritage adaptive reuse, does not use a specific set of indicators to assess the impacts of diverse conservation, regeneration and reuse projects. It still remains difficult to assess the contribution of cultural heritage to territorial development (Lykogianni et al. 2019). It was therefore considered timely and useful to develop a set of criteria and indicators that, according to the CLIC circularity framework, can practically suggest sustainable directions for cultural heritage adaptive reuse, and contribute to enhance the quality of interventions (ICOMOS 2019) and the accountability of management bodies that "care" for cultural heritage as common good.

Based on the research conducted, a particular issue was raised about skills in cultural heritage accounting. The public and private sector managers interviewed were rarely aware of environmental impacts of cultural heritage adaptive reuse, and in some cases they were not able or not interested in accounting for social impacts. On the other side, third sector/grassroots organisations and more "spontaneous" groups of citizens who "take care" of cultural heritage as common good in the frame of the FARO Convention (Council of Europe 2005) have usually less capacity for business modelling, financing and economic accounting, resulting in difficulties to develop sustainable business and management models starting from bottom-up initiatives. On this base, a call should be raised to policy makers to develop new training and educational programmes to enhance skills in the cultural heritage sector, including skills related to circularity management. Specialised and multi-sectorial competences are needed to implement the proposed impacts assessment framework for circular adaptive reuse of cultural heritage, as well as higher collaboration and cooperation capacity between diverse specialists, and between diverse territorial organisations. These new capacities are strongly needed to enhance heritage-led local innovation ecosystems based on "collaborative attitude and capacity" at district/regional, national and European level. Moreover, issues of data availability and data management can be overcome through the inclusion of the right skills which are currently lacking in the cultural heritage sector.

The circularity indicators matrix for ex-post evaluation used for the analysis of case studies represents a tool to create evidence-base of cultural heritage impacts, allowing comparison and benchmark between diverse adaptive reuse practices, thus providing useful information to enhance choices for new adaptive reuse interventions. Heritage sites managers and owners, in the public, private and third sector, can use the indicators to orient adaptive reuse circular solutions and monitor results in a systematic way. However, the assessment of some indicators requires specific expertise and skills to be conducted, calling for multidisciplinary knowledge and new capacities of managers that should be able to coordinate diverse aspects of circularity, from environmental measures for the "circular building" to sensitiveness for historic-cultural and "intrinsic" values of cultural heritage, to economics and financial assessments, to aspects related to social cohesion, participation, engagement, inclusion.

References

Abastante F, Lami IM, Mecca B (2020) How to revitalise a historic district: a stakeholders-oriented assessment framework of adaptive reuse. In: Mondini G, Oppio A, Stanghellini S, Bottero M, Abastante F (eds) Values and functions for future cities, Green energy and technology series. Springer, pp 3–20. https://doi.org/10.1007/978-3-030-23786-8_1

Angrisano M, Fusco Girard L, Bianchi A (2019) A literature review about life cycle assessment as a tool to support circular economy innovation in the built environment sector. BDC Bollettino Del Centro Calza Bini 19(1):125–143. https://doi.org/10.6092/2284-4732/7064

ARUP (2016) Circular economy in the built environment. ARUP, London. Available at: https://www.arup.com/perspectives/publications/research/section/circular-economy-in-the-built-environment

Bandarin F, van Oers R (2014) Reconnecting the city: the historic urban landscape approach and the future of urban heritage. Wiley-Blackwell

Bosone M, De Toro P, Fusco Girard L, Gravagnuolo A, Iodice S (2021) Indicators for ex-post evaluation of cultural heritage adaptive reuse impacts in the perspective of the circular economy. Sustainability 13(9):4759. https://doi.org/10.3390/SU13094759

Bottero M, Lerda M (2019) Valutazione circolare degli interventi di riuso adattivo: il caso della città di Torino. BDC Bollettino Del Centro Calza Bini 19(2):497–513. https://doi.org/10.6092/2284-4732/7279

Bourdieu P (1986) The forms of capital. In: Richardson JE (ed) The handbook of theory of research for the sociology of education. Greenword Press, pp 241–258

Bullen PA, Love PED (2010) The rhetoric of adaptive reuse or reality of demolition: views from the field. Cities 27:215–224

CHCfE Consortium (2015) Cultural heritage counts for Europe. International Cultural Centre, Krakow. Available at: http://blogs.encatc.org/culturalheritagecountsforeurope//wp-content/uploads/2015/06/CHCfE_FULL-REPORT_v2.pdf

Circular economy network (2021) The third circular economy report. Focus: the role of circular economy in the transition to climate neutrality

Costanza R, de Groot R, Sutton P, van der Ploeg S, Anderson SJ, Kubiszewski I, Farber S, Turner RK (2014) Changes in the global value of ecosystem services. Glob Environ Chang 26(1):152–158. https://doi.org/10.1016/j.gloenvcha.2014.04.002

Council of Europe (2005) Convention on the value of cultural heritage for society (Faro Convention, 2005). Available at: https://www.coe.int/en/web/culture-and-heritage/faro-convention

Douglas J (2006) Building adaptation, 2nd edn. Butterworth-Heinemann, Woburn

Ellen MacArthur Foundation (2012) Towards the circular economy, vol 1. Ellen MacArthur Foundation. https://doi.org/10.1162/108819806775545321

Ellen MacArthur Foundation (2013) Towards the circular economy. Ellen MacArthur Foundation

Ellen MacArthur Foundation (2015a) Circularity indicators. An approach to measuring circularity. Project overview. https://www.ellenmacarthurfoundation.org/assets/downloads/insight/Circularity-Indicators_Project-Overview_May2015.pdf

Ellen MacArthur Foundation (2015b) Delivering the circular economy a toolkit for policymakers

Ellen MacArthur Foundation (2015c) Growth within: a circular economy vision for a competitive Europe

Ellen MacArthur Foundation & ARUP (2019) Circular economy in cities: project guide. ARUP, London. Available at: https://ellenmacarthurfoundation.org/circular-economy-in-cities

European Commission (2019) Sustainable & circular re-use of spaces & buildings. Handbook. https://ec.europa.eu/futurium/en/sustainable-land-use/handbook-sustainable-and-circular-re-use-spaces-and-buildings-now-available

EUROSTAT (2021) Indicators circular economy Eurostat. https://ec.europa.eu/eurostat/web/circular-economy/indicators

Foster G (2020) Circular economy strategies for adaptive reuse of cultural heritage buildings to reduce environmental impacts. Resour Conserv Recycl 152:104507. https://doi.org/10.1016/j.resconrec.2019.104507

Foster G, Kreinin H (2020) A review of environmental impact indicators of cultural heritage buildings: a circular economy perspective. In: Environmental adaptive reuse of cultural heritage letters. https://doi.org/10.1088/1748-9326/ab751e

Foster G, Kreinin H, Stagl S (2020) The future of circular environmental impact indicators for cultural heritage buildings in Europe. Environ Sci Eur. https://doi.org/10.1186/s12302-020-00411-9

Fusco Girard L (1987) Risorse architettoniche e culturali: valutazioni e strategie di conservazione. Franco Angeli

Fusco Girard L (2013) Toward a smart sustainable development of port cities/areas: the role of the historic urban landscape approach. Sustainability 5:4329–4343

Fusco Girard L (2014) The role of cultural urban landscape towards a new urban economics: new structural assets for increasing economic productivity through hybrid processes. HOPUE 1(1):3–27

Fusco Girard L (2019a) The human-centred city development and the circular regeneration. In: Fusco Girard L, Trillo C, Bosone M (eds) Matera, città del sistema ecologico uomo/società/natura il ruolo della cultura per la rigenerazione del sistema urbano/territoriale. Giannini Publisher

Fusco Girard L (2019b) Implementing the circular economy: the role of cultural heritage as the entry point. Which evaluation approaches? BDC Bollettino Del Centro Calza Bini 19(2):245–277. https://doi.org/10.6092/2284-4732/7269

Fusco Girard L (2020) Metropolitan cities: which development strategies? Which new perspectives for improving productivity? Which governance tools? In: Urban empires cities as global rulers in the new urban world, 1st edn. Routledge, New York

Fusco Girard L (this volume) The circular "human-centred" adaptive reuse of cultural heritage: theoretical foundations. In: Fusco Girard L, Gravagnuolo A (eds) Adaptive reuse of cultural heritage. Springer

Fusco Girard L, Gravagnuolo A (2018) Il riuso del patrimonio culturale religioso: criteri e strumenti di valutazione. BDC Bollettino Del Centro Calza Bini 18(2):237–246

Fusco Girard L, Nijkamp P (1997) Le valutazioni per lo sviluppo sostenibile della città e del territorio. FrancoAngeli, Napoli

Fusco Girard L, Vecco M (2021) The "tntrinsic value" of cultural heritage as driver for circular human-centered adaptive reuse. Sustainability 13/6. https://doi.org/10.3390/su13063231

Fusco Girard L, Gravagnuolo A, Nocca F, Angrisano M, Bosone M (2015) Towards an economic impact assessment framework for Historic Urban Landscape conservation and regeneration projects. BDC Bollettino Del Centro Calza Bini 15(2):1–29

Fusco Girard L, Nocca F, Gravagnuolo A (2019) Matera: city of nature, city of culture, city of regeneration. Towards a landscape-based and culture based urban circular economy. Aestimum 74:5–42. https://doi.org/10.13128/aestim-7007

Gaballo M, Mecca B, Abastante F (2021) Adaptive reuse and sustainability protocols in italy: relationship with circular economy. Sustainability 13(14):8077. https://doi.org/10.3390/su13148077

Geissdoerfer M, Savaget P, Bocken NMP, Hultink EJ (2017) The circular economy – a new sustainability paradigm? J Clean Prod 143:757–768. https://doi.org/10.1016/j.jclepro.2016.12.048

Getty Conservation Institute (1998) Economics and heritage conservation, a meeting organised by the Getty Conservation Institute (edited by Maria de la Torre), December 1998, Getty Center, Los Angeles

Ghisellini P, Cialani C, Ulgiati S (2016) A review on circular economy: the expected transition to a balanced interplay of environmental and economic systems. https://doi.org/10.1016/j.jclepro.2015.09.007

Gravagnuolo A, Fusco Girard L (2021) Database of indicators and data in pilot cities. Deliverable D2.4, Horizon 2020 CLIC.

Gravagnuolo A, Fusco Girard L, Ost C, Saleh R (2017) Evaluation criteria for a circular adaptive reuse of cultural heritage. BDC Bollettino Del Centro Calza Bini 17(2/2017):185–216

Gravagnuolo A, Angrisano M, Girard LF (2019a) Circular economy strategies in eight historic port cities: criteria and indicators towards a circular city assessment framework. Sustainability (Switzerland) 11(13). https://doi.org/10.3390/su11133512

Gravagnuolo A, Angrisano M, Girard LF, Fusco Girard L (2019b) Circular economy strategies in eight historic port cities: criteria and indicators towards a circular city assessment framework. Sustainability 11(13):3512. https://doi.org/10.3390/su11133512

Gravagnuolo A, Angrisano M, Bosone M, Buglione F, De Toro P, Fusco Girard L (2023) Participatory evaluation of cultural heritage adaptive reuse interventions in the circular economy perspective: a case study of historic buildings in Salerno (Italy). J Urban Manag, forthcoming

Gupta J, Hurley F, Grobicki A, Keating T, Stoett P, Baker E, Guhl A, Davies J, Ekins P (2019) Communicating the health of the planet and its links to human health. Lancet Planet Health 3(5):e204–e206. Elsevier B.V. https://doi.org/10.1016/S2542-5196(19)30040-3

Hosagrahar J, Soule J, Fusco Girard L, Potts A (2016) Cultural heritage, the UN sustainable development goals, and the new urban agenda. BDC Bollettino Del Centro Calza Bini 16(1):37–54

ICOMOS (2019) European quality principles for EU-funded interventions with potential impact upon cultural heritage. ICOMOS International, Paris

Keeney R, Raiffa H (1976) Conflicting objectives in decisions'. Wiley

Kirchherr J, Reike D, Hekkert M (2017) Conceptualizing the circular economy: an analysis of 114 definitions, 127 Resources. Conserv Recycl 221(2017). https://doi.org/10.1016/j.resconrec.2017.09.005

Labadi S (2011) Evaluating the socio-economic impacts of selected regenerated heritage sites in Europe. Project report. European Cultural Foundation; Riksbankens Jubileumsfond

Le Moigne R (2014) L'économie circulaire. Comment la mettre en œuvre dans l'entreprise grâce à la reverse supply chain? Dunod. http://www.decitre.fr/livres/l-economie-circulaire-9782100600083.html

Lemille A (2017) Optimizing circular value. Ensuring circular economy projects and solutions delivers on genuine value creation. Medium – online. Available at: https://alexlemille.medium.com/optimizing-circular-value-5ee670ef1b9c

Lupacchini R, Gravagnuolo A (2019) Cultural heritage adaptive reuse: learning from success and failure stories in the city of Salerno, Italy. BDC Bollettino Del Centro Calza Bini 19(2):353–377. https://doi.org/10.6092/2284-4732/7273

Lykogianni E, Mobilio L, Procee R, Airaghi E, Kern P, Le Gall A, Krohn C, Vanhoutte C (2019) The MCH as a strategic territorial development resource: mapping impacts through a set of common european socio-economic indicators. Synthesis Report. ESPON

Marston FJ (1982) Historic preservation: curatorial management of the built world. McGraw-Hill, New York

MEA (2003) Ecosystems and human well-being: a framework for assessment. World Resources Institute. Available at: https://www.unep.org/resources/report/ecosystem-and-human-well-being-framework-assessment

Moraga G, Huysveld S, Mathieux F, Blengini GA, Alaerts L, Van Acker K, de Meester S, Dewulf J (2019) Circular economy indicators: what do they measure? Resour Conserv Recycl. https://doi.org/10.1016/j.resconrec.2019.03.045

Moreau V, Sahakian M, van Griethuysen P, Vuille F (2017) Coming full circle: Why social and institutional dimensions matter for the circular economy. J Ind Ecol 21(3):497–506. https://doi.org/10.1111/jiec.12598

Nijkamp P (1977) Theory and application of environmental economics. Elsevier, Amsterdam

Nijkamp P (2012) Economic valuation of cultural heritage. In: Licciardi G, Amirtahmasebi R (eds) The economics of uniqueness. World Bank, Washington, DC

Nocca F (2017) The role of cultural heritage in sustainable development: multidimensional indicators as decision-making tool. Sustainability 9(10):1882. https://doi.org/10.3390/su9101882

OECD (2021) OECD inventory of circular economy indicators. Available at: https://www.oecd.org/cfe/cities/InventoryCircularEconomyIndicators.pdf

Ost C (2009) A guide for heritage economics in historic cities. Values, indicators, maps, and policies. Getty Conservation Institute, Los Angeles

Ost C (2016) Innovative financial approaches for culture in urban development. In: UNESCO (ed) Global report on culture for sustainable urban development

Parchomenko A, Nelen D, Gillabel J, Rechberger H (2019) Measuring the circular economy a multiple correspondence analysis of 63 metrics. J Clean Prod 210:200–216. https://doi.org/10.1016/j.jclepro.2018.10.357

Potting J, Hekkert M, Worrell E, Hanemaaijer A (2017) Circular economy: measuring innovation in the product chain. PBL Netherlands Environmental Assessment Agency, The Hague

Saidani M, Yannou B, Leroy Y, Cluzel F, Kendall A, Saidani M, Yannou B, Leroy Y, Cluzel F, Kendall A (2019) A taxonomy of circular economy indicators to cite this version: HAL Id: hal-01954800. J Clean Prod 207:542–559

Stanik N, Aalders I, Miller D (2018) Towards an indicator-based assessment of cultural heritage as a cultural ecosystem service – a case study of Scottish landscapes. Ecol Indic 95(December 2017):288–297. https://doi.org/10.1016/j.ecolind.2018.07.042

TEEB (2010) The economics of ecosystems and biodiversity ecological and economic foundations (Kumar P (ed)). Earthscan.

van Buren N, Demmers M, van der Heijden R, Witlox F (2016) Towards a circular economy: the role of Dutch logistics industries and governments. Sustainability 8(7):647. https://doi.org/10.3390/su8070647

Vecco M (2007) L'evoluzione Del Concetto Di Patrimonio Culturale. Franco Angeli, Milano, Isbn: 978-88-464-8789-6

Watts N, Amann M, Arnell N, Ayeb-Karlsson S, Belesova K, Berry H, Bouley T, Boykoff M, Byass P, Cai W, Campbell-Lendrum D, Chambers J, Daly M, Dasandi N, Davies M, Depoux A, Dominguez-Salas P, Drummond P, Ebi KL et al (2018a) The 2018 report of the Lancet countdown on health and climate change: shaping the health of nations for centuries to come. The Lancet 392(10163):2479–2514. Lancet Publishing Group. https://doi.org/10.1016/S0140-6736(18)32594-7

Watts N, Amann M, Ayeb-Karlsson S, Belesova K, Bouley T, Boykoff M, Byass P, Cai W, Campbell-Lendrum D, Chambers J, Cox PM, Daly M, Dasandi N, Davies M, Depledge M, Depoux A, Dominguez-Salas P, Drummond P, Ekins P et al (2018b) The Lancet Countdown on health and climate change: from 25 years of inaction to a global transformation for public health. Lancet. https://doi.org/10.1016/S0140-6736(17)32464-9

Watts N, Amann M, Arnell N, Ayeb-Karlsson S, Beagley J, Belesova K, Boykoff M, Byass P, Cai W, Campbell-Lendrum D, Capstick S, Chambers J, Coleman S, Dalin C, Daly M, Dasandi N, Dasgupta S, Davies M, Di Napoli C et al (2021) The 2020 report of The Lancet countdown on health and climate change: responding to converging crises. Lancet. https://doi.org/10.1016/S0140-6736(20)32290-X

Wijkman A, Skånberg K (2015) The circular economy and benefits for society: jobs and climate clear winners in an economy based on renewable energy and resource efficiency. The Club of Rome, p 59

World Health Organization Regional Office for Europe (2018) Circular economy and health: opportunities and risks. WHO Press. Available at: https://apps.who.int/iris/handle/10665/342218

Yung EHK, Chan EHW (2012) Implementation challenges to the adaptive reuse of heritage buildings: towards the goal of sustainable, low carbon cities. Habitat Int 36:352–361

Zamagni S, Venturi P, Rago S (2015) Valutare l'impatto sociale. La questione della misurazione nelle imprese sociali. Impresa Sociale 6:77–97

Zeleny M (1991) Measuring criteria: weights of importance. Hum Syst Manag 10(4):237–238. https://doi.org/10.3233/HSM-1991-10401

Zeleny M (2006) From knowledge to wisdom: on being informed and knowledgeable, becoming wise and ethical. Int J Inf Technol Decis Mak 5(4):751–762

Chapter 9
A Multicriteria and Multi-scalar Decision Support System to Implement Circular Economy in Cultural Heritage Adaptive Reuse

Simona Panaro, Salvatore Greco, and Alessio Ishizaka

1 Introduction

Although sustainability approaches to the urban environment have been recognised as needed, implementing them is not an easy process. Indeed, the current environmental, social, economic, and health crises are proving the inability of cities to endure those pressures in the long term. On the other hand, effects of climate change (e.g. heat waves, floods, and droughts, etc.) and other threats such as the Covid-19 outbreak make evident the need to re-think the city environment and the urban life (housing, mobility, working, education, free time, health, and wellbeing).

In this context, the circular economy approaches, aimed to prolong materials life cycle and reduce waste and pollution, have sparked a new interest. Circular systems are always more frequently associated with policies for climate change, energy, reduction of waste. However, the circular approaches have shown several limitations in their applications and request more accurate research for including sustainability issues in a comprehensive manner (Nakajima 2000). Indeed, the circular economy has seemed more focused on economic and environmental issues, ignoring often social inclusion, cultural dimensions and decision-making processes (Geissdoerfer et al. 2017). The most recurrent applications of circular principles

S. Panaro (✉)
Science Policy Research Unit, University of Sussex, Falmer, Brighton, UK

S. Greco
Department of Economics and Management, University of Catania, Catania, Italy
e-mail: salgreco@unict.it

A. Ishizaka
Department of Information Systems, Supply Chain and Decision Making, Neoma, Rouen, France
e-mail: alessio.ishizaka@neoma-bs.fr

L. Fusco Girard, A. Gravagnuolo (eds.), *Adaptive Reuse of Cultural Heritage*,
https://doi.org/10.1007/978-3-031-67628-4_9

concerned changes of forms of consumption (e.g. sharing, second hand, and delivery services) rather than the way to produce and manage resources. The hardest task to accomplish seems to start up interrelated cycles to reclaim the value of unused resources (materials already headed for the landfill, abandoned spaces, etc.).

The recent Mission Board for Climate-neutral and Smart Cities of the EU commission (European Commission 2020) underlines the need to innovate the forms of participatory governance and economic and funding models and promote integrated urban planning.

In that sense, in the CLIC project, the adaptive reuse of cultural heritage offered a testbed to consider the circular economy principles in the larger perspective of urban sustainability (including social inclusion, cultural heritage dimensions, and citizens' wellbeing). Indeed, cultural heritage has been identified as a resource for the sustainable development of Europe (Council of the European Union 2014) and has been added to urban agendas and local development strategies for sustainability (Blake 2000). Several studies have argued that actions on cultural heritage are able to generate economic, environmental, cultural and social benefits altogether. However, those benefits are not easy to identify and quantify and often require high financial costs of restoration or restrictive regulatory requirements, which did not encourage the adaptation and the reuse of cultural sites or historical urban areas.

In this respect, the development of methodologies and procedures able to support more informed decisions on the reuse of cultural heritage could be beneficial.

Indeed, the Multiple Criteria Decision Aiding (MCDA) methods (Greco et al. 2016; Ishizaka and Nemery 2013) have been already used in the cultural heritage sector (Hong and Chen 2017; Dutta and Husain 2009; Giove et al. 2010). What came out from those experiences is that the complexity of the heritage problem requested often the integration of methods and appropriate frameworks.

A versatile framework for the evaluation of alternative regeneration interventions of the cultural sites has been proposed by Ferretti and Comino (2015). The authors considered qualitative and quantitative dimensions to support the development of urban strategies and stressed the importance to interact with different stakeholders to improve the transparency and inclusive choices. Indeed, in line with national and international recommendations, it is needed to explore and include the points of view of the different actors involved in the transformation/reuse process of cultural sites.

On the other hand, the need to reduce public expenditure and make better use of scarce resources available suggested considering the adaptive reuse of the heritage also as a resource allocation problem. This kind of problems usually has been associate with the Portfolio Decision Analysis (PDA), used in situations with budget limits to direct better financial resources. In particular, to respect budget constraints and multidimensional criteria, Nesticò et al. (2018) have integrated a linear programming model with multi-criteria analysis (weighted sum of criteria) to identify the subset of interventions to be made starting from a larger list of projects.

In this perspective, the Portfolio Decision Analysis can be integrated with many other Multi-Criteria methods to analyse a larger spectrum of heritage problems. The methodology of the Decision Support System (DSS) developed within the CLIC project experimentation entails a new procedure to consider several criteria, budget restrictions, restrictive constraints, and interlinked issues to implement integrated strategies. In addition, it can be implemented in multi-stakeholder environments, aiding to make informed decisions and shared visions.

Indeed, the interactive approach, the opportunity to use different multi-criteria methods and considering both qualitative and quantitative information make it very flexible in different contexts and stages of the reuse process. It aids the development of well-structured comprehensive strategies also in deficiency of information as at preliminary phases of the reuse process. In the next section we provide an overview of the issues of adaptive reuse of cultural heritage. We then present the DSS methodology developed in the CLIC project, accompanied by a brief description of its implementation in the Italian city of Salerno and the Croatian city of Rijeka (Boxes 9.1 and 9.2).

Box 9.1: Implementation of DSS in Salerno (Italy)

1. *Context*

The city of Salerno has four historical buildings called "Edifici mondo" complex in its uptown, significant for the city's history, which are nowadays abandoned and run-down.

Within the Heritage Innovative Partnerships (HIP) participatory process activated in the CLIC project, the City of Salerno has opened an exchange with the local community and stakeholders in order to identify local needs and objectives and develop circular strategies for the reuse of those buildings.

In this context, an idea generation process for the reuse of "Edifici mondo" was opened looking at the circular economy prospective. In this direction, it was organised the "CLIC Business Model Workshop" by the Municipality of Salerno, ICHEC Brussels Management School, and CNR. In this workshop, four different reuse proposals have been generated.

2. *DSS implementation*

In that context, the implementation of the DSS aimed to explore in a structured way relevant points of view for the adaptive reuse of the historical complex "Edifici Mondo" and to analyse the four reuse proposals in terms of capacity to regenerate cultural, human, environmental, and social capital, to promote cooperation and synergies, and to improve the territorial vitality.

(continued)

Box 9.1 (continued)

The assessment of the reuse proposals aimed to inform the future design of a circular reuse strategy for "Edifici mondo" complex.

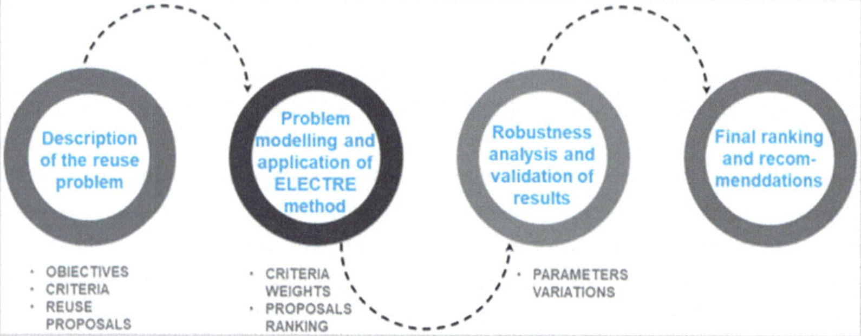

3. *Output*

The output of the DSS is the ranking of the four reuse proposals and identification of strong/weak points of each one providing useful elements for the reuse of "Edifici mondo" complex in a circular economy perspective.

Box 9.2: Implementation of DSS in Rijeka (Croatia)

1. *Context*

During the Heritage Innovative Partnerships (HIP) participatory process activated in the CLIC project, experts of the city of Rijeka and local stakeholders have developed the concept of a Cultural Corridor for the old industrial area of the city, located on the edge of the city centre and back-boned by the Rječina River, aiming to improve the liveability of this area and the city.

During the HIP meetings, the following objectives have been identified: improving the liveable and accessible of the Rječina River and the city, reusing existing cultural assets of the area, preserving the historic cultural heritage of the city. In addition, in the "CLIC Urban Seeding workshop" organised by the University of Nova Gorica in collaboration with the University of Rijeka, several projects and interventions to develop the Cultural Corridor strategy have been developed.

(continued)

Box 9.2 (continued)

2. *DSS implementation*

In that context, the implementation of the DSS aimed to support the definition of a complex adaptive reuse problem, helping to consider and manage at the same time elements, conditions and constraints listed in the figure below. It shows how the identification of satisfying portfolios of projects/interventions can be carried out in the step by step procedure of the DSS.

3. *Output*

The output of the DSS is the identification of a portfolio of projects and interventions more suitable to contribute to the definition of a circular regeneration strategy for the area of the Cultural Corridor. Those have been identified through the exchange with the group of local experts involved in the study and taking into account all the conditions and the budget scenarios explored with them.

2 Adaptive Reuse of Cultural Heritage from a Circular Perspective

Nowadays the cities begin to look at the possibilities offered by the circular economy in order to try to reduce their negative impacts on the environment. Indeed, the built environment has a prominent role in the consumption of global resources and generation of waste. In the last decades, the urban sprawl has determined an impoverishment of urban landscape, generating land consumption and abandoned areas. For its high impacts, the construction industry should contribute more to the development of urban sustainability strategies. Several studies have underlined the benefits to shift from new-build to reuse or refurbishment vision.

Indeed, the renewed use of buildings could address the need of new places for activities, without increasing energy consumption, land use, and waste for new constructions. In enhancing spaces and changing their functions, the adaptive reuse allows obstructing buildings obsolescence, recovering economic value and use the embodied energy still available in the existing building stock.

However, the adaptive reuse of cultural heritage has several limitations (e.g. historical significance) to be respected and, for this reason, it has seemed, sometimes, more difficult to realize (Hein and Houck 2008). On the other hand, effecting on cultural values and places identity, the reuse of cultural heritage is able to attract many actors and interests as proved from several ongoing experimentations in which different stakeholders are directly involved in the reuse process (Mangialardo and Micelli 2018).

Indeed, in the last few years, the spending capacity of public authorities has completely changed. More and more the public owners have not been able to manage and fund alone the reuse of cultural sites. Consequentially, actions of organizations and groups interested in reusing abandoned property have become more recurrent, showing as the cooperation among public, private and non-government sectors is crucial not only to start and carry out reuse projects but also to sustain the reuse over time (Macdonald and Cheong 2014).

In this perspective, to make the most of existing resources, adaptive reuse should be considered in a larger urban vision adopting more strategic and integrated approaches. Indeed, the current urban challenges (urgent needs, quick changes, scarce resources) cannot be dealt with traditional planning.

Circular and sustainable solutions should consider a bigger scale than a single intervention, exploring possible interactions among new uses, context, current, and future city programs. Therefore, the conceptualizing of circular reuse actions/strategies of cultural heritage must meet several issues and becomes a very articulated process that needs specific support.

In this direction, the DSS can help the reuse process of cultural heritage in different stages as shown in the next paragraph.

3 Decision Support System Methodology

The DSS developed in the CLIC project proposed a step-by-step procedure to help to make transparent and informed decisions at different stages of the reuse process (defining proposals, projects assessing, interventions programming, monitoring of reuse actions). It has been developed to take into account the diverse points of view of actors involved (public authorities, experts, civic society, citizens), and integrated several approaches to manage the complexity of the circular reuse challenges.

Firstly, multicriteria approach has been used to take into account the different aspects of the problem (economic, environmental, cultural, social) including those intangible and difficult to quantify (e.g. heritage perceptions and significances).

Indeed, trough the multicriteria analysis, economic aspects (Wang and Zeng 2010) can be analysed together with environment and territorial issues (Veldpaus et al. 2013), and qualitative dimensions of the heritage, as the place identity, intrinsic values, well-being and life quality of citizens (Tweed and Sutherland 2007).

Nowadays, there are several multicriteria methods and the one selected should be in relation to the features of the specific problem. In Sect. 3.1.2 it has been discussed some arguments to support the identification of opportune methods in heritage reuse problems.

However, because currently decisions are often made in situations marked by the lack of resources, a multicriteria perspective alone is not sufficient to make effective decisions. In that case, it could be beneficial to consider also budget limitations and the larger benefits of combined reuse interventions. In other words, the reuse actions should not be analysed individually, but in relation to the synergies that are able to generate with other actions.

Indeed, the adaptive reuse problem from a circular perspective needs to be analysed as a portfolio problem. This kind of problem is explored thanks to a portfolio decision analysis that aims to identify the set of actions to be implemented with respect of several constraints as the budget limitations. In the DSS, a multimethodological approach has been developed to combine the portfolio analysis with several multicriteria methods. In addition, different kinds of constraints have been modelled to explore possible synergies among actions in an urban context.

In the DSS, both multi-criteria approach and portfolio approach has been designed to be implemented in an interactive way. It represents an additional innovation and has been planned to facilitate the contribution of several actors in the decision process for the adaptive reuse of cultural sites. Indeed, according to the recent recommendations on the heritage (Council of Europe 2005), the activities that regards the cultural heritage need to take into account the diverse points of view and to promote inclusive processes (Dutta and Husain 2009). In this direction, procedures oriented to facilitate the interactions with stakeholders are of particular interest (Fusco Girard et al. 2014). The literature on the stakeholder engagement is very rich and different methods have been tested in participatory processes. However, the current interest in multi-stakeholders environments is focused on co-management processes of cultural and natural assets. In this kind of processes, the different actors are affected in a direct way by the decisions. In that case, the interactive process should aim not only to explore the diverse points of view (as in traditional participatory processes), but also to include them in the decision, keeping high interest of people involved and improving transparency of decisions. In this direction, the DSS used simple interactions to create a mutual-learning environment for the local actors and improve their awareness on adaptive reuse opportunities thanks to exchanging among the expert and common knowledge. During the process, those interactions helps to clarify the problem, discuss and validate results, and in some cases to generate new solutions in a collaborative way. The interactive process of the DSS has been described more in depth in the Sects. 3.1.1, 3.1.2, 3.1.3, and 3.1.4.

Compared with other procedures, the DSS, combining different approaches and methods, allows making a comprehensive decision on several reuse actions, identifying a satisfying project or portfolio of actions with the actors involved.

3.1 How to Implement the DSS

The DSS is a procedure that helps to identify a satisfying reuse project or set of reuse projects. Through a step by step process, several points of view and impacts of the new uses are explored, developing an integrated knowledge on the reuse opportunities of specific cultural sites. The DSS consists of four main stages:

1. Description of the reuse problem.
2. Problem modelling and application of a multicriteria method.
3. Generation and selection of portfolios of actions.
4. Robustness analysis and validation of results.

The procedure has been designed to be interactive and in each stage, one or more interactions with the actors involved can be conducted. These interactions are important to explore because key information of a specific decision problem are collected and allow the system to be adaptable at different contexts.

It requires expert knowledge for managing process, conducting interactions with local actors, collecting and processing the data. The team should include at least one specialist on the cultural heritage, one analyst, and one facilitator.

3.1.1 Description of the Reuse Problem

This stage regards the identification and description of the reuse problem through its key elements: stakeholders involved, aims and points of view to be considered, proposals and criteria to assess them, and eventual constraints to reuse cultural sites (e.g. requirements/performance to be respected, etc.). It is a crucial phase because it regards the problem statement and strongly influences subsequent stages of the procedure.

As underlined in the literature, many mistakes in the choice and application of decision support tools came from a non-right definition of the problem (Cinelli et al. 2022). Usually, to address messy or wicked problems, Problem Structuring Methods (PSMs) are suggested to improve situations characterised by uncertainty, conflict, and complexity. Fundamentally, PSMs (for example Stakeholder Analysis, SWOT analysis, Scenario Analysis, Soft Systems Methodology, Strategic Options Development, and Analysis, etc.) requires to analyse a problematic situation from different perspectives, facilitating the synthesis and good organisation of information collected (Mingers and Rosenhead 2004).

It has been proved that exploring in a structured way the perspectives, values, and preferences of actors responsible for and impacted by the decisions facilitates the right identification of the problem especially in the multi-stakeholders environments, and consequentially allows to carry out more robust analyses.

In any cases, to have an effective stakeholder engagement in the problem definition and description stage, several techniques and recommendations can be used. The literature on group involvement is large, many studies underlined as effective communication is important for managing discussion or conflicts while the engagement approach to be used (e.g. Pull communications, Push communications, Consultation, Participation, Partnership) should be selected according to the features of the specific situation.

In the CLIC project the exploration of objectives and issues relevant for the local actors have been conducted within the Heritage Innovation Partnerships process (HIPs) where several interactions with them took place. The main outputs of this process (e.g. identification of local adaptive reuse issues, general and specific objectives, exploration of points of view of diverse actors involved) have contributed to define the local problems for the CLIC pilots cities.

Another important point in the definition of adaptive reuse problem is the scale of analysis, which can interest a single building, a set of buildings considered as an unique element/project, or a portfolio of buildings/cultural sites to be analysed as distinct elements/projects. In the first two cases, the DSS supports the definition of a satisfying reuse project taking into account benefits for a larger community and the several issues explored. In the third case, the DSS supports the definition of a satisfying portfolio of projects to be implemented together. This could be the case of the design/development of regeneration/reuse strategies of large historical urban areas in which different kinds of sites/projects/interventions and synergies among the different actions need to be considered.

3.1.2 Problem Modelling and Application of a Multicriteria Method

At this stage, the problem is modelled according to the elements defined in the previous stage. Firstly, it is important to identify the type of problem (choice, ranking or sorting problem) and measurement scale of criteria. The identification of the problem also infer how the decision recommendations system should be built (e.g. elicitation of preference, features of the aggregation, etc.).

In general, the cultural heritage features and the adaptive reuse issues explored in the CLIC project suggest choosing a multi-criteria method that allows:

- to work with the qualitative and/or quantitative criteria,
- to handle criteria with heterogeneous scales,
- to take into consideration imperfect knowledge of data,
- to model and detect the preferences through exchange with the actors involved.

With respect to the last point of this list, the DSS procedure uses the Simos-Roy-Figuiera method (SRF) (Figueira and Roy 2002) to explore the relative importance of criteria thanks to direct interaction with actors or decision makers. Indeed, the SRF method works in this way: firstly it is asked to order the criteria according with their relative importance and then to identify the smaller or bigger differences of importance among them. The method has been tested in the CLIC pilot cases. These tests gave a good response in relation to the interaction with actors involved and outputs provided.

In case the scale of the reuse problem is a single building, the results of multi-criteria analysis provide a solution and the robustness analysis should be conducted immediately after.

3.1.3 Generation and Selection of Portfolios of Actions

This stage needs to be considered only in case the reuse problem has been defined as a portfolio problem. It means that the diverse actions must be considered as a part of a comprehensive program of interventions. In that case, the output of the DSS is represented by a portfolio of reuse actions able to respect the recommendations identified in the previous step (ranking/sorting of the actions conducted with multi-criteria analysis) and a coherent set of constraints explored (e.g. synergies among the actions, budget limitations, etc.).

More in depth, in order to identify a satisfying solution, several scenarios, different for budgets and constraints configurations are generated and a binary linear programming model is applied. For each scenario, a portfolio of actions is identified that respects the ranking/classification of the actions and do not violate constraints. In this way, more portfolios are generated which should be shown to the actors involved in the process. After the discussion with them, a satisfying portfolio of actions should be identified. In case no solution is accepted, the set of constraints should be reconsidered and new solutions generated in order to support the identification of a compromise solution.

The opportunity to integrate different ranking/sorting multi-criteria procedure into a portfolio decision analysis (PDA) approach represent an original contribution of the DSS along with the integration of an interactive process in PDA. Indeed, usually, in PDA the solution is not determinate through the exchange with participants involved or decision-makers. In this perspective, the DSS allows to explore and solve a portfolio problem in very complex situations integrating different approaches.

3.1.4 Robustness Analysis and Validation of Results

This stage regards the stability of the DSS recommendation, testing their adherence to the preferences system of actors involved. Basically, it is verified if the results obtained are sufficiently stable with respect to the variations of some parameters (e.g. the importance of the criteria, the formulation of some of the constraints, etc.). In this way, several scenarios are considered to realize whether the solution is consistent and the actors still agree with it.

4 Conclusions

The DSS developed in the CLIC project is a multimethodological procedure that allows using of the scarce resources available in a more accurate way improving transparency to the choices related to the transformation of the natural, built and historic environment.

The procedure helps to explore the opportunities to reuse cultural heritage through the interaction with different actors in order to consider larger benefits for the city, the citizens and the stakeholders. Indeed, reuse processes interest a multiplicity of cultural, economic, environmental and social features and heterogeneous objectives. For that, the procedure proposed is strongly interactive in order to take adequately into account the plurality of points of view involved in the decision process.

During the different stages of the DSS, different interactions are carried out to identify and model the key elements of a specific reuse problem. For example, it is suggested implementing a multicriteria method in a constructive way for the ranking or sorting of actions. In case of a portfolio problem, conditions, budget constraints and satisfying solutions can be defined in an interactive way, promoting a dialogic and learning perspective and facilitating the contribution of different actors at the design of the most appropriate solutions.

In this way, during the process, new knowledge is developed that helps the formulation of useful justifications and argumentations for acknowledging the goodness of the proposed solutions and supporting the adopted decisions towards a third party and public opinion.

From the technical point of view, the integration of different approaches and methods represents the most innovative part of the work.

References

Blake J (2000) On defining the cultural heritage. Int Comp Law Q 49:61–85

Cinelli M, Kadziński M, Miebs G, Gonzalez M, Słowiński R (2022) Recommending multiple criteria decision analysis methods with a new taxonomy-based decision support system. Eur J Oper Res 302(2):633–651. https://doi.org/10.1016/j.ejor.2022.01.011

Council of Europe (2005) Council of Europe Framework Convention on the value of cultural heritage for society. Faro, Lake Mary

Council of the European Union (2014) Council conclusions of 21 May 2014 on cultural heritage as a strategic resource for a sustainable Europe

Dutta M, Husain Z (2009) An application of multicriteria decision making to built heritage. The case of Calcutta. J Cult Herit 10(2):237–243

European Commission (2020) Proposed Mission: 100 Climate-neutral Cities by 2030 – by and for the Citizens. Report of the Mission Board for climate-neutral and smart cities.

Ferretti V, Comino E (2015) An integrated framework to assess complex cultural and natural heritage systems with multi-attribute value theory. J Cult Herit 16(5):688–697

Figueira J, Roy B (2002) Determining the weights of criteria in the ELECTRE type methods with a revised Simos' procedure. Eur J Oper Adapt Reuse Cult Herit 139(2):317–326

Fusco Girard L, Cerreta M, De Toro P (2014) Integrated assessment for sustainable choices. Ital J Reg Sci 13:111–142
Geissdoerfer M, Savaget P, Bocken NM, Hultink EJ (2017) The circular economy-a new sustainability paradigm? J Clean Prod 143:757–768
Giove S, Rosato P, Breil M (2010) An application of multicriteria decision making to built heritage. The redevelopment of Venice Arsenale. J Multi-Criteria Decis Anal 17(3–4):85–99
Greco S, Figueira J, Ehrgott M (2016) Multiple criteria decision analysis: state of the art surveys. Springer, Berlin
Hein MF, Houck KD (2008) Construction challenges of adaptive reuse of historical buildings in Europe. Int J Constr Educ Res 4(2):115–131. https://doi.org/10.1080/15578770802229466
Hong Y, Chen F (2017) Evaluating the adaptive reuse potential of buildings in conservation areas. Facilities 35(3/4):202–219
Ishizaka A, Nemery P (2013) Multi-criteria decision analysis: methods and software. Wiley
Macdonald S, Cheong C (2014) The role of public-private partnerships and the third sector in conserving heritage buildings, sites and historic urban areas. The Getty Conservation Institute, Los Angeles
Mangialardo A, Micelli E (2018) From sources of financial value to commons: emerging policies for enhancing public real-estate assets in Italy. Pap Reg Sci 97(4):1397–1409. https://doi.org/10.1111/pirs.12310
Mingers J, Rosenhead J (2004) Problem structuring methods in action. Eur J Oper Res 152(3):530–554. https://doi.org/10.1016/S0377-2217(03)00056-0
Nakajima N (2000) A vision of industrial ecology: state of the art practices for a circular and service-based economy. Bull Sci Technol Soc 20:54–69
Nesticò A, Morano P, Sica F (2018) A model to support the public administration decisions for the investments selection on historic buildings. J Cult Herit 33:201–207
Tweed C, Sutherland M (2007) Built cultural heritage and sustainable urban development. Landsc Urban Plan 83(1):62–69. https://doi.org/10.1016/j.landurbplan.2007.05.008
Veldpaus L, Pereira Roders AR, Colenbrander BJ (2013) Urban heritage: putting the past into the future. Hist Environ Policy Prac 4(1):3–18
Wang H-J, Zeng Z-T (2010) A multi-objective decision-making process for reuse selection of historic buildings. Expert Syst Appl 37(2):1241–1249. https://doi.org/10.1016/j.eswa.2009.06.034

Chapter 10
Social Sustainability Framework for Measuring Socio-cultural Impacts of Cultural Heritage Adaptive Reuse

Agata Żbikowska, Magdalena Roszczyńska-Kurasińska, Bartosz Ślosarski, and Anna Domaradzka

1 Introduction

In this chapter, we argue that the successful implementation of circularity frameworks, especially in cultural heritage projects that relate to local identity, shared history, and place attachment, might require long-term monitoring of the socio-cultural context in general and social sustainability in particular (Roszczyńska-Kurasińska et al. 2021). This importance stems from two primary reasons: first, certain aspects of the socio-cultural context affect the adaptation process that aims to introduce change in built environment as well as in a mindset of local society; second, the adaptive reuse of cultural heritage impacts the socio-cultural context of local communities by offering new solutions and information. This feedback loop can operate in both directions. It might either strengthen the local community and make adaptive reuse a successful endeavor, or it could lead to insignificant changes or even negative impacts on the neighborhood.

Our past research shows that many investments in circular re-adaptation have resulted in unforeseen positive outcomes and spillover effects that increased the well-being of local communities (Roszczyńska-Kurasińska et al. 2019). However, there have also been projects that failed to meet the needs of local communities. In these cases, existing habits, attitudes towards change, and traditions (Eagle 1999) were often not carefully considered during the implementation of circular solutions, leading to lack of acceptance by the community. An unrecognized social context has

A. Żbikowska · M. Roszczyńska-Kurasińska (✉) · A. Domaradzka
Robert Zajonc Institute for Social Studies, University of Warsaw, Warsaw, Poland
e-mail: agata.zbikowska@uw.edu.pl; m.roszczynska@uw.edu.pl; anna.domaradzka@uw.edu.pl

B. Ślosarski
Robert Zajonc Institute for Social Studies/Faculty of Sociology, University of Warsaw, Warsaw, Poland
e-mail: b.slosarski@uw.edu.pl

L. Fusco Girard, A. Gravagnuolo (eds.), *Adaptive Reuse of Cultural Heritage*,
https://doi.org/10.1007/978-3-031-67628-4_10

caused these projects to grow in isolation from the local community, hindering their potential positive impacts. This issue is particularly visible in projects which are driven by single goals, such as economic or cultural, where a more complex long-term goal, like social sustainability, is not considered. While socio-cultural factors are crucial for the overall success of the circular economy, they are also the most elusive and understudied. This presents an interesting challenge, which we address by defining the scope of potential social impacts and identifying specific features of local communities that may be decisive for successful circular project implementation.

This chapter focus on combining existing theoretical framework for social sustainability (Missimer et al. 2017) with well-being indicators to propose a valid approach for assessing the socio-cultural impact of adaptive reuse investments in the cultural heritage field. We begin with a review of the existing and widely used macro level measures of well-being to identify potential indicators related to cultural heritage. Further examining the specificity of adaptive re-use, we focus on the relation between adaptive reuse projects and local communities, keeping in mind that the prosperity of any community depends on its ability to adapt to an ever-changing environment. We explore the potential features of sustainable community and test possibilities of measuring them.

2 Cultural Heritage in the Well-Being Indexes

The importance of cultural heritage in shaping local identity, place attachment, and economic development is well established. However, its role in the formation of well-being is less recognized, partly due to the way in which well-being is defined and measured. In this chapter, we explore this issue from a historical perspective. The concept of the life well-lived and the overall happiness of an individual has been important to philosophers since ancient times. There have been two main trends in the conceptualization of well-being (Oishi 2010). One trend, deriving from Aristotle, defines well-being as encompassing "not only life satisfaction and positive affect, but also purpose in life, a sense of autonomy, self-acceptance, connectedness, and a psychological sense of vitality" (after Oishi 2010, p. 37), the other trend is more closely associated with the idea of happiness derived from pleasure and the absence of pain, thefulfillment ones' wishes, and the attainment of good health and relationships.

However, the measurement of well-being is a relatively recent development, emerging at the beginning of the 20th century and gaining momentum after World War II. Initially, well-being was primarily assessed through economic indicators of living standards. One of the first tools used for this purpose was Gross Domestic Product (GDP), developed around 1934 in the United States by Kuznets (1934).

Around the same time, Maslow (1943) offered a somehow different approach to understanding of well-being. This renowned psychologist view well-being as a combination of both physical and psychological dimensions. In his understanding, the feelings of physical and economic security, good health, and lack of pain must be complemented by opportunities for self-development, a sense of connectedness, and autonomy. However, his work did not lead to the creation of any particular measurement tool.

In social sciences, one of the first indexes developed for measuring well-being at the national level was the Physical Quality of Life Index (PQLI), proposed in 1970 by Morris Davis Morris. This index considered statistics such as literacy rate, infant mortality and life expectancy at age one. The PQLI employed a rather utilitarian approach relying on quantitative data, and not accounting for individuals' subjective impressions. However, later developments in social science sought to accommodate the subjective perception. In 1985, Diener, Emmons, Larsen and Griffin published a paper introducing a tool called Satisfaction With Life Scale (SWLS). This scale measured life satisfaction by asking respondents to report their level of agreement with five statements on a seven-point response scale. The SWLS, which was later used as a national index (Diener 2000), included statements such as "So far I have gotten the important things I want in life" or "If I could live my life over again, I would change almost nothing" which focused on the respondent's subjective views. These two approaches—objective, relating to health and living conditions and subjective, relating to individual feelings of happiness remain the main frameworks for creating tools and indicators for measuring life satisfaction.

In the earliest discussion on wellbeing, there was almost no reference to cultural heritage, however some developments have been made in the recent years. Our investigation into the most popular and innovative well-being indexes revealed that cultural heritage, although infrequently, is sometimes included in the measurement of well-being. Altogether, we analysed 15 indexes: Canadian Index of Wellbeing, Human Development Index, Happy Planet Index, World Happiness Index, Gallup World Survey, Genuine Progress Indicator, Index of Individual Living Conditions, JFS Sustainability Vision and Indicators, UN Sustainable Development Goals, Better Life Index, UK National Well-being, Social Progress Indicator, Quality of Life (Eurostat), BES Equitable and Sustainable Wellbeing, and Gross National Wellness. The selection of wellbeing indexes was based on the number of scientific publications citing each index from 2000 till 2019. The search was conducted using the Google Scholar search engine. The main aim of this analysis was to verify whether cultural heritage is represented in wellbeing indexes and, if so, to identify which aspects of cultural heritage have been most recognized as important factor.

The analysis revealed that very specific aspects of cultural heritage are measured in the listed indexes: loss of cultural heritage, public spending on cultural heritage, cultural participation, and total production of traditional crafts. Other aspects of cultural heritage, like accessibility or its importance to local or national community, are not reflected in the indexes. Our analysis shows that only 4 out of 15 well-being indexes analysed and 9 out of 890 indicators that compose those indexes, took into account indicators related to the cultural heritage field. Therefore, we focused our reserch on these four indexes that acknowledge the cultural heritage: United Nations' Sustainable Development Goals, Canadian Index of Wellbeing, Italian BES index, and JFS Indicators of Sustainability.

The United Nations Sustainable Development Goals (UNSDG) index was introduced by United Nations as a part of their worldwide Agenda 2030, which aims to achieve a sustainable future for the global population and eradicate global poverty. The index includes seventeen factors of well-being, encompassing structural, economic, environmental, and social indicators of life quality. Despite covering a broad range of aspects of contemporary human life, only one indicator within the entire set directly pertains the cultural heritage field. This indicator was defined as "direct economic loss to cultural heritage damaged or destroyed attributed to disasters", and it is measured by the United Nations in millions of current United States dollars. This indicator focuses on the economic aspects of cultural heritage sites, quantifying the economic loss due to armed conflict, climate change, and natural disasters in a given area. While this can be informative to planning of sustainable reconstruction (Guarnacci 2012) it does not fully capture the broader issue of the destruction of cultural heritage sites. Moreover, the losses resulting from top-down decisions are not accounted for.

The UNSDG index is frequently used in adaptive reuse of cultural heritage and within the public policy field. However, the index is more often applied as a guide for setting goals and directing the transformation of cultural heritage assets rather than as an analytical tool or a basis for cultural heritage studies. Several studies have demonstrated this purpose. For example, Fatorić and Egberts (2020) introduced the term UNSDG "climate change actions" in an explorative study of perceived heritage benefits for leveraging climate change policy in the Netherlands. Similarly, Wong (2019) discussed opportunities to achieve UN sustainability goals within the Malaysian heritage industry by using a self-developed cultural heritage refurbishment framework.

The second index was Canadian Index of Wellbeing (CIW), a composite index predominantly focused on the Canadian population. The CIW covers eight areas related to both tangible/objective and intangible/cultural factors, such as community vitality, democratic engagement, education, environment, health, leisure, living standards, and time use. In the context of cultural heritage and associated fields, the CIW can be used to measure culture-related activities of citizens through cultural indicators such as the "average percentage of time spent on the previous day in arts and culture activities" and the "average number of hours volunteering for culture and/or recreation organizations". The index also references historic heritage sites, specifically in terms of "average visitation per site to all National Parks and National Historic Sites". While the index provides valuable insights into the importance of culture in the everyday life of the studied population, it offers a limited information about the role and significance of specific heritage sites.

The third index – BES index (Benessere Equo e Sostenibile = Equitable and Sustainable Well-being) is a national-scale composite subjective well-being index associated with the Italian National Institute of Statistics. The BES index is broader in terms of cultural heritage and its structural background. Inspired by the Happy Planet Index, the BES index comprises eight dimensions that describe three different areas: personal well-being, social well-being, and well-being at work (Iacus et al. 2020). It covers issues of cultural policy, such as "current expenditure of municipalities in the cultural sector (€ per capita)" and "cultural participation" understood as a number of active users of cultural goods and assets. It also includes the "density and relevance of museum heritage (values for 100 km^2)". Among all the reviewed indexes, the BES index most significantly addresses cultural heritage positing it as a common good of a non-renewable nature that is crucial to general well-being (Costanzo and Ferrara 2015).

The last index, which we included in our review as a national index that addresses issues of tradition and cultural heritage is the Japan for Sustainability (JFS) index, called "Indicators of Sustainability." Regarding cultural heritage, the JFS index primarily focuses on traditional skills that are considered part of intangible cultural heritage. One of its indicators is strictly centred on the "production volume of traditional crafts" which measures the economic potential of traditional manufacturing areas.

The presented review of well-being indexes reveals that cultural heritage has not yet been widely recognized as an important indicator of wellbeing. Only the most recent indexes, such as the Italian BES index and the Canadian Index of Wellbeing, consider the value of cultural heritage as a potential contributor to well-being. Notably, the intangible effect of cultural heritage on local community are often overlooked. The important relationship between built cultural heritage, particularly the circular adaptive reuse of it, will be the focus of the next section.

3 The Social Aspect of Adaptive Reuse of Built Heritage

The adaptive reuse of built heritage process, which involves reorienting production and consumption patterns towards sustainability through Circular Economy models, requires focused action at all levels, from local communities to organizations. This is an enormous challenge that entails identifying and sharing common values (the 'mission' or 'purpose') as the change-enabling factor. As pointed out by LeMille: "Value is important or lasting beliefs shared by the members of a culture about what is desirable or undesirable" (Lemille 2017). Value is therefore recognized as a key element in driving the development and success of "Circular Economy 2.0", where profit-oriented businesses are replaced by purpose-oriented ones (Lemille 2017), creating shared value for all stakeholders (Kramer and Porter 2011).

Closing the loops, changes the traditional logic of both economic and social development towards one driven by sustainability. The 2030 Agenda for Sustainable Development (Desa 2016) highlights the role of cultural heritage as an enabler of social cohesion and inclusion by developing places and spaces that foster the building of a shared identity, therefore enhancing place attachment and the integration of newcomers. In its documents, the ICOMOS Task Force underlines the importance of cultural heritage and creativity as a driver for equity and inclusive economic development in urban context (e.g. heritage places as incubators of creativity, cultural capital and sustainable tourism). It also underlines the role of cultural heritage in improving the quality of life, sustainability, and resilience in urban areas through implementation of practices such as walkability and compactness, the adaptive reuse of existing built fabric, and the embodiment of traditional skills and knowledge (ICOMOS 2017).

Similar inspiration comes from the recent works on the concept of the human-centred city (Landry 2019; UNESCO 2016). As stated in the UNESCO Global Report on Culture for Sustainable Development (2016):

> Culture lies at the heart of urban renewal and innovation. (…) Creativity and cultural diversity have been the key drivers of urban success. Cultural activities can foster social inclusion and dialogue among diverse communities. Tangible and intangible heritage are integral parts of a city's identity, creating a sense of belonging and cohesion. Culture embodies the soul of a city, allowing it to progress and build a future of dignity for all. (…) A human-centred city is a culture-centred space. (…) We must strengthen the cultural assets of cities, the heritage that provides a sense of meaning and identity to their inhabitants, and the creative opportunities that enhance the vitality, liveability and prosperity of our cities.

In their latest report, Landry (2019) emphasizes the need for fostering innovation through the physical urban setting, where "there is history, character, distinctiveness, diversity and vitality, with high levels of liveability and all the necessary support facilities, from health and education to culture and public spaces." All of these factors have the potential to generate a rich civic life and make our cities more human-centred, where "citizens become city-makers and shapers, makers and co-creators of their evolving urban development" (p. 18).

Scientists are increasingly recognizing that profit-maximization actions taken by business entities often have negative impacts on communities, cities and regions. This is because business activities frequently infringe upon basic principles of social, environmental, ethical, and human rights. As a result, there is growing interest within the public, private and social economy sector in conducting business in a more sustainable way, with a focus on improving community quality of life and generating positive social impacts through for-profit actions (Gidron and Domaradzka 2021). We propose linking indicators related to the social economy and impact investing with the final measurement tools of community well-being to reflect the shared value and inspire a different approach to corporate social responsibility practices.

We also believe that effort to measure the socio-cultural impacts of cultural heritage adaptive reuse can draw inspiration from the Theory of Change, which is widely applied in evaluation studies. Theory of Change focuses on providing a comprehensive description of how and why a desired change is expected to happen, particularly by mapping out the "missing middle" between activities and achievement of desired goals. In looking for a comprehensive framework for conducting circular interventions in the cultural heritage field, it is essential to understand the features that make a community more predisposed to desirable change. **Marginalization of the role of society in the circular economy** could severely impede the implementation of this new economic logic. Communities should not be seen merely as end users or beneficiaries of a transition but rather as active players who can enforce change, either by pressuring companies to adopt new technologies or by introducing change themselves. In many cases, the adoption of new ideas, products, and solutions starts with local activists who demonstrate to their neighbours that alternative forms of conduct are possible and more beneficial in the long term (Brandsen et al. 2016).

A similar problem was diagnosed earlier in the sustainability domain (Colantonio et al. 2009). Although many interventions aiming at achieving sustainability have social consequences, the social dimension of sustainability remains vaguely defined and barely examined (Dempsey et al. 2011; Vallance et al. 2011). Numerous adaptive reuse projects have called for an investigation into which measures should be employed to support an effective transition toward sustainability (Johnston et al. 2007; Marsden et al. 2010). These projects, particularly, strived to assess which solution work under specific conditions and what makes a social system sustainable.

In response to these challenges, Missimer, Robèrt, and Broman (Missimer et al. 2017) proposed a universal framework for studying social sustainability. Building on the study of **complex adaptive systems**, they developed a distictive profile of a sustainable social system – one that is capable to prosper amid uncertainty and constant change. They list altogether five key characteristics of a social system that are essential for achieving sustainability: **diversity, common meaning, trust, capacity for learning**, and **capacity for self-organization**.

The first aspect of the adaptive capability of social systems is **diversity** (Norberg and Cumming 2008). This concept encompasses diversity of knowledge, skills, opinions, beliefs, and values. Anything that contributes to the variety within a community enhances its preparedness for the unknown (Folke et al. 2005). Heterogeneity, which reates creative tension and encourages out-of-the-box thinking, is believed to facilitate innovation. The adaptive reuse of cultural heritage has a significant potential to strengthen this aspect of local communities. Therefore, measuring the level of diversity is crucial for assessing both the opportunities for embracing the change and the impact of the projects on the diversity of the local community.

The inner heterogeneity of society interacts with the second aspect of resilient social systems—**common meaning**. Common meaning is the ability of people to make sense of their situation and actions (Cacioppo et al. 2005). It helps them to set goals, establish rules of conduct, and agree on values as a community. In the context of a circular economy, the existence of common meaning an facilitate transition, but only if the shared understanding within a community aligns with the principles of the circular economy. For instance, if a local community perceives its environment as an unlimited resource (what could be their common meaning) they might be reluctant to invest extra time and money in recycling, refurbishing, or reusing. In such circumstances, the likelihood of grassroots circularity emerging spontaneously is very low. The introduction of adaptive reuse of cultural heritage in communities with low ecological consciousness can have a particularly significant impact. Circular adaptive reuse of built heritage can resonate within the local community, teaching them new practices. Over time, a society can even learn to value circular solutions, although the path to this goal may be challenging.

What helps a community to achieve a new understanding of reality are **trust** and the **capacity for learning**, two additional aspects of a sustainable social system.

Communities vary in their ability to learn; some adjust quickly to changing environments; while others require more time to adapt. Societies with a high capacity for learning are considered more resilient because they sense change fairly quickly, gaining extra time to react when the change actually occurs (Scheffer et al. 2001). The diverse experience of society members enhances their learning potential, particularly through learning-by-doing, which is an effective way to aquire skills and knowledge (Schank et al. 1999). The circular economy demands **innovation and experimentation**, which can be more easily achived in a society that is well-prepared for attaining new skills and knowledge.

In a social environment, the safety and ease of introducing new projects can also be described by the **level of trust** among its members. When people trust each other, they focus on potential opportunities rather than perceived threats, and they do not waste time and effort on verifying others' trustworthiness (Roszczynska-Kurasinska and Kacprzyk 2013). Instead of protecting themselves against potential mistrust, they can concentrate on developing shared meaning and creating better solutions (Nowak et al. 2019).

A high level of trust facilitates open-minded interaction and the exchange of opinions, which can serve as foundation for self-organization. The **capacity for self-organization** is the final aspect of a resilient social system highlighted by Missimer et al. (2017). To respond quickly and adequately to unpredictable shocks and changes in the environment, communities need to process the potential for self-organization at different levels and across varying scope (Levin 1998). This capability allows them to create new structure or enhance existing ones, enabling collective action, disseminating of information, and collaborating to ensure sustainability of a given project.

4 Measuring Social Sustainability

Dividing the concept of social sustainability into five clearly defined dimensions enhances strategic planning and enables the introduction of innovations within the field of the circular economy. It is also a crucial step in developing a better measurement framework for assessing the social impacts of the adaptive reuse intervention. To fully leverage this concept, policymakers and decision-makers need to be equipped with tools that can effectively capture these five dimensions. In this context, we examine the composition of existing well-being indexes from the perspective of social sustainability to test their usefulness for impact assessment. To acomplish this, we extracted all the indicators related to any of the dimensions of social sustainability by assigning them specific codes. Our analysis included all five dimensions of social sustainability.

The results revealed that five dimensions of social sustainability are not equally represented in existing well-being indexes (Fig. 10.1). Moreover, not all indexes take into account all dimensions in the same proportion. For example, Gallup World

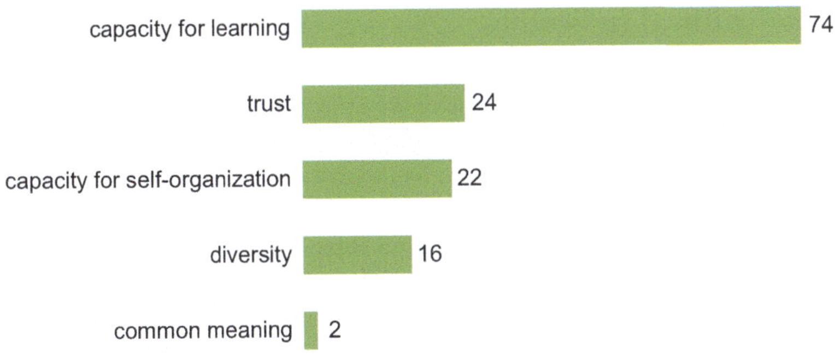

Fig. 10.1 Number of indicators measuring individual dimensions of social sustainability based on 890 indicators from 15 indexes of well-being

Survey covers four dimensions – capacity for learning, common meaning, trust, and capacity for self-organization – while Better Life Index only one – capacity for learning. The most highly represented dimension in well-being indexes is the **capacity of learning**, with 8% of indicators measuring this aspect. In contrast, common meaning is the least measured dimension of social sustainability. Other dimensions, like diversity, trust, and capacity for self-organizations, are only superficially present in the indexes.

In the studied indexes, the measurement of capacity for learning primarily focuses on the level of education. The education rate has been a part of many well-being indexes for some time. Is is often measured by mean years of schooling or proficiency in different areas of education. It also considers inequalities in education, such as gender, socio-economic status, and rural to urban disparities.

Indicators measuring **diversity, capacity for self-organization and trust** are much less visible in these indexes with their representation being three times lower than that of capacity for learning. Out of 890 indicators, only 24 are dedicated to trust, 22 to self-organization, and 16 to diversity. Trust in well-being indexes is often measured by the level of trust in local and national authorities (such as the judicial system, members of parliament, etc.). The capacity for self-organization is understood as the ability to act in the non-private sphere, expressed through indicators such as past experiences of volunteering and involvement in environmental organizations. Diversity is captured in two specific ways: first, economic inequalities measured by income and outcome indicators or income gaps expressed by the Gini coefficient, and second, gender inequalities based on different types of power asymmetries in private, public, and professional sphere.

The least represented factor is **common meaning**, with only two indicators out of 890 classified as capturing this dimension. The aspect considered under common meaning include voicing one's opinion publicly and having a sense of belonging to the community.

5 Socio-cultural Indicators

The above analysis shows that questions related to sustainability in social systems are scarce in global and local well-being indexes. Therefore, in this part of the article, we propose more adequate measurements of socio-cultural impact in the context of cultural adaptive reuse. We have created questionnaires designed to evaluate these aspects for the proposed project within three groups: users (all people who visit or will visit the renovated site), the local community (people leaving in the vicinity of the site), and stakeholders (public and private actors included in the process of adaptive reuse). Ideally, the questionnaires might be used before renovation as a means of evaluating the direction of changes in planned work and after the project is completed to assess the influence of the adaptation process on a given neighbourhood.

We propose to measure the aforementioned aspects in two main areas—at a personal or grassroots level, where the needs and opinions of individuals can be assessed; and at the institutional level, where the influence of actions taken by local governments or institutions can be estimated. Some aspects will be more easily measured at the individual level, while others require a broader view, thus necessitating the inclusion of questions that allow for both types of assessment.

As mentioned above, we assumed that diversity, capacity for self-organization, common meaning, learning and trust are key elements in creating a sustainable society – one that is resilient and capable of change. Therefore, the measurements proposed in our questionnaires are designed to account all these elements.

5.1 *Diversity*

In the context of adaptive reuse of cultural heritage, it is important to assess the diversity of the local community as well as ability of the transformed built heritage to not only accommodate but also enhance this diversity. The diversity of local community can be measured by asking respondents how they perceive others who live in the area. For example, a statement like "Our community is highly diverse in terms of skills/professions" can provide insights into perceived diversity. Alternatively, rather than measuring the current level of diversity in the community, one could assess the community's potential to embrace a higher level of diversity. This potential can be indirectly measured by statements such as "I see myself as someone who enjoys meeting new people" or "I see myself as someone who enjoys new ideas" as openness to new people and ideas is essential for embracing diversity. The assessment of a built heritage site's ability to accommodate a diverse population typically involves measuring its accessibility and safety with statements like "This is a place accessible for everybody", "This is a place where people feel safe". However, these questions alone are insufficient, as various groups of people may perceive diversity, safety, and accessibility differently due to factors such as age, sex, or health. Therefore, we propose that the questionnaire also include statements that assess the project's influence on social diversity in the area. Such measures could provide valuable information to key stakeholders, including directors of local institutions and municipal officials involved in the project. For instant, asking whether the access to certain public services has improved or worsened can directly indicate whether the project has (or is likely to have in the future) a positive or negative influence on diversity in the area. Specifically, questions about public spaces (e.g. squares, patios, pavements), green and blue areas (parks, bodies of water), social services, public transport, recreation and culture sites are very informative. From an economic diversity perspective, monitoring and systematically analysing rental and property prices can provide a good approximation of inequalities in the neighbourhood. If prices are very low or very high, it is more likely that the area is less varied. Since adaptive reuse can, in some cases, contribute to higher recognition of the area and potentially lead to gentrification, it is important for policymakers to consider measures to counteract these effects where possible.

5.2 *Capacity for Self-Organization*

The capacity for self-organization within a local community can be captured either by asking direct questions such as "How many of your friends are active in civil society organizations/volunteering?" or "Did you vote in the last local elections?". It can also be assessed through analysis of existing data on political or social activity, like voter turnout in general elections, levels of volunteerism, participation in local decision-making, and engagement in public consultations. Additionally, the number of Civic Society Organizations (CSOs) operating in the region serve as a good approximation of self-organization. An increase in the number of CSOs following the adaptive reuse of cultural heritage can indicate that a new investment has sparked collective agency, encouraging the use of social ties to achieve shared goals. Questions such as "If there is a problem in this community, people who live here can get it solved" or "In this neighbourhood, there are many people who participate in community/civic activities" might provide a good indication of the level of perceived agency or collective efficacy (Sampson 2013). However, if self-organization is a desirable outcome of the investment, the leaders of the project should ensure there is space where people can easily meet, preferably free of charge. Therefore, a question such as "If you wanted to organise a community meeting, is there a space in your neighbourhood you could use free of charge?" should be included in the questionnaire.

5.3 *Common Meaning*

For an adaptive reuse project to succeed, it is crucial to properly identify the elements that local residents find particularly important and consider them when planning the adaptive reuse process. Institutions responsible for the change should have a comprehensive understanding of the place's significance in the context of the local community, as this significance can be both positive and negative. A place attachment scale can be used to evaluate the importance of the place to the residents. We propose using statements such as: "This place is special, one of a kind", "People who live here are/should be proud of this place", "This place means a lot to the history and culture of the community" (see also Lewicka 2011).

When planning an adaptive reuse process, it is also important to understand whether residents feel part of a community and share similar values. Statements such "I feel I am part of a community, not just a resident" or "My vision of the neighbourhood and community is similar to the vision of my neighbours" are examples of questions that should be asked to assess the level of common meaning within the area. Common activities help to develop a sense of shared reality and meaning within the community. Therefore, it is advised to assess how many cultural events are organized in the neighbourhood and how many residents participate in these events. Example questions include: "Have you participated in the cultural/social event organised in your neighbourhood in the last year (concert, picnic, sport event)?" or "Do you regularly search for information regarding your neighbourhood, the place you live e.g., on Facebook, in Wikipedia or community newsletter?" However, it is important to not only determine whether the local community shares values but also to assess whether these values are connected to sustainability, ecology, circularity, and cultural heritage. We suggest asking questions such as "I see myself as someone who preserves nature for future generations" and "This community is conscious about saving natural resources" to assess this aspect. A community that is not particularly interested in environmental issues but is interested in history, may be less inclined to appreciate circular solutions in the adaptive reuse of cultural heritage, especially if these investments alter the original appearance of the building.

5.4 *Capacity for Learning*

There are two main aspects to capacity for learning in the context of cultural adaptive reuse projects. The first aspect relates to the willingness and potential of individuals to learn and develop. The more people enjoy learning new things, the more likely they are to adopt new ideas and embrace change. Capacity for learning can be measured with questions such as: "How often do your work or other activities require learning new things?", "Have you gained any new skills during the last year?", "It's easy to bring new ideas to this community", "Members of this community learn from each other", "This is a place to learn something new", and "I see myself as someone who likes learning new things (yes/no)." However, as in economics, the supply of learning opportunities needs to meet the demand. It is not enough for people to be willing to learn; there also need to be learning opportunities available to them. An adaptive reuse project should provide such opportunities for diverse groups of people while also being closely connected to the needs of the particular community. Local institutions, before staring an adaptive reuse project, should consider whether it would create new opportunities for education and lifelong learning, enhance the development of new skills (technical, traditional etc.), and promote healthy, ecological lifestyle and behaviours. Conducting such evaluation before and after the implementation of the project would allow for verification of whether it broadened the opportunities for development and growth in the neighbourhood, thereby contributing to its sustainability.

5.5 Trust

The process of change requires trust, making trust between citizens and authorities crucial for the smooth adaptive reuse of cultural heritage initiatives. When the level of trust between these two groups is low, introducing new solutions becomes much more difficult. In some societies and neighbourhoods, trust level may generally be low due to historical, cultural, and economic factors. However, a low level of trust is not a reason to abandon adaptive reuse projects. Instead, measuring trust before the project is launched allows for better preparation for potential difficulties and the design of counter-strategies to address distrust among local actors. Trust can be measured with following questions: "Can local authorities be trusted?", "Are local authorities skilful and competent?, "Can local business owners be trusted?", "Are local business owners skilful and competent?". Additionally, understanding the level of trust within the neighbourhood might prove to be very important for planning change. For example, if trust among residents is high but trust in institutions is low, finding a community leader to help implement change in a bottom-up manner might be an effective strategy. Trust within a group also affects the sense of security; people who feel unsafe are less likely to embrace change because they already feel they have little control. A set of additional trust related questions might include: "Is this a place where people feel safe?", "Can people in this place be trusted?", "Can people in this neighbourhood/community be trusted?", "Are people in this neighbourhood skilful and competent?", "Do you believe this community is a good place for kids to grow up (yes/no)?", "Would you recommend this place to entrepreneurs looking for a place to set a business?/others?".

To evaluate existing projects, we propose asking whether the place enables greater social integration (new relations within the community, people getting to know each other), higher trust within the local community, and increased cooperation between local government and entrepreneurs (e.g. public-private partnerships).

6 Summary

In CLIC project we strived to define a framework for assessing the socio-cultural impact of cultural heritage adaptive reuse. Our starting point was the assumption that to facilitate the transition to a circular economy, we need to understand how the introduction of this new model may influence the wellbeing of people and communities.

First, by revising existing measures of well-being, we were able to indicate the blind spots in current quality of life indicators. Second, we explored the issue of social sustainability and applied it to adaptive reuse projects in the cultural heritage field, to better understand the factors behind their positive and negative impacts.

We conclude by proposing practical tools to evaluate the social sustainability potential of specific communities and stakeholder groups. We began by gathering the most widely used methods used for measuring wellbeing and analysed their potential for capturing the social and cultural impact of circular economy implementation, particularly in adaptive reuse. We investigated the presence of five dimensions of social sustainability in highly regarded indexes of well-being: diversity, common meaning, capacity for learning, trust, and capacity of self-organization. Our analysis revealed that while all these dimensions play an important role in community formation and the successful implementation of adaptive reuse projects, the existing quantitative indexes are not well-suited for the job of impact measurement. Therefore, we proposed measures to collect data on the sustainability of local community. We explored various parameters of community relations, including the strength and density of ties between members, the level of trust (understood as general trust, in-group and out-group trust), openness, and diversity of members and their competences.

Our basic assumption is that people are inherently social beings, and their identity is formed not only through internal cognitive processes but also by external influences, including the places and spaces where they spend time and live their lives. For example, in a recent study conducted in Warsaw (Realizacja 2017), Warsaw residents indicated that their sense of identity is shaped by their place of residence (75%) a similar extent as by gender (77%), family (76%), and hobbies (76%). Interestingly, only 62% of respondents identified their job as a significant source of identity. This data illustrates high capacity of built cultural heritage to impact the identity formation of local residents. In other words, we must recognize that changes to cultural heritage or the landscape can have a profound impact on the people living in the area.

Special attention has to be paid to the long-term impact of the intervention on the local community. Understanding the social aftermath may help to assess the hidden qualities that make a place attractive or significant, going beyond aesthetics or physical quality to capture the intrinsic value and meaning attached to a specific place, building, or landscape.

References

Brandsen T, Cattacin S, Evers A, Zimmer A (2016) Social innovations in the urban context. Springer Open

Cacioppo JT, Hawkley LC, Rickett EM, Masi CM (2005) Sociality, spirituality, and meaning making: Chicago health, aging, and social relations study. Rev Gen Psychol 9(2):143–155

Colantonio A, Dixon T, Ganser R, Carpenter J, Ngombe A (2009) Measuring socially sustainable urban regeneration in Europe

Costanzo L, Ferrara A (2015) Well-being indicators on landscape and cultural heritage: the experience of the BES project. In: A new adaptive reuse of cultural heritage agenda for improvements in quality of life. Springer, pp 1–15

Dempsey N, Bramley G, Power S, Brown C (2011) The social dimension of sustainable development: defining urban social sustainability. Sustain Dev 19(5):289–300

Desa UN (2016) Transforming our world: the 2030 agenda for sustainable development

Diener E (2000) Subjective well-being: the science of happiness and a proposal for a national index. Am Psychol 55(1):34

Diener ED, Emmons RA, Larsen RJ, Griffin S (1985) The satisfaction with life scale. J Pers Assess 49(1):71–75

Eagle M (1999) Why don't people change? A psychoanalytic perspective. J Psychother Integr 9(1):3–32

Fatorić S, Egberts L (2020) Realising the potential of cultural heritage to achieve climate change actions in the Netherlands. J Environ Manag 274:111107

Folke C, Hahn T, Olsson P, Norberg J (2005) Adaptive governance of social-ecological systems. Annu Rev Environ Resour 30:441–473

Gidron B, Domaradzka A (2021) The new social and impact economy: an international perspective. Springer

Guarnacci U (2012) Governance for sustainable reconstruction after disasters: lessons from Nias, Indonesia. Environ Dev 2:73–85

Iacus SM, Porro G, Salini S, Siletti E (2020) An Italian composite subjective well-being index: the voice of Twitter users from 2012 to 2017. In: Social Indicators Adaptive reuse of cultural heritage, pp 1–19

ICOMOS (2017) ICOMOS action plan: cultural heritage and localizing the UN sustainable development goals (SDGs)

Johnston P, Everard M, Santillo D, Robèrt K-H (2007) Reclaiming the definition of sustainability. Environ Sci Pollut Adapt Reuse Cult Herit Int 14(1):60–66

Kramer MR, Porter M (2011) Creating shared value, vol 17. FSG

Kuznets S (1934) National Income, 1929–1932. In: National Income, 1929–1932. NBER, pp 1–12

Landry C (2019) Advanced introduction to the creative city. Edward Elgar Publishing

Lemille A (2017) Circular human flows. In: Enhancing humans as an integral part of circular economic flows

Levin SA (1998) Ecosystems and the biosphere as complex adaptive systems. Ecosystems 1(5):431–436

Lewicka M (2011) Place attachment: how far have we come in the last 40 years? J Environ Psychol 31(3):207–230

Marsden G, Kimble M, Nellthorp J, Kelly C (2010) Sustainability assessment: the definition deficit. Int J Sustain Transp 4(4):189–211

Maslow, A. H. (1943). A theory of human motivation. Psychol Rev 50(4):370–396. https://doi.org/10.1037/h0054346

Missimer M, Robèrt K-H, Broman G (2017) A strategic approach to social sustainability–Part 1: exploring the social system. J Clean Prod 140:32–41

Norberg J, Cumming G (2008) Complexity theory for a sustainable future. Columbia University Press

Nowak A, Vallacher R, Rychwalska A, Roszczynska-Kurasinska M, Ziembowicz K, Biesaga M, Kacprzyk-Murawska M (2019) Target in control: social influence as distributed information processing. Springer. https://doi.org/10.1007/978-3-030-30622-9

Oishi S (2010) Culture and well-being: conceptual and methodological issues. In: International differences in well-being, vol 34. Oxford University Press, New York, p 69

Realizacja Sp. z o.o (2017) Badanie tożsamości Warszawy i jej mieszkańców

Roszczynska-Kurasinska M, Kacprzyk M (2013) The dynamics of trust from the perspective of a trust game. In: Complex human dynamics. Springer, pp 191–207

Roszczynska-Kurasinska M, Ann D, Bartosz Ś, Agata Ż (2019) Embracing circularity in adaptive reuse – grassroots perspective. In: STS conference Graz 2019. Critical issues in science, technology and society studies, pp 363–383

Roszczyńska-Kurasińska M, Domaradzka A, Ślosarski B, Żbikowska A (2021) Facebook data as part of cultural heritage investments toolbox: pilot analysis of users interests and preferences concerning adaptive reuse. Sustainability 13(4):2410

Sampson RJ (2013) Great American city. University of Chicago Press

Schank RC, Berman TR, Macpherson KA (1999) Learning by doing. In: Reigeluth CM (ed) Instructional-design theories and models: a new paradigm of instructional theory. Routledge, pp 161–181

Scheffer M, Carpenter S, Foley JA, Folke C, Walker B (2001) Catastrophic shifts in ecosystems. Nature 413(6856):591

UNESCO (2016) Culture: urban future; global report on culture for sustainable urban development. UNESCO

Vallance S, Perkins HC, Dixon JE (2011) What is social sustainability? A clarification of concepts. Geoforum 42(3):342–348

Wong PF (2019) A framework of sustainability refurbishment assessment for heritage buildings in Malaysia. In: IOP conference series: earth and environmental science, vol 268. IOP Publishing, p 012011

Chapter 11
Grassroots Social Sustainability Framework and Intrinsic Value Measures for Cultural Heritage Adaptive Reuse Projects

Anna Domaradzka and Magdalena Roszczyńska-Kurasińska

1 Introduction

This chapter illustrates the results of pilot implementation of the grassroot social sustainability framework (GSS) to study specific cases of adaptive reuse projects. The framework is based on the ideas described in detail in another chapter of this book (Żbikowska et al. 2024, Chap. 10, this volume). Here, we present the results of three pilot studies to demonstrate how concepts of social sustainability and intrinsic value can be linked to create a practical methodological framework. Its primary purpose is to enable the analysis of the social surroundings of cultural heritage sites as well as the qualities of these sites, to identify opportunities and threats for achieving long-term grassroots sustainability.

As there is a growing literature on the topic of sustainability (Nosratabadi et al. 2019) and social sustainability in particular (Afshari et al. 2022; Cope et al. 2022; Kyttä et al. 2016), we used these concepts as our starting point to develop practice-oriented tools for the evaluation and planning of adaptive reuse investments. Our focus was mainly on the potential of local communities to embrace and uphold changes concerning cultural heritage, with focus on circular economy approach (see Roszczynska-Kurasinska et al. 2019). Learning from existing conflicts around similar projects (Roszczynska-Kurasinska et al. 2019), we wanted to highlight the fundamental importance of community response and engagement in such investments. We believe that a careful diagnosis of strengths and weaknesses from a social sustainability perspective allows for better planning, faster implementation, and long-lasting positive effects of similar project in the future. By introducing new functionalities to a built heritage, it is possible to stimulate specific traits of the local community that are necessary for attaining and maintaining resilience and long-term social sustainability. As a result, we propose a new concept of grassroots social

A. Domaradzka · M. Roszczyńska-Kurasińska (✉)
Robert Zajonc Institute for Social Studies, University of Warsaw, Warsaw, Poland
e-mail: anna.domaradzka@uw.edu.pl; m.roszczynska@uw.edu.pl

L. Fusco Girard, A. Gravagnuolo (eds.), *Adaptive Reuse of Cultural Heritage*,
https://doi.org/10.1007/978-3-031-67628-4_11

sustainability that underlines the importance of local actors and their relations for obtaining desired impacts of adaptive reuse projects. We also highlight the intrinsic social value of heritage sites as an important aspect to assess when designing successful interventions (Roszczynska-Kurasinska et al. 2021). The intrinsic value of the place sets certain constrains on the range of potential interventions in the built heritage. Due to the specificity of the place, not all kinds of interventions are equally welcomed by members of the local community.

Using three pilot cases, we will also suggest how the tools described in Chap. 10 for diagnosing the interplay between the grassroots potential of local communities and intrinsic value of the site could help to plan and execute further interventions that engage and strengthen the community, creating positive feedback loop of social sustainability. Moreover, conducting studies in three different locations allowed us to test the universality of the approach and interpret the results in relation to different national contexts and site functions.

Our empirical pilot study was divided into two lines. Firstly, we wanted to learn about different qualities of the communities living in close proximity of the cultural heritage sites: what are the residents' experiences of living in this particular neighbourhood, how they perceive their neighbours, and how they assess the opportunities for fulfilling their needs in the area. The information gathered in this line of adaptive reuse of cultural heritage allowed us to compile the grassroots social sustainability index.

Secondly, we wanted to explore the intrinsic value of the built heritage. This involved seeing and evaluating the site through the users' eyes. Therefore, we asked visitors of each site to describe how they perceive the site, what are the motives for visiting it, and what changes they would recommend to increase the social value of the space.

As a result of both studies, we were able to explore the intrinsic value and test the grassroots social sustainability framework, which we see as a useful diagnostic tool for all adaptive reuse interventions.

2 Description of Cases

We conducted our study in three local communities surrounding cultural heritage locations in Italy, Poland, and Sweden. The sites differ in terms of history and character and include: (A) a botanic medical garden in a southern Italian city, (B) a wooden housing settlement surrounded with greenery in the centre of a Polish city, (C) a post-industrial site in northern Sweden. Each of these heritage sites has undergone adaptive re-use process in recent years, changing its function and opening it up to visitors, activists, and artists, who are engaged in shaping the new future of previously neglected places. Sites A and B are located in urban environment, while site C is situated in a rural/wooded setting. In sites A and B, the greenery/landscape is an important aspect of the cultural heritage and local engagement, while sites B and

C are home to civil society organizations and regular residents. All three sites host artistic or educational initiatives on a regular basis.

3 Method

The study was carried out among community members residing in the neighbourhoods of three cultural heritage sites, as well as with visitors to these sites, in 2020. We used two questionnaires: one concerning the quality of life around heritage site (for both community members only) and one about the heritage site itself (for community members and outside visitors). The respondents' opinions on different aspects of the community and the cultural heritage site are presented on the scale from 1 to 5, with average scores noted in brackets throughout the text.

Due to the Covid-19 pandemic, we decided to conduct surveys online. We sent emails to people familiar with the studied places and advertised the study online. Only in the case of post-industrial site (C), it was possible to conduct the survey face to face[1].

Altogether we gathered data from 119 residents and 143 visitors across three locations. While the samples in each location differ (Table 11.1), they represent the density of the local population and the popularity of the site. The density of the local population is lowest at site C and highest at site B.

The demographical structure of our respondents also reflects the different characteristics of the local communities. In Italy, the majority of respondents in group of residents were adults above 55 years, while in Sweden, the most represented group were adults aged between 35 and 54 years. Only 9% of respondents in Sweden were under 34 years old (Fig. 11.1a).

Studied places vary in terms of visitors: A is an attractive tourist location in a small town, B is centrally located in a pedestrian area in a big city, and C is located in a forested area, a 20-minute drive from the closest city. The demographic profile of surveyed visitors shows additional differences among sites (Fig. 11.1b). In the Swedish case the most represented group of visitors and users of the place were adults between 34 and 54 years. The group of adults over 55 years was the most

Table 11.1 Number of respondents per study

	(A) Botanic garden in Italy	(B) Urban garden in Poland	(C) Post-industrial site in Sweden
Local community	33	64	22
Visitors	75	47	18

[1] We would like to thank Antonia Gravagnuolo and Vera Telemo for conducting questionnaires in Italy and Sweden. Moreover, we would like to thank Luciano Mauro and the organisation of Garden of Minerva for supporting us in process of data collection.

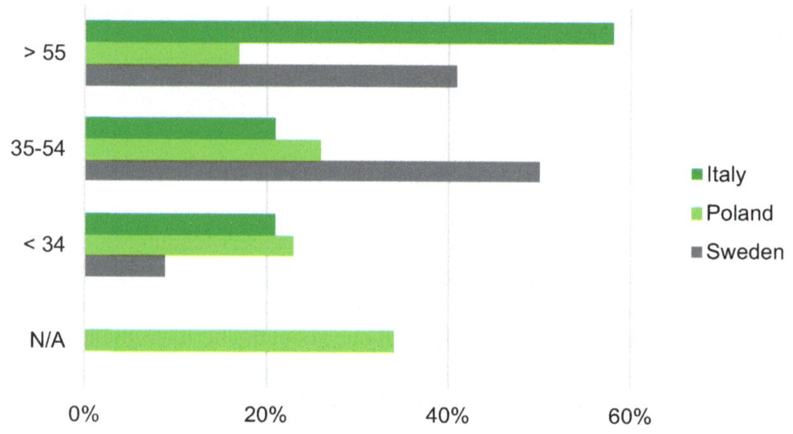

Fig. 11.1a Age of respondents—community members

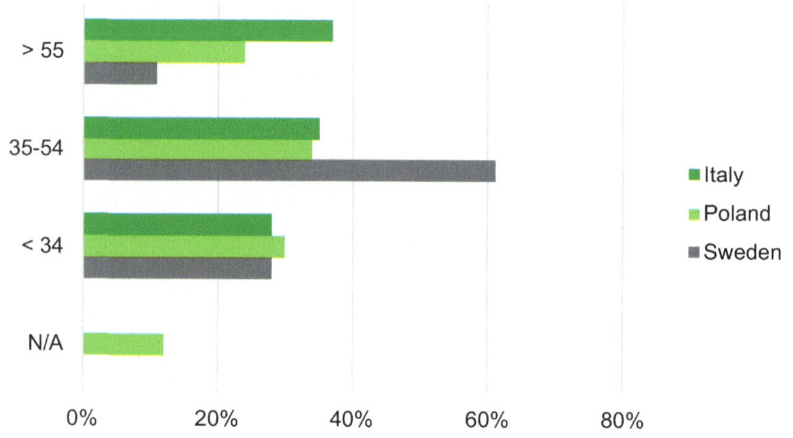

Fig. 11.1b Age of respondents—users of the site

numerous in the Italian location. The Polish urban garden attracted visitors of all age groups at the similar level.

In all three cases, the respondents were relatively well-educated, at least half of the visitors to each site had a Bachelor's, Master's, or doctoral degree. This may be a by-product of the fact that all three sites organize cultural and educational activities popular with certain target groups. Our data collection method could also have led to a certain self-selection bias, resulting in lower participation of less educated users. However, this type of bias seems unavoidable in most adaptive reuse of cultural heritage projects, and the Covid-19 pandemic made it impossible to engage directly with on-site users. In future adaptive reuse projects, sample characteristics should be carefully studied to ensure representativeness. For the purpose of these pilots, however, we considered our sample to be sufficient.

4 Grassroots Social Sustainability Framework

Data collected in all three locations allowed us to test our grassroots social sustainability framework approach and compare our sites on each of the six dimensions: diversity, openness, common vision, trust, capacity for learning and capacity for self-organization. Those dimensions constitute the GSS Index, based on the theoretical review conducted by Missimer et al. (Missimer et al. 2017) and developed in the work of Roszczyńska-Kurasińska et al. (2019).

One of the main aims of the study was to assess the grassroots social sustainability potential in three communities surrounding re-adapted heritage sites. We strongly believe that social sustainability is measurable as a combination of several characteristics concerning the people living in the area and their relations with each other, such as diversity of residents, trust between people and organizations, common understanding of the community's challenges and goals, ability to learn and ability to self-organize. According to the literature (Missimer et al. 2017) this combination of characteristics is a good predictor of the social readiness to embrace changes. Therefore, the socio-economic impact of cultural heritage sites interventions should take them into account.

As a result, we formulate recommendations for counteracting potential gaps and risks and enhancing local communities' potential to embrace cultural adaptive reuse projects. This pilot study also allowed us to further adapt our measures to ensure they build a universal tool for evaluating communities and sites from a social sustainability perspective.

4.1 *Diversity and Openness*

According to the literature (Curşeu and Schruijer 2017; Folke et al. 2002; Missimer et al. 2017), the potential for social sustainability in a given neighbourhood lies in the diversity of its community members and their openness towards other people and experiences. In a diverse community, the distinctive needs of its members can be addressed locally by people with different skills and knowledge. Residents of such communities do not have to spend much time and effort searching for providers of know-how, services, or goods they need.

In case of low diversity, rich social relations outside of the neighbourhood can compensate for the lack of skills and knowledge within the community itself. The needed knowledge can be easily brought into the community through the personal links of its residents, as long as the community is open enough to accept new information and ideas.

The analysis of gathered data revealed that respondents in all three studied locations felt that they live in communities where the diversity of skills is high and everybody has rather sociable attitudes. However, the differences were found in openness to newcomers and the breadth of social relations.

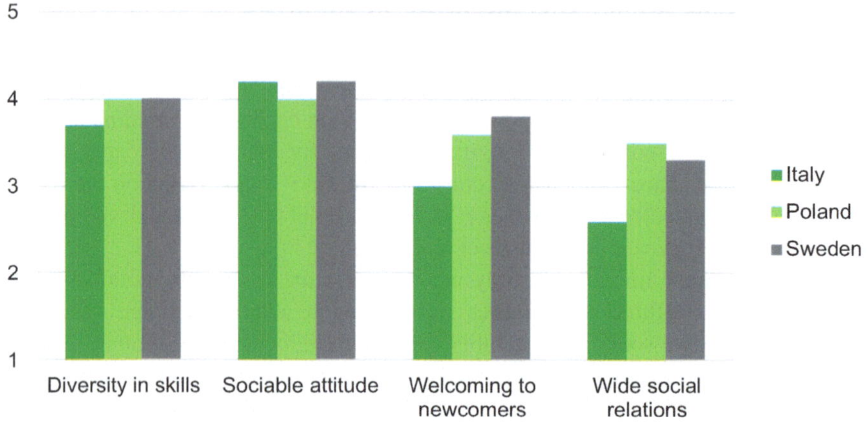

Fig. 11.2 Diversity and openness

As illustrated by Fig. 11.2[2], the majority of respondents in Italy (70%) agreed that their community is diverse in terms of skills (M = 3.7). They have a rather sociable attitude towards others when it comes to meeting new people—76% of respondents enjoys meeting new people (M = 4.2). However, they perceive their neighbourhood as less welcoming to newcomers—only 24% of respondents declared that the newcomers might feel welcomed in their neighbourhood (M = 3). People living in this community are perceived by respondents also as rather lacking wide social relations (55%); only 18% of respondents declared that people there have many social relations (M = 2.6). This result suggests that there is relative agreement among respondents that while they enjoy meeting new people, and have diverse skills and expertise from which the local community can benefit, they perceive their neighbours as not having very dense social relations.

Similar to the Italian case, members of the community living around the urban garden in Poland see themselves mostly as sociable people, who enjoy meeting new people (M = 4) and find their community diverse in terms of skills (M = 4.0). Contrary to Italian respondents, Poles perceive their neighbourhood as rather welcoming to newcomers (M = 3.6). People living in this community are also believed to have fairly wide social relations (M = 3.5).

Respondents from the Swedish location also thought that their neighbourhood community is diverse in terms of skills (M = 4). They have a sociable attitude towards others when it comes to meeting new people (M = 4.2), and they perceive others in the community as open to newcomers (M = 3.8). However, people living in this community are considered to have weaker social relations—survey participants reported that the members of the local community have some links with others but not many (M = 3.2).

[2] The average values for each indicator are shown in brackets and represent the mean value of the indicator for the respective community (M) on a scale from 1 to 5.

In our view, the observed differences may reflect important cultural differences among locations.

4.2 Common Vision

For a diverse community to generate a socially sustainable reaction to change, it must have the ability to develop a common vison among its members (Missimer et al. 2017; Roszczynska-Kurasinska et al. 2019). Effective cooperation and engagement requires a clear vision that is shared by all involved. Without a common vision, members of local communities tend to focus on their own interests, which can often be contradictory. In such a situation, members of the community will find it difficult to collaborate in the long run. We used five measures to evaluate the common meaning in our three communities: shared vision, shared values, sense of belonging to the community (see Fig. 11.3).

Additionally, to see how much respondents feel part of the local community and are attached to the place they live in, we measured their place attachment and place identity. Place attachment informs about the emotional bond between respondents and their place of residence, how much they like living in their neighbourhood and how much they feel "at home" there (Lewicka 2008). Place identity is related to the concept of community formation and indicates how much respondent feels part of his/her community (Hernández et al. 2007).

The data from site A shows that residents of the botanic garden area felt that they do not share a unified vision of their neighbourhood with other residents (M = 2.6). Only 15% of respondents agreed with the statement that they had a similar vision of the neighbourhood with their neighbours, while 48% did not agree with the

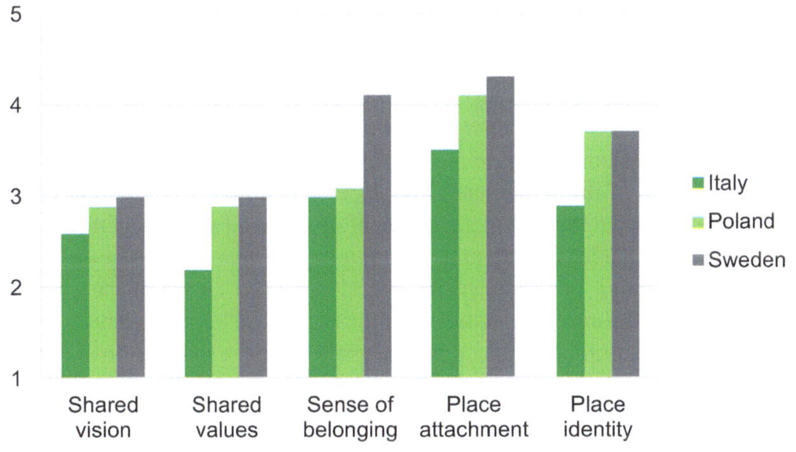

Fig. 11.3 Common vision, place attachment, and place identity

statement. Moreover, only 9% of respondents agreed that their neighbours share the same values (M = 2.2). Despite the perceived low similarity of views among neighbours, they seem to feel somewhat like part of the local community (M = 3). However, one-third of respondents did not agree with the statement that they are part of the local community. These findings are consistant with results concerning place attachment and place identity. On average, respondents declared being only slightly attached to the neighbourhood they were living in (M = 3.5) and did not seem to feel strongly that they are a part of this neighbourhood (M = 2.9). In other words, living in the area of the botanic garden did not significantly influence their sense of belonging or identity.

In the community B, 31% of the residents felt that they share common vision of their neighbourhood with other residents (M = 2.9) and 27% of respondents agreed with the statement that their neighbours shared similar values as their neighbours (M = 2.9). Moreover, respondents did not feel very much like a part of the local community (M = 3.1).

Despite the differences in values and visions, respondents declared being attached to the neighbourhood they were living in (M = 4.1), and they identified with the area (M = 3.7). This suggests that the bond might be formed more on the relationship with the character of the place rather than with people living there.

According to the respondents from community C, the residents of the studied area seldom share the same values (M = 3) or a common vision (M = 3). Half of the respondents felt that the community neither agrees nor disagrees on common visions, while 27% of respondents felt that people in the community shared a common vision. Moreover, only 19% of respondents felt that the residents have the same values. Vision and values appear not to be a strong bonding agent of this community. Despite this, the sense of belonging to the community itself was relatively high (M = 4.1), as were place attachment and place identity. Respondents reported feeling strongly attached (M = 4.3) to the neighbourhood where they lived, and they expressed a sense of belonging and identification with the area (M = 3.7).

These results should be interpreted in relation to the characteristics of the sites. Site A is closed to visitors after operational hours and requires an entry fee, which may create a sense of detachment between the residents and the site. Meanwhile, visiting sites B and C is free of charge and the events organised there are often open to visitors as well as community members. Site B is also where grassroots protests against demolition of the settlement took place, which increased awareness of the place's value and strengthened the attachment of involved persons, although more to the site than people leaving there.

While investigating common vision, we also looked closer at the communities' attitude toward nature. We assume that the successful implementation of circular economy and sustainable development goals requires that the members of a community are sensitive to matters connected with ecology and nature. Therefore, their common vision should reflect the need to protect the environment as well as a shared perception of the benefits stemming from preservation of natural resources.

As presented in Fig. 11.4, the studied communities differed in terms of attitudes towards nature and its benefits. According to respondents from the Italian site, the

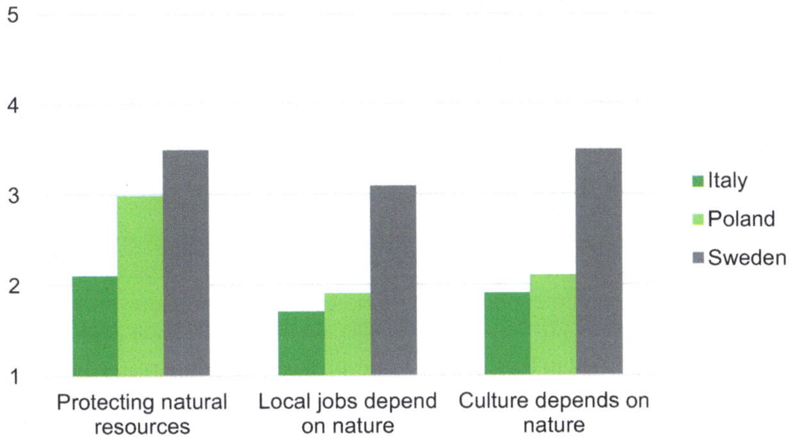

Fig. 11.4 Attitude toward nature

community's awareness of the need to protect natural resources is relatively low (M = 2.1) (Fig. 11.4). This might be explained by the fact that local jobs are not seen as dependent on natural resources (M = 1.7), and the local culture does not seem to relate to natural resources such as rivers, forests, and local animals (M = 1.9). This shows a certain disconnect between the botanic garden's natural resources and the general perception of nature's benefits in an otherwise not-so-green neighbuorhood.

The Polish site's community's awareness of the need for protecting natural resources is average (M = 3). The variation of answers is high, with 38% of respondents disagreeing with this statement and 33% agreeing with it. The ambivalent attitude toward nature preservation might be an aftermath of the fact that local jobs are not seen as dependent on natural resources (M = 1.9), and the local culture does not seem to relate to natural resources such as rivers, forests, and local animals (M = 2.1). This can be explained by the modern urban setting in which green areas have mainly recreational or decorative value.

In contrast, the Swedish site's community is quite conscious about protecting natural resources (M = 3.5). The majority of respondents (55%) declared that the community is aware of the need to protect natural resources, while only 9% of survey participants held the opposite view. This might stem from the fact that respondents perceive local culture as being somehow related to natural resources such as rivers, forests, and wildlife (M = 3.5). Moreover, 55% of respondents felt that the local culture is directly related to local nature. The situation is less clear when it comes to the issue of dependence of local jobs on natural resources. Only one-third of respondents felt that the local jobs are related to natural resources, 27% did not see such a relation between jobs and nature (M = 3.1). The high level of recognition of the cultural value of nature may result in Nordic cultural patterns, which are very deeply rooted in respect and symbolism of nature. The location of site C is also close to the lake region, with many forests and waterways. While the history of the postindustrial heritage is very much related to the proximity of those resources, they are not crucial for local economy at the moment.

4.3 Trust

Trust is one of the crucial aspects of social processes as it ensures smooth and fast interactions between people (Hardin 2002; Sztompka 1999). It makes things work without the need to implement costly and time-consuming measures of control. In that way, trust strongly contributes to social sustainability (Missimer et al. 2017). In this context it is important to look at both in-group and out-group trust, which allows us to grasp the potential strengths and weaknesses of relations within local community, which is seldom homogenous (Roszczynska-Kurasinska et al. 2019). Therefore, we asked respondents to comment on their level of trust in particular groups and institutions (other residents, local businesses and local authorities). Figure 11.5 illustrates the differences in trust among studied locations.

In the studied community A, the level of trust is comparatively low—almost half of the respondents (43%) do not believe that people in this neighbourhood can be trusted, compared to 27% that claim people from their community can be trusted. As much as 30% could not decide whether other members of their community could be trusted or not (M = 2.8). The trust towards local business owners is also low—36% of respondents do not perceive owners of local business as trustworthy, 33% declare that they neither can be trusted or not, and 30% agree that local business owners can be trusted (M = 3).

The perception of trustworthiness of local administration is similarly low. Almost half of the respondents (42%) do not agree that local authorities can be trusted, and only 15% judge them trustworthy (M = 2.5).

Fig. 11.5 Trust in neighbours, business owners, and local administration

This result can be better understood when we compare it to the data from the World Value Survey Wave (WVS) 7 (2017–2020) conducted in Italy (N = 2282) (Haerpfer et al. 2022). When respondents were asked to indicate whether they trust their neighbours, 73% declared that they trust their neighbours completely or somehow; while only 26% indicated that they do not trust their neighbours very much or at all their neighbours. However, when respondents of WVS in Italy were asked if they trust people they meet for the first time, only 27% of them declared that they trust people they just met a lot or somehow; the majority (72%) do not trust strangers very much or at all. The results from our site A indicate that either the trust level there is lower than the average for Italy due to some past conflicts, or that the residents of this area perceive other residents more as strangers than neighbours.

In case of site B, the level of trust in the studied community was average (M = 3.4). Over half of the respondents (52%) believe that people in their neighbourhood can be trusted, while 27% claim that people from their community can't be trusted. The remaining respondents could not decide whether other members of their community could be trusted or not. This is in line with WVS results according to which 23% of Poles do not trust their neighbours, while 70% do not trust people who they meet for the first time.

The trust towards local business owners is at the average level (M = 3.5), similar to trust in neighbours, but the perception of trustworthiness of local administration is much lower (M = 2.5). Half of the respondents do not agree with the statement that local authorities can be trusted, with only 21% judging them trustworthy. Distrust towards local government might origin from the plans to demolish the site, which was championed by the local government at one point and stopped thanks to the collective efforts and protests of local residents and supporting organizations. Difficult relations with the municipality continue as the site is located on the high-value land, which could reap great profits for investors and the administration.

In the Swedish case, the level of trust is high in the studied community, specifically when it comes to the trust in other residents. On average, respondents declared that people in the community are trustworthy (M = 4.3). This is in line with WVS results according to which only 10% of Swedes do not trust their neighbours, and 25% do not trust people they met for the first time. The trust in local business representatives is similarly high. On average, respondents reported that businesses owners in the community can be trusted (M = 4.4). The perception of the trustworthiness of the local administration is relatively lower than other groups in the community, but significantly higher than in other locations. On average, respondents agreed that the local government can be trusted (M = 3.5). This high level of trust is typical for Nordic countries.

4.4 Capacity for Learning

According to several sources (Folke et al. 2002; Missimer et al. 2017; Nelson 2007), social sustainability means that a society is capable of adapting to changing conditions. The adaptation cannot happen without acquiring new knowledge and skills, making the capacity for learning (Olejniczak et al. 2014) an important indicator in assessing the potential for social sustainability. We found that there are differences between individual learning capabilities and structural as well as social constraints that shape learning capacities in local communities (Fig. 11.6).

Respondents from the Italian site A declared that they enjoy developing new ideas (M = 4.2) and learning new things (M = 4.9). However, they rated the learning opportunities in their neighbourhoods rather low (M = 2.4)—which could indicate either a lack of such opportunities or that residents are unaware of them. The matter requires further investigation, as rare learning opportunities could affect the neighbourhood's opennes to new ideas. Indeed, on average, respondents declared that it does not seem to be easy to introduce new ideas to the neighbourhood (M = 2.2), and they do not perceive people in their neighbourhood as being willing to learn new things (M = 2.0).

In the case of Polish urban garden, respondents declared that they like developing new ideas (M = 4.0) and learning new things (M = 4.6). They also view their neighbourhood as a conductive place for learning new things—they rated the learning opportunities in their neighbourhoods highly (M = 4.4), and stated that it is not too difficult to bring new ideas to the neighbuorhood (M = 3.1). Other community members are also seen by respondents as fairly eager to learn (M = 3.3), with 40% of respondents believing that people in the neighbourhood willingly learn new things, while only 17% disagreed with this statement.

Respondents from the Swedish site also declared that they very much like learning new things (M = 4.5) and rather enjoy developing new ideas (M = 3.4). The structural opportunities for learning were rated as average (M = 3.4), with 45% of

Fig. 11.6 Capacity and opportunities for learning

respondents reporting that there are opportunities to learn. Only 18% felt that learning opportunities in the community were poor. The same is true for bringing new ideas to the community. On average, respondents agreed that it is relatively easy to introduce new concepts into the community (M = 3.4). Particularly interesting is the positive assessment of other community members in terms of their willingness to learn from each other (M = 3.8), which is higher than in other studied locations.

Comparing results for all three communities, we can see that individual learning capabilities were evaluated highly in all three locations, while on the community level this assessment was not as optimistic and univocal. The results indicate that Swedish respondents have a better vision of their neighbours than Polish or Italian respondents and assess members of local community more optimistically. The discrepancy between self-description and the description of others is lowest at the Swedish site. When it comes to opportunities for learning, the Polish site received the highest rating, reflecting the character of grassroots activism that is developing the site as a place for workshops, meetings, and new social and artistic endeavours.

4.5 *Capacity for Self-Organization*

The final component of social sustainability is the capacity for self-organization. This factor relates directly to social capital (Putnam 2001) and civil society organizing capabilities, which are crucial for collective efforts and community efficacy (Sampson 2012). Figure 11.7 illustrates significant differences in terms of self-organization capabilities among studied sites.

Site A's respondents rated opportunities for social activity and civic self-organization relatively low compared to other locations. Most notably, respondents evaluated opportunities for social activism in the neighbourhood as very low (M = 2.0). Additionally, half of the respondents indicated that other

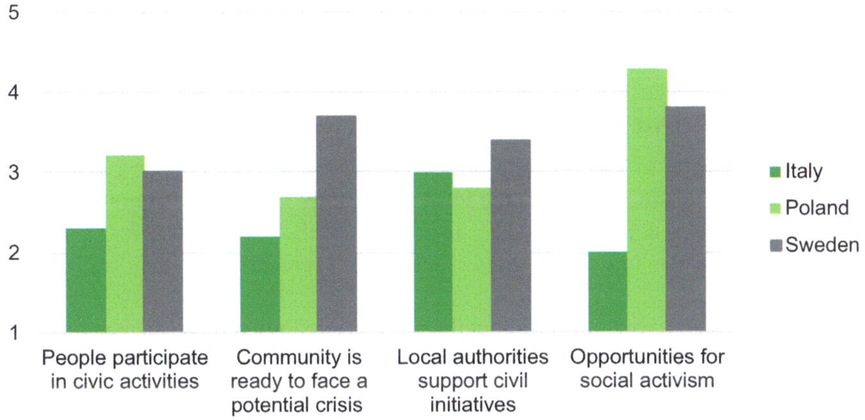

Fig. 11.7 Capacity and opportunities for self-organization

community members do not participate in civic activities (M = 2.3). Respondents also lacked confidence in the local community's ability to address potential difficulties in the near future, rating the community resilience as rather low (M = 2.2). More than half of the respondents (58%) denied the existence of local authorities' support for local organizations and civil initiatives, while only 21% view local authorities as supportive (M = 3).

In site B, respondents rated opportunities for social activism in the neighbourhood rather high (M = 4.2). They perceived the level of volunteering among their friends and acquaintances as average (M = 3.2), and 40% of respondents agreed that other community members participate in civic activities (M = 3.2). However, the respondents were not convinced of the local community's ability to address potential difficulties in the near future (M = 2.7). Only 23% of respondents believed in the local community's ability to solve problems, while 42% of respondents was convinced otherwise. One third of respondents (35%) did not agree that local authorities supported local organizations and civil initiatives, while 31% saw local authorities as supportive (M = 2.8).

In the case of site C, respondents rated opportunities for social activity as rather high (M = 3.8), and 67% of respondents hold a believe that the local community has the ability to solve potential future problems (M = 3.7). Opportunities for environmental activities were particularly well-rated (M = 3.8). Respondents also declared that their friends are active in volunteering (M = 4). Participants believed that local authorities support local initiatives (M = 3.4) with one-third of respondents holding a strong positive view on this issue.

To summarize, community resilience in the face of problems was rated highest in the Swedish location, which may stem from higher trust and perceived support for civic initiatives from local authorities in Sweden. This aligns with Sweden's high level of social engagement and developed policies supporting civil society. However, it was the Polish site that scored highest in terms of opportunities for social activism. As mentioned above, this is likely a result of the social mobilization related to protests against demolishing of the settlement, which evolved into continued social engagement.

5 Socio-cultural Impacts of Cultural Heritage

Our study of the characteristics of the local community where adaptive re-use projects took place was complemented by an analysis of the sites themselves. The aim was to diagnose the traits of the sites and users' perception of their value and friendliness.

In terms of traits, we inquired about the place accessibility, their openness to different groups of users, their functionalities, and the activities available at the site. The personal evaluation of the site focused on its perceived value to users and wider community, as well as the emotions evoked by the site. By assessing users experiences at cultural heritage adaptive reuse sites, we aim at gathering information that should guide the further development of site functions, open them up to new groups and activities, and ultimately increase their value to society – while respecting the current functions and needs they satisfy.

As with previous measures, respondents' opinions concerning the sites were measured on a scale from 1 to 5, with average value presented in brackets and on graphs.

5.1 Accessibility and Diversity

All three sites were evaluated as being rather accessible and important for the diversity of the neighbourhood (Fig. 11.8). The Italian site A was considered a place that enhances the diversity of what the city has to offer (M = 4.5). At the same time, it was perceived as a place only moderately accessible to everyone (M = 3.3). This is connected to the fact that the garden is open six days a week, and only during working hours (from 9:30 am till afternoon or evening, depending on the season). Additionally, parts of the garden are not accessible by wheelchair (due to stairs), and educational activities requires a fee for both tourists and community members. Despite these limitation, the potential activities visibly contribute to the diversity of the neighbourhood's character.

In case of site B, the urban garden and settlement of wooden houses are considered to be a place very much adding to the diversity of local offer (M = 4.3). Simultaneously, it was perceived as a place accessible to everyone (M = 4.2). This is the result of community's collective efforts to open the area for artistic and civil society endeavours, leading to the plethora of potential activities happening in the garden free of charge and accessible for anyone.

The post-industrial site C in Sweden is considered as a place that substantially expands the diversity of local offer (M = 4.9). At the same time, the site is perceived as a place only moderately accessible to everyone (M = 3.4). Although the site

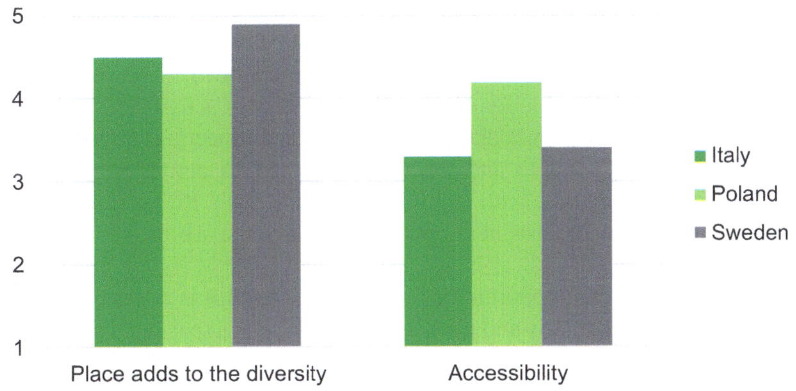

Fig. 11.8 Perceived accessibility and added value of the place

offers numerous interesting experiences, including post-industrial architecture, art and artists' workshops, organised tours, a handicraft shop, and a café, it is still perceived as not easily accessible. This perception might stem from the site's primary function of providing artists with space for their projects, making the main user group relatively small and requiring specific skills.

5.2 *Preferred Activities on the Site*

Heritage sites are associated with different forms of activity: some are seen as place of work and business meetings, while others are perceived as cultural spots where friends meet. To capture the character of each site, we asked visitors to list the activities they usually engage in at the site (Fig. 11.9).

The Italian garden is appreciated mostly for two aspects and available activities—it allows users to experience nature and scenic beauty (83%) and it creates opportunities for education and learning (64%). Respondents also cherished the fact that it provides a safe haven for animals and birds that can be observed there (59%), as well as offering fresh air and opportunities to gaze at bodies of water. The garden is also considered as a good place to relax. Meanwhile, the garden is not perceived as destination for recreation or spending time with friends, nor for engaging in hobbies or discussing community-related issues. Apart from educational activities, respondents see it as a place to enjoy solitary activities.

It is worth analysing these results in light of the grassroots social sustainability index. Considering that opportunities for learning, capacity for self-organization and trust are not the highest at the location A, it might be important from a resilience perspective to create more opportunities for citizens to interact at the site.

The polish urban garden and wooden settlement area are appreciated mostly for aspects connected to community activities and social meetings (72%), experiencing nature and scenic beauty (70%), enjoying fresh air (66%), relaxation (66%) and leisure activities (62%). Respondents also liked educational activities offered and the opportunity to observe animals and birds (54%). Many respondents thought it was a good place to meet with friends and family (50%). The area was also considered a good place to engage with cultural heritage (44%). However, the site was rarely perceived as suitable for work or business-related activities.

The Swedish site is appreciated mostly for three aspects: community activities associated with discussions and social meetings (83%), activities related to work (78%), and spending time with family and friends (72%). Respondents also valued the fact that it provides business opportunities and a space to indulge in hobbies during leisure time.

The site offers dual opportunities—on the one hand it is a space for relaxation and socializing with family and friends, on the other it is a space to work and develop own businesses by resident artists. In both cases, the place provides opportunities for networking and building social relationships with other people.

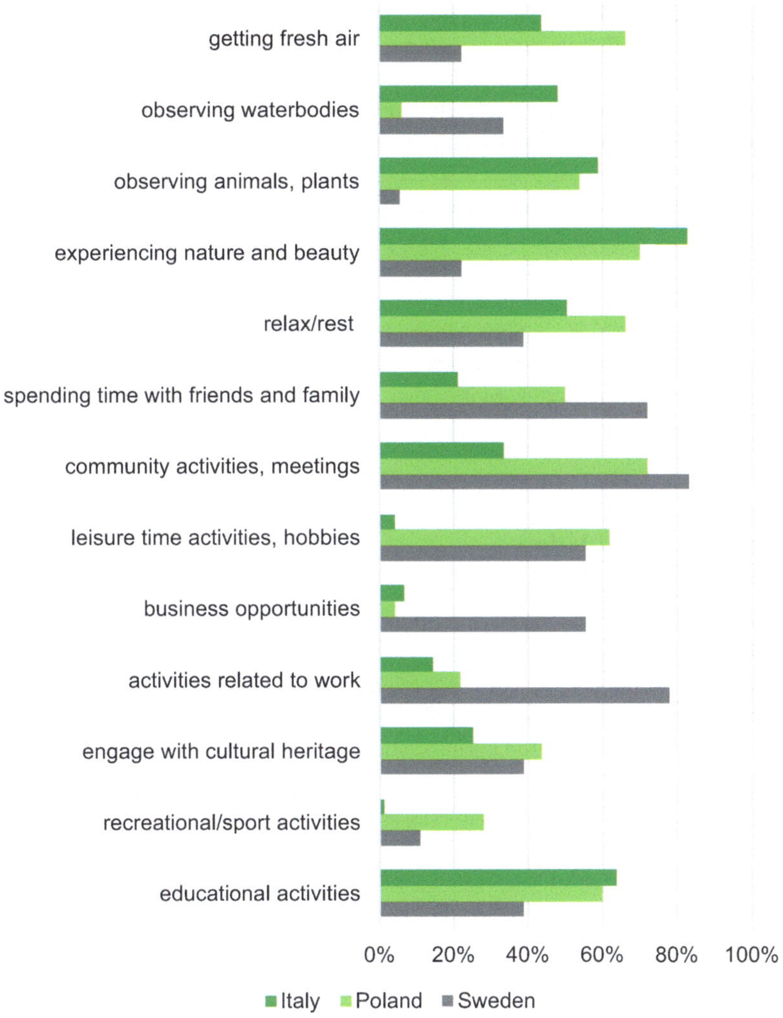

Fig. 11.9 Activities

5.3 Learning Opportunities

The individual capacity for learning is strongly connected with the previous and current experience of learning. However, for learning to occur, people first need to find themselves in a situation where acquiring and processing knowledge is possible. We believe that heritage sites are unique places where learning can take place, so we decided to ask visitors to evaluate the educational efforts of these heritage sites.

Most respondents evaluated the botanic garden as a good place for learning. It was considered an area that inspires creativity (M = 4.5) and is intellectually stimulating (M = 4.7). In the case of the Polish urban settlement, most of the respondents evaluated it as an area that inspires creativity (M = 4.5), is intelectually stimulating (M = 4.3) and in general is a good place for acquiring new knowledge (M = 4.4). In Sweden, the post-industrial site was also recognized as a good place for learning (M = 4.6), one that greatly inspires creativity (M = 4.9) and is intellectually stimulating (M = 4.3). All three sites rated highly in terms of offering users positive stimuli for inspiration and learning. The Swedish site was assessed as most creativity-inducing, reflecting the fact that it is mainly the cultural centre with artists in residence and their art on display. Meanwhile, the Italian site offered the most intellectual stimulation, which relates to the botanical aspects of its flora and existing information and educational activities that allows users to increase their knowledge on different plants.

5.4 *Intrinsic Social Value*

While cultural heritage sites can be important to people for different reasons, we wanted to establish which aspects of the places make them valuable to users. We assumed that while some people would appreciate the beauty of nature or buildings, others might focus on their ecological value or economic potential. Therefore, we asked our respondents whether the heritage site had a special value for them and if yes, why?

In the case of the botanic garden (Site A), respondents considered it both beautiful and special (Fig. 11.10). The garden's significance in their personal lives was evaluated quite high (M = 4), possibly because important events had occured there, giving them reason to value it as a special place in their lives. The garden certainly made people proud (M = 4.4) and for some it was an important part of their own

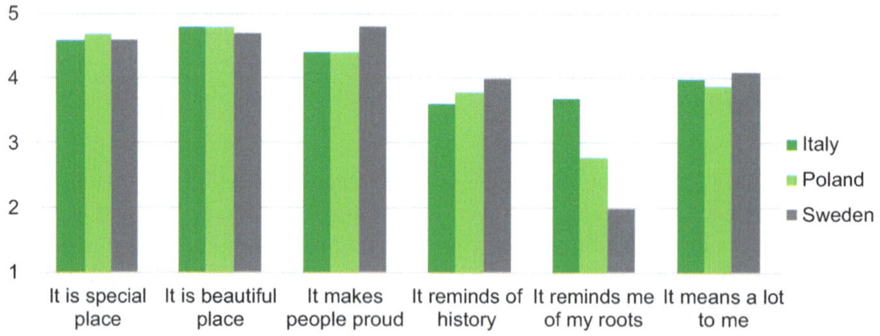

Fig. 11.10 Social value of the sites

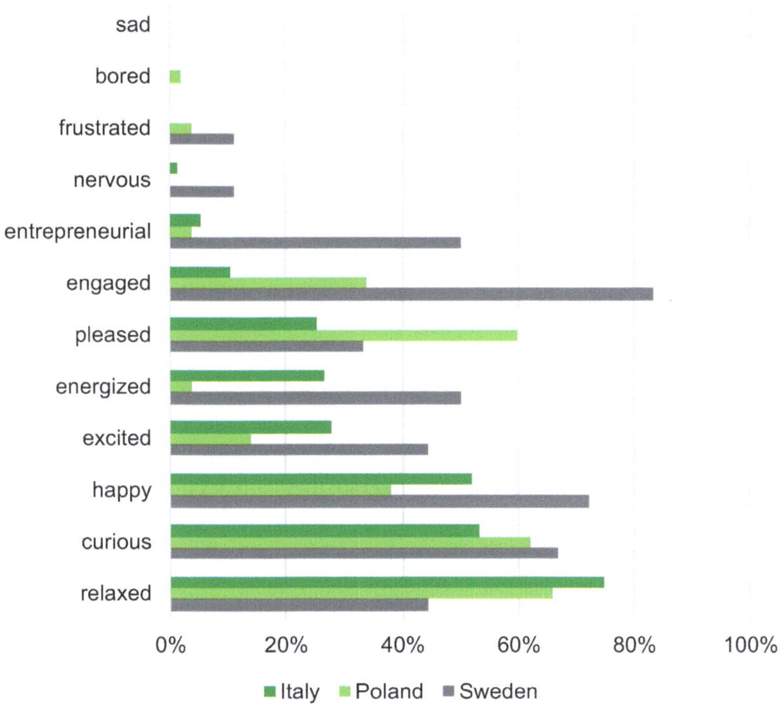

Fig. 11.11 Emotions evoked by the sites

history—they felt that the site connected them to their roots (M = 3.7). The garden also seemed to inspire reflection, with respondents often indicating that it is a spiritual place (M = 4.3). Moreover, the garden reminded people of the history of the neighbourhood (M = 4.4), though less so the history of the country (M = 3.6), which is not surprising, considering that it is a lesser-known local landmark. In general, our respondents felt that the garden is a place worth protecting and that it evokes positive feelings. As shown in the Fig. 11.11, very few people felt uneasy there.

Respondents visiting the Polish settlement and garden area (site B) considered the place beautiful and special as well. The meaning of this cultural heritage in their personal lives was quite high (M = 3.9), possibly because local activists involved many users in actions to preserve the place. The site was making people proud (M = 4.4), and for some, the place was an important part of their own history (33%). Spending time in the area seemed to make people reflective; many considered it as a spiritual place (M = 3.6). The area reminded people mostly of the history of the neighbourhood (M = 4.4) and less so of the national history (M = 3.8), which is not surprising, giving that it is a lesser-known cultural heritage.

Visitor of the Swedish post-industrial site C declared that the place strongly affects their sense of local pride (M = 4.8), which may be related to the fact that they recognize this place as both beautiful (M = 4.7) and special (M = 4.6). On average, survey participants emphasized that the place meant a lot to them (M = 4.1). Apart from its aesthetics and meaning, this place is also important to users for historical reasons—it reminds them of both the history of the local community (M = 4.2), and also brings back history in a general sense (M = 4). However, this historical significance does not translate to individual or family history (M = 2). As many as 83% of respondents said that the site does not remind them of their roots, which may be related to the recent renovation or the fact that many users come from outside the area.

5.5 Emotions

Generally, people reported feeling positive emotions in each of the sites (Fig. 11.11), although some differences among the sites can be captured.

A visit to the botanic garden site elicited mostly positive emotions in our respondents. As shown in the chart, people usually reported feeling relaxed, curious, and generally happy when visiting the place. For many, the visit was exiting or energizing, and they felt pleased. Almost none of the respondents experienced negative emotions.

Visiting the garden and wooden settlement area also generated mostly positive emotions in respondents. People felt relaxed, curious, and happy when visiting the place. They more frequently reported feeling pleased and engaged at this site compared to others. Almost none of the respondents experienced negative emotions like boredom or frustration.

A visit to the post-industrial site also elicited mostly positive emotions in respondents. When visiting the place, people felt mostly engaged, happy, curious, and entrepreneurial. For many, the visit increased their energy level, while others felt more relaxed. Compared to other sites, location C visibly triggered an entrepreneurial attitude and high levels of engagement, which may relate to the fact that the site showcases the work of artists-entrepreneurs and invites others to engage in a similar way. Almost none of the respondents experienced negative emotions, but some mentioned feeling nervous or frustrated.

5.6 Things to Enjoy

By asking visitors what they enjoy most about the heritage site, we aimed to identify the unique added value of the place. We believe it is one of the best questions to reveal differences in the character of the sites.

At site A, respondents declared that they mostly enjoyed trees, greenery, sights, and quietness. Other enjoyable characteristics of the place included its history and landscape. As mentioned before, the generally pleasant atmosphere and historical character of the garden contributed to its positive reception (Fig. 11.12).

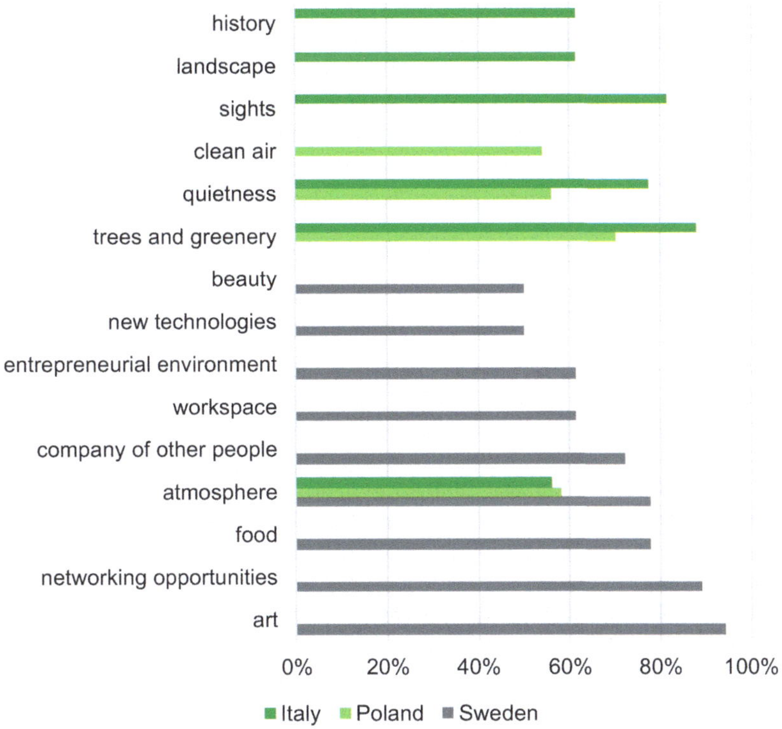

Fig. 11.12 Thinks that people enjoy the most on the site. Only the items that were indicated by more than 50% of respondents are listed on the graph

Respondents declared that the most enjoyable characteristics of the place B were trees, greenery, atmosphere, quietness, and clean air. The landscape and shade were also considered important, but by less than 50% of respondents. Even fewer respondents appreciated the company of other people and networking opportunities. The sights and contact with art were considered important by a few, as well as the fact that this is a place with history.

Visitors to site C declared that art, networking opportunities, food, and general atmosphere were the most enjoyable characteristics of the place. The company of other people, as well as entrepreneurial environment and space for work were also considered important. The generally pleasant atmosphere and the opportunity for networking and leisure added to the positive reception of this post-industrial site.

5.7 Recommendations

In the final part of the survey, we asked the users of the sites for their recommendations for the future development of the cultural heritage functions.

Most respondents felt that the botanic garden (site A) could be further improved by offering new social and cultural activities open to a wider public. Adding services such as places where visitors could eat or drink in or around the garden would be welcomed. Respondents suggested that the space could be expanded to allow for scientific activities and cultural events. Opening the site to the local community and involving its members in various educational and integrating activities (e.g., neighbourhood gardening, study areas) were also often mentioned as ideas for improvement.

Importantly, respondents underlined the need to remove architectural barriers to enable access for older citizens or people with disabilities. Some respondents mentioned that the garden would be a good place for wellness-related activities such as tasting herbal teas and healthy products, participating in yoga classes, or attending seminars on health and the value of nature.

Upgrading the garden in terms of the quality of flora (making sure that plans are watered and taken care of), improving the paths design, and the clearly signage (possibly with translation into English) were also recommended. Some respondents suggested developing cooperation with schools (for educational projects for children) and the Faculty of Medicine to make the garden more vibrant and attractive to the community.

For the urban garden surrounding the wooden settlement (site B), most respondents believed it could benefit from expanded offer of social and cultural activities open to all. They recommended renovating the houses in the area (many which are in poor shape) and better maintaining of the whole area (with more cleanliness, pavements, and benches, etc.). Upgrading the site with additional plants and ensuring that information on activities is clear and well-displayed were also mentioned.

Adding services, like places where visitors could eat or drink (for example vegetarian food) in or around the area, would be welcomed at site B. Respondents suggested that more information about both the history of the place and the events happening there should be provided. Some respondents felt the space should host more cultural and educational events for different age groups, facilitating opportunities to meet new people.

The question of finance stability was also raised by some respondents, who suggested development of a new financial model for social activities or allowing more tax revenue to be directed to the area.

Visitors to the post-industrial site C wanted the site to be further improved by offering new entrepreneurial and networking spaces. Adding services, like small shops, studios, or places where visitors and community members could eat or drink would be welcomed. Respondents would also welcome improvements to the space itself, especially workspaces, offices, and workshops—both in terms of heating and appliances (including a refrigerator). More importantly, respondents felt that the availability of the space could be significantly improved.

Many respondents commented on the infrastructural aspects of site C, some pointing out that new opportunities for businesses to invest in the premises through long-term contracts would be beneficial for all parties. Others focused on the greenery, suggesting upgrades to both the outdoor space and the residential areas for artists. Finally, respondents felt that the design of the old building should be enhanced, to emphasize its artistic values and increase the aesthetics of the place.

6 Summary

We conducted two studies at each of the three pilot sides to test the grassroots social sustainability framework and intrinsic value measures developed within the CLIC project.

The first study focused on the social context of heritage adaptive reuse and collected data on the perception of community strengths and weaknesses in various aspects of social sustainability. We discovered that the sites differed in how community members perceived newcomers and each other. In some cases, national patters were reflected on the local scale— for example the high levels of trusts and capacity for organization observed in the Swedish location, compared to medium trust levels in the Polish location, which were paired with wider social relations.

All three studied communities exhibited a high diversity of skills and a generally sociable attitude, which can be seen as key to enabling the development of the site and attracting new ideas and people. However, the lower levels of openness to newcomers and trust in the Italian community could hinder cooperation and self-organization, lowering the community's potential to work together on developing the site's capacity to respond to various needs. Therefore, the adaptive reuse of the Italian site should be designed in a way that helps to boost grassroots social sustainability elements such as openness, trust, and self-organization.

Our study highlights how many of the communities' struggles can be either facilitated or hindered by the space characteristics. For example, the fact that the site A is closed to visitors after opening hours and requires a fee creates a sense of detachment between the residents and the site, which may exacerbate problems with reaching out to newcomers and building trust. In contrast, site B which is free to visit and hosts many events, has managed to create a community that extands beyond just the residents, serving as a hub for diverse initiatives. Site B is also the location where grassroots protests protecting the settlement took place, which both increased awareness of the site's value as well as strengthened the attachment of those involved. In the case of site C, the perception of accessibility is just average, meaning that the accessibility of the place still has some potential to be fully realized. Taking into account that residents rated the opportunities for learning in the area as average, enhancing services in this domain seems like a promising direction. Although the local community here is aware of its diversity and shares a common vision only to a slight degree, they still trust each other a lot and look toward the future with optimism. Our additional observations reveal that despite the site's relative openness to visitors, there is a division between the heritage site community and the surrounding village community, reflecting the divide between newcomer urbanites and rooted locals. While popular with external visitors coming from further away, site C faced the challenge of connecting with the local community and realigning different stakeholders' visions for the sites' future.

While investigating common vision, we studied the community's attitude toward nature, as an important part of circular adaptive reuse projects. For the successful implementation of circular economy and sustainable development goals, a certain openness and sensitivity to environmental and ecological matters are essential. Therefore, we examined whether local community's vision reflects the need to protect the natural environment and if there is a shared understanding of the benefits stemming from natural resources. We found that our sits varied in terms of community's awareness of the need for protecting natural resources, which might be explained by the level of economic and cultural dependence on natural resources. At site A there is a certain disconnect between the botanic garden's natural resources and a general perception of nature's benefits in otherwise not-so-green neighbourhood. In contrast, site C, which is close to lakes, forests, and waterways, we noted a high level of recognition for the cultural value of nature. Visitors to site B expressed an ambivalent attitude toward nature, which can be explained by the modern urban setting where green areas are primarily valued for recreational or decorative purposes.

Last but not least, we studied the trust levels within the communities to identify patterns that facilitate cooperation and found that two of our sites could greatly benefit from increased trust. Site A's residents were generally careful of each other and seemed unaware of opportunities to develop positive local bonds. In case B, the main challenge related to the lack of trust towards city authorities, stemming from the negative experience with demolishment plans. These trust deficits were mentioned as significant barrier to further developing the potential of the site. Therefore, we identified trust-building efforts as key to building grassroots sustainability in both these locations.

Moreover, we posit that trust relations strongly influence community resilience in the face of challenges. While resilience was rated highest at our Swedish location, we suggest that it correlates with higher generalized trust and perceived support for civic initiatives from local authorities. This aligns with Sweden's high level of social engagement and well-developed policies supporting civil society. However, it was the Polish site that scored highest in terms of opportunities for social activism. As mentioned earlier, this is likely a result of social mobilization related to protests against demolishing of the settlement, which evolved into continued social engagement. However, this engagement has not yet translated into empowerment and optimism that the local community has enough resources to face a potential crisis. We believe that the Italian site has the highest potential to develop new functions and attract new users, which should meaningfully increase grassroots social sustainability.

Learning from our pilot studies allowed us to delve deeper into various aspects of what makes adaptive reuse projects more succesful and how they relate to the characteristics of both the site itself and the local community. Our key takeaways from applying the grassroots social sustainability framework are as follows:

First, we emphasize the importance of studying both the "social health" of the neighboring community and the "social perception" of the site itself. Only by combining these two perspectives we can fully anticipate reactions to change and potential conflicts around new investments. Second, we see the proposed questionnaires and the GSS framework as easy-to-use tool for all potential stakeholders, enabling better-informed planning and implementation processes. Third, we believe that the proposed framework can be used not only to evaluate cultural heritage adaptive reuse projects but also a wide range of investments in public places and spaces. Its greatest potential, however, lies in the empowering the local community by placing it at the centre of various planning and evaluation processes.

We believe that the joint study of community characteristics and the perceived intrinsic value of the site allowed us to elevate the evaluation framework to a different level. By underling the interaction between the value attributed to a place and grassroots social sustainability, we sought to highlight the relationship between openness to change in a given location and the grassroots sustainability of the local community, particularly in places with high intrinsic value. As part of the CLIC project, we developed an approach to adaptive reuse projects that views not only as opportunities to preserve the historical or artistic value of the cultural heritage but also as a means to enhance sustainability and resilience within the local community.

Our methodology underlines the responsibility that stakeholders planning and performing adaptive reuse project bear, not only for the material value of the heritage site but also for less tangible factors like the wellbeing of its users and the social resilience of its neighbours. If hastily planned and executed, adaptive reuse projects may merely preserve or even deteriorate important aspects of community potential and social sustainability. However, if done correctly, by strategically addressing the strengths and weaknesses of the grassroots context, such interventions can bring numerous social, cultural and economic benefits.

References

Afshari H, Agnihotri S, Searcy C, Jaber MY (2022) Social sustainability indicators: a comprehensive review with application in the energy sector. Sustain Prod Consum 31:263–286. https://doi.org/10.1016/j.spc.2022.02.018

Cope MR, Kernan AR, Sanders SR, Ward C (2022) Social sustainability?: Exploring the relationship between community experience and perceptions of the environment. Sustainability 14(3):1935

Curşeu PL, Schruijer SG (2017) Stakeholder diversity and the comprehensiveness of sustainability decisions: the role of collaboration and conflict. Curr Opin Environ Sustain 28:114–120

Folke C, Carpenter S, Elmqvist T, Gunderson L, Holling CS, Walker B (2002) Resilience and sustainable development: building adaptive capacity in a world of transformations. AMBIO A J Hum Environ 31(5):437–440

Haerpfer C, Inglehart R, Moreno A, Welzel C, Kizilova K, Diez-Medrano J, Lagos M, Norris P, Ponarin E, Puranen B (2022) World values survey: round seven – Country-Pooled Datafile Version 3.0. JD Systems Institute & WVSA Secretariat, Madrit/Vienna

Hardin R (2002) Trust and trustworthiness. Russell Sage Foundation

Hernández B, Hidalgo MC, Salazar-Laplace ME, Hess S (2007) Place attachment and place identity in natives and non-natives. J Environ Psychol 27(4):310–319
Kyttä M, Broberg A, Haybatollahi M, Schmidt-Thomé K (2016) Urban happiness: context-sensitive study of the social sustainability of urban settings. Environ Plan B Plan Design 43(1):34–57
Lewicka M (2008) Place attachment, place identity, and place memory: restoring the forgotten city past. J Environ Psychol 28(3):209–231
Missimer M, Robèrt K-H, Broman G (2017) A strategic approach to social sustainability–Part 1: exploring the social system. J Clean Prod 140:32–41
Nelson PJ (2007) Human rights, the Millennium Development Goals, and the future of development cooperation. World Dev 35(12):2041–2055
Nosratabadi S, Mosavi A, Shamshirband S, Kazimieras Zavadskas E, Rakotonirainy A, Chau KW (2019) Sustainable business models: a review. Sustainability 11(6):1663
Olejniczak K, Kupiec T, Estelle R (2014) Brokerzy wiedzy. Nowe spojrzenie na rolę jednostek ewaluacyjnych
Putnam R (2001) Social capital: measurement and consequences. Can J Policy Adap Reuse Cult Herit 2(1):41–51
Roszczynska-Kurasinska M, Domaradzka A, Ślosarski B, Żbikowska A (2019) Embracing circularity in adaptive reuse – grassroots perspective. In: STS conference Graz 2019. Critical issues in science, technology and society studies, pp 363–383
Roszczynska-Kurasinska M, Domaradzka A, Wnuk A, Oleksy T (2021) Intrinsic value and perceived essentialism of culture heritage sites as tools for planning interventions. Sustainability 13(9):5078
Sampson RJ (2012) Great American city. In: Great American City. University of Chicago Press
Sztompka P (1999) Trust: a sociological theory. Cambridge University Press
Żbikowska A, Roszczyńska-Kurasińska M, Ślosarski B, Domaradzka A (2024, this volume) Social sustainability framework for measuring socio-cultural impacts of cultural heritage adaptive reuse. Springer

Chapter 12
Multidimensional Spillovers of Cultural Heritage Through Regional Development and Circular Economy

Jermina Stanojev and Christer Gustafsson

1 Introduction

In the phase of the advancement of conservation as a science and process, defined as Conservation 3.0, conservation of historic environments is observed as a source for development and shall be envisaged as a production factor, and as investments that are anticipated to be central to the returns in the future (Ost 2009). In line with that, an enlarged significance has been determined in evaluating social environmental and economic spillover effects of cultural heritage that could be understood as a vector with focus on development and continuity (Janssen et al. 2017). This indicates that the consequence, or the spillover effects, from investing in cultural heritage are comprehended as contributors to sustainable development and are used as starting points for planning the priorities for cultural heritage management (Gustafsson 2019). The importance for the cultural heritage sector is no longer solely on preservation and protection- to be able to identify new actions to take place in historic buildings has become more essential. Adaptive re-use is defined as "any building work and intervention aimed at changing its capacity, function or performance to adjust, reuse or upgrade a building to suit new conditions or requirements" (Douglas 2006). These activities and actions are expected to bring spilovers at the regional level.

Investments in cultural heritage are not only planned out from heritage values. Instead, it has the demand from the society in general, as expresses in the sustainable development perspective and smart specialisation at regional level in particular as a starting point (Gustafsson 2019). On the other hand, region's specific needs, problems, opportunities and challenges are presented and cultural heritage is mainstreamed into these (Fig. 12.1). The spillover effects from investment in cultural heritage as well as from the new activities in the historic buildings are understood as

J. Stanojev (✉) · C. Gustafsson
Department of Art History, Uppsala University, Visby, Sweden
e-mail: jermina.stanojev@gmail.com; christer.gustafsson@konstvet.uu.se

L. Fusco Girard, A. Gravagnuolo (eds.), *Adaptive Reuse of Cultural Heritage*,
https://doi.org/10.1007/978-3-031-67628-4_12

Fig. 12.1 Region's specific needs, problems, opportunities and challenges

contributing to the aims presented in, for example, sustainable development goals and smart specialisation strategies. The adaptive reuse could therefore be planned to also improve and strengthen innovation-driven growth in the field of creative force and creative clusters, inclusiveness and social cohesion, as well as sustainability.

2 Regional Development and Cultural Heritage in the Context of Global Challenges

The global society of today is facing three major global challenges: climate change; global economic competition; poverty and social exclusion. The Global Risks Report 2021 has identified Climate action failure, Human environmental damage and Natural resource crises as three out of seven Top Global Risks by Impact (World Economic Forum 2021) all of them directly impacting cultural heritage too. The political response in various organizations to these challenges has been synthesized with the comprehensive concept of sustainable development through different milestones.[1] On the other hand, a failure of national governance (e.g. failure of rule of law, corruption, political deadlock) has been identified as one of Global Risks 2020 and the regional development and regional governance have been identified as model to turn to. These two trajectories implies that solutions from global risks for cultural could be defined at the regional level.

In literature, and in debate, the concepts of region and regional development have been interpreted in many ways based on several epistemological frameworks. Often, regional development has been understood in terms of economic growth and employment. From the perspective of welfare economics, however, the regional development may be understood as increased welfare and this includes qualities influencing well-being, such as regional identity, democracy and ecology. Thus, it seems impossible to promote regional development without creating one single job,

[1] Such as the Brundtland Report (WCED 1987) the Rio Declaration (UNCED 1993) the Habitat Agenda (UN—Habitat 1996) the UN World Summit 2002 (UN—Habitat 2002) and the World Urban Forum (UN—Habitat 2012, 2009) etc.

without attracting one single business or without any person moving into a region, but simply by stimulating identity and creativity within a region.

In the last decades, methods of territorial development and consequent physical and entropic transformations have represented a fertile field of research, e.g. for urbanists, architects, environment scientists, economists, geographers, anthropologists, sociologists etc. A creative relation between economy, society and territory has emerged from the analyses operated in these different research fields: this relation configures itself as a complex system able to generate growth and development (Hubbard 2006). Depending on consciousness about this linkage, it is obvious that the dimension of competitiveness increasingly shifts from the micro-level of single economic operators to the macro-level of territorial systems and this applies to cultural heritage too—from a micro-level of single cultural heritage adaptive reuse to the macro-level of the regional territorial system. Here, the organization of the systemic logic of resources and the creation of cooperation networks fed by huge levels of social capital become a necessary requirement for productivity and potential for territorial growth.

When applying this method of regional development, Sacco et al. draw on three different paradigms. Firstly, the creativity-based attraction model of Richard Florida, which emphasizes the role of quality of life and of technological infrastructure in the creation of a critical mass for the emergence of a knowledge-orientated economy (Florida 2002). Secondly, the competitiveness-based urban renovation model of Michael Porter, which is focused on the transition from an investment-based industrial orientation toward a self-sustaining innovation-based economy (Porter and Porter 1998). Finally, the capability-based model of Amartya Sen has been used, which underlines the central role of general social involvement in capability-building activities as a prerequisite for viable economic development (Ferilli et al. 2019). With the exception of Florida, none of these models was created to explain cultural and cultural heritage driver phenomena, since they rather highlight post-industrial growth as an interesting field of application.

New approaches might be decentralized policy-making and decision-making, as well as elucidating the roles of politicians and civil servants, to promote the development of innovations, entrepreneurship and cross-sectorial collaboration in relations to smart specialisations and cultural heritage, and to develop education and training programmes in relations to businesses, such as those promoted by the concept of the KIC (Knowledge and Innovation Communities of the European Institute of Innovation and Technology). It might also imply the creation of horizontal multi-problem-oriented networks at the regional level, the development of project design and project management and the development of a systematic approach for process-oriented knowledge, as well as enabling better understanding of the role and possibilities for the cultural heritage and its spillover effects in Europe and beyond.

3 Smart Specialisations as the Framework for Regional Development

Smart specialisations are playing a key role in the concept of regional development, representing a new policy measure to overcome disorganised and weak parallel activities in countries, and offer support to those areas that have research, technological and production capacity to carry out particular activities. Smart specialisations, therefore, follow the socio-economic situation in countries and their integrated technological, institutional and business processes. This approach has a significant potential as an alternative measure to many currently present horizontal policies, mostly not orchestrated and even competitive. The latter may create a conflict between public institutions and programmes producing low level collaboration and therefore effectiveness of the legislative documents, slow implementation and development of new strategies and finally low impact on securing technology transfer and economic growth. In respect to the principles of the Europe 2020 strategy, smart specialisations are constituted to ensure the most effective results through rational and strategic spending of public capital. The idea of smart specialisation is based on the notion that regions cannot achieve everything in science technology and innovation, and therefore it is crucial to follow a thoughtful process of prioritisation, concentrating resources in certain domains of expertise based on the needs and available resources of each region.

A creative and cultural sector forms are most important growth factor in many cities and even regions. Integration of these cultural and creative sectors in innovative regional strategies for smart specialisations represents the challenge. By that means, regions need to understand and take into account complexity of interconnections between traditional cultural assets such as cultural institutions, services and heritage, and on the other side the improvement of creative industries or tourism. A Guide to Regional Innovation Strategies for Smart Specialisation, published by the European Commission and drafted by the Smart specialisation platform—S3Platform, has a purpose to support Member States and regions in creating, executing and revising regional smart specialisation strategies, where one part is devoted to the cultural and creative industries.

Therefore, integration of cultural heritage into smart specialisations is mainly conducted through cultural and creative industries and cultural tourism (Stanojev and Gustafsson 2021). Some EU regions have recognised very well this vast potential in promoting socio-economic development supported by the EU structural funds. On the other hand, it seems that majority of others have not realised this potential (Stanojev and Gustafsson 2021).

4 Cultural Heritage in Relations to Smart Specialisations and Circular Economy Strategies

The first EU action plan for the Circular Economy advocates that waste management should not be the only issue tackled within the concept of circular economy, but that rather should be considered as a broader sustainable development strategy that also should "support Member States and regions to strengthen innovation for the circular economy through smart specialisations" (European Commission, Joint Research Centre 2015) (Fig. 12.2). Circular economy is now explicitly put forward in the European Commission's legislative proposals for the 2021–2027 programming period, and the European Commission has adopted a new Circular Economy Action Plan (European Commission 2020)—one of the major strengths of the European Green Deal (European Commission 2019). Having in mind the importance of regions in the EU policy it is essential to explore the regional level of circular economy implementation and the role of cultural heritage in these processes.

While concepts of adaptive reuse and circular economy are growing, circular economy design processes in the building sector are not advanced yet, neither well-articulated, despite the fact that circular economy models could greatly fit for cultural heritage assets and construction sector in regions and cities. The concept of circular economy is rising its interest in different sectors, but also in different fields, in particular the ones that are not initially linked with circular economy, meaning fields beyond waste management, resource efficiency, low-carbon investments etc. (Bosone et al. 2021; Fusco Girard and Vecco 2021; Stanojev and Gustafsson 2021). This link opens up many niches of research for fields and sub-fields that could be linked and fit under circular economy and that could be considered as a recourse for its implementation. Cultural heritage and its adaptive reuse shall be considered as one of them. Having in mind enormous number of cultural heritage sites and building stock in Europe this subfield could be considered even as one of circular economy pillars. A large and growing body of literature (Pickerill 2021; Gravagnuolo et al. 2021; Acri et al. 2021) has investigated adaptive reuse and sustainability of buildings—definitions of adaptive reuse, the interest in adaptive reuse of buildings as an alternative to demolition for the benefit of the society, an analysis of the

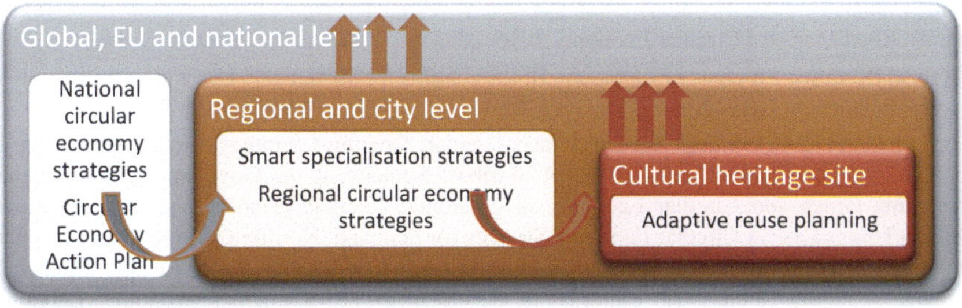

Fig. 12.2 Smart specialisation strategies in relation to global and local challenges

renovation process in terms of quality of intervention and investments, also in terms of impact on the environment, but some of them considering adaptive reuse as a strategy towards conservation of cultural heritage.

Although a circular economy monitoring tool has been set at the level of the EU, there have been circular economy related actions taking place not only at the EU level, but also at national and regional levels. Different European countries and regions have developed their own guidelines and documents related to circular economy and its actions. At the national level, the EU Member States have developed and adopted different strategic documents regarding circular economy, taking diverse approaches toward its understanding. Several countries developed their circular economy strategic frameworks, roadmaps or national plans[2] (Greece, Italy, Denmark etc.) while some countries integrate circular economy aspects into their national strategies through waste management[3] (Germany, Romania, Slovakia etc.), and Sweden does it through its bio-based economy. However, regions and cities (NUTS 2 and NUTS 3) have rather identified their circular economy strategies instead of spreading circular economy actions through different plans.[4]

In order to understand the state of the art regarding circular economy and cultural heritage at national and regional levels, the search was performed related to the following concepts in national, regional and city strategic frameworks, roadmaps or national plans related to "cultural heritage", "cultural", "culture", "creative", "heritage", "adaptive reuse", "historic buildings".

Although a small number of countries, regions and cities have officially adopted circular economy strategies and roadmaps, it should be taken into consideration that other national and local governments have also started implementing circular economy principles through other actions (a full list of good practices, European Circular Economy Networks and events is available at the European Circular Economy Stakeholder Platform—a joint initiative by the European Commission and the European Economic and Social Committee).

[2] National Action Plan on Circular Economy of Greece; Roadmap towards the Circular Economy in Slovenia; Towards a Model of Circular Economy for Italy—Overview and Strategic Framework; Leading the Transition: Circular Economy Action Plan for Portugal; Circular Economy Roadmap of France: 50 Measures for a 100% Circular Economy; A circular economy in the Netherlands by 2050; Danish Circular Economy Strategy; Spain—Circular Spain 2030; Leading the Cycle—Finnish Road Map to a Circular Economy 2016–2025.

[3] German Resource Efficiency Programme II: Programme for Sustainable Use and Conservation of Natural Resources; Luxembourg's National Waste and Resources Management Plan.

[4] Extremadura 2030: Strategy for green and circular economy; Strategy for the Transition to Circular Economy in the Municipality of Maribor; Strategy of the Government of Catalonia: Promoting Green and Circular Economy in Catalonia; Circular Flanders Kick-off Statement; Making Things Last: A Circular Economy Strategy for Scotland; Northern Irish Region—Circular Economy Strategy; Regional Plan for Circular Economy, Brussels Capital Region; Roubaix's Circular Economy Route Map; Päijät-Häme Roadmap toward a Circular Economy; London's Circular Economy Route Map; Circular Amsterdam, A vision and action agenda for the city and metropolitan area; White Paper on the Circular Economy of Greater Paris.

Circular economy became an umbrella assembling strategies, but also practical solutions at different levels regarding economic transformations. However, at regional levels circular economy is also directed at the green and bio-economic sectors, which implies that agriculture and biotechnology are prioritized, as is the case in Germany, Sweden and Portugal. On the other hand, some countries, such as Spain, France and Romania, integrate circular economy principles into their national strategies through waste management. Waste reduction and conversion is an essential part of circular economy; however, it should not be the only possible way to implement the circular model.

A search performed for "cultural heritage", "cultural", "culture", "creative", "heritage", "adaptive reuse", "historic buildings", showed the results that "reuse" has been applied the most, but mainly in the context of "reuse of building materials", "material reuse", "waste reuse", "reuse by enabling reallocation of materials". Even Amsterdam city, which manages the Seventeenth-Century Canal Ring Area of Amsterdam inside the Singelgracht designated as the UNESCO World Heritage Site, does not recognise the term "cultural heritage" or "adaptive reuse" in context of "historic buildings" in their document "Circular Amsterdam, A vision and action agenda for the city and metropolitan area".

On the other hand, the "Regional Plan for Circular Economy, Brussels Capital Region" implemented through four sectors (construction, resource and waste, logistics, retail business), in its construction sector clearly recognises "making use of the building stock—urban mining" as one of the main strengths, as well as "occupying empty buildings" and "building conservation". In Brussels, with between 15,000 and 30,000 buildings standing empty and with increasing numbers of people looking for an affordable place to live or to carry out a wide variety of activities, the local government renovates these buildings and makes them temporarily available for social initiatives, with the idea to bring about a proliferation of urban activities and a laboratory illustrating creative potential which can intermix social, economic and charitable activities, while also accommodating cultural gatherings.

Päijät-Häme region in Finland included circular economy in its regional innovation strategy for smart specialisation, thus defining circular economy as a priority sector for the region, but still not including cultural heritage as a part of its implementation.

No other region declared "cultural heritage adaptive reuse" in the context of their circular economy strategies and its understanding mainly remains in domains "Constructions" and "Waste management".

As the European Commission's monitoring frameworks on circular economy has not offered instruments for following the implementation of circular economy concepts and effects at the local and regional level, the research of Stanojev and Gustafsson 2021 contributes to setting up a baseline for the monitoring of including cultural heritage in circular economy processes at regional levels through smart specialisations.

Taking into account previously discussed, that study "Smart Specialisation Strategies for Elevating Integration of Cultural Heritage into Circular Economy" (Stanojev and Gustafsson 2021) evaluated the impact that cultural heritage and culture have on each specific region/city toward circular economy through smart specialisation strategies (Figs. 12.3 and 12.4).

These analyses have elaborated on main regional tendencies and to understand which of the regions have potentially have developed spilovers. This research also explores what are the existing and registered priorities and tendencies of the CLIC regions in order to give a support to how these priorities could serve to ensure economic slipovers at regional level, being transformed and/or linked to adaptive reuse activities, which were explored in the development of Local Action Plans within CLIC.

None of the CLIC pilot cases includes cultural heritage and culture as a comprehensive part of their priorities and strategies.

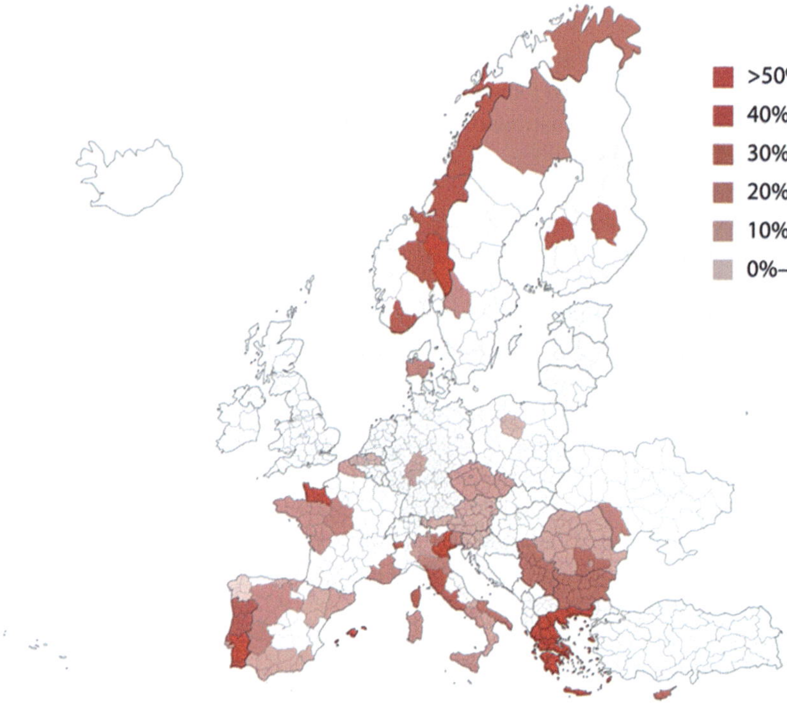

Fig. 12.3 Graphical presentation of regions prioritising culture and cultural heritage with color-coding. (Stanojev and Gustafsson 2021)

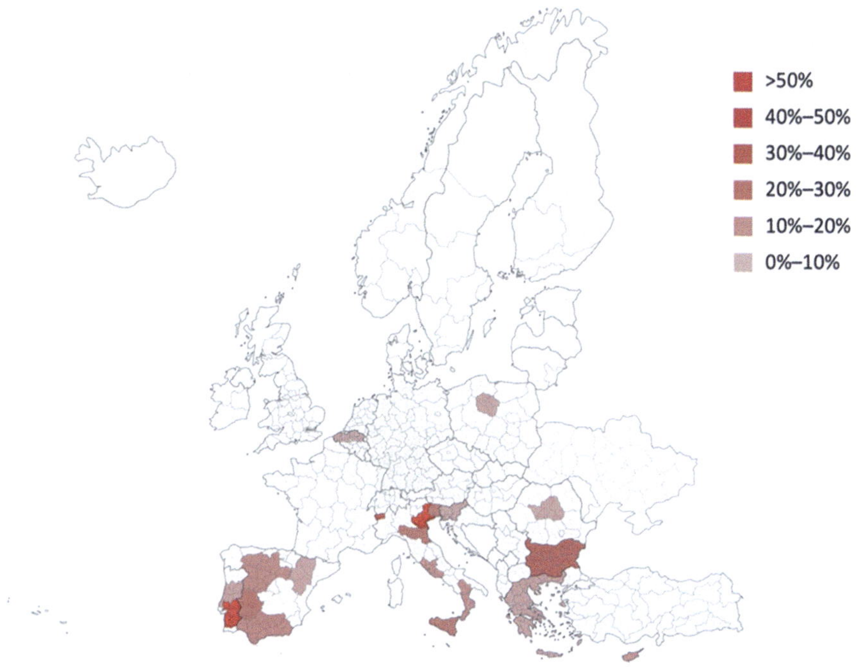

Fig. 12.4 Graphical presentation of regions prioritising cultural heritage with color-coding. (Stanojev and Gustafsson 2021)

5 Cultural Heritage Spillover Effects

The process that leads from the disclosure to the full cultural heritage service is rather straightforward, probably more than it may seems. Nowadays, regarding this aspect, there are two perspectives that make impact on cultural heritage policies in the EU. First one considered cultural heritage as a "territorial capital" that represents a potential for sustainable development of places taking into account institutional setting, socio-economic organisations, knowledge, cultural and natural heritage and infrastructure (Camagni 2009, 2019). Territorial capital efficiency is assessed through the quantity of new investments, income and new jobs created. Moreover, specific creation of place image and identity may positively influence the creativity and spirit of new entrepreneurs across various sectors within local economy, for instance architecture, construction, fashion design, crafts, tourism, hospitality, catering, etc. Nevertheless, according to the Richard Florida, who created the idea that commercial success happens in those cities that have embraced so-called "creative class", the biggest influence on the creation of "creative city" have people of action, knowledge and energy who recognise issues and find solutions, who develop innovative products and amenities, and at the same time enrich and civilise the environment by the means of creativity and artistic sense, and finally define society development to follow their visions. Yet, these people usually choose to live in an environment that has a beneficial cultural, social and technological influence on them (Florida 2002, 2004). The argument is found in the necessity of special devotion to the uniqueness of places that attract representatives of the creative class. This view of the cultural heritage as an element for commercial success is part of the general movement of the re-framing culture in EU and understanding that cultural heritage also involves innovation and improves the long-term competitiveness of the European economy (Gustafsson and Lazzaro 2021).

Preservation and adaptive reuse of cultural heritage have become a pro-active process, or a production factor, where historic buildings and related activities, can be used as an infrastructure for innovative initiative in the creative industries and in specific field of creative powers or platforms of innovations. The process targets long-term benefits, addresses economic spillovers and non-use values which are not immediately related to the use of cultural properties, but may give larger benefits to local systems in terms of increase of human and relational capital.

The European Commission, put spillover effects of the culture on the political agenda (COM 2012 537) by stating that "Being at the crossroads between arts, business and technology, cultural and creative sectors are in a strategic position to trigger spill-overs in other industries" and "The cultural and creative sectors need multi-disciplinary environments where they can meet with businesses from other industries. Any public intervention aiming to further develop them calls for cross-sectoral fertilisation. This requires the development and testing of better business support instruments and policies that aim to facilitate cross-sectoral linkages and spill-overs."

Furthermore, one priority of the Agenda Europe 2020 was to promote spillovers from the cultural and creative sectors as well as A New European Agenda for Culture claims that "Culture and creative industries also have the power to improve lives, transform communities, generate jobs and growth, and create spill over effects in other economic sectors."

However, research into our fundamental understanding of spillover effects are deficient (Vickery 2014). There are research gaps about the term and concept, as well as about methods of evaluations, both qualitative and quantitative as well as about their translation regarding adaptive reuse and local economy. Nowadays, the term "spillover" is mostly understood and used in the context of "impact".

One definition of spillover was presented by the KEA as "(A) process by which the interactions between artists, creative professionals and industries and/or cultural organisations contribute to economic and/or social innovation in other sectors of the economy or society. The spillover process takes place when creativity originating from culture and creative professionals and industries influences innovation in sectors where culture and creative professionals do not usually evolve." Moreover, "Spillover is defined as benefits arising from the activities of CCIs including artists and creative professionals, which determine positive effects on other sectors of the economy or society. Those positive externalities result from processes through which culture-based creativity spreads out from the CCIs, across economic sectors and industries, thus contributing to innovation in the wider economy."

Jonatan Vickery provided one definition in the first issue of the publication series "to be debated "by ecce, 2015 as following "Spillover might involve: Complex interactions/effects/influences operating on different registers – not simply "cause effects"... A process of dialogue, interaction and engagement that might be place specific or place sensitive or optimised by drawing on the resources of place and contributing to the broad economic development of place... Crossing boundaries—informal as well as formal jurisdictions, questions of agency and legitimacy."

"Creative clusters and innovation" published by Nesta in 2010 tried to understand innovation spillovers from a co-location basis of creative businesses to the wider economy perspective by mapping UK's creative clusters, showing their location, related sectors, and role within systems of innovation where they are embedded.

Another report "Creative Industry spillovers – understanding their impact on the wider economy", published by Frontier Economics and prepared for the Department of Culture, Media and Sport in England, developed the necessary knowledge, product and network economic framework that was massively used within the area by many.

On the other hand, in a 2015 study "The Ecology of Culture", John Holden argues the "concentric circle" model of spillovers proposes that spillovers ineffectively describe the processes because: "The notion of spillover defines a cultural "expressive" core that is then commercialised through the creative industries. As this report makes clear, no such division should be drawn—creativity and expression flourish throughout the cultural ecology and can be exploited for economic gain anywhere within it."

However, scientific literature identified the issue of spillovers in a more systemic way that also corelates with adaptive reuse of cultural heritage and regional development. The term "spillover", often replaced with synonyms such as externalities, cross-overs, value-added or subsumed within a wider set of outcomes, impacts or values, etc., originates from the economic geography and cluster theory (such as Jacobs 1969 and Porter 1990). Its usage has later started spreading across fields, among others culture, cultural heritage and creative industries, and yet no common definition is adopted within scientific and professional community. Literature on spillovers shows two main cases, investigating growth and knowledge spillovers (Audretsch and Feldman 2004). However, in both cases, the process is unintentional and can be classified as externalities (Maier and Sedlacek 2005).

Extensive empirical studies explored multiple aspects of spillovers such as relationship between knowledge spillovers and regional growth (Döring and Schnellenbach 2006), estimating the effects of spillovers (e.g. Fritsch and Franke 2004, Van Stel and Nieuwenhuijsen 2004) and their characteristics e.g. geographical area of influence, the maximum distance beyond which their effects are no longer significant (which Bottazzi and Peri 2003, estimate as 300 km). For the magnitude and quality of local spillovers, characteristics of internal networks play a big role. On a regional level, these internal networks differ as they are organised through different, more informal channels, e.g. governance, social capital, collective learning, productive interdependence and labour market integration to name a few of all the channels that are crucial for local spillovers to emerge.

The amplification of agglomeration economies[5] leads to "localized knowledge spillovers" effect that has been studied for its magnitude and importance (Breschi and Lissoni 2001). Significant case studies on US high-tech clusters (Saxenian 1994), Italian industrial districts, and local innovation systems (innovative milieux) in Europe and elsewhere (Markusen 1996; Keeble and Wilkinson 1999) have demonstrated importance of innovation opportunities as a locational factor. A claim that knowledge relevant for innovative and entrepreneurial activities is mostly tacit and that therefore knowledge transmission is a matter of face-to-face contacts and labour mobility appears as a linking theme for these contributions. Furthermore, as the access to relevant knowledge requires co-location, the expectation that the tendency for innovative and entrepreneurial activities for spatial clustering will be highest in industries in the early stages of their industry's life cycle and where tacit knowledge plays an important role (Audretsch and Feldman 1996) which nowadays co-relates with the concept of a limited number of innovation hubs across Europe called "co-location centres" of KICs (Knowledge and Innovation Communities of the European Institute of Innovation and Technology), established also for thr EIT Culture & Creativity with cultural heritage playing an important role in it.

Spatial clustering of firms and their economic activities are one of the fruitful areas of research and interest over the last two decades by the economists working on all levels applicable to cultural heritage. It is believed that agglomeration economies, network effects and knowledge spillovers are the main factors for entrepreneurship and consequently territory growth and key engines of cluster development in the economic literature (Audretsch 2002; Acs and Audretsch 2003; Audretsch and Keilbach 2004).

Information and knowledge spillovers occur from the proximity to the knowledge hubs, expertise and skilled labour which are critical for companies of all sizes, but especially for small ones that are more dependent on resources in their local environment (Feldman and Francis 2004). The growth of the regional and national economies relies on the knowledge spillovers and multi-regional and inter-regional linkage, as emphasized by the endogenous growth theory. Moreover, spillovers are responsible for emergence of dynamic externalities and agglomeration effects leading to the faster growth (Glaeser et al. 1992; Griliches 1992; Smolny 2000, Fritsch 2000).

Technological externalities as defined by Hanusch and Cantner (1993) are the sum of intended and unintended spillovers, while broader definition of externalities involves aspects of public regulation, social costs of innovations and generally allocation problems. Moreover, Woeckener (1993) noticed that externalities are far more present than simply the positive spillovers from an innovator to potential imitators thus becoming foundation for analyses both in evolutionary and in neoclassical models of economic growth.

[5] Agglomeration economies imply positive externalities, because bringing additional people or firms to an urban area increases the productivity of other individuals or firms in that area.

Bakshi et al. 2013 reported "Creative clusters and innovation", and outlined subjective understanding of spillovers: "In addition to contributing directly to regional innovation processes through the innovative activities in which they engage, they could also do so indirectly, by generating spillovers that benefit the wider economies of the places where they are located." These aspects are of extreme importance for the adaptive reuse processes, as they might introduce new functions and following recognition from which types of activities different spillovers may occur, could facilitate a decision-making process on that (e.g. if the growth of the regional and national economies relies on the knowledge spillovers that occur from the proximity to the knowledge hubs, expertise and skilled labour that shall be an indication related to new functions expected to produce spillovers).

Starting from the report "Cultural Heritage Counts for Europe" (CHCfE 2015) the concept of cultural heritage as providing economic good can raise external factors that need to be considered. These are unintended benefits or costs (spillovers) generated by an economic good. They affect people who are not its direct consumers and are not accounted for by market transactions (hence the name -they are external to the workings of the market). As benefits generated by cultural heritage are difficult to capture through conventional market mechanisms, there might be a tendency to undersupply this type of goods. There is also the "free-rider" problem related to cultural heritage, occurring in situations when people have access to a good yet do not pay for it. Private for-profit providers are not to be expected to supply enough of such goods. Therefore, it is the role of the government (or sometimes non-profit organisations) to provide the society with them (the same goes for other public goods, such as clean air).

When examining spillover effects of cultural heritage adaptive reuse and interventions, previous theories are applicable in a scaled and adjusted context, but it is fundamental to understand the significance of a holistic approach regarding the economic value embodied in cultural heritage and activities around it, including adaptive re-use. This approach considers, both, use (value derived from actual "using", consuming a given good) and non-use values (ascribed to goods that might not ever be "used" by a given individual), therefore two types of use values can be distinguished: direct and indirect (CHCfE 2015).

The direct use value generated by immovable heritage is usually lower than the financial return generated by a given object. In an era of tightly constrained public finances, this may lead to the conclusion that cutting financial support in this area is justified, as it only generates costs (Pūķis 2011).

Economic impact may be seen as an influence or as a result of changes brought about by either implementing a special programme, policy or a project, or the existence of a certain object or project. Various authors, including Bowitz and Ibenholt (2009), distinguish between direct and indirect impact. The direct impact refers to the effects of the cultural heritage project or a site itself and may be measured in sales (however, they tend to be inflated and difficult to interpret), added value or employment. The authors draw our attention to the methodology of measurement; namely, that it is important to include only fees from visitors from outside the region where a given heritage site is located (if the residents spend more money on cultural heritage, it is at the expense of their other expenditure that if spent in the place of residence would contribute to the local economy anyway). Additional spending by residents could only be considered if it was possibly to prove that the spending would otherwise have been taken outside the community. Indirect impact refers to the influence of the heritage on the wider environment or to spillover effects. These are more difficult to measure, since it is not always clear whether there is a causal relation between the cultural heritage site, its adaptive reuse and the impact.

Effects arising from cultural heritage adaptive reuse may also be divided into direct, indirect and induced impacts. The induced impact concerns a situation where different cultural heritage benefits materialise when various associations to cultural heritage are used in other economic sectors, not related to heritage (that includes culture and creative industries, crafts and other production). To understand the effect cultural heritage might have on its environment one must also remember that a euro incurred on cultural heritage may stimulate actions and flows of financial resources in other areas or sectors bringing additional income or development to a given place. For cultural heritage to have impact on its local community through the multiplier effect, it is important to bear in mind that the effect is stronger when goods and services for the cultural heritage project or maintenance of a heritage site are bought locally and when it uses its local labour.

According to Seaman (2003), traditional economic impact studies usually focus on increases in short-run net local income and as such they are incomplete and could be misleading, especially if one tried to use them in public policy towards arts and heritage.

Finally, European Commission recognized the spillover effects of the arts, culture and the creative industries for the first time as "The process by which activities in the arts, culture and creative industries has a subsequent broader impact on places, society or the economy through the overflow of concepts, ideas, skills, knowledge and different types of capital."

Following previous definitions and considerations, within the CLIC project, spillover effect is considered as the process where an adaptive reuse of cultural heritage in one area has a subsequent broader impact on places, society or the economy through the overflow of concepts, ideas, skills, knowledge and different types of capital. Spillovers exist over fluctuating timeframes, intentionally or unintentionally, planned or unplanned, direct or indirect, negative as well as positive. In the framework of the CLIC research, the focus was on those spillover effects that emerge as a consequence of investment by public or private stakeholders in the adaptive reuse of cultural heritage buildings.

The findings from literature suggest that there are several types of economic spillovers related and implied from the culture and cultural heritage, where following three are the most represented typologies with rooting from economic literature, and featured in the European statistical system network on culture final report (ESSnet Culture 2013) and later used by (among others) Bakshi et al. (2013) and O'Connor and Gibson (2014).

The most common and persuasive of the three is the knowledge spillover considering the role of culture in developing social capital (OECD 2005), benefits to individuals of long-term engagement with arts organisations (CEBR 2013, and Cuypers et al. 2011), the wide impact of large-scale cultural events (Rutten 2006), the importance of culture in improving cross-border co-operation (Interact 2014), the spillover between publicly funded and commercially funded arts and cultural heritage (Albert et al. 2013, and Tafel-Viia et al. 2011) and the linkages between cultural heritage creative industries and innovation (Rutten 2006).

Provided evidence is modest, implicating it falls short of proving causality, but still provides a straightforward argument while promoting the necessity for additional research, of the role of arts and culture in improving national productivity (CEBR 2013), the role for festivals in boosting professional development (BOP 2011) the importance of heritage in connecting communities (Dümcke and Gnedovsky 2013), in the role of culture in boosting transferable skills (CEBR 2013) and social innovation (KEA 2009) and the importance of arts and cultural organisations in innovating new forms of organisation and ways of working (ECCE 2013). The role of culture in boosting academic attainment (CEBR 2013, and ACE 2014) falls within the same category, as well as the importance of culture as a form of participatory democracy (Rutten 2006), the role of culture in promoting social cohesion, the importance of the cultural capital of place (Krynica 2012) and the importance of cross-sector knowledge exchange as a driver for innovation (Tafel-Viia et al. 2011).

It links with new ideas, innovations and processes developed within cultural heritage, culture and creative business, which spillover into the wider economy and society. However, it falls short of proving causality to scientifically accepted standards and evidence of knowledge spillovers that could be improved by exploring how experiencing and practising "creativity" in one sphere transfers into bringing a more creative approach to other spheres of activity. Includes also transfer of (culture and cultural heritage) skills, e.g. through labour flows, spillover effects of culture-related education on young people etc. Moreover, it could be reinforced by studies that track long-term engagement with culture and cultural heritage and its importance in personal development, a role of volunteering in developing social capital, impact and value of large-scale cultural events, the value of cross-border networks, and the impact of creativity throughout the value chain and beyond manufacturing.

The most evident cases representing industry spillovers are boosts of communications within organisations (Antal and Strauss 2013), cross-fertilisation occurs between commercial and non-commercial sectors (OCE 2014), culture-led regeneration has a positive impact (Rutten 2006), spillovers play a role in boosting uptake of new technology (KEA 2006), investment in design has an impact (Sternö and Nielsén 2013), and networks are important in spreading innovation (Schopen et al. 2008).

Further evidences might be visible if there was more analysis of the two-way relationship between culture and cultural heritage adaptive reuse and the wider economy in terms of innovation and entrepreneurship. Similarly, research covering the value of public sector investment in stimulating risk-taking, as well as exploring the role of social media and spillover effects that occur without the benefits of physical proximity through clusters, would be highly relevant. Industry spillovers relate to effects for the economic performance, for example, when activities from one sector influence performance in another through a value chain between or within sectors (e.g. productivity, competitiveness or practice). Examples of evidence of the importance of cultural heritage in inspiring competitive markets in border regions is visible but less clearly articulated (Interact 2014), as in the reciprocal connection between technology and creative clusters (Chapain et al. 2010), in the case of the positive role that improved facilities can have on the property market (WWC 2014) and in the role of creative industries in the innovation process (Schopen et al. 2008). The evidence of reverse spillovers is weaker between productive industries and creative industries (Interact 2014), of the impact of large-scale events on the local economy (WWC 2014), of the negative impacts of culture-led regeneration (Evans 2005, and Slach/Boruta 2013) and of the pathways that exist for spillovers between public and commercial culture. These spillovers are divided into more categories such as how culture and cultural heritage stimulate business cultures and entrepreneurship, property markets, private and foreign investment, productivity, profitability and competitiveness, and innovation and digital technology.

The most structured and developed evidence of network spillovers is represented through numerous reports and papers as following: on the way that the process of social cohesion occurs (Goodlad et al. 2002), in the impact of culture on social cohesion (KEA 2009, and BOP 2011) and community cohesion (Dümcke and Gnedovsky 2013), on the association between cultural activity and perceived health and satisfaction with life (Cuypers et al. 2011, and Billington 2010), on the individual benefits of visiting museums (Fujiwara 2013), on the "creative milieu" effect and on the importance of creative entrepreneurs, on the role of culture in place-making and city-branding (ICC 2010, and Rutten 2006), among many others. Network spillovers are linked to the impact and outcome to the economy and society that spill over from the presence of a high density of cultural heritage and cultural activities in a specific location- effects associated with clustering.

Yet, improvement of the network spillover evidence might be improved by exploring complex relationship between art, culture, cultural heritage and wellbeing, and by analysing complex interplay of factors through an ecosystem approach to back our understanding of the role of the cultural heritage for place attractiveness. Negative outcome, as exclusive gentrification, are also common.

Analysis and definitions of spillovers, represents one of the contributions to the CLIC understanding of multidimensional spilovers of the cultural heritage adaptive reuse. This discussion is also valuable for the current scientific and political debate. However, there is a lack of structured evaluations of the causality in the cultural and creative sectors against scientific standards such as Bradford Hill Criteria (Bakshi et al. 2013, and Cuypers et al. 2011). There is a growing need for more methods originated from the social sciences, particularly those testing hypotheses using qualitative research methods. Indicators for assessing the role and contribution of the heritage community to cultural heritage adaptive reuse are still in their infancy and only few attempts have been made to structure robust evaluations (Bosone et al. 2021).

6 Learning from the Halland Model and Spillover Effects Through Trading Zone(S)

The case study of Halland model is still one of the best cases to demonstrate the spillover effects of cultural heritage adaptive reuse involving a complex network of partners, collaborators and co-creators. One of the reasons for this, is the timeframe that has ended and that has given us the closure of the complete cycle to understand all elements that brought spillovers at the regional level as well as detailed data collection that has enabled clear spillover distinction.

The low unemployment rates traditionally enjoyed by Sweden have often been attributed to the country's extensive system of active labour market programmes, and thus have often been regarded as a model for other countries to emulate (Sianesi 2001). Research in abundance has been directed towards understanding the causes of unemployment. Unemployment implies economic costs and welfare losses on a societal level. On an individual level, unemployment entails immediate income losses and reduced future earnings capacity, as well as decreased physical and mental well-being. The evaluation of macro-effects on labour market policy is to a great extent about studying how the supply of labour forces, employment and the mobility of the labour force are affected by the scope of and the direction of labour market policy (Hemström and Martinson 2002). According to Kenneth Carling and Katarina Richardson, labour market action programmes, in which the participants obtain subsidized work experience and training provided by firms, have better outcomes than classroom vocational training (Carling and Richardson 2001). These effects can have intrinsic values or be instrumentally valuable. Everything of instrumental value is related directly or via other instrumentally valuable agents to something of intrinsic value.

The cooperation carried out with the different conservation objects may be described in different ways, depending on the angle from which the project was observed. For example, the projects might be regarded as: "Hallandic", "labour market", "cultural heritage", "historic environment", "building conservation", "education", "regional development", "environment", "youth", "tourism", "culture", "regional identity" or "democracy" projects which also shows diversity of potential spillover effects rising from the project.

The process starts with planning procedures within each sector. In the trading zone, negotiations about different policies, values, facts, resources and activities result in the selection of conservation projects. Consequently, feasibility studies are realized, which provide the foundation for project planning, including surveys of quality and quantity, estimation of costs etc. After the finished conservation work, new activities were made available to take place in the improved premises. These activities were decided in the trading zone, while spillover effects have happened in related sectors (Fig. 12.5).

The Halland Model was organized to make priorities of specific meanings and needs. These were of cultural and local identity, cultural history, employment, training needs and the overall importance of sustainable development. These specific

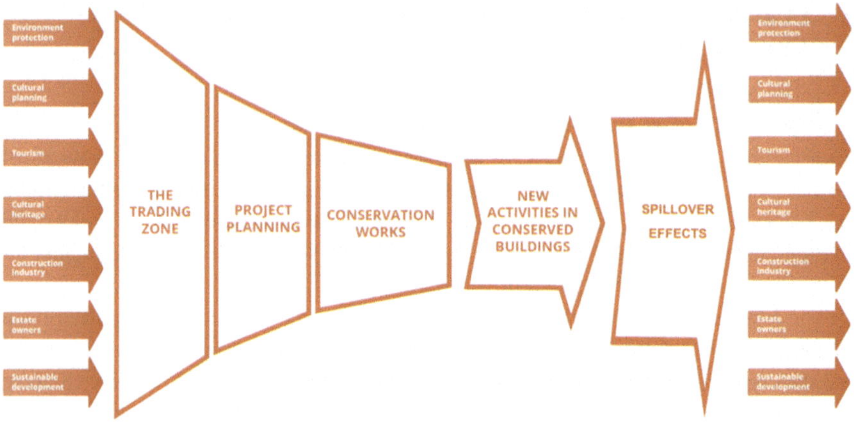

Fig. 12.5 Flow chart of projects of the Halland Model and spillover effects

meanings were discussed and negotiated during the feasibility studies, where key words for the success of conservation projects, as well as for cross- sector and multi-problem-oriented approaches, were formulated as "flexibility among stakeholders, trust for the partners, and transparent methods".

How to cooperate in comprehensive projects was a new kind of experience for the representatives of the cultural heritage. The knowledge about the comprehensive structure in which the Halland Model was operating and the role of the historic environment in this context had led to the understanding of the importance of cultural heritage to regional sustainable development.

From this perspective, the role of the Historic Environment Sector was not just building conservation and cultural tourism. The conservation and restoration works also improved the knowledge among the conservation officers in the region about traditional building techniques and the use of traditional building materials as well as craftsmanship in general creating knowledge spillover, indirect and non-monetary. The financial assets together with well-developed organization and increased knowledge among the participants implied that the quality of the conservation could be higher than in ordinary conservation works taking place at the same time in Halland (Fig. 12.6).

The objective was to increase the amount of construction projects in progress in the region, which would spread the public investments in the Halland Model to enterprises, and their sub-contractors and material producers and suppliers from the whole construction industry, also including consultants. One important aim was that, through this strategy, Halland would gain a rejuvenated labour force ready for construction works during the next period of prosperity (Fig. 12.7).

The labour market sector had the financial assets and available unemployed construction workers, but was lacking in terms of working places. The construction industry offered training programmes for construction workers, and estate owners offered objects to be conserved. The main objective of training construction workers

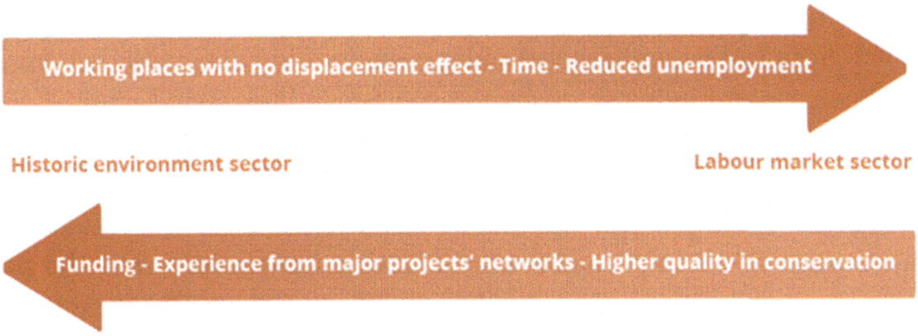

Fig. 12.6 Trading between the historic environment and labour market sectors

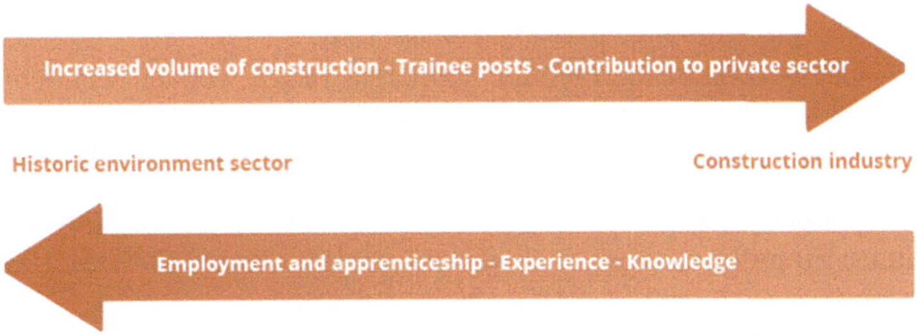

Fig. 12.7 Trading between cultural heritage and the construction industry

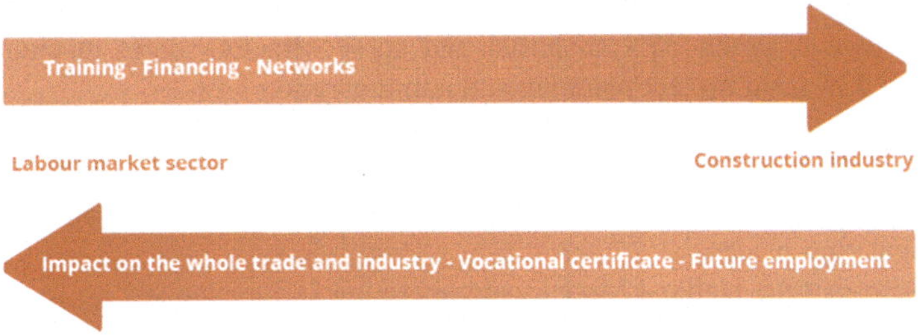

Fig. 12.8 Trading between the labour market sector and the construction industry

was to prepare them to fit the expected demands of industry in future. Together, the labour market sector and the construction industry had the capacity to train a new generation of construction workers (Fig. 12.8).

In the Halland Model unemployed construction workers were trained in traditional building techniques by practising these skills on historic buildings at risk. It started 1993 as a building conservation and labour market policy project, but soon it developed to a regional cross-sectorial joint action network focused on adaptive re-use of historic buildings at risk aiming at sustainable growth.

The Halland Model was organized as a joint venture between construction industry companies and the labour market and cultural heritage sectors together with other actors at national, regional and local levels, aiming at sustainable development and regional growth. The multi-stakeholder collaboration also included representatives from business and industry, academic society and the civic sector.

Halland is a region with approximately 300,000 inhabitants on the Swedish West coast. The Halland Model has had a considerable regional impact, e.g. on the employment and training in the construction industry and for the cultural heritage sector in the region. Of importance for the successful projects was the partners' agreement for a common objective for the cooperation, their clear role in the partnership, trust between them as well as the flexible and multi-problem-oriented approach—"try to solve several problems within one project". The co-operation has also implied considerable impact on other sectors of the society, i.e.:

- 100 historic buildings were saved and conserved
- approximately 1200 of the region's 3600 constructor workers were employed and trained in traditional building techniques
- 350 new jobs were created on the improved premises

The labour market sector was the biggest financier and contributed with appr. 80% of the almost 50 million euros which was the project's total turn-over. 10% came from the heritage sector and property owners respectively.

The total turnover in the Halland Model was approximately SEK 375 million. Of this amount the County Labour Market Board contributed with three quarters, the property owners with 15% and the cultural heritage sector with 10%. Each SEK from the cultural heritage sector had, in other words, been tenfold. The cost was divided between labour 37%; material and sub-contractors 28%; machines and equipment 22%; and project planning and administration 13%.

Before the conservation within the Halland Model half of the conserved buildings had no function. After completed conservation one-third was used for arts and cultural functions, one sixth as a conference centre in various sizes, and one-tenth as local meeting-places.

The actors of the Halland Model were operating simultaneously on several levels, trying to solve specific conservation matters according to conservation principles, as well as designing conservation projects according to all-embracing regional development policy. The actors had to judge within the Halland Model, as well as within their own sectors. The Halland Model can be compared with a trading zone, where different actors present their values and goods to achieve the established goal. All the team members of the comprehensive consortium of the Halland Model had to understand and agree that historic buildings at risk were potential conservation objects, and as such are conveyors of intended and desirable meanings.

The conservation interventions, and the entire conservation processing, could be understood as circular, since conservation is about caring for existing resources from a long-term perspective, instead of degrading and demolishing them. Further, environmentally friendly construction materials were used together with renewable sources of energy in the conserved buildings.

The Halland Model projects are considered to be economic since they have resulted in an obvious return on the investments and guaranteed future income covering the costs of maintenance, and they have moreover contributed to regional growth. Concerning social aspects, the conservation process involved regional cohesion and developed cross-sectoral networks and a multi-problem- oriented approach. Further, they increased the knowledge level and strengthened the local identity and democracy, as well as creating new jobs in the region. Finally, the process was sustainable from an environmental viewpoint since conservation is about caring for existing resources instead of demolishing them, but also since environmentally friendly construction materials were used together with renewable sources of energy in the conserved buildings. The results from the Halland Model have had a considerable impact, e.g. on employment, capability building, increased proficiency of traditional building techniques and the preservation of historic buildings at risk, as well as planning the use of historic buildings, regional cohesion and regional sustainable development in general.

7 Discussion

A simple comparison between case studies, regions, communities and cultures cannot be easily made as they are unique, dynamic and constantly changing. This constant change of the cultural heritage systems represents a reality of the process of interaction and collaboration between regions, people and communities. Despite the difficulties in taking individual measurements, the main goal of taking these indicators is to track their change over time. Cultural indicators usually work on the principles of cultural participation and asses, being similar to the common mainstream measurement systems that assumes culture, like some other sectors, behaves like an independent eco-system, having boundaries and static existence. This is applicable in case of spillovers of cultural heritage adaptive reuse as evident from elaboratin on the complex natures of spilovers.

Overall, it is possible to state that only a very small number of regions include cultural heritage and adaptive reuse through smart specialisation in their regional strategies and this state of the art was very much needed in order to understand current situation at the regional level. Findings from both the qualitative and the quantitative analysis, suggest the following: some NUTS entities have included circular economy in their regional innovation strategies for smart specialisation, thus defining circular economy as a priority sector for the region, but still not including cultural heritage adaptive reuse as a part of its implementation. Adaptive reuse in relation to cultural heritage, as a term, has not been mainstreamed in circular economy strategies, neither in smart specialisation strategies. Nevertheless, the research shows that the significance and representation of cultural heritage and/or culture through circular economy strategies and smart specialisations strategies is very low. This data, with accurate percentages shall offer a starting point for further improvements at both, academic research and policy improvements and act as a benchmark for set of indicators.

Overall, these results indicate that, although creative and cultural sector forms are an important growth factor in many cities and regions, integration of these aspects in circular economy and regional development domains remains a challenge that national and supranational governments will have to address in the future. Regions need to recognize and take into account the complexity of interconnections between the traditional cultural assets, such as cultural heritage adaptive reuse, on the one hand, and the improvement of circular economy and regional development, on the other.

8 Conclusions

As previously highlighted, the society of today is facing major global challenges and a failure of national governance (e.g. failure of rule of law, corruption, political deadlock) has been identified as one of Global Risks 2020. The point of departure for this paper was differentiating main alternative option for governing cultural heritage at different levels, such as smart specialisations, and how that relates to cultural heritage spillovers for unlocking the potential of adaptive reuse through circular economy. By focusing on distinguishing regional development, circular economy and different spilovers of cultural heritage adaptive reuse, the chapter presented various aspects through which the cultural heritage and its sectors and environments can address these matters at different levels also pointing out some criticalities. Through the review of the scholarly and other literature, one of the most relevant conclusions is how cultural heritage spillovers are difficult to be caught by simple indicators or quantifications when they are allowed to happen spontaneously. They are important drivers of economic innovation, but it highlights the interest in moving towards a cultural heritage approach that can catalyse spillover effects in various economic and social contexts.

Another conclusion of this chapter is the importance of public policy in supporting cultural heritage impact through circular economy and regional development offering an overview of latest major policy initiatives at the EU, national, and regional levels, with the focus on smart specialisation strategies. While the theoretical background of this instrument is known, research about the assessment of its application to circular economy and cultural heritage in the context of regional development is still too scarce, though promising.

Overall, the chapter stressed how cultural heritage has the potential to stimulate innovation, growth, sustainable development, welfare, jobs, income, and liveability of regional settings.

References

Acri M, Dobričić S, Debevec M (2021) Regenerating the historic urban landscape through circular bottom-up actions: the urban seeding process in Rijeka. Sustain For 13:4497. https://doi.org/10.3390/su13084497

Acs ZJ, Audretsch DB (2003) Innovation and technological change. In: Handbook of entrepreneurship research: an interdisciplinary survey and introduction. Springer, Boston, pp 55–79

Albert A, Bakhshi H, Mitchell S, Smithies R (2013) Publicly-funded arts as an R&D Lab for the Creative Industries? Arts Council England

Antal A, Strauss A (2013) Artistic interventions in organizations: finding evidence of value-added. Creative Clash

Arts Council England (2014) The value of arts and culture to people and society – an evidence review

Audretsch DB (2002) The dynamic role of small firms: evidence from the US. Small Bus Econ 18:13–40

Audretsch DB, Feldman MP (1996) R&D Spillovers and the Geography of Innovation and Production. Am Econ Rev 86(3):630–640. Published by: American Economic Association Stable. http://www.jstor.org/stable/2118216

Audretsch DB, Feldman MP (2004) Knowledge spillovers and the geography of innovation. In: Handbook of regional and urban economics, vol 4. Elsevier, pp 2713–2739

Audretsch DB, Keilbach M (2004) Entrepreneurship and regional growth: an evolutionary interpretation. J Evolut Econ 14:605–616

Bakshi H, Edwards J, Roper S, Scully J, Shaw D, Morley L, Rathbone N (2013) Creative credits: a randomised controlled industrial policy experiment. Nest, Palo Alto

Billington J (2010) An investigation into the therapeutic benefits of reading in relation to depression and well-being. University of Liverpool

Bosone M, De Toro P, Fusco Girard L, Gravagnuolo A, Iodice S (2021) Indicators for ex-post evaluation of cultural heritage adaptive reuse impacts in the perspective of the circular economy. Sustain For 13:4759

BOP (2011) Edinburgh festivals impact study final report May 2011. https://www.efa-aef.eu/media/12317-110520edinburghfestivalsfinaloverallreportbop_final.pdf, https://www.efa-aef.eu/media/12317-110520edinburghfestivalsfinaloverallreportbop_final.pdf

Bottazzi L, Peri G (2003) Innovation and spillovers in regions: evidence from European patent data. Eur Econ Rev 47(4):687–710

Bowitz E, Ibenholt K (2009) Economic impacts of cultural heritage–research and perspectives. J Cult Heritage 10(1):1–8

Breschi S, Lissoni F (2001) Knowledge spillovers and local innovation systems: a critical survey. Ind Corp Change 10(4):975–1005

Brundtland G (1987) Report of the world commission on environment and development: our common future. United Nations General Assembly document A/42/427

Camagni R (2009) Territorial capital and regional development. In: Handbook of regional growth and development theories. Edward Elgar Publishing

Camagni R (2019) Territorial capital and regional development: theoretical insights and appropriate policies. In: Handbook of regional growth and development theories. Edward Elgar Publishing, pp 124–148

Carling K, Richardson K (2001) The relative efficiency of labor market programs: Swedish experience from the 1990's, IFAU-Office of Labour Market Policy Evaluation. Working Paper 2001: 2

Centre for Economics and Business Research (2013) The contribution of the arts and culture to the national economy. Arts Council England

Chapain C, Cooke P, Propis L, MacNeill S, Mateos-Garcia J (2010) Creative clusters and innovation. Putting creativity on the map. Nesta

Cultural Heritage Counts for Europe (2015) https://www.europanostra.org/our-work/policy/cultural-heritage-counts-europe/

Cuypers K, Krokstad S, Holmen T, Knudtsen M, Bygren L, Holmen J (2011) Patterns of receptive and creative cultural activities and their association with perceived health, anxiety, depression and satisfaction with life among adults: The HUNT study, Norway. JECH online 66:698–703
Douglas J (2006) Building adaptation, 2nd edn. Butterworth-Heinemann, Woburn
Döring T, Schnellenbach J (2006) What do we know about geographical knowledge spillovers and regional growth?: a survey of the literature. Reg Stud 40(03):375–395
Dümcke C, Gnedovsky M (2013) The social and economic value of cultural heritage: literature review. European Expert Network on Culture
ECCE (2013) Forum d'Avignon Ruhr. Culture is the Key European centre for creative economy
ESSnet Culture (2013) Creative industries spillovers ESSnet-Culture final report, pp 365–396
European Commission (2012) Communication from the commission to the European Parliament, the Council, the European economic and social committee and the committee of the regions promoting cultural and creative sectors for growth and jobs in the EU. https://www.europarl.europa.eu/registre/docs_autres_institutions/commission_europeenne/com/2012/0537/COM_COM%282012%290537_EN.pdf, https://www.europarl.europa.eu/registre/docs_autres_institutions/commission_europeenne/com/2012/0537/COM_COM%282012%290537_EN.pdf
European Commission (2015) Communication from the commission to the European Parliament, the council, the European economic and social committee and the Committee of the regions closing the loop—an EU action plan for the circular economy. European Commission, Brussels
European Commission (2017) Strengthening innovation in Europe's regions: strategies for resilient, inclusive and sustainable growth (Doc. 11426/17 + ADD 1). European Commission, Brussels
European Commission (2019) The European green Deal (COM (2019) 640 final). European Commission, Brussels
European Commission (2020) A new circular economy action plan for a cleaner and more competitive Europe (COM (2020) 98 final). European Commission, Brussels
European Commission, Joint Research Centre (2015) Mapping innovation priorities and specialisation patterns in Europe. European Commission, Seville
Evans G (2005) Measure for measure: evaluating the evidence of culture's contribution to regeneration. Urban Stud 42(5/6)
Feldman MP, Francis JL (2004) Homegrown solutions: fostering cluster formation. Econ Dev Q 18(2):127–137
Ferilli G, Gustafsson C, Buscema M, Sacco PL (2019) The complex dynamic evolution of cultural vibrancy in the region of Halland, Sweden. Int Reg Sci Rev 43(3):1–44. https://doi.org/10.1177/0160017619849633
Florida R (2002) Bohemia and economic geography. J Econ Geogr 2(1):55–71
Florida R (2004) Cities and the creative class. Routledge, New York
Forum d'Avignon Ruhr (2013) Culture is the Key. european centre for creative economy
Fritsch M (2000) Interregional differences in R&D activities—an empirical investigation. Eur Plan Stud 8(4):409–427
Fritsch M, Franke G (2004) Innovation, regional knowledge spillovers and R&D cooperation. Res Policy 33(2):245–255
Fusco Girard L (1987) Risorse Architettoniche e Culturali: Valutazioni e Strategie di Conservazione. Franco Angeli, Milano
Fusco Girard L (2013) Toward a smart sustainable development of port cities/areas: the role of the historic urban landscape approach. Sustain For 5:4329–4348
Fusco Girard L (2014) The role of cultural urban landscape towards a new urban economics: new structural assets for increasing economic productivity through hybrid processes. Hous Policies Urban Econ 1:3–27
Fusco Girard L, Nijkamp P (1997) Le Valutazioni per lo Sviluppo Sostenibile Della Città e Del Territorio. Franco Angeli, Milano

Fusco Girard L, Vecco M (2021) The "intrinsic value" of cultural heritage as driver for circular human-centered adaptive reuse. Sustain For 13:3231. https://doi.org/10.3390/su13063231

Fusco Girard L, Gravagnuolo A, Nocca F, Angrisano M, Bosone M (2015) Towards an economic impact assessment framework for historic urban landscape conservation and regeneration projects. BDC Boll Cent Calza Bini 15:1–29

Glaeser EL, Kallal HD, Scheinkman JA, Shleifer A (1992) Growth in cities. J Polit Econ 100(6):1126–1152

Goodlad R, Hamilton C, Taylor P (2002) Not just a treat: arts and social inclusion – a report to the Scottish Arts Council. Centre For Cultural Policy Research, University of Glasgow

Gravagnuolo A, Micheletti S, Bosone M (2021) A participatory approach for "circular" adaptive reuse of cultural heritage. Building a heritage Community in Salerno, Italy. Sustain For 13:4812. https://doi.org/10.3390/su13094812

Griliches Z (1992) Introduction to "output measurement in the service sectors". In: Output measurement in the service sectors. University of Chicago Press, pp 1–22

Gustafsson C (2009) The Halland model—a trading zone for building conservation in concert with labour market policy and the construction industry, aiming at regional sustainable development. Chalmers University of Technology, Göteborg

Gustafsson C (2019) Conservation 3.0 – Cultural heritage as a driver for sustainable growth. Scientific Research and Information Technology Ricerca Scientifica e Tecnologie dell'Informazione 9(1):21–32. e-ISSN 2239-4303. https://doi.org/10.2423/i22394303v9n1p21

Gustafsson C, Lazzaro E (2021) The innovative response of cultural and creative industries to major European societal challenges: toward a knowledge and Competence Base. Sustain For 13:13267. https://doi.org/10.3390/su132313267

Hanusch H, Cantner U (1993) Neuere Ansätze in der Innovationstheorie und der Theorie des technischen Wandels—Konsequenzen für eine Industrie-und Technologiepolitik. In: Innovationsökonomie und Technologiepolitik: Forschungsansätze und politische Konsequenzen. Physica-Verlag HD, pp 11–46

Hemström M, Martinson S (2002) Att följa upp och utvärdera arbetsmarknadspolitiska program, No. 2002: 1. IFAU-Institute for Evaluation of Labour Market and Education Policy

Hubbard G (2006) Business knowledge and global growth capitalism and society. J Capital Soc 1(3). https://doi.org/10.2202/1932-0213.1013

Interact (2014) Inspiring creativity promoting culture and creative industries across Europe. https://cercles.diba.cat/documentsdigitals/pdf/E130343.pdf, https://cercles.diba.cat/documentsdigitals/pdf/E130343.pdf

International Cultural Centre Krakow (2010) Culture and development – 20 years after the fall of Communism in Europe

Jacobs J (1969) The economy of cities. Routledge, New York

Janssen J, Luiten R, Renes H, Stegmeijer E (2017) Heritage as a sector, factor and vector: conceptualizing the shifting relationship between management and spatial planning. Eur Plan Stud 25(9):1654–1672

KEA (2006) The economy of culture in Europe. European Commission

KEA (2009) The impact of culture on creativity. European Commission

Keeble D, Wilkinson F (1999) Collective learning and knowledge development in the evolution of regional clusters of high technology SMEs in Europe. Reg Stud 33(4):295–303

Maier G, Sedlacek S (2005) Spillovers and innovation, environment and space: an introduction. Spillovers and innovations. Springer, Vienna, pp 1–18

Markusen A (1996) Sticky places in slippery space: a typology of industrial districts. Econ Geogr 72:293–313

O'Connor J, Gibson M (2014) Culture, creativity, cultural economy: a review. Australian Council of Learned Academies, Acton

Observatory for Cultural Economics (2014) Evidence for spillover effect: audiovisual cluster. European Cultural Foundation

Organisation for Economic Co-operation and Development (2005) Culture and local development. Organisation for Economic Co-operation and Development, Paris

Ost C (2009) A guide for heritage economics in historic cities. Values, indicators, maps, and policies. Getty Conservation Institute, Los Angeles

Pickerill T (2021) Investment leverage for adaptive reuse of cultural heritage. Sustain For 13:5052. https://doi.org/10.3390/su13095052

Porter M (1990) The competitive advantage of nations. Free Press, New York

Porter ME, Porter MP (1998) Location, clusters, and the "New" microeconomics of competition. Bus Econ 33(1):7–13. http://www.jstor.org/stable/23487685

Pūķis M (2011) Kultūras mantojuma sociālā un ekonomiskā loma. Valsts Kultūras pieminekļu aizsardzības inspekcijas, Riga

Rutten (2006) Cultural activities & creative industries, a driving force for urban regeneration, Online, https://d1wqtxts1xzle7.cloudfront.net/31747931/PRutten-E-libre.pdf?1392425546=&response-content-disposition=inline%3B+filename%3DCulture_and_urban_regeneration_Findings.pdf&Expires=1724024521&Signature=La~XBtNXk7OcrhejgM2if3ch5Ilrbw0bp399aNBVNE0nuRy7pYBqfw6IFeo8Yn-FLy9kpMWdgheALP6bPKgpVKKM14wiGNtlSM0VxHL4hhLUjLiMBPPDL9yIGe0NR1mja741R29uVYdG~ie2JXwHDMnaL4yAMANtqY5HeTAcQXvuld62GADWBMuCN0WZAhoQ0G4aUETU59ZMMUpj-4sIbGj2d9pislno898aQ4Hxd~OkVmMxxM3Wdwj57nV6AcrG3TbfP941wpDuHe2W4Dez-XMWiMn1d~pq2qA0ke~JN7KMtqPUfSOqwBlPL0dZEp51blJTbN54LNdMa6It7~Tw__&Key-Pair-Id=APKAJLOHF5GGSLRBV4ZA

Saxenian A (1994) Book note regional advantage: culture and competition silicon valley and route 128

Schopen K, Rammer C, Trüby J (2008) The role of creative industries in industrial innovation. Centre For European Economic Research (ZEW)

Seaman BA (2003) Cultural and sport economics: Conceptual twins? J Cult Econ 27:81–126

Sianesi B (2001) An evaluation of the active labour market programmes in Sweden (No. 2001: 5). Working Paper

Slach O, Boruta T (2013) What can cultural and creative industries do for urban development? Three stories from the post-Socialist industrial city of Ostrava Quaestrones geographicae 31(4):99–112

Smolny W (2000) Endogenous innovations and knowledge spillovers: a theoretical and empirical analysis; with 78 Tables, vol 12. Springer Science & Business Media

Stanojev J, Gustafsson C (2019) Circular economy concepts for cultural heritage adaptive reuse implemented through smart specialisation strategies. Proceedings of the STS Conference Graz

Stanojev J, Gustafsson C (2021) Smart specialisation strategies for elevating integration of cultural heritage into circular economy. Sustain For 13:3685. https://doi.org/10.3390/su13073685

Sternö J, Nielsén T (2013) Kreativa Krafter – Stockholm-Mälarregionen Mälardalsrådet

Tafel-Viia K, Viia A, Purju A, Terk E, Keskpaik A, Lassur G (2011) Creative industries spillovers in the frame of the ESSnet Culture project. Tallinn University Estonian Institute for Future Studies

UN HABITAT (1996) Participation in shelter strategies at community level in urban informal settlements. UN Habitat

UN-Habitat (2009) Global report on human settlements 2009: planning sustainable cities, Earthscan, London/Sterling

Un-Habitat (2012) Meeting development goals in small urban centres: water and sanitation in the worlds cities 2006. Routledge

United Nations Conference on Environment and Development Rio de Janeiro, Brazil (1993) Agenda 21: Programme of action for sustainable development; Rio declaration on environment and development; Statement of forest principles: the final text of agreements negotiated by governments at the United Nations Conference on Environment and Development (UNCED), 3–14 June 1992, Rio de Janeiro. United Nations Department of Public Information

United Nations Human Settlements Programme (2002) Global urban indicators database: version 2. UN-HABITAT

Van Stel AJ, Nieuwenhuijsen HR (2004) Knowledge spillovers and economic growth: an analysis using data of Dutch regions in the period 1987–1995. Reg Stud 38(4):393–407

Vickery J (2014) To be debated SPILLOVER. european centre for creative economy

What Works Centre (2014) What works? Centre for economic growth – evidence review – sports and culture. What Works Centre

Woeckener B (1993) Innovation, externalities, and the state: a synergetic approach. J Evolut Econ 3:225–248

World Economic Forum (2021) The global risks report 2021, 16th edn. World Economic Forum. ISBN: 978-2-940631-24-7

XV11 Economic Forum Krynica (2012) Culture as a stimulus for creativity and innovation in the region – a summary of the panel discussion

Chapter 13
Towards Participatory, Dynamic, Co-evolutionary Evaluation for Circular Adaptive Reuse of Cultural Heritage: The Experimentation of Salerno Circular City of Health

Antonia Gravagnuolo, Martina Bosone, Serena Micheletti, Mariarosaria Angrisano, and Luigi Fusco Girard

1 Introduction

Adaptive reuse describes the process of giving "new life to old buildings" through new uses and functions compatible with the existing building/site (Schmertz 1982; Marston 1982; Douglas 2006; Bullen and Love 2010). This process can be implemented in cultural heritage sites that have lost their functionality for many reasons, and lay in a state of abandonment or underuse (Fusco Girard 2021a, b). The recovery of the rich cultural heritage in abandonment, from religious heritage buildings not used anymore in today's society, to ancient fortifications, civic buildings, lighthouses, commercial sites, can be a leverage for the regeneration of urban areas and rural landscapes, enhancing the attractiveness of cities and regions (Fusco Girard and Gravagnuolo 2017; Gravagnuolo et al. 2019, 2021a; Gravagnuolo 2021).

A. Gravagnuolo (✉)
CNR ISPC Institute of Heritage Science, National Research Council, Naples, Italy
e-mail: antonia.gravagnuolo@cnr.it

M. Bosone
Department of Architecture, University of Naples Federico II, Naples, Italy
e-mail: martina.bosone@unina.it

S. Micheletti
Administration and Finance Directorate, Italian Space Agency, Rome, Italy
e-mail: serena.micheletti@asi.it

M. Angrisano
Engineering Department, Pegaso Telematic University, Naples, Italy
e-mail: mariarosaria.angrisano@unipegaso.it

L. Fusco Girard
Emeritus Professor at University of Naples Federico II; CNR IRISS, National Research Council, Institute for Research on Innovation and Services for Development, Naples, Italy

L. Fusco Girard, A. Gravagnuolo (eds.), *Adaptive Reuse of Cultural Heritage*,
https://doi.org/10.1007/978-3-031-67628-4_13

Moreover, as discussed in this volume, the functional reuse of cultural heritage can be an opportunity to recover unused resources implementing the principles of the circular economy: reuse, recovery, refurbishing, restoration, regeneration (Gravagnuolo et al. 2017). Sustainable and circular adaptive reuse of cultural heritage should focus on ecosystems regeneration, social and cultural regeneration, and economic benefits for communities (Fusco Girard 2021a, b). Attention to the formation of "heritage communities" caring for cultural heritage and contributing to its maintenance and conservation, a careful knowledge of the urban and territorial system on which the assets insist, the sustainable valorisation of the intangible heritage as a tool for understanding the tangible one, represent tools that can reasonably orient the transformation processes in line with a virtuous operational practice (Gravagnuolo et al. 2021a). Consistent with these premises is the "circular" adaptive reuse model of cultural heritage, developed within the CLIC project (Fusco Girard this volume, Chap. 2). In particular, the circular adaptive reuse of cultural heritage is oriented towards the construction of an ecological, safe, inclusive and multifunctional "spatial dimension", in a circular economy perspective consistent with both the European Action Plan for the Circular Economy (European Commission 2020) and the Action Plan for the Social Economy (European Union 2021).

The study on "Cultural Heritage Counts for Europe" identified diverse domains in which cultural heritage conservation and regeneration contributes to sustainable development. Also, this study identified different evaluation tools used to assess cultural heritage projects in the ex-ante and ex-post evaluation phases, considering also participatory tools to engage stakeholders in the decision-making process. Multicriteria evaluation was described as a group of methods and tools able to support decisions in cultural heritage projects, as it allows to consider multiple objectives, criteria and indicators, however it was not considered as a potential tool to support participatory processes (CHCfE consortium 2015). According to scientific literature, different studies support the use of multicriteria evaluation for participatory decision-making, through the use of "weighted" criteria expressing the preferences of different stakeholders, elicited through different methods (Nijkamp 1977; Fusco Girard 1987; Fusco Girard and Nijkamp 1997; Garmendia and Gamboa 2012; Seghezzo et al. 2017; Barinaga-Rementeria et al. 2019; Dean 2022). However, participatory processes for cultural heritage adaptive reuse can be particularly complex due to conflicting interests and objectives, and they can proceed through different phases and feedbacks that make it difficult to choose between alternative projects/solutions in a transparent and collaborative way. The process of identification of suitable circular solutions for cultural heritage adaptive reuse can become a "learning journey" for all stakeholders involved in co-design and co-planning, in the search for new creative circular "win-win-win" solutions for the public, private and social sectors (Gravagnuolo et al. 2021b).

In the CLIC project, participatory co-design of circular adaptive reuse solutions was experimented, supported by multicriteria evaluation tools. The aim of this Chapter is to present the CLIC dynamic and co-evolutionary evaluation methodology for participatory circular adaptive reuse of cultural heritage. The methodology was applied in Salerno, Italy to support the adaptive reuse of four large historic buildings in a state of abandonment. The following sections present the methodology adopted in Salerno and the results obtained in the different phases of the participatory process. A critical discussion and conclusions highlight the innovativeness of the proposed methodology compared to the state of the art in cultural heritage studies, as well as the possible barriers and further research needs.

2 Methodology

2.1 Dynamic and Co-evolutionary Evaluation for Participatory Decision-Making in Cultural Heritage Adaptive Reuse Processes

Cultural heritage represents a "common good" able to enhance the attractiveness of cities and regions and generate positive impacts in multiple dimensions (European Commission 2014a, 2015; Licciardi and Amirtahmasebi 2012; Council of Europe 2005; Rojas 2016; UNESCO 2016a, b). As common good, decisions on cultural heritage adaptive reuse should be taken ensuring engagement and participation of local communities, promoting participatory decision-making processes to enhance wellbeing. Adaptive reuse implies the identification of new functions for "old" buildings and heritage sites. In the circular perspective, the new uses/functions should promote people's wellbeing, reaching multiple objectives such as cultural values conservation, ecological regeneration, new employment, urban and social regeneration, human capital enhancement. Circular and "human-centred" adaptive reuse of cultural heritage is focused on the "wellbeing" of people and local communities, considering not only present generations but also future ones (Fusco Girard 2021a, b, this volume, Chap. 2). Thus, it aims to reduce negative environmental impacts such as climate-altering emissions and natural resources consumption as a way of preserving resources and ecosystems ensuring equal opportunities for present and future generations (Foster 2020).

Circular adaptive reuse of cultural heritage can be a very complex process, as it aims to reach multiple goals and combine often conflicting interest towards the "common good". Multidimensional and multicriteria evaluation methods and tools can support the process of adaptive reuse, enabling participation of diverse stakeholders in decision-making. Participation is not only expressed through assignment of specific weights to criteria and indicators, but should represents a dynamic, co-evolutionary process in which stakeholders have the opportunity to discuss, share ideas and visions, learn and change their mind at each step. Through participatory processes, stakeholders can learn about new concepts, examples, proposals, and develop a enhanced capacity to cooperate and collaborate towards a satisfying choice.

The CLIC evaluation methodology proposed aims to promote stakeholders' engagement, cooperation and dialogue, showing how transparent and participatory decision-making supported by multicriteria tools can substantially contribute to generate viable and effective circular solutions for cultural heritage adaptive reuse.

Starting from the CLIC theoretical framework of circular and human-centred adaptive reuse of cultural heritage, the methodology adopted in Salerno was defined into four main phases (Fig. 13.1):

Phase 1. Engagement and public debate.
Phase 2. Ideation of alternative adaptive reuse solutions.
Phase 3. Co-design of adaptive reuse solutions.
Phase 4. Satisfying solution identification.

According to the CLIC theoretical framework, the circular model in the territory is oriented towards the ability to regenerate different forms of capital (manufactured capital, natural capital, social capital, human capital, economic-financial capital). Circular reuse is aimed at transforming abandoned places into 'living' systems (Fusco Girard 2021a, b, this volume, Chap. 2). In this way, it is able to generate positive effects in the context and contribute to the resilience of the city/territory system over time.

The circular reuse is geared towards generating a community, in particular a 'heritage community' as described in the FARO Convention (Council of Europe 2005) that in turn cares for heritage and regenerates its meanings, in a virtuous circular process, increasing social capital and strengthening cohesion and cooperative and collaborative capacity in communities. In fact, circular reuse is characterised by

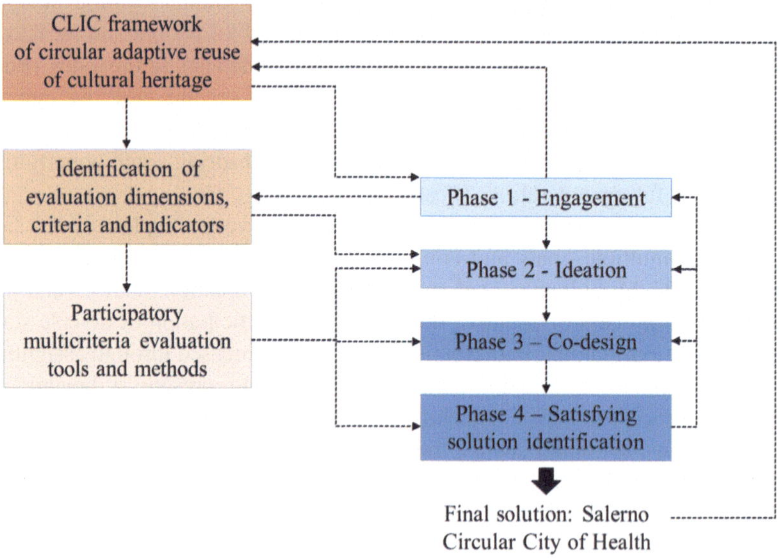

Fig. 13.1 Methodological phases of the participatory adaptive reuse process in Salerno

the search for synergies/synergies and cooperative activities between actors in the area that increase overall productivity.

The circular adaptive reuse of cultural heritage also re-produces intangible values: cultural values (Hosagrahar et al. 2016; Fusco Girard and Vecco 2019, 2021). This reuse process should also be able to regenerate cultural values /meanings/horizons, generating new meanings and new values linked to the original meanings and value. In summary, the circular reuse of cultural heritage aims at regenerating the tangible and intangible cultural, natural, social and economic resources of the territory, promoting symbioses/synergies and cooperation between public, private and civil society actors, and generating net positive economic, social, environmental and cultural impacts in the territory.

The ex-ante evaluation methodology experimented in CLIC was based on the theoretical framework and the dimensions, criteria and indicators developed through the research (see previous Chapters in this book). Main circularity goals/objectives are defined through the 3 dimensions framework: auto-poietic capacity, symbiotic capacity, generative capacity (Fusco Girard this volume, Chap. 2). According to these three dimensions, a set of evaluation criteria and indicators was defined, discussed and revised through stakeholders' engagement and iterative feedback loops (Bosone et al. 2021). As showed in Fig. 13.1, the phases of the methodology are interconnected and recurrent: the final preferable solution results from a series of project alternatives development, assessed and discussed to search for further new creative solutions towards the "ideal" point. Through the phases of engagement and discussion, ideation, co-design and iterative evaluation processes, a final "satisfying" solution (Simon 1959) can be found, which was not even visible at the start of the process.

The CLIC methodology was implemented in the city of Salerno, through the case study of the so-called "Edifici Mondo", a group of four large abandoned historic buildings waiting for an adaptive reuse solution since more than 30 years. The following section presents the case study in the city of Salerno.

2.2 The Case Study: Edifici Mondo Abandoned Historic Buildings in Salerno, Italy

The CLIC evaluation methodology was experimented for the identification of a circular adaptive reuse solution of four historic buildings located in a large abandoned area within the city center of Salerno, in Southern Italy.

Four buildings were selected for the experimentation: the ex-Convent of "San Pietro e San Giacomo", the ex-Convent of "San Francesco d'Assisi", the ex-Convent of "Santa Maria della Consolazione", and a historic noble palace called "Palazzo San Massimo", all in state of complete abandonment and neglect. The four buildings lay in a large area of the historic city centre, and are so-called "Edifici Mondo", for their size and complexity.

The buildings and the entire urban area were in a state of degrade since 30 years despite a long series of tentative initiatives for their recovery and reuse (Lupacchini and Gravagnuolo 2019). Shows the size of the buildings and their current state of degrade, requiring a high level of investment. However, their localisation in the city centre could generate a high attractiveness, thus a solution able to blend financial and social returns, as well as environmental and urban regeneration, could be identified.

The challenge of the CLIC project in the city of Salerno was to identify a possible solution for the adaptive reuse of those iconic buildings in the historic city centre, which could generate social benefit for the local community, ecological regeneration, adequate financial return of the investment, while recovering a symbolic cultural heritage for the city currently at risk of being completely lost for the future generations (Fig. 13.2—Historic architectural and arts details of Palazzo San Massimo).

The CLIC circular adaptive reuse model, expressed through well-defined circularity goals/objectives, dimensions, criteria and indicators, represented a clear orientation towards the identification of a satisfying adaptive reuse solution.

The participatory evaluation process experimented aimed at overcoming the barriers to the adaptive reuse of the historic buildings, supporting the development of financially and socially viable project alternatives, which should also preserve cultural values and regenerate environmental resources (multiple objectives and criteria), engaging stakeholders in a structured co-evaluation process supporting participatory decision-making.

To support the co-development of project alternatives in Salerno, a process of envisioning and community engagement was implemented, starting from a phase of knowledge building and local stakeholders involvement. The following section presents the result of the evaluation phases towards the definition of a satisfying project/solution in the perspective of circularity.

Fig. 13.2 Historic architectural and arts details of Palazzo San Massimo, one of the Edifici Mondo buildings

3 Results of Dynamic, Co-evolutionary Evaluation Process

3.1 Phase 1—Engagement and Public Debate

The first phase of the participatory adaptive reuse process consisted in a series of public meetings, workshops, surveys and debates on the future of the Edifici Mondo buildings, in which more than 50 local organisations participated, invited by Salerno municipality (Garzillo et al. this volume, Chap. 15). During the six Heritage Innovation Partnerships workshops, four Peer-review meetings, an exploratory survey and additional focus groups with key interested stakeholders, the CLIC framework for circular adaptive reuse of cultural heritage was presented, discussed and adapted to the specific case study thanks to the feedbacks provided. This phase was particularly relevant to build trust, open dialogue and understand between public, private and civic stakeholders, acknowledging different levels of interest and power in the decision making process, as well as possible conflicting interests and objectives. As cultural heritage is a "common good", its management requires the interaction of community actors (Ostrom 1990) to take decisions oriented to wellbeing. Thus, the phase of knowledge sharing, goals definition, and common vision identification represents a needed starting point for the subsequent phases of co-design supported by co-evaluation processes.

This phase resulted in the definition of specific objectives of circular adaptive reuse of cultural heritage in Salerno, particularly in the area of the "Edifici Mondo". Thus, the municipality of Salerno adopted the "Objectives of adaptive reuse of cultural heritage in a circular economy perspective"[1]:

[1] http://www.comune.salerno.it/allegati/30257.pdf

> *The adaptive reuse of cultural heritage in the perspective of circular economy is oriented towards the ability to regenerate different forms of capital (manufactured capital natural capital social capital human capital economic-financial capital). Circular reuse is aimed at transforming abandoned places into "living" systems, and as such regenerative. In this way it is able to generate positive effects in the context and contribute to the resilience of the city/territory system over time. Circular reuse is configured as the regenerative reuse that helps to implement the transition to a local de-carbonized economy (green economy). It minimizes waste, negative environmental impacts, and ecological footprint; it reuses/recycles waste, turning it into resources for new production cycles. It derives most of its resources from the surrounding area: it uses renewable energies as much as possible; it reuses rainwater and grey water; it contributes to regenerating the ecosystem services on which human activities and people's well-being depend; it promotes the use of green and nature-based solutions. It contributes to transform linear metabolism (extraction-production-consumption-waste/emissions) into circular (reuse, recovery, recycling…), imitating the wisdom of nature. In addition, circular reuse is characterized by seeking the ability to regenerate financial resources for its operation over time, minimizing subsidies from public/private sources. Circular reuse is a promoter of economic impacts in terms of locating new activities, also generating new direct, indirect, induced jobs. From the social point of view, circular reuse is oriented to generate a community, a "community of heritage" convention of lighthouse that in turn takes care of the heritage itself, in a virtuous circular process. Circular reuse is characterized by the adaptive reuse of cultural heritage*

This general description was further systematized through evaluation criteria, related to the three dimensions of circularity as expressed in the CLIC evaluation framework and adapted to the specific case study (Tables 13.1 and 13.2).

Then, 14 macro-criteria were detailed in 28 specific criteria, on the basis of criteria used in Public Consultation.

The following phase was focused on the ideation of alternative solutions for the adaptive reuse of the buildings.

3.2 Phase 2—Ideation of Alternative Adaptive Reuse Solutions

Based on the shared definition of the "circular adaptive reuse of cultural heritage in Salerno", the Municipality of Salerno launched a Public Consultation for the elaboration of adaptive reuse proposals for the "Edifici Mondo" in the circular economy perspective.

The public consultation was considered an operational experiment and contributed first of all to the identification of the needs expressed by the local community about regeneration processes/projects, providing an opportunity to express ideas, points of view, opinions and proposals. The public consultation allowed individual or associated subjects, including private ones (companies, foundations, cooperatives, associations, individual citizens, etc.) to present and motivate their adaptive reuse proposals in a initial stage.

As per the public consultation call, the overall objective of the adaptive reuse project should be the activation of a circular and regenerative local economy centred on the valorisation of the historical and cultural heritage as a lever for local sustainable development: attraction of investments, creation of new jobs, increase in entrepreneurship, increase and qualification of human capital, environmental, social and cultural regeneration.

The Public Consultation resulted in 14 different proposals of adaptive reuse, submitted by interdisciplinary teams. The 14 project ideas included functions oriented at increasing the attractiveness of the city through the enhancement of the intangible heritage of the ancient "Salerno Medical School", introducing functions related to medical and pharmaceutical activities. Other project ideas were focused on new forms of innovative living (from co-housing to solidarity condominiums), musical, theatrical and artistic cultural productions, craftsmanship, including digital

for synergies/synergies and cooperative activities between the subjects of the territory that increase the overall productivity of the intervention. The reuse of cultural heritage also reproduces intangible values: cultural values. The reuse of cultural heritage is also able to regenerate values cultural horizons and meaning, generating new contemporary meanings and new values related to the original meanings and value. In summary, the circular reuse of cultural heritage aims to regenerate tangible and intangible, natural, social and economic cultural resources of the territory, to promote synergies/synergies and cooperation between public, private and civil society actors, and to generate net positive economic, social, environmental and cultural impacts in the territory.

Table 13.1 The 14 macro-criteria defined for the evaluation of the adaptive reuse of the Edifici Mondo in Salerno

N.	Macro-criteria	Description
1	Regeneration of historical and cultural values	Ability to regenerate historical and cultural values through compatible interventions capable of transmitting the identity values of the heritage and of positively affecting the perceptual quality of the landscape.
2	Regeneration of human capital	Capacity to regenerate human capital through the provision of learning opportunities and the promotion of activities aimed at the recovery of traditional knowledge and the development of entrepreneurial skills.
3	Regeneration of environmental resources	Capacity to regenerate environmental resources through the use of circular technical and technological solutions (e.g. solutions for the recovery/reuse of water, reduction of waste and the use of non-renewable energy, nature-based solutions, ecosystem services) compatible with the historical fabric of the area.
4	Economic and financial self-sustainability	Ability to sustain itself over time with its own resources.
5	Activation of partnerships and collaborative networks	Ability to activate partnerships and collaborations with different stakeholders.
6	Local Symbiosis	Ability to activate or promote a local network for the extension of the life cycle of materials and resources.
7	Citizens' participation	Ability to involve citizens in the implementation of the proposal.
8	Contribution to the implementation of the RIS3 strategy of the Campania region	Consistency of proposed uses and functions with the smart specialization strategy of the Campania region.
9	Generation of jobs	Ability to create full-time jobs.
10	Contribution to the cultural vitality of the city (cultural vibrancy)	Ability to contribute to the cultural vitality of the area through the promotion of arts and crafts, cultural activities and events, and the establishment of creative and cultural enterprises.
11	Contribution to the attractiveness of the adaptive reuse of cultural heritage, development and innovation sectors	Ability to attract innovative/technology startups/enterprises.
12	Contribution to the attractiveness of the cultural tourism sector	Ability to attract activities in the field of cultural tourism and related services.
13	Contribution to the liveability of the area	Ability to improve the liveability of the area through the improvement of pedestrian routes and perception of safety, the maintenance of local stores and different social groups in the historic Centre.
14	Contribution to perceived health and wellbeing	Ability to enhance health and wellbeing

Table 13.2 The 28 specific criteria for the evaluation of the adaptive reuse of the Edifici Mondo in Salerno

Macro-criteria	ID_ Criterion	CRITERION
Regeneration of historical and cultural values	C1	Compatibility with the historical and cultural value of the asset
	C2	Capacity to regenerate and transmit the identity value of the cultural heritage
	C3	Ability to positively affect the perceptive quality of the landscape
Regeneration of human capital	H1	Ability to provide formal and informal learning opportunities for the improvement of competences and skills
	H2	Ability to promote activities aimed at recovering traditional knowledge
	H3	Capacity to promote entrepreneurial skills
Regeneration of environmental resources	N1	Capacity to implement circular solutions to extend the life cycle of materials and reduce waste production
	N2	Capacity to implement circular solutions for the recovery/reuse of water
	N3	Capacity to implement circular solutions to reduce the use of energy from non-renewable sources
	N4	Capacity to implement natural-based solution
	N5	Potential to generate environmental and/or health risks
Economic and financial self-sustainability	E1	Ability to sustain itself over time with its own resources
Activation of partnerships and collaborative networks	P1	Ability to activate partnerships and collaborations with different stakeholders
	P2	Ability to include marginalised groups
Local symbiosis	P3	Ability to activate or promote a local network for the extension of the life cycle of materials and resources
Citizen participation	PE1	Ability to involve citizens in the implementation of the proposal
Contribution to the implementation of the RIS3 strategy of the Campania region	R1	Coherence with the smart specialisation strategy of the Campania region
Generation of jobs	J1	Capacity to create full-time jobs
Contribution to the cultural vitality of the city (cultural vibrancy)	CV1	Capacity to attract creative and cultural enterprises
	CV2	Capacity to promote arts and crafts
	CV3	Capacity to promote cultural activities and events

(continued)

Table 13.2 (continued)

Macro-criteria	ID_ Criterion	CRITERION
Contribution to attractiveness for adaptive reuse of cultural heritage, development and innovation (R&D) sectors	RD1	Attractiveness of attracting innovative/ technological start-ups/enterprises
Contribution to attractiveness for the cultural tourism sector	CT1	Attractiveness of attracting cultural tourism activities and services
Contribution to the liveability of the area	L1	Ability to implement walkability in the historic Centre
	L2	Ability to preserve local shops in the area
	L3	Ability to favour the perception of security in the area
	L4	Ability to contribute to the maintenance of different social groups within the historic Centre
Contribution to perceived health and wellbeing	HW1	Ability to enhance health and wellbeing

craftsmanship, and commercial activities linked to local typical products, as well as tourist and social functions, guaranteeing new and more functional services to citizens, in terms of accessibility, sustainability and better liveability. Below a synthetic description of the project ideas submitted (Gravagnuolo et al. 2024) (Table 13.3).

The proposals were analysed and assessed according to the multiple evaluation criteria defined, in line with Multi-Attribute Utility Theory (MAUT) (Zeleny 1982, 1983).

The TOPSIS method (Technique for Order Preference by Similarity to Ideal Solution) multicriteria evaluation tool was used in this phase. TOPSIS is a MCDA (multi-criteria decision analysis) method used in different fields of scientific research to support decision-makers to choose between different alternatives (Assari 2012; Łatuszyńska 2014; Roszkowska 2011). This method was chosen for the participatory decision-making process in Salerno because it allows to measure the relative performance of the alternatives through both quantitative and qualitative criteria and indicators, while its concept is easy to understand for non-experts (Kabir and Hasin 2012; Zanakis et al. 1998).

The application of multicriteria evaluation in the initial stage of public consultation was able to support the identification of the most valuable proposals, despite few and undefined information. In fact, multicriteria evaluation can be effective also with qualitative information (Zeleny 1983).

It is important to highlight that multicriteria evaluation methods are oriented to "support" decisions but not to "take" decisions instead than the decision-maker. They support stakeholders, owners and managers of cultural heritage by making explicit the rationale behind choices, through the discussion about objectives, criteria, and their relative weights, thus contributing to the transparency of

Table 13.3 The 14 proposals of adaptive reuse in Salerno. (Adapted from Gravagnuolo et al. 2024)

Title	Brief Description
A. Hotel complex 'Plajum Montis'	Hotel and accommodation facility that pursues the objectives of adaptive reuse of cultural heritage in the perspective of a circular economy.
B. SALERNO (re) STARTS	Widespread hotel to regenerate the historic Centre for the development of tourism and youth entrepreneurship to overcome the tourism seasonality through activities such as residences and offices, co-working spaces and a student residence.
C. Reggia di Salerno	Cultural hub and museum capable of creating synergies with local stakeholders to develop a cultural and educational institution aimed at rediscovering the traditional values of handicrafts and typical local products.
D. The identity between tradition and innovation	Integrated redevelopment of the buildings for cultural purposes through a research and vocational training Centre for the management of cultural heritage and the enhancement of sustainable tourism, a museum Centre, a Centre for artistic and cultural activities, and a Centre for the promotion of agrifood excellence linked to the Mediterranean diet.
E. Solidarity condominium	A social housing project based on the culture of solidarity, brotherhood, and mutual respect with the aim of intensifying relations between the public and private sectors by strengthening the relationship with the neighbourhood.
F. Hippocratica Hills health heritage hub	A Centre of excellence dedicated to research and experimentation in the field of health and wellbeing in the tradition of the Salerno medical school.
G. Academy ASSE3 T	Academy for the environment and the empathetic-ethical-ecological-economic sustainable development of the territory—a campus providing continuous training, work strategies, innovative and sustainable services for local stakeholders.
H. Tourism learning based	Experiential tourism Centre based on a bottom-up regeneration process carried out by citizens and local businesses for the valorisation of typical products and traditional crafts.
I. School hotel	Advanced training project in tourism that promotes accommodation, educational, catering and territorial promotion activities in a school-hotel project.
J. Discreet interventions for the reactivation of 'Edifici Mondo'	Small-scale interventions that generate a new culture by setting up a socio-cultural Centre with spaces for exhibitions, musical and artistic performances, workshops for handicrafts, student residences, and a botanical research Centre.
K. The awakening of the senses	An innovative and technological museum for the establishment of both an experiential archive of places and cultural exchanges.
L. The house of music	The project aims at promoting strategies for the reuse of cultural, natural, social and economic resources by rethinking cultural heritage as a creative system where music is the main activity.

(continued)

Table 13.3 (continued)

Title	Brief Description
M. Water paths	Hub for the cultural, social and economic regeneration of Salerno historic Centre. The project envisages a thermal Centre powered by geothermal systems, an academy for architects with a student residence, a hotel, and green paths with botanical gardens providing also a sustainable mobility system.
N. Creative reuse of abandoned buildings under an artistic key	A Centre for artistic and cultural activities, training courses and workshops, artist residencies, events and music festivals.

Table 13.4 Ranking of preferability of adaptive reuse proposals. (Adapted from Gravagnuolo et al. 2024)

Ranking	Adaptive reuse alternatives
1	M. Water paths
2	F. Hippocratica Hills Health Heritage Hub
3	D. The identity between tradition and innovation
4	L. The house of music
5	E. Solidarity condominium
6	J. Discreet interventions for the reactivation of 'Edifici Mondo'
7	N. Creative reuse of abandoned buildings under an artistic key
8	H. Tourism Learning Based
9	G. Academy ASSE3 T
10	C. Reggia di Salerno
11	A. Hotel Complex 'Plajum Montis'
12	K. The Awakening of the Senses
13	I. School hotel
14	B. SALERNO (re)STARTS

decision-making processes that is clearly particularly important in public participatory processes (Gravagnuolo et al. 2021b).

Through the support of multicriteria analysis, it was possible to identify the most valuable adaptive reuse solutions to be discussed and revised in the subsequent stage of co-design. An order of preferability was identified (Table 13.4).

Participants were thus invited to take part in a series of co-design workshops, to refine and revise the adaptive reuse proposals assessing their financial and social sustainability, as well as the impacts, in search of a fully circular business and management model for the adaptive reuse solution.

3.3 Phase 3—Co-design of Adaptive Reuse Solutions

In the advanced phase of co-design, a group of 31 stakeholders took part in a series of workshops (Saleh and Ost this volume, Chap. 16) to define the adaptive reuse proposals in more detail. Through the facilitation of the workshops, and the support of the multicriteria evaluation, a set of four adaptive reuse solutions was defined as a synthesis of the main features of "best" solutions. The four alternative solutions for the adaptive reuse of Edifici Mondo were defined starting from the "top ranked" proposals submitted in the public consultation stage. A series of exercises and interactions were conducted with participants to define the main new uses/functions for the buildings taking into account elements of cultural values conservation especially compatibility of the new uses with heritage values, financial sustainability (potential costs/revenues, financing sources), energy self-sustainability and environmental resources regeneration, social impacts and the overall circular business model viability (Saleh and Ost this volume, Chap. 16). The proposals elaborated were described in detail, providing quantitative and qualitative information and data to be used in the following stage of co-evaluation. A synthetic overview of the four resulting alternative solutions is provided in Table 13.5.

Table 13.5 The four adaptive reuse alternatives developed through the co-design workshops in Salerno

Alternative solutions	Description and main uses/functions proposed
Water paths (merged with) Hippocratica Hills health heritage hub (5H)	The main functions in the **Hippocratica Hills health heritage hub and water paths** reuse project include spaces for advanced pharmaceutical research and training (laboratories and classrooms for nutraceuticals, herbal medicine), residences, conference rooms, wellness (spas, thermal plants, sensory pathways), crafts and trade—In synergy with a private foundation for medical and pharmaceutical research active in Salerno since 10 years, and located nearby the Edifici Mondo buildings. Particularly, the proposal focuses on the recovery and regeneration of the ancient Salerno medical school, that represents the most important intangible cultural heritage of the city, recognised by all citizens as a symbol of local identity, history and culture. The proposal creates particular synergies with the "garden of Minerva", a historic botanic garden nearby the buildings to be reused, that was used since the middle ages as didactic garden and for production of officinal plants, exploiting the water paths springing from the top of Salerno hills. The spaces of the 'Edifici Mondo' would thus host a cluster of activities in which the main resources are locally sourced (a nearby water system for the main facilities and thermal plants, food and herbs harvested from the gardens for restaurants and research). This hybrid model of functions guarantees a circular dimension from an economic point of view, as well as cultural values conservation and regeneration, urban ecosystem regeneration, and social benefits with the generation of high skilled jobs in pharmaceutical and medical research. The innovation is in the technologies used (renewable energy, etc.) for the renovation and conservation of the old structures. This proposal aims to develop a heritage-driven circular ecosystem composed of incubators, laboratories and services specialised in health-related research and dedicated to the creation of craft products and cultural activities related to health and Well-being; a complex system integrated in the urban context that enhances the cultural and territorial potential.

(continued)

Table 13.5 (continued)

Alternative solutions	Description and main uses/functions proposed
Identity between tradition and innovation	The adaptive reuse project in the **identity between tradition and innovation** proposal is based on five main function: 1. Observatory and Centre for research and specialised vocational training on tourism and cultural heritage. 2. Salerno museum Centre (contemporary, modern, street and civic art) with open innovation activities and environments in the traditional and digital art sectors and transdisciplinary co-working/co-factory spaces and workshops. 3. Food and wine hub for the promotion of indigenous Agri-food and marine products, as part of the Mediterranean diet, with exhibitions, showrooms and events. 4. Intelligent garden city: a green network of paths and terraced systems with herbal, botanical and shared urban gardens, linked to the 'Schola Medica Salernitana' and to the resources and traditions of the Amalfi coast. 5. Improved interconnection, mobility and accessibility by creating a network of different historical routes with a better information/signalling system, mechanised carriers and parking spaces sized to the context. This adaptive reuse project is a multifunctional/disciplinary cultural Centre intended to host a large number of socio-cultural, artistic and research subjects as well as events and activities, open to local residents but also to a national and international audience. All the planned activities are strongly interconnected and intended for a wide range of functions; the various buildings are interconnected through external courtyards, terraced and urban gardens, pathways, which guarantee not only effective communication routes but also the complementarity of the buildings by exploiting the spaces to increase local self-sufficiency. All physical interventions on the 'Edifici Mondo' include optimising natural light, equipping them with solar panels, using recycled rainwater and a wastewater treatment system. More than 50% of the structures are based on renewable energy and recycled materials. The reuse project contributes to the resilience of the city-territory system through the transformation of abandoned places by encouraging synergies and cooperation between public and private actors and the involvement of the local community.

(continued)

Table 13.5 (continued)

Alternative solutions	Description and main uses/functions proposed
House of music	The **house of music** project provides spaces where musicians can experiment, record or perform in an alternative and sustainable economic environment, but also many public spaces where music is the main activity. The building complex has the following spaces: Hospitality rooms and conservatory rooms. Functional spaces for musical activities including classrooms, recording studios, recycling workshop for damaged or disused musical instruments, lutherie workshops in the area of repair/maintenance of damaged instruments, courses to learn and/or improve artistic skills related to traditions, music courses related to local traditions, library, kindergarten. Events include: Open-air concert hall, exhibition and performance space, instrument museum. Commercial activities include: Restaurant and bar, musical instrument shop. Adaptive reuse in the house of music proposal aims to promote territorial synergies, rethinking the 'Edifici Mondo' complex as the central node of a corridor of public green spaces that, linked together, form a green mesh of reactivated areas. The inclusion of several functions minimises the transformation of material and immaterial cultural resources and adapts the building to the needs of new users. The functional choice is based on a naturalistic approach to regenerate the building. Green solutions are adopted in the consolidation of the building envelope. Roofs are differentiated by type: Sloping or flat. For flat roofs, green roofs are proposed to reduce the building's energy requirements. In the second case, the use of photovoltaic tiles that can be integrated into the roof, which are less invasive and more maintainable than panels, is considered. Finally, green solutions are hypothesised that push the linear metabolism of the building's waste towards the circularity of waste reuse: Aiming to achieve an almost zero impact of the building, aerobic mini-composters have been placed in the perimeter rooms. The energy obtained is used for the central heating of the building. To enhance the strategy, pyrolysis mini-plants could be also installed to treat the remaining organic waste. Circularity from a social point of view refers to the sharing of the skills of individuals (to repair instruments, to give lessons to citizens, to spread their knowledge about the local musical culture, etc.).

(continued)

Table 13.5 (continued)

Alternative solutions	Description and main uses/functions proposed
Solidarity condominium Hippocratica Civitas	The **solidarity condominium** is a social housing project. The purposes of the reuse project are to improve the quality of life, create accessible housing, develop an alternative to loneliness and marginalisation, improve interpersonal relations, and activate the culture of solidarity, coexistence and mutual aid. The adaptive reuse of the 'Edifici Mondo' envisages their transformation through an organisation of spaces that allows for housing, services and common areas. The social housing project consists of modular housing from 35 to 95 square metres accommodating families of one, two, three or more persons. The shared spaces are the gardens and the social vegetable garden enjoyed by the members of the social housing and the local community. The objective of the garden is the promotion of the local Mediterranean diet. The services offered include craft activities. The objective of the craft activities is to rediscover local traditions (shoemaker, umbrella repairer, tailor, chair maker, etc.) connecting them with twenty-first century technologies through the FabLab methodology. Thus, synergies are created between the old and the new, where specialised craftsmen teach the younger generation and working members of the community cooperative handicrafts. These activities offer the design and production of parts needed to repair/modify household utensils or anything useful for the members and citizens of Salerno. The reuse project envisages spaces for training activities based on an interweaving of the Salerno medical school, the Mediterranean diet and territorial excellence, a stimulus for new forms of sustainable and responsible tourism and an engine of economic development for the territory and social welfare. Services to the neighbourhood consist of shared shopping services, regular house maintenance, provision of food on a social basis by restaurants, caring services, babysitting services, organisation of recreational activities.

Once defined the four alternative solutions for the Edifici Mondo, the final phase of the participatory adaptive reuse process was aimed at identifying a "satisfying" solution according to circularity objectives, criteria and indicators, as described in the following section.

3.4 *Phase 4—Satisfying Solution Identification*

After detailing and improving the four proposals, the TOPSIS method was used again to obtain a new ranking of preferability considering the four alternatives, evaluated according to circularity objectives, criteria and indicators.

Three groups of stakeholders were engaged to re-discuss and adapt the criteria and indicators of circularity, as well as their relative weights: public officers of Salerno municipality; civil society organisations; sustainability experts and researchers. As described in Panaro et al. (this volume, Chap. 9), the Simos-Roy Figueira method of "cards" was used to assess the relative importance of the different criteria for each stakeholder group, and assign numeric weights to be used in multicriteria evaluation. In this stage of the ex-ante evaluation process, thanks to the more detailed information provided through the co-design workshops, both qualitative and quantitative indicators were used. Qualitative indicators were expressed using a five-point Likert scale (5 = very high performance; 4 = high performance; 3 = medium performance; 2 = low performance; 1 = very low performance), while quantitative indicators were expressed through diverse units of measure (e.g. estimated Return on Investment, Net Present Value of the investment, energy self-sufficiency levels).

The process of re-discussing and refining the specific evaluation criteria at each subsequent evaluation stage was particularly important to trigger a structured reflection and evolution of ideas and beliefs of all stakeholders involved in the participatory actions (Fig. 13.3).

Through a series of feedbacks and interactions with stakeholders, criteria and weights were adjusted to fit with their evolving preferences, needs and objectives.

The four alternatives were evaluated through the TOPSIS method, according to the adapted criteria and weights. The ranking was obtained considering their relative proximity to the "ideal" solution, which should maximise all circularity objectives. The alternative with the greatest relative proximity (the highest value) to the ideal solution is preferred (Table 13.6).

"Hippocratica Hills and Water Paths" resulted the preferable solution (value 0.764) as it was capable of attracting entrepreneurial, training and social activities in the medical, pharmaceutical and wellness sectors, enhancing the tangible and intangible cultural heritage of the Salerno ancient Medical School. The "Identity between tradition and innovation" was in second position (value 0.512), aimed at

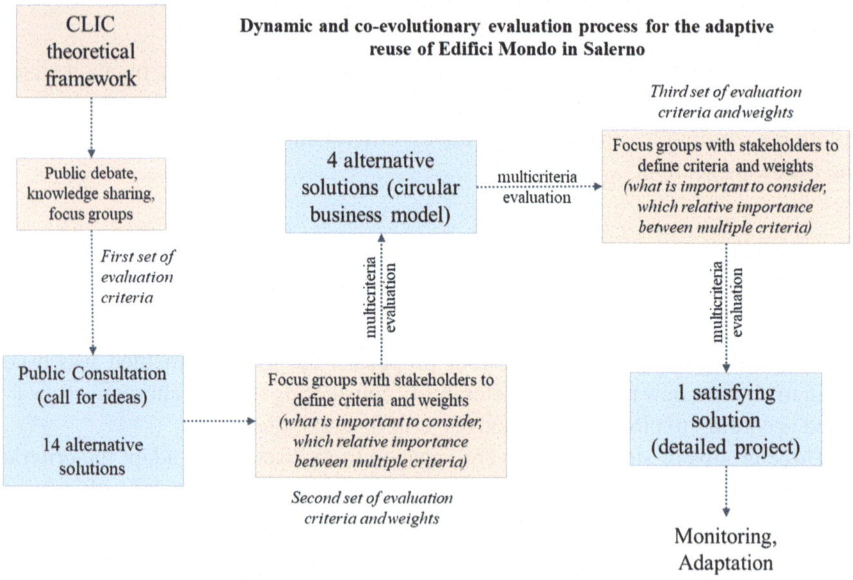

Fig. 13.3 Dynamic and co-evolutionary evaluation process in Salerno

Table 13.6 Ranking of preferability of four proposals through the TOPSIS method

Relative proximity to the ideal solution (TOPSIS evaluation method)			
Solidarity Condominium	The identity between tradition and innovation	House of music	Hippocratica Hills and Water Paths
4 (value = 0.181)	2 (value = 0.512)	3 (value = 0.347)	1 (value = 0.764)

involving the productive fabric of the entire province of Salerno also in relation to agri-food and the Mediterranean Diet. "House of Music" resulted the third preferable solution (value 0.347), as it aimed to stimulate local and international cultural productions and make Salerno a regional/national cultural attraction. Finally, "Solidarity Condominium" promoted sociality, health and intergenerational well-being, however its contribution to the overall circularity objectives was lower (value 0.181). Details on the implementation of the TOPSIS evaluation method, specific criteria and weights assigned, are described in Gravagnuolo et al. (2024).

The last and final stage was the definition of the adaptive reuse "satisfying" solution (Simon 1959) for the Edifici Mondo buildings, as described in the following section.

3.5 *Elaboration of a Defined Satisfying Solution for the Adaptive Reuse of Edifici Mondo: Salerno Circular City of Health*

According to the participatory evaluation process conducted in Salerno, a final adaptive reuse proposal for the Edifici Mondo historic buildings was defined. The final proposal referred to the realisation of a demonstrative circular adaptive reuse project involving the entire urban area towards ecological and urban regeneration: "Salerno Circular City of Health" (Fig. 13.4). Through the classification, reorganization and reshaping of the functions of the initial project proposals, the most viable mix of functions was identified. This process is intended as an iterative and interactive process which, starting from the evaluation of the project alternatives, improves the decision-making and co-design processes through continuous and circular feedback mechanisms, shifting towards the "ideal" solution through the search of a "satisfying solution" (Simon 1959).

The overall project found to be most consistent with the proposed circular model and most satisfactory according to the preferences expressed by the stakeholders was a functional mix centred on the recovery and enhancement of the Salerno Medical School, referring to the top ranked proposal (Hippocratica Hills and Water Paths). The project developed from the results of the co-design and evaluation thus aims to create a centre of adaptive reuse of cultural heritage, innovation and entrepreneurship of excellence for the development of Salerno as a 'Circular City of Health'. The final project elaborated is inspired by the best European practices of

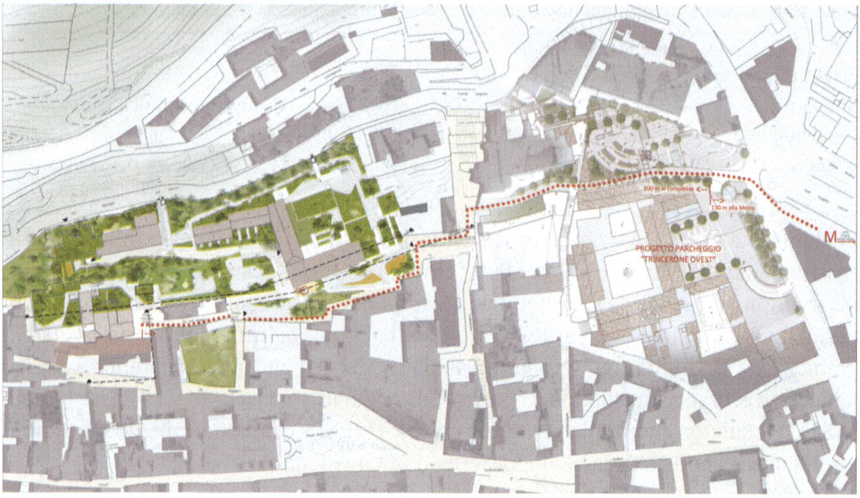

Fig. 13.4 The masterplan of the historic urban area for the Circular City of Health. (Authors: Salerno municipality, arch. Silvia Napoli)

adaptive 'circular' reuse of buildings and sites of historical and cultural value, analysed by the CLIC project, aiming not only at the realisation of multi-functional spaces for adaptive reuse of cultural heritage excellence, but also at the minimum environmental impact: energy efficiency, renewable energies, recovery and reuse systems for rainwater and grey water, and the use of 'nature-based' solutions for the re-naturalisation of the urban area and the improvement of air quality and microclimate. Energy analysis (through assessment of energy classes of buildings based on the retrofit interventions foreseen) was conducted to ensure that the higher investment initially required to realise 'near-zero energy' buildings will be compensated over time thanks to the reduced impacts in terms of carbon equivalent emissions and reduced energy and water consumption. The final design solution thus contributes to addressing climate change both in terms of combating climate-changing emissions and better use of materials and energy, and in terms of adapting to the effects of rising average temperatures. This aspect is extremely important in Southern Italy, considering that the impacts of climate change are particularly evident in the Mediterranean area.

The final circular adaptive reuse project of the Edifici Mondo buildings was elaborated with the goal to become an exemplary ("demonstrative") intervention in the city of Salerno, showing how to synergistically obtain positive environmental, social and economic impacts together with the regeneration of the cultural identity and symbolic value of cultural heritage, thus offering a "cultural" reference perspective for sustainability and circularity. The assessment of impacts in the ex-ante design phase supported the identification of the most satisfactory functional mix based on circularity criteria.

The positive impacts assessed include the revitalisation of the entire urban area through the adaptive reuse of abandoned cultural heritage assets, the activation of cultural and creative activities, which becomes an attractor for visitors/tourists and new residents, the increase of commercial activities and neighbourhood services, the generation of highly qualified and specialised employment, as well as financial self-sustainability contributing to operation and maintenance in the long term without need of additional public resources.

The functional mix was thus carefully designed to generate a circular business model capable of guaranteeing a high degree of economic and financial self-sustainability, which is necessary to avoid further abandonment phenomena after the redevelopment. In addition to the direct employment impact, externalities related to the economic added value produced by the activated sectors, i.e. adaptive reuse of cultural heritage, tourism and production, must be considered. Further synergetic and indirect economic impacts concern the infrastructural intervention. The wholesomeness and beauty of the urban area and buildings was considered a strong point of the intervention, in line with the New European Bauhaus initiative (European Commission 2021) (Fig. 13.5).

Finally, an important role should be played by civil society organisations, already active in the area, also through the activation of Collaboration Pacts for Shared Administration, adopting a "circular" governance model (LabSus 2021). The Collaboration Pact was proposed in Salerno[2] as a possible tool for the regeneration of the "culture of citizenship", an instrument enabling citizens to take care of urban spaces and cultural heritage, and thus to regenerate democratic vitality. This regulatory tool could be adopted in diverse contexts to promote the capacity to generate intangible values, a "civic culture" that activates potential resources, collaboration, trust, which triggers social and cultural regeneration through the conservation of cultural heritage. Pacts of Collaboration were integrated into the adaptive reuse solution in Salerno, with the aim of enhancing heritage care, civic engagement and responsibility.

4 Discussion

The important aspect of the dynamic, co-evolutionary participatory evaluation methodology experimented in Salerno is the capacity to trigger the evolution of knowledge, vision and objectives of the involved stakeholders. Through recurring "circular" stages of alternative proposals definition, co-evaluation, discussion, and further refinement, stakeholders are enabled to enhance their capacity to address circularity in the adaptive reuse of cultural heritage, cultivating cooperation and synergies where there was conflict of ideas and interests, using their knowledge and

[2] The mandate for the elaboration of the Collaboration Pacts Regulation was approved in the City Council.

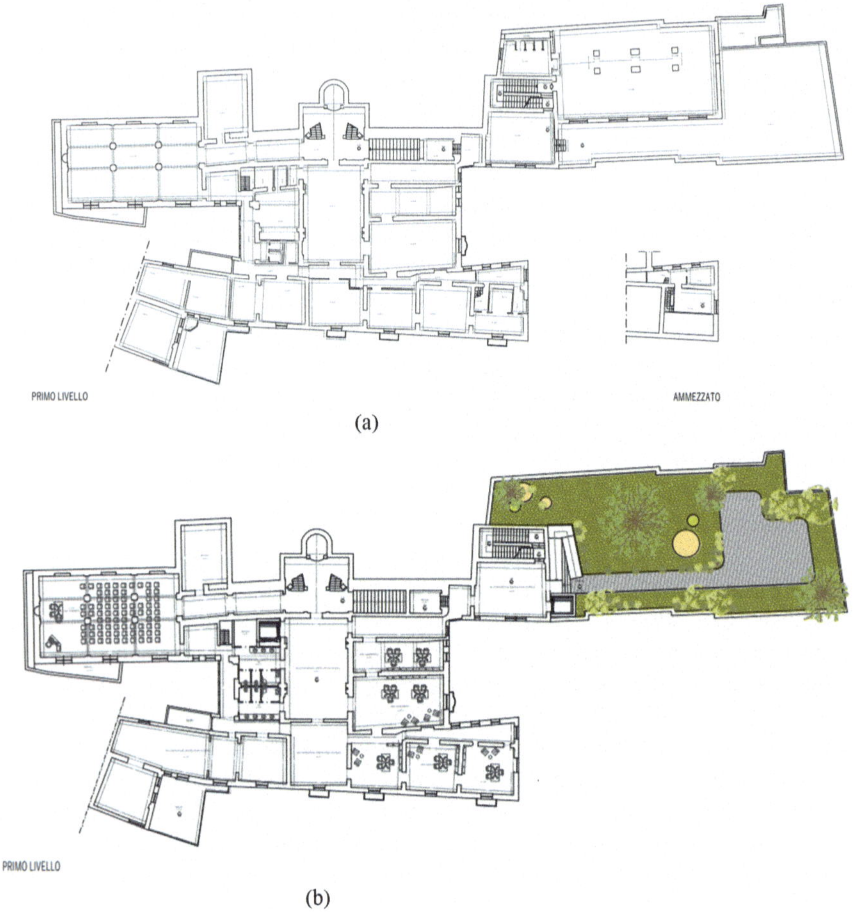

Fig. 13.5 Functional reuse of the Palace San Massimo, state of the art and project. (**a**) Current state, (**b**) Adaptive reuse proposal

resources to contribute to the collective process in search of a satisfying solution, and finally entering a learning and educational journey in which each participant comes out with new knowledge and perspectives. Through the evolutionary, dynamic evaluation process the objectives, criteria and indicators do not remain the same in each iteration, but they evolve together with the collective consciousness of the stakeholders' group, reflecting their change in priorities, needs and knowledge. This aspect becomes extremely important not only for the identification of a circular, "self-sustainable" solution for cultural heritage adaptive reuse, but also to enhance communities capacity to cooperate, collaborate and trust each other, contributing to ensure longer term care, maintenance, and conservation of cultural heritage—as well as contributing to social cohesion and new forms of participatory democracy through active engagement of citizens and communities. In this way, cultural heritage can regain a central role in contemporary and future society, generating social and cultural value, as well as ecological and urban regeneration, becoming an engine of circular development.

In the city of Salerno, the co-planning and co-design process was conducted in line with national Laws and regulations. In fact, the co-planning is promoted by Italian Law 117/2017 (art. 1–5, 55), however it is rarely implemented as possible instrument to enhance the quality and effectiveness of projects impacting local communities. Other regulatory acts are oriented to promote the engagement and participation of citizens and stakeholders in decision-making processes, particularly when decisions directly affect their wellbeing, such as the Italian Constitutional Court Judgement 131/2020 and Italian Law on Public Debate. Participatory processes can thus be supported by multi-criteria and multi-dimensional evaluation tools. “Public debate” and evaluation was strengthened by the Constitutional Court Sentence 235/2018, focusing on the need of comparing project alternatives and implement participation as fundamental instrument of democracy. Public debate is mentioned also by art. 22 of the Public Contracts Code, while a National Commission on Public Debate was established with DM 627/2020. Moreover, the PNRR (Italian National Resilience and Recovery Plan/EU Next Generation Funds) guidelines for projects implementation (DL 77/2021, becoming Law 108/2021) foresees the assessment of alternative scenarios and the engagement of stakeholders in co-design and co-programming, according to the Guidelines for assessing investment projects of the European Commission (European Commission 2014a). To reach public interest, the public sector can also take advantage of cooperation with the Third Sector. In last years, new tools for shared governance, administration and collective care of common goods are emerging in the Italian context, such as the Pacts of Collaborations (LabSus), adopted by more than 200 cities in Italy, in line with art. 118 of Italian Constitution (subsidiarity principle). The Law 117/2017 and the Constitutional Court Sentence n.131/2020 focus on the possibility for public institutions to cooperate with third sector actors to reach objectives of public interest. However, despite the regulatory orientation towards higher participation of communities in decision processes, and use of decision-support and evaluation tools, which implementation is still limited.

The project discussed in Salerno assumed that circular economy models can be applied at territorial level to avoid “waste” of cultural and natural resources, re-activating areas challenged by abandonment, depopulation and underuse. Regenerating vulnerable and discarded cultural heritage and landscape means in particular to reconnect human beings to nature, and re-building the symbiotic relationships among them into a systemic perspective (Fusco Girard 2021a, b). In this context, the human capital, social cohesion, solidarity, common identity, communities participation, the feeling of confidence in the future, in interpersonal relations and between citizens and institutions emphasize the role of “culture” as a driver for change and development. Integrating this “human-centred” perspective into a circular/eco-centric perspective allows to consider landscape regeneration as a “multiplier of values” able to regenerate the vitality of vulnerable and discarded cultural heritage and landscapes, transforming them into "living systems". Considering the multidimensional and multi-objective nature of the circular human-centred economy paradigm, evaluation tools play a fundamental role in its implementation.

In circular adaptive reuse of cultural heritage, a key aspect is the ecological and environmental impact of the reuse. The reuse/regeneration of cultural heritage is currently considered from a conservation perspective, that does not take into account sufficiently the ecological transition and the problem of climate change. As a result, the evaluation tools available in the heritage sector can be seen as incomplete (ICOMOS 2011; SoPHIA consortium 2021). The circular perspective allows to integrate the ecological dimension introducing the concept of cultural heritage as complex "ecosystem", a "living system" able to regenerate itself and generate new resources for the territory over time, in symbiotic relationship with the social, environmental and economic context.

The implementation of participatory processes for circular cultural heritage adaptive reuse through dynamic, co-evolutionary multicriteria evaluation tools could enhance adaptive reuse projects and ensure that cultural and natural heritage is conserved and transmitted to future generations, enhancing wellbeing and health of local communities (WHO 2018). However, participatory processes require also specific skills and competences that are not always available especially in small and peripheral cities. Usually, human and financial resources are sought for participatory processes in case of large infrastructural projects, while cultural heritage sites remain separated and mostly managed through public authorities and top-down approaches led by conservation experts. As cultural heritage is a sensitive "object" that requires very specific skills and knowledge, it is also widely acknowledged that it has a relevant impact on people's wellbeing, thus structured participatory processes including diverse competences and stakeholders can contribute to enhance the quality of conservation projects, identifying creative solutions thanks to the collective knowledge and exchanges between diverse actors. Clearly, participatory processes engaging multiple stakeholders can be complex, therefore the skills related to complex evaluation, facilitation, diplomacy and negotiations should be more and more developed and requested at all levels.

Finally, the "ease" of understanding evaluation tools should not be underrated. In participatory processes like that implemented in Salerno, with non-expert stakeholders coming from diverse backgrounds, it is of utmost importance that transparency, clarity and openness is maintained. Evaluation tools should not become a "black box" through which decisions are taken through software and algorithms, on the contrary they should be used to support dialogue and understanding between stakeholders, clarifying the implications of different choices and alternatives, helping to produce evidence-base of results, supporting reliable estimations based on observation of case studies. In this way, evaluation tools can support "human-centred" adaptive reuse of cultural heritage, centred on wellbeing and responsibility of the stakeholders involved, as well as stimulating creativity in the search of possible satisfying solutions more and more near to the "ideal" circular project. In fact, evaluation tools should support the "tension" towards the ideal circular project, enabling evidence-based discussion and understanding of the multiple implications of diverse choices, not only enhancing knowledge and capacity of stakeholders, but also promoting a shift in cultural values towards a circular development model. This "shift" in values is the most important contribution of circular cultural heritage adaptive reuse to today's and future society.

5 Conclusions

The CLIC evaluation methodology was tested in Salerno to experiment how a dynamic, co-evolutionary approach to multicriteria assessment can work to support better decisions towards circular and human-centred circular adaptive reuse of cultural heritage. Community engagement and participation was the focus of the experimental approach, addressing initial issues and barriers. The experience of Salerno showed that multicriteria evaluation tools can be successfully used to support participatory decision-making processes for cultural heritage adaptive reuse.

In the case of the Edifici Mondo buildings in Salerno, the preliminary assessment of impacts in the ex-ante phase guided and oriented the design process from an initial phase of uncertainty (public consultation) to the feasibility design phase. A series of subsequent in-depth studies and co-assessments based on the acquisition of more and more detailed data on the projects, in relation to stakeholder preferences and foreseeable impacts, made it possible to identify the most "satisfying" project solution with respect to the territorial and social context. In this way, it was possible to share project choices at each stage and trigger a process of innovation that involved the Municipality of Salerno itself in a process of 'co-learning'. This project process, oriented towards the territorial circular economy model, and hinged within the framework of participatory governance, can realistically be transferred and implemented in other contexts, since it does not offer 'ready-made' project solutions, but a processual, dynamic and adaptive model, based on sharing, co-designing and co-assessment (Fusco Girard 2021a, b). According to Patrick Geddes (1918), planning/layout should create organic relationships among people, places, and work activities: the "Triad" between the environment, functions, and organism should be the approach to design. Evaluations in a circular economy perspective as proposed in this work can contribute to implement a synergistic vision of the city as "organism", integrating stakeholders' perspectives and ecological perspectives, towards an integrated ecological, social, cultural, perceptual "planning balance sheet".

The implementation of the circular adaptive reuse model in Salerno showed how cultural heritage can become a "circular cultural ecosystem", configured as a dynamic "living lab" in which people and communities continuously learn. Through participatory processes supported by dynamic and co-evolutive evaluation for the adaptive reuse, the role of cultural heritage expands and becomes a "laboratory of citizenship", contributing to social cohesion, responsibility, collaboration, dialogue and cultural and social development of cities and communities.

References

Assari A (2012) Role of public participation in sustainability of historical city: usage of TOPSIS method. Indian J Sci Technol 5(3). https://doi.org/10.17485/ijst/2012/v5i3.2

Barinaga-Rementeria I, Erauskin-Tolosa A, Lonzano PJ, Latasa I (2019) Individual and social preferences in participatory multi-criteria evaluation. Sustainability 11(20):5746. https://doi.org/10.3390/su11205746

Bosone M, De Toro P, Fusco Girard L, Gravagnuolo A, Iodice S (2021) Indicators for ex-post evaluation of cultural heritage adaptive reuse impacts in the perspective of the circular economy. Sustainability 13(9):4759. https://doi.org/10.3390/SU13094759

European Union (2021) Building an economy that works for people: an action plan for the social economy. Publications Office of the European Union, Luxembourg. https://doi.org/10.2767/12083. Available at: https://ec.europa.eu/social/main.jsp?langId=en&catId=89&newsId=10117&furtherNews=yes#navItem-1

Bullen PA, Love PED (2010) The rhetoric of adaptive reuse or reality of demolition: views from the field. Cities 27:215–224

CHCfE Consortium (2015) Cultural heritage counts for Europe. International Cultural Centre, Krakow. http://blogs.encatc.org/culturalheritagecountsforeurope//wp-content/uploads/2015/06/CHCfE_FULL-REPORT_v2.pdf

Council of Europe (2005) Convention on the value of cultural heritage for society (Faro Convention, 2005). Available at: https://www.coe.int/en/web/culture-and-heritage/faro-convention

Dean M (2022) Including multiple perspectives in participatory multi-criteria analysis: a framework for investigation. Evaluation 28(4):505–539. https://doi.org/10.1177/13563890221123822

Douglas J (2006) Building adaptation, 2nd edn. Butterworth-Heinemann, Woburn

European Commission (2014a) Guide to cost-benefit analysis of investment projects. Economic appraisal tool for cohesion policy 2014–2020. European Commission, directorate-general for regional and urban policy. Publications Office of the European Union, Luxembourg

European Commission (2014b) Communication from the commission to the European Parliament, the council, the European economic and social committee and the Committee of the Regions – towards an integrated approach to cultural heritage for Europe, COM (2014) 477 final, Brussels, 22.7.2014

European Commission (2015) Getting cultural heritage to work for Europe. Report of the Horizon 2020 expert group on cultural heritage

European Commission (2020) A new Circular Economy Action Plan. For a cleaner and more competitive Europe. COM (2020) 98 final, Brussels, 11.3.2020. Available at: https://eur-lex.europa.eu/legal-content/EN/TXT/?uri=COM%3A2020%3A98%3AFIN

European Commission (2021) New European Bauhaus – beautiful, sustainable, together. Available at: https://new-european-bauhaus.europa.eu/index_en

Foster G (2020) Circular economy strategies for adaptive reuse of cultural heritage buildings to reduce environmental impacts. Resour Conserv Recycl 152:104507

Fusco Girard L (1987) Risorse architettoniche e culturali: valutazioni e strategie di conservazione. Franco Angeli

Fusco Girard L (2021a) CLIC framework of circular human-centred adaptive reuse of cultural heritage, Deliverable D2.7 Horizon 2020 CLIC, Napoli

Fusco Girard L (2021b) The evolutionary circular and human cantered city: towards an ecological and humanistic "re-generation" of the current city governance. Hum Syst Manag 40(6):753–775

Fusco Girard L (this volume) The circular "human-centred" adaptive reuse of cultural heritage: theoretical foundations. In: Fusco Girard L, Gravagnuolo A (eds) Adaptive reuse of cultural heritage. Springer, forthcoming

Fusco Girard L Gravagnuolo A (2017) Circular economy and cultural heritage/landscape regeneration. Circular business, financing and governance models for a competitive Europe. BDC Bollettino Del Centro Calza Bini 1/2017(1):35–52

Fusco Girard L, Nijkamp P (1997) Le valutazioni per lo sviluppo sostenibile della città e del territorio. FrancoAngeli, Napoli
Fusco Girard L, Vecco M (2019) Genius loci: the evaluation of places between instrumental and intrinsic values. BDC. Bollettino Del Centro Calza Bini 19(2):473–495. https://doi.org/10.6092/2284-4732/7278
Fusco Girard L, Vecco M (2021) The "Intrinsic Value" of cultural heritage as driver for circular human-centered adaptive reuse. Sustainability 13(6):3231. http://www.mdpi.com/2071-1050/13/6/3231
Garmendia E, Gamboa G (2012) Weighting social preferences in participatory multi-criteria evaluations: a case study on sustainable natural resource management. Ecol Econ 84(C):110–120
Garzillo C, Wildman A, Balenciaga I, Izulain A (this volume) A new approach for cultural heritage adaptive reuse: circular governance. In: Fusco Girard L, Gravagnuolo A (eds) Adaptive reuse of cultural heritage. Springer, forthcoming
Gravagnuolo A (2021) Il modello di economia circolare nella città e nel territorio. Documenti Geografici 2021:451–457
Gravagnuolo A, Fusco Girard L, Ost C, Saleh R (2017) Evaluation criteria for a circular adaptive reuse of cultural heritage. BDC Bollettino Del Centro Calza Bini 17(2/2017):185–216
Gravagnuolo A, Angrisano M, Fusco Girard L (2019) Circular economy strategies in eight historic port cities: criteria and indicators towards a Circular City assessment framework. Sustainability 11
Gravagnuolo A, Fusco Girard L, Kourtit K, Nijkamp P (2021a) Adaptive re-use of urban cultural resources: contours of circular city planning. City Cult Soc 26:100416
Gravagnuolo A, Micheletti S, Bosone M (2021b) A participatory approach for "circular" adaptive reuse of cultural heritage. Building a heritage Community in Salerno, Italy. Sustainability 13(9):4812. https://doi.org/10.3390/SU13094812
Gravagnuolo A, Angrisano M, Bosone M, Buglione F, De Toro P, Fusco Girard L (2023) Participatory evaluation of cultural heritage adaptive reuse interventions in the circular economy perspective: a case study of historic buildings in Salerno (Italy). J Urban Manage, forthcoming
Gravagnuolo A, Angrisano M, Bosone M, Buglione F, De Toro P, Fusco Girard L (2024) Participatory evaluation of cultural heritage adaptive reuse interventions in the circular economy perspective: a case study of historic buildings in Salerno (Italy). J Urban Manag 13(1):107–139. https://doi.org/10.1016/j.jum.2023.12.002
Hosagrahar J, Soule J, Fusco Girard L, Potts A (2016) Cultural heritage, the UN sustainable development goals, and the new urban agenda. BDC. Bollettino Del Centro Calza Bini 16(1):37–54
ICOMOS (2011) Guidance on heritage impact assessments for cultural world heritage properties. ICOMOS, Paris
Kabir G, Hasin MAA (2012) Comparative analysis of Topsis and fuzzy Topsis for the evaluation of travel website service quality. Int J Qual Res 6(3)
Labsus (2021) Rapporto Labsus 2021. Available online: https://www.labsus.org/rapporto-labsus-2021/
Łatuszyńska A (2014) Multiple-criteria decision analysis using Topsis method for interval data in research into the level of information society development. Folia Oeconomica Stetinensia 13(2). https://doi.org/10.2478/foli-2013-0015
Licciardi G, Amirtahmasebi R (2012) The economics of uniqueness. Lnvesting in Historic City cores and cultural heritage assets for sustainable development. The World Bank, Washington, DC
Lupacchini R, Gravagnuolo A (2019) Cultural heritage adaptive reuse: learning from success and failure stories in the city of Salerno, Italy. BDC. Bollettino Del Centro Calza Bini 19(2):353–377
Marston FJ (1982) Historic preservation: curatorial management of the built world. McGraw-Hill, New York
Nijkamp P (1977) Theory and application of environmental economics. Elsevier, Amsterdam
Ostrom E (1990) Governing the commons: the evolution of institutions for collective action. Cambridge University Press, Cambridge. ISBN: 0-521-40599-8

Panaro S, Greco S, Ishizaka A (this volume) A multicriteria and multi-scalar decision support system to implement circular economy in cultural heritage adaptive reuse. In: Fusco Girard L, Gravagnuolo A (eds) Adaptive reuse of cultural heritage. Springer, forthcoming

Patrick Geddes (1918) Town planning towards city development: a report to the durbar of Indore. Holkar State Printing Press, Indore

Rojas E (2016) Urban heritage for sustainable development, in UNESCO (2016). Global Report on Culture for Sustainable Urban Development

Roszkowska E (2011) Multi-criteria decision making models by applying the topsis method to crisp and interval data. Int Sci J 6(1)

Saleh R, Ost C (this volume) Circular business model for cultural heritage adaptive reuse. An iterative journey. In: Fusco Girard L, Gravagnuolo A (eds) Adaptive reuse of cultural heritage. Springer, forthcoming

Schmertz MF (1982) New life for old buildings: an architectural record. McGraw-Hill, New York

Seghezzo L, Venencia C, Catalina Buliubasich E, Iribarnegaray MI, Volante JN (2017) Participatory, multi-criteria evaluation methods as a means to increase the legitimacy and sustainability of land use planning processes. The case of the Chaco region in Salta. Argent Environ Manag 59:307–324

Simon HA (1959) Theories of decision-making in economics. Am Econ Rev 49(3):253–283

SoPHIA consortium (2021) Final version of the SoPHIA impact assessment model. Deliverable D2.3. SoPHIA Social Platform for Holistic Heritage Impact Assessment

UNESCO (2016a) Global report on culture for sustainable urban development

UNESCO (2016b) The HUL guidebook. Managing heritage in dynamic and constantly changing urban environments A practical guide to UNESCO's Recommendation on the Historic Urban Landscape

WHO World Health Organization, Regional Office for Europe (2018) Circular economy and health. Risks and opportunities. Report. Available online: https://apps.who.int/iris/handle/10665/342218

Zanakis SH, Solomon A, Wishart N, Dublish S (1998) Multi-attribute decision making: a simulation comparison of select methods. Eur J Oper Res 107(3). https://doi.org/10.1016/S0377-2217(97)00147-1

Zeleny M (1982) Multiple criteria decision making. McGraw-Hill, New York

Zeleny M (1983) Qualitative versus quantitative modelling in decision making. HSM 4(1983):39–42

Part III
Barriers, Bottlenecks and the Need of a New Governance

Introduction

The adaptive reuse of cultural heritage faces numerous challenges that can impede the successful transformation of historic cultural sites into functional assets for modern use. This section delves into the barriers and bottlenecks that often arise in such projects, ranging from regulatory constraints and financial limitations to societal resistance and the complexities of balancing preservation with innovation. These obstacles highlight the need for a new governance model that fosters collaboration among stakeholders, streamlines processes, and integrates flexible policies to support adaptive reuse initiatives. By addressing these challenges, new circular governance frameworks can help unlock the full potential of cultural heritage within sustainable development strategies.

Chapter 14
Adaptive Reuse of Cultural Heritage: Barrier Assessment and Policy-Related Recommendations

Deniz Ikiz Kaya, Nadia Pintossi, and Caroline A. M. Koot

1 Introduction

As an economic and cultural asset, cultural heritage boosts economic growth, enhances urban liveability, and contributes to environmental adaptability in cities. The reuse of abandoned and underused historic buildings and sites is a practical substitute to demolition and reconstruction (Bullen and Love 2011a). Adaptive reuse of cultural heritage is thus an essential instrument to achieve circularity in the built environment and support transition to circular economy (Fusco Girard 2020). First, heritage reuse prolongs the cultural heritage lifespan by bypassing the wasteful processes of demolition and new construction (Foster and Kreinin 2020). Second, it stimulates the reduction of waste, environmental costs, raw material use, energy consumption while favouring the use of renewable sources of energy, the water recovery and the reuse of materials and construction elements (Foster 2020). Third, it brings forth substantial economic, social, and cultural advantages of reusing historic buildings (Conejos et al. 2016; Architects' Council of Europe 2018; Mohamed et al. 2017; Bullen and Love 2010; Bosone et al. 2019). Hence, by curbing carbon emissions, implementing climate change adaptation and mitigation, and enhancing urban liveability, the adaptive reuse of cultural heritage contributes to achieving both circular economy and sustainable development goals (SDGs) (United Nations General Assembly 2015).

The adaptive reuse of cultural heritage is a heritage conservation and urban development strategy, which has become an increasing trend within the built environment in the past two decades. Previously, reuse was considered within the scope of function and material/structural change of derelict buildings, associated with the terms of building rehabilitation and renovation (Shahi et al. 2020). With a growing

D. Ikiz Kaya (✉) · N. Pintossi · C. A. M. Koot
Department of the Built Environment, Eindhoven University of Technology, Eindhoven, The Netherlands
e-mail: d.ikiz.kaya@tue.nl; caroline.koot@agentschapnl.nl

L. Fusco Girard, A. Gravagnuolo (eds.), *Adaptive Reuse of Cultural Heritage*,
https://doi.org/10.1007/978-3-031-67628-4_14

interest towards instrumentalization of heritage with its inherent cultural and economic values, adaptive reuse has now been better acknowledged and employed as an environmentally sustainable and financially feasible mean of regeneration and heritage conservation (Fusco Girard 2019). According to the 2013 ICOMOS Burra Charter, the goal of the adaptive reuse is sustaining the values attached to buildings and sites while ensuring their future usefulness (Australia ICOMOS 2013). The 2011 UNESCO Recommendation on the Historic Urban Landscape (HUL) suggests a "conservation through transformation" approach to conserve the manifold values of cultural heritage, engaging local communities and stakeholders in conservation, transformation and adaptation choices (UNESCO 2011). The Leeuwarden Declaration also calls for a smart and quality-based heritage transformation process (Architects' Council of Europe 2018). Adaptive reuse can thus be better implemented in the perspective of circular economy as it plays a significant role in the transition towards circularity by relying upon a regenerative and reusable resource. This is fully embraced and investigated by the CLIC project framing the research presented in this chapter.

The CLIC project systemically explored how the adaptive reuse of cultural heritage has the potential to stimulate growth, social regeneration, welfare, jobs, income, and liveability of urban / territorial settings: to implement the circular economy model and sustainable development. A human-centred circular city approach has thus been developed that places cultural heritage adaptive reuse at the core of the transition to circular economy that takes into account complex values inherent in cities and their heritage as part of the circular economy agenda (Fusco Girard 2020). To better understand the challenges to adaptive reuse processes and practices in order to come up with the best solutions and recommendations, it was important to 'build knowledge', which formulated the basis for the innovative circular governance, financial and business models that were to be tested and validated in the CLIC pilot cities. For this purpose, the local, regional and national stakeholders involved in the decision-making processes for adaptive reuse policies and practices were consulted to identify the main barriers to adaptive reuse, suggested solutions and to formulate a multi-level toolkit and policy enablers to facilitate and accelerate adaptive reuse in local contexts. This knowledge base thus has a dual purpose: to inform the formulation of the CLIC circular instruments and models, and to constitute the first step to formulate policy-related enablers and strategies to tackle these barriers at local, regional, national and global levels.

While the economic, environmental, social and cultural benefits of cultural heritage adaptive reuse have been widely acknowledged in the last decade by scholars and experts (Bullen and Love 2010; Yung and Chan 2012; Kurul 2007; Gravagnuolo et al. 2017), the barriers and challenges of undertaking adaptive reuse practices have been covered in a limited number of studies. The decisions regarding whether to demolish or reuse an existing building, and the planning, design, execution and operation stages of the reuse policies and practices entail a complex set of considerations and issues to be addressed at multiple levels of decision making. Despite the worldwide acceptance of the socio-economic and environmental benefits of cultural heritage adaptive reuse, local administrators, developers and building owners may still regard the reuse of historic buildings and sites as an unviable option due to the complexity of barriers, and limitations of existing tools and strategies to cope with them (Pintossi et al. 2020).

This chapter thus investigates stakeholders' experiences and views on adaptive reuse of cultural heritage within the context of circular economy from a multidimensional and multifocal perspective. The stakeholders' reflection on barriers, solutions and policy-related enablers to support cultural heritage adaptive reuse are identified using a mixed multistep methodology. This multidimensional barrier assessment is based on a multiscalar case study analysis conducted in the CLIC pilot cases; namely, the cities of Amsterdam in the Netherlands, Rijeka in Croatia, and Salerno in Italy as well as the region of Västra Götaland in Sweden. The assessment was multiscale as it considered the site, urban, national, and international levels. As part of this multiple case study analysis, workshops and an online survey were conducted, recruiting a wide range of stakeholders from the public, private and civic sectors, including representatives of local administrators, NGOs, developers, and researchers.

This research thus enhances the understanding of barriers in adaptive reuse and provides a set of toolkit and enablers that can facilitate and accelerate adaptive reuse practices in the transition to circular economy based on the perspectives of multiple stakeholders, reflecting their plurality of views on different local contexts. The critical examination of these barriers contributed to the identification of the underlying parameters that guided the formulation of policy-related strategies and enablers that will facilitate building a roadmap for circularity. The outcomes support the numbering-up of adaptive reuse practices by providing solutions and tools to overcome the identified challenges. By raising awareness and building consensus on barriers among relevant actors involved in adaptive reuse, this research also promotes the transition towards a proactive attitude in adaptive reuse practices worldwide.

1.1 PESTEL-CA Evaluation Framework

The adaptive reuse of cultural heritage is a multidisciplinary practice with a complex set of indicators necessary to evaluate the performances of adaptive reuse practices at strategic and operational levels (Gravagnuolo et al. 2017). Thus, the evaluation framework for the barrier assessment needs to embrace a multi-criteria, multi-level and multi-stakeholder approach embodied in the barrier assessment methodology holistically in an integrated manner.

The evaluation tools for adaptive reuse have developed substantially in recent years, introducing a multi-dimensional and multi-actor perspective to the analysis (Gravagnuolo et al. 2017; Throsby 2016). These tools refer to the four pillars for the evaluation of sustainability indicators, namely environmental, economic, social, and cultural. However, these four pillars are limited to address all the wide range of barriers encountered to implement adaptive reuse policies and practices. Hence, for the designation and categorisation of barriers, solutions and policy-related enablers to adaptive reuse, the evaluation framework conveyed in this chapter was elaborated using the political; economic; social; technical and technological; environmental; legal, legislative, and regulatory; cultural; and administrative (PESTEL-CA) framework.

The PESTEL-CA framework was drawn from the PESTEL framework initially used in the business management field (Yüksel 2012). The PESTEL framework was used for the analysis of political, economic, social, technological, environmental, and legal factors in strategic management (Witcher and Chau 2010). The PESTEL categorization is usually employed to analyse operational factors of business organisations to identify the opportunities and minimise threats to maximise their business activities. To cover the wide variety of dimensions impacting cultural heritage adaptive reuse, this framework was further expanded to also consider the cultural and administrative factors that affect adaptive reuse policies and practices. This newly elaborated framework, i.e. the PESTEL-CA framework, introduces a more holistic approach for evaluation of adaptive reuse implementations. The eight dimensions of PESTEL-CA are introduced and defined in Table 14.1.

1.2 Barriers to the Adaptive Reuse of Cultural Heritage: Literature Review

In the realm of built environment practice, the factors influencing the decision making, design and implementation phases of adaptive reuse practices, and the barriers encountered in these processes can be further understood and addressed to be fully grasped. The literature on adaptive reuse encompasses a limited number of scholarly contributions that bring up the issues and barriers affecting the adaptation of reuse projects, and they usually have limited scope, scale and geographical distribution (Conejos et al. 2016; Witcher and Chau 2010; Ikiz Kaya et al. 2021a; Issa et al. 2010; Oxford College of Marketing n.d.).

The barriers to the adaptive reuse that have already been identified in the existing literature fall under one or more of the multiple PESTEL-CA domains. These barriers and the sources referring to them are presented in Table 14.2, categorized based on PESTEL-CA dimensions and the level of decision-making they entail.

This semi-systematic review on existing literature shows that most of the barriers to adaptive reuse are concentrated on the economic, environmental, and technical issues to be tackled at building and local scales (Table 14.2). Parallel to the incorporation of sustainability framework into the adaptive reuse discourse, there had been a shift towards a more integrated approach that also embodies cultural and social dimensions in terms of issues to be addressed. Almost half of the barriers are associated with decisions taken at local level (48%), followed by barriers concerned with the building scale (20%). The multilevel distribution of barriers to adaptive reuse is followed by regional, national, and urban levels, respectively. Specifically, the environmental, technical/technological, and cultural barriers are tended to be concerned

Table 14.1 Definition and explanation of PESTEL-CA framework (Ikiz Kaya et al. 2021a)

Factor/category	Keywords	Scope	Key references
Political	Government, authority, policy, democracy, transparency	It is concerned with governments, governmental policies, and regulations that institutions have to comply with. It includes political policy and stability as well as national trade, fiscal and taxation policies.	Issa et al. (2010) and Oxford College of Marketing (n.d.)
Economic	Investment, funding, grants, loans, financial incentives, partnerships, economic activities	It is based upon the theoretical framework of urban and ecological economics. It involves the following themes: market forces, land use, urban transportation, public funding policies, housing, local government expenditures and taxes	Fusco Girard and Gravagnuolo (2017) and Verhoef and Nijkamp (2004)
Social	Social equity, social inclusion, social cohesion, wellbeing, quality of life	It stands for the combination of social principles for basic societal needs, i.e. housing and health, equality and social justice with concepts associated with wellbeing, such as a sense of place, happiness and quality of life	Lami and Mecca (2020), Santi et al. (2019), Polese and Stren (2000), and Colantonio (2010)
Technical/ technological	Sustainable buildings, building components and materials, design, techniques / methods, technology, innovation, tools, accessibility	It focuses on physical characteristics and technological aspects related to the built environment. It integrates both building and urban related factors, and encompasses a wide range of disciplines from design and engineering.	Foster and Kreinin (2020) and Blagojević and Tufegdžić (2016)
Environmental	Environment, climate change, natural hazards, energy efficiency, eco-friendly, green areas	Concerned with protecting the natural environment (particular ecosystems in and around properties), environmental factors involve gradual changes due to geological, climatic or other environmental factors, threats and protection from natural hazards, pollution, efficiency and improvement of natural resources, environmental friendly interventions, etc.	Guzmán et al. (2017)
Legal/ legislative/ regulatory	Legislation, legal acts, regulations, buildings codes, health, and safety	It involves all legal-related topics and issues. Factors include zoning, land regulations, heritage legislation, building codes, local policies and strategies, health, and safety regulations	Oxford College of Marketing (n.d.)

(continued)

Table 14.1 (continued)

Factor/category	Keywords	Scope	Key references
Cultural	Cultural heritage, cultural values, intrinsic values, significance, sense of belonging, attractiveness	Set of cultural, spiritual, material, intellectual and emotional features of society or a social group, and that it encompasses, in addition to art and literature, lifestyles, ways of living together, value systems, traditions and beliefs	Torggler et al. (2015)
Administrative	Administration, governance, stakeholder engagement, citizen participation, collaboration	It is concerned with holding the balance between economic and social goals and between individual and communal goals. The aim is to align as nearly as possible the interests of individuals, of cultural heritage, and of society.	UNESCO (n.d.)

Adapted by authors from the given sources

Table 14.2 List and categories of barriers derived from the literature

Dimensions	Barriers	Levels	Key references
Political	Transparency and accountability		Yung and Chan (2012)
Economic	High costs of energy retrofitting	Building scale	Yung and Chan (2012), Shipley et al. (2006), and Ellison et al. (2007)
	High maintenance costs	Building scale	Douglas (2006), Bullen and Love (2011b), Ellison et al. (2007), Kohler and Yang 2007, and Remøy and Van Der Voordt (2014)
	Commercial risk and uncertainty	Local level	Bullen and Love (2011b), Shipley et al. (2006), Remøy and Van Der Voordt (2014), Reyers and Mansfield (2001), and Bruce et al. (2015)
Social	Human resources—lack of skilled tradesmen	Local level	Remøy and Van Der Voordt (2014) and Reyers and Mansfield (2001)
	Inability to estimate social viability	Local level	Bullen and Love (2011c)
	Community value of existing buildings	Local level	Bullen and Love (2011c)
	Lack of social services and transportation	Urban and regional level	Yung and Chan (2012)
	Public awareness of adaptive reuse	Local, regional, and national level	Bullen and Love (2011c)
	Meeting the needs of all relevant stakeholders		Lufkin et al. (2005)

(continued)

Table 14.2 (continued)

Dimensions	Barriers	Levels	Key references
Technical/ technological	Longevity of building materials (durability of external fabric and finishes etc.)	Building scale	Bullen and Love (2011b), Ball (1999), Erik Bradley and Kohler (2007), and Lützkendorf and Lorenz (2005)
	Flexibility of buildings to accommodate new use	Building scale	Bullen and Love (2011b), Ellison et al. (2007), and Bruce et al. (2015)
	Complexity and technical difficulties	Building scale	Shipley et al. (2006), Bruce et al. (2015), Ball (1999), and Kronenburg (2007)
	Limitation of knowledge and data	Building scale	Remøy and Van Der Voordt (2014) and Fox (2003)
	Health and safety requirements	Local level	Conejos et al. (2016)
Environmental	Contamination and high remediation costs	Local level	Douglas (2006), Bruce et al. (2015), Wilkinson et al. (2009), and Leadbeter 2013)
	High energy performance requirements	Local and national level	Douglas (2006), Shipley et al. (2006), Ellison et al. (2007)
Legal/legistative/ regulatory	Inertia of urban development criteria	Local level	Bullen and Love (2011b) and Bromley et al. (2005)
	Zoning	Local and urban levels	Douglas (2006) and Langston and Shen (2007)
	Compliance with local building codes	Local and regional levels	Conejos et al. (2016), Douglas (2006), Shipley et al. (2006), and Cooper (2001)
	No sustainable tourism measures	Local and regional levels	Tweed and Sutherland (2007)
	Building regulations / planning restrictions	Local and national level	Bullen and Love (2011b), Bruce et al. (2015), and Leadbeter (2013)
	Supportive governmental policies and strategies	Local, regional, and national levels	Bullen and Love (2011c), Steinberg (1996), and Zhang (2011)
Cultural	Culture perceptions	Building level	Kurul (2007)
	Balancing cultural significance and economic viability	Building level	Yung and Chan (2012) and Murtagh (2005)
	Intangible dimensions—difficulty of assessing intangible heritage values	Local level	Yung and Chan (2012) and Department of Environment and Heritage (2004)
	Sense of place and identity	Local level	Tweed and Sutherland (2007), Rodwell (2003), and Stubbs (2004)
	Significance assessment and changing perceptions of heritage	Local and urban level	Ellison et al. (2007), Ball (1999), Leadbeter (2013), and Gregory (1997)

(continued)

Table 14.2 (continued)

Dimensions	Barriers	Levels	Key references
Administrative/ Institutional	Lack of participatory processes in decision making	Local level	Harnack (2016)
	Conflict of priorities of different actors	Local level	Harnack (2016)
	Community involvement	Local level	Yung and Chan (2012) and Pendlebury et al. (2004)
	Market opportunity due to location and site	Local level	Bullen and Love (2011c) and Murtagh (2005)
	Incentive schemes	Regional and national levels	Shipley et al. (2006) and Barber (2003)

with building and local scales, whereas political, regulatory, and administrative barriers address issues of wider scale.

The review of the literature on factors influencing adaptive reuse policies and practices also reveals certain limitations and knowledge gaps in the normative literature that can be summarized as: limitations of scale and scope, limited geographical distribution, and lack of a holistic sustainability and circular economy framework. Firstly, there is clearly lack of a wider and multilevel approach to barriers of implementing adaptive reuse. The barriers that are already defined in the normative literature usually have a limited scope focusing on the individual building and site scale. This is associated with the earlier focus on costs of operation for reused buildings and their environmental performances. Even though the scope of the recent publications has expanded to a certain extend to include regulatory, political, and administrative concerns to be tackled at wider urban, regional, and national scales, the approach is still sporadic and fragmented. Secondly, while there are certain national level problems discussed specifically on regulatory and legal-basis in several academic publications, they are mostly case-specific and represent non-European contexts. For instance, there are a number of papers that examine the regulatory and legislative framework in Australia (Bullen and Love 2011b; Leadbeter 2013), in addition to papers that investigate the barriers associated with the construction sector in the United Kingdom (Kurul 2007), and in Canada (Shipley et al. 2006; Tam and Hao 2019). In addition, some papers focus only on specific building typologies in certain countries, such as adaptive reuse of religious buildings (Velthuis and Spennemann 2007) and office buildings in the Netherlands (Remøy and Van Der Voordt 2014). Consequently, there is lack of an integrated vision that investigates and identifies barriers to adaptive reuse at European scale. Finally, the limited scale, scope, and geographical representation of the scholarly contribution to the field of adaptive reuse fails to provide a holistic consideration on the economic, social, environmental and the cultural concerns which constitutes the four fundamental pillars in a solid sustainability framework. The barriers already identified in the existing knowledge mostly focus on issues that are to be addressed during the formative stages of the design process so that necessary actions can be taken towards more sustainability efforts. However, its implementation is still sporadic, not fully aligned with the circular economy framework, and unframed in the adaptive reuse and regeneration policies and practices at local level.

It is thus clear that there is a need for a more holistic framework that will integrate sustainability and circular economy agendas. Addressing only the economic and environmental concerns is not sufficient. All the other domains of sustainability elaborated with the PESTEL-CA dimensions also significantly contribute to the extent to which heritage buildings, sites and landscapes adaptively reused can be sustainable and contribute to circular economy at local, national, and European levels.

2 Materials and Methods

This research employed a mixed methodology combining qualitative and quantitative research techniques to convey the multi-vocal stakeholders' views (Fig. 14.1). A multiple-case study analysis was conducted in the four CLIC pilot cases, namely the cities of Amsterdam, Rijeka, and Salerno, and the region of Västra Götaland. The multi-step methodology included a semi-systematic literature review, thematic analysis, post coding using the PESTEL-CA analytical framework, and network analysis, depicted by complexity mapping, to identify and group the barriers to adaptive reuse. Additional steps of the methodology entailed a thematic analysis of the solutions suggested by stakeholders involved in the research. Finally, a semi-systematic literature review and an online questionnaire were used to identify and assess European, national, and local policy enablers to facilitate the adaptive reuse of cultural heritage by solving some of the barriers identified.

2.1 Data Collection

2.1.1 Semi-systematic Review

In the multi-step methodology adopted in the present research, two semi-systematic literature reviews were performed (Snyder 2019). This method allowed to determine the identification of themes and categories derived from secondary sources (Ward et al. 2009). The first review aimed to identify the barriers to the adaptive reuse of cultural heritage mentioned in the literature. The second review focused on policy-related strategies and enablers that support adaptive reuse.

The semi-systematic review on barriers was conducted using the Scopus and Web of Science databases. The keywords searched included: "adaptive reuse" and "cultural heritage", "adaptive reuse" and "heritage", in addition to search items added to "adaptive reuse", as such "landscape", "heritage sites", "urban spaces", "cultural landscapes", "historic centre", "heritage cities", "heritage areas". This

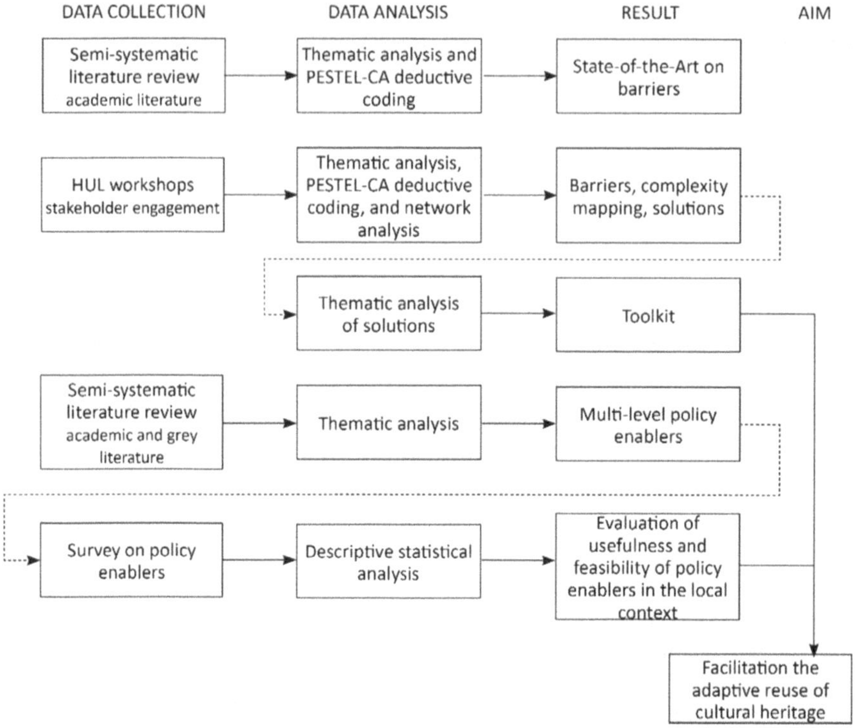

Fig. 14.1 Overview of the methodology

search identified a total number of 151 unique journal papers published until November 2018 on the topic of adaptive reuse. An overview of the abstract of this high number of published sources shows that the main focus has been on the adaptive reuse historic buildings and infrastructures, rather than landscapes and wider urban contexts. The corpus of 151 documents was further analysed focusing on "barriers", "challenges" and "problems" of adaptive reuse, resulting in a selection of 33 papers.

The second semi-systematic literature review was conducted to identify policy enablers for the adaptive reuse of cultural heritage (Ikiz Kaya et al. 2021a). For this review, both the academic literature and the grey literature, including policy documents and reports adopted by public, private and non-governmental organizations and bodies were consulted (Schöpfel 2010). The literature search used the databases Scopus and Web of Science gathering sources available until December 2020. On both databases, the search was conducted with the following search strings: "Adaptive reuse" OR Reuse Or Regeneration OR Adapt* AND polic*, "Adaptive reuse" OR Reuse Or Regeneration OR Adapt* AND Driv* OR Factor OR Enabl*". This sample of literature resulted in 69 publications. This semi-systematic literature review was complemented by an additional literature step focusing on the so grey literature to identify multi-levelled policy enablers to be assessed. The overall selection process resulted in a corpus of grey literature constitute by 10 local, 5 national, and 12 European between policy documents and reports.

2.1.2 The Historic Urban Landscape Workshops

The Historic Urban Landscape (HUL) workshop is a data collection methodology applied in the CLIC pilot cities through participatory engagement of all the relevant stakeholders concerned with adaptive reuse practices and policymaking in focus groups. We employed purposive sampling to recruit the stakeholders involved in the workshops (Tongco 2007). Together with the CLIC city/region partners, we sought stakeholders representing governmental agencies, administrators, experts, small or medium-sized enterprises and NGOs at local, regional, or national level that are directly or indirectly involved in adaptive reuse practices (Fig. 14.2). Each case study hosted one workshop. The series of workshops took place between May 2018 and September 2019. The workshops in Amsterdam, Rijeka, and Salerno aimed to identify barriers and bottlenecks at city, regional, national, and EU level. The workshop in Västra Götaland was concerned with the barriers identified in the HIP process (Ikiz Kaya et al. 2021b).

In Amsterdam, Rijeka, and Salerno, the identification of the barriers was guided by using the six steps of the Historic Urban Landscape approach (HUL steps) (WHITRAP 2016), which was introduced by the UNESCO 2011 Recommendation on the Historic Urban Landscape (UNESCO 2011). These steps were also previously used to develop local heritage management plans, e.g. in Ballarat (Fayad 2019), and was also employed in HUL workshops to

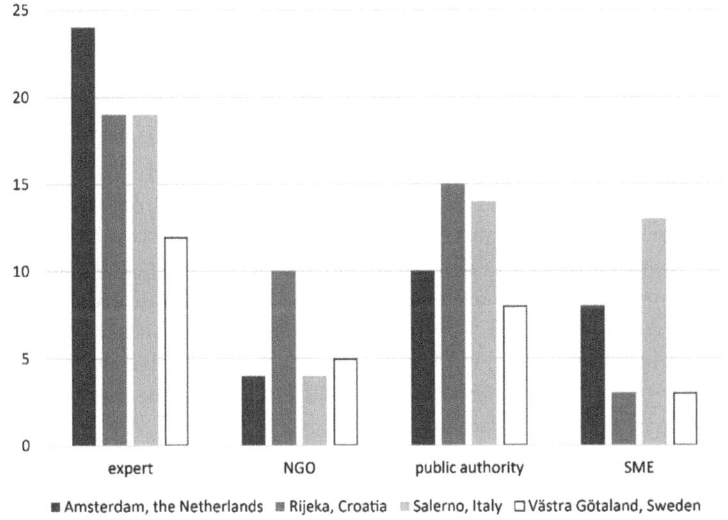

Fig. 14.2 Overview of the stakeholders participating in the HUL workshops

determine the needs to improve the integration of conservation and planning, e.g. in Zanzibar (Van Oers 2013). They were thus adopted to this study to assess the adaptive reuse processes and to identify the barriers hampering it. During the workshops, participants reflected on the process of adaptive reuse heritage and their related experiences from the perspective of the HUL steps, and the administrative scale associated with the barriers identified, namely the site scale, the urban scale, or elsewhere for those issues referring to other context or scales, such as the national level. This multiscalar identification of the barriers allows to consider the impacts of practices and measures at the various levels (Wilkinson 2018). During these three HUL workshops, participants also reported or proposed solutions to cope with the barriers.

In Västra Götaland, the barriers identified during the first meeting of the Heritage Innovation Partnership (HIP) were clustered by the authors forming a list of 15 barriers. These barriers were classified under five PESTEL-CA categories: social (elderly population, depopulation, lack of job), economic (i.e. lack of attractiveness, seasonality, restraining regulation), administrative (bureaucracy, lack of involvement), environmental (contamination, degradation and decay, lack of evidence), and cultural (access to cultural facilities, lack of understanding, different perceptions). The participants individually evaluated these barriers via a digital questionnaire hosted on the platform Lime Survey during the workshop (Lime Survey n.d.). This evaluation aimed to gather insights on the participants' opinion about the validity of these barriers after 15 months.

2.1.3 Survey

An online questionnaire was conducted to evaluate the usefulness and feasibility of the policy enablers identified with the semi-systematic literature review. For this survey, a list of 49 different stakeholder groups were identified from the four pilot cases in consultation with the local or regional administrators, and an online survey was distributed. In the survey, policy enablers were presented in sets based on their level of governance of reference, namely European, national, and local, and were asked to rank the enabler usefulness and feasibility considering their local context. Usefulness is defined as the "functionality and practicality of the policy enablers and instruments, and their quality of being useful at local scale" and feasibility is defined as the "degree of being conveniently and effectively implemented" (Ikiz Kaya et al. 2021a) (p. 3). For this evaluation, the respondents were provided a 5-point Likert scale ranging from 1 to 5, where 1 represented "least" or "strongly disagree" and 5 "most" or "strongly agree". We gathered responses from stakeholders in all four CLIC pilot cases who represented various levels of governance (Ikiz Kaya et al. 2021a). Responses were gathered from November 2019 to March 2020 and counted 23 fully completed responses.

2.2 Data Analysis

The data analysis included multiple steps employing a range of qualitative and quantitative research methods. To analyse the big set of qualitative data from the semi-systematic reviews and the HUL workshops, we conducted reflexive thematic analysis through a semantic approach where coding and theme development are drawn on the explicit content of the papers (Braun et al. 2019). We used coding and generated themes defining the barriers and the enablers, and then categorized them based on the PESTEL-CA categorization. Afterwards, to explore the interlinkages between the themes, the workshop data were further analysed by network analysis (Hevey 2018). In cases where traditional barrier analysis techniques fail to encompass the complex nature of adaptive reuse practices, this methodology portrays a more holistic image of all the relevant processes and issues. Descriptive statistical analysis was performed to analyse the quantitative data collected from the survey.

3 Results

The multi-step data analysis methodology resulted in findings on barrier assessment, solutions and toolkit recommendations and policy enablers that support adaptive reuse in the transition to circular economy.

3.1 Barrier Assessment

Barriers to adaptive reuse processes were identified based on stakeholders' experiences gathered from the HUL workshops. Based on the barriers identified in all the four CLIC cases, a comparative analysis was conducted to examine the common trends in adaptive reuse, as well as to compare the diverse tendencies, vulnerabilities, strengths, threats and opportunities as part of the multiple case study analysis. The bar chart in Fig. 14.3 shows the overview of the barrier distribution based on the PESTEL-CA categories for each case city/region. According to this analysis, the administrative barriers were predominant for the three city cases (between 30% and 41%), whereas in the region case they accounted for 15% of the overall barriers. In Västra Götaland, the most predominant barriers were economic (28%), the ratio of economic barriers to the overall list was 7–13% in the three cities. Similar differences in overall weight also prevailed for legal, legislative, and regulatory barriers. These barriers had almost one fourth weight in the city of Amsterdam, while they counted for around a tenth in other cases. Nevertheless, all the four cases presented a similar trend for the environmental barriers, which were limited to 5% of the barriers per case. Technological/technical and political barriers had a relatively low proportion. Notably, no political barriers were identified for the region case.

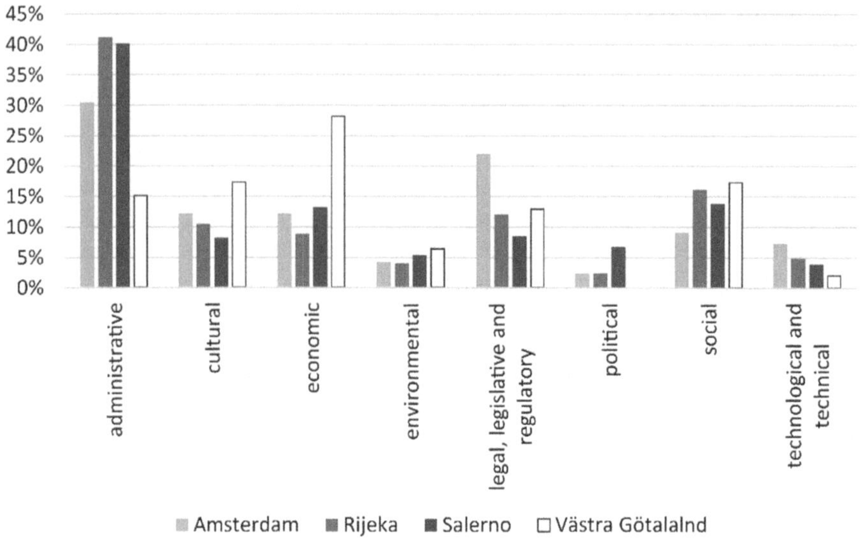

Fig. 14.3 Comparison of number of barriers per each PESTEL-CA dimension

The thematic analysis of the qualitative data gathered from the HUL workshops resulted in 111 main barriers to adaptive reuse of cultural heritage (Fig. 14.4). These barriers were clustered through network analysis and presented in the complexity mapping depicted in Fig. 14.5. This mapping was derived from the integrated assessment of the multiple case study and categorization of the barriers for each CLIC pilot based on the PESTEL-CA framework. This network analysis also demonstrates the frequency of references to these barriers with the dimensions of the nodes, i.e. the bigger the node the more frequently a barrier was identified in the multiple-case comparison.

The findings show that administrative barriers were the most frequently mentioned barrier group. The city cases, namely Amsterdam, Salerno and Rijeka, the predominant barriers to adaptive reuse were related to administrative and governance issues, based on the high frequency of barriers clustered under this category. Lack of cooperation (n = 37), collaboration and communication (n = 4), and public participation (n = 32) have been the main concerns in relation to decision making, emphasizing the necessity of open dialogue and participatory processes of governance and decision making to tackle these administrative barriers.

Regarding economic barriers, lack of funding and financial resources was a common challenge for all cases. These economic barriers included problems associated with limited funding and financial resources for the adaptive reuse of cultural heritage (n = 28). However, the context of these limitations varied based on the size, scale and governance model of the cities. In Amsterdam, the focus had been on the lack of cooperation and communication between different public and private parties, resulting in problems arising in public-private partnerships. In Salerno and Rijeka, limited public investments from the national authorities have been articulated as the biggest concern. Similarly, lack of public investments had also been an issue in Västra Götaland, paired with lack of interest from entrepreneurs and private investors. It was only in Västra

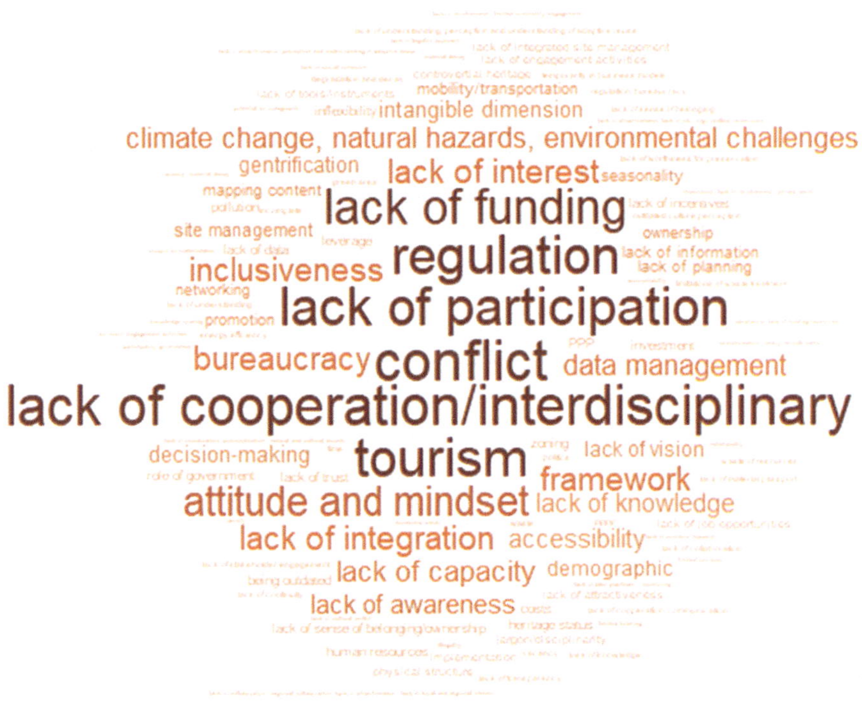

Fig. 14.4 Word cloud showing the densities for barrier assessment

Götaland where the economic barriers were mentioned more than the administrative barriers (Fig. 14.3). It is associated with the small scale of the residential areas and lack of major investments and financial support from the regional authorities, in addition to national and European funding for rural settlements.

Lack of awareness and knowledge came forth in the social and cultural domains. In all the four cases, social and cultural barriers followed similar trends, coming as third or fourth most predominant category of barriers to adaptive reuse (Fig. 14.3). One of the most articulated problem in this context was the lack of awareness on cultural heritage, its significance and potential for adaptive reuse and circular economy ($n = 14$). In addition, tourism ($n = 32$) was addressed as a major issue, as shown in Fig. 14.5, even though it followed different trajectories in each case. In Amsterdam, overtourism has been a major concern due to regulatory, economic, social, and cultural impacts on adaptive reuse. In Salerno and Västra Götaland, on the other hand, seasonality of tourism activities was presented as a barrier, resulting in seasonality in business and economic activities and temporality in the job and real estate markets ($n = 7$).

The thematic analysis revealed a limited reference to environmental threats and issues. In all the four cases, the references to environmental concerns including the impact of climate change, natural hazards, as well as pollution and environmental degradation and decay of buildings and sites were limited. Only in Salerno, the natural threats, such as risk of volcanic eruption and earthquakes, were mentioned. This limited reference to environmental threats and issues could be caused by lack

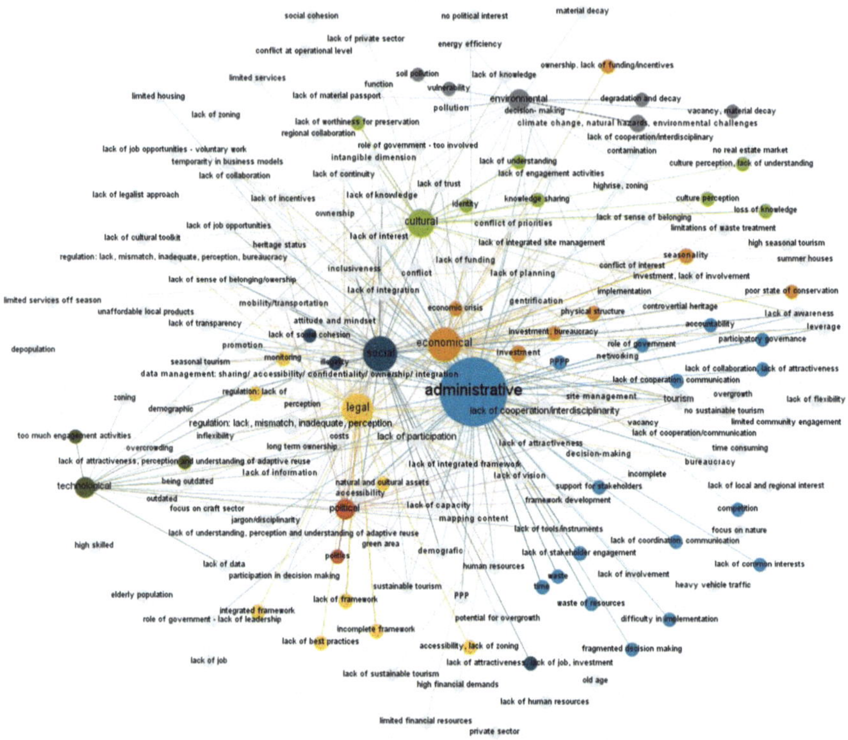

Fig. 14.5 Complexity mapping of all main barriers to adaptive reuse

of awareness on environmental concerns, and the limited knowledge on the impact of climate change on cultural heritage.

Technological and technical issues were also less referred by the stakeholders. The barriers identified in this domain mainly related to mapping and data management. Notably, in the normative literature, barriers related to the physical structure of the historic buildings and issues concerned with its adaptation played a significant role. However, in the HUL workshops, these issues were barely indicated (n = 4). This is likely due to the holistic HUL approach introduced in the data collection methodology that extended the context of cultural heritage beyond the individual building and site scale.

3.2 Solutions and Toolkit Recommendation

During the HUL workshops, the local stakeholders provided insights into possible solutions and recommendations to overcome the challenges posed by the adaptive reuse barriers. These suggestions from stakeholders offered future solutions that can be instrumentalized in a number of tools.

In addition to a set of barriers, the stakeholders also identified a list of 159 solutions to address these challenges. The solutions suggested by the stakeholders presented a wide variety of strategies, tools, models, and policy-related recommendations. Through thematic analysis, these large set of solutions were categorized into 12 themes that were further grouped under six categories of tools to facilitate adaptive reuse. These six categories and associated themes included: (i) Knowledge and planning instruments—mapping, mobility, visitor management; (ii) regulatory systems; (iii) financial tools and structures; (iv) environmental considerations; (v) governance-related—reaching consensus, decision-making, partnership, civic engagement; (vi) education—educational tools, raising awareness. These suggestions contributed to the formation of a toolkit and recommended instruments within.

The Historic Urban Landscape Recommendation (UNESCO 2011) provides toolkit that is classified under four categories of tools to facilitate management of urban heritage and adoption of the HUL approach in various local contexts: knowledge and planning tools, regulatory systems, financial tools and civic engagement tools. It was emphasised in the HUL Guidebook (WHITRAP 2016) that the HUL toolkit provides an ever-expanding set of innovative and multi-disciplinary tools, policies, and actions. The findings derived from the barriers and solution analysis reveal that the existing HUL toolkit is limited in context because its offer is limited to four categories of tools and actions to facilitate the local adaptation process. They address the administrative, regulatory, and financial aspects of the normative framework to a large extent but fail to provide effective solutions to overcome governance-related and environmental barriers, as well as socio-cultural problems. As it has been emphasised in the HUL Recommendation that the toolkit provided is continually evolving, thus a more elaborated toolkit with additional categories and tools were introduced as part of this study.

In terms of administrative and governance-related issues, the key tools provided by the HUL Toolkit have been limited to community engagement tools. In the barrier assessment, lack of collaboration, communication, and cooperation, coupled with lack of partnership between a wide range of relevant stakeholders were indicated as one of the major concerns, in addition to lack of public participation. Hence, a broader context of governance-related tools needs to be developed to tackle the broader context of administrative barriers. In parallel to the growing interest towards climate-heritage topics and climatic adaptation of cultural heritage, there is also a necessity to include environmental tools to build resilience, mitigate natural and human-induced risks, and support climatic adaptation of historic buildings, sites and landscapes. In addition, to address the social and cultural barriers associated with lack of awareness and interest towards adaptive reuse and cultural heritage, it is also important to include educational tools to raise awareness among a wide variety of interest groups ranging from private investors to local community groups and youth.

Building from these knowledge gaps and the solutions suggested by local stakeholders as part of the multiple-case study analysis, an extended multi-level toolkit with examples of associated tools to facilitate adaptive reuse policies and processes within the circular economy perspective was developed and presented as follows (Table 14.3).

Table 14.3 Multi-level toolkit for adaptive reuse and related instruments (tool suggestions derived from solutions are presented in italic)

Tool categories	Sub-categories	Examples of tools
Knowledge and planning tools	Mappings	Mappings—*perception mapping*
	Impact assessments	Heritage, social and environmental impact assessments—*impact assessment for cultural heritage adaptive reuse*
	Mobility	*Smart mobility plans* *Measurement tools for local accessibility*
	Visitor management	*Sustainable tourism management plans* *ICT-based destination plans and mapping* *Supporting tools for disabled people*
Regulatory systems	Laws, legislations, regulations	*Flexible land use regulations*
	Policies and strategies	*Governmental priorities for heritage-related strategies, bottom-up policy development*
	Plans	*Local action plans*
Governance-related tools	Participatory decision-making tools	*Policies for national clusters, Decision support system*
	Consensus and partnership	*Multi-stakeholder platforms, local strategic partnerships*
	Citizen engagement tools	Public consultations, *community workshops*
Financial tools		Global and governmental funds—*urban heritage development fund* Micro-credit and incentives Public-private-*people* partnerships *Business improvement districts* *crowdfunding*
Environmental tools	Circular built environment	*Circular environmental strategies* *Material passports*
	Environmental and climatic adaptation	Environmental impact assessment *Resilience building tools (seismic retrofitting, drainage systems etc)* *Local ambassadors for sustainability*
	Risk management	*Vulnerability assessment* *Risk mitigation plans*
Educational tools	Education	*Local history and storytelling* *Craftsmen training courses* *Educational programs at schools*
	Raising awareness	*Data sharing platforms for best practices* *Heritage awareness campaigns*

This toolkit introduces six key categories of strategies and instruments that meets the need for a more elaborated and holistic set of tools. It involves knowledge and planning tools, regulatory systems, governance and administration-related tools, financial tools, environmental strategies and actions and educational tools. The solutions suggested by the stakeholders were also categorized through thematic analysis and were associated to the relevant key categories. For instance, vulnerability assessment and adoption of risk mitigation plans for vulnerable buildings and sites were classified as risk management instruments embedded into environmental tools. With this extended toolkit, all the key pillars of the PESTEL-CA framework would be covered, and the challenges mentioned for different local contexts would be addressed to a certain extend.

3.3 Analysis on Policy Enablers

Going beyond the toolkit derived from the suggestions of stakeholders, the research was further expanded to identify policy-related strategies and enablers that support adaptive reuse with a semi-systematic review of policy enablers deduced from academic and grey literature, and their assessment on usefulness and feasibility in the local contexts of the CLIC case studies (Ikiz Kaya et al. 2021a). Based on the results of the semi-systematic review, we identified 19 drivers and associated policy-related instruments. These instruments were concentrated on administrative, regulatory and financial tools. This analysis led to the identification of 18 policy enablers that can be adopted at three levels: European, national and local. These key enablers contribute to the acceleration of adaptive reuse practices and replication of effective instruments to promote wider and systemic changes toward circular economy.

The identified European, national and local policy enablers are listed and defined with content description and key references in Table 14.4 as follows:

These multi-level enablers supporting policies were presented to local stakeholders through an online survey and were assessed in terms of their usefulness and feasibility at local contexts. The findings show that all enablers were evaluated as useful and feasible to a certain extent (Fig. 14.6). A greater mean value signified assessment of usefulness over feasibility. Almost all the enablers thus reported higher scores for their usefulness compared to their feasibility. Half of the policy enablers were regarded as very useful and the other half as moderately useful. In comparison, only one policy enabler (EU funding) was evaluated to be very feasible, whereas twelve of them were evaluated moderately feasible.

Among the European-level policy enablers, EU Funding was assessed to be both very useful (mean = 4.90) and very feasible (mean = 4.50). It was also the only enabler that was evaluated to be both very useful and very feasible with the highest scores. Another European enabler regarded to be very useful was the EU directives, despite its assessment as moderately feasible. Regarding the national policy enablers their usefulness values were mainly higher than their feasibility. The ones that were very useful were evaluated moderately feasible, and the moderately useful enablers were regarded slightly feasible. The very useful national-level enablers were bottom-up approach, national subsidies and market-based incentives, and national public funding, respectively. As for the local-level enablers, they scored high in terms of usefulness. Flexible land use regulation was evaluated to be the only slightly feasible enabler, whereas remaining were considered moderately feasible with

Table 14.4 Multi-level policy enablers and their descriptions (Ikiz Kaya et al. 2021a)

Policy enabler	Description	Key references
EU Funding and Grants	The EU provides funding to support research and innovation on heritage-related projects through programmes such as Horizon2020 and Horizon Europe. The European Regional Development Fund, European Structural and Investment Funds and the Cohesion Fund also provide support in the transition towards circular economy, and to promote economic and social cohesion across Europe.	European Regional Development Fund (n.d.)
EU Directives	Regulatory measures that support the Circular Economy agenda and the European Framework for Action on Cultural Heritage also act as facilitators for relevant actions to be adopted at local contexts.	European Commission (2015, 2019)
Support coming from Development Banks	As another European financial resource, European Investment Bank and the European Bank for Reconstruction and Development may also provide financial support.	Allegro and Lupu (2018)
EU Action Plan for the Circular Economy	Adopted in 2015, this Action Plan aims to transform the economy of the EU and its Member States in favour of the circular economy.	European Commission (2015)
Pact of Amsterdam	In this EU Urban Agenda revised 2019, the role of social dimension and employment benefits of the adaptive reuse of cultural heritage are indicated.	European Commission (n.d.)
Historic Urban Landscape approach	Adopted by UNESCO in 2011, the HUL Recommendation provides an interdisciplinary and continually evolving toolkit to support conservation through transformation approach	European Commission (n.d.)
Bottom-up approach to policy development	In order to enhance greater citizen engagement in policy making and implementation, bottom-up approaches that engage local communities and social groups horizontally in the decision making process is supported.	Zhang et al. (2019) and Othman and Elsaay (2018)
National subsidies and market-based incentives	Public financial mechanisms and supplementary tools, such as subsidies, tax relief and other market-based incentives, leverage investments in adaptive reuse.	Ellison et al. (2007), Kohler and Yang (2007), Remøy and Van Der Voordt (2014), Zhang et al. (2019), and Muminović et al. (2020)
National public funding and special budget	Public funding provided through grants, special budgetary allocations and loans also provide financial support.	Allegro and Lupu (2018) and Tanrıkul and Hoşkara (2019)

(continued)

Table 14.4 (continued)

Policy enabler	Description	Key references
Policies in favour of key national clusters	To foster cooperation and innovation between public and private bodies, agglomeration of certain economic entities can be promoted to support collaboration for effective adaptive reuse practices.	Abastante and Lami (2020) and Darchen and Tremblay (2013)
Governmental circular economy and heritage priorities for smart specialization	Lack of partnerships and prioritisation of circular economy and heritage-related strategies were addressed as challenges by local stakeholders. National smart specialization strategies aiming to support sustainability and circularity goals through adaptive reuse can be adopted through public-private and –people partnerships	Fusco Girard and Gravagnuolo (2017), Remøy and Van Der Voordt (2014), Gustafsson (2019)
Awareness raising campaign and education tools	Awareness raising campaigns focusing on action-based initiatives, coupled with educational tools are essential to raise awareness among different stakeholder groups on the impact of adaptive reuse in the transition to circularity.	Yung et al. (2017) and Prada et al. (2019)
Multi-stakeholder platforms and citizen engagement	Establishing collaboration between governments, local administrators, private and public bodies, developers, local organizations and citizens are essential in reaching consensus to push local circularity agendas forward.	Prada et al. (2019), Langston et al. (2008), and Torrieri et al. (2019)
Support for the development of sustainable tourism and mobility plans	Dedicated support from local administrative bodies for the development of sustainable tourism and mobility plans are important to enhance sustainable tourism development in the cities and regions.	Ross et al. (2016) and Vecchio and Arku (2020)
Environmental impact assessments and risk mitigation plans	With rising impact of climate change and other natural and human-induced threats, it is important to enforce the conduct of environmental impact assessment, and adoption of risk preparedness and mitigation plans at multiple levels.	Abastante and Lami (2020) and Torrieri et al. (2019)
Scaling up public procurement for adaptive reuse	As public procurement decisions are predominantly based on economic concerns, often without considering the environmental costs, it is important to build and ensure a closer relationship between public procurement and circular economy agenda.	Zhang et al. (2019) and Foster (2019)
Enhancement of policy communication and enforcement	Lack of communication between policy makers and stakeholders is an issue to be tackled with enhancement of policy communication for better implementation and promotion of reuse activities.	Abastante and Lami (2020)
Flexible land use regulations	Flexibility in land-use plans can facilitate culture- and tourism-led regeneration practices with temporary or permanent land use policies.	Langston et al. (2008), Della Spina (2020), and Morandotti et al. (2019)

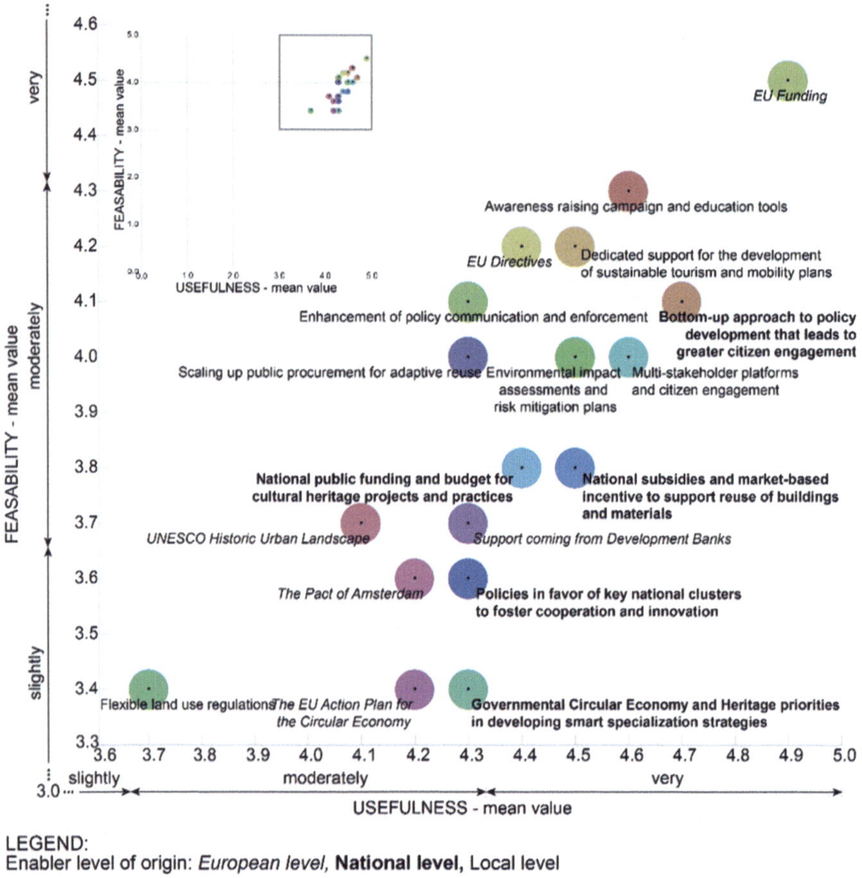

Fig. 14.6 Distribution of the policy enablers per mean value of usefulness and feasibility resulting from the descriptive statistical analysis of the evaluation results. The bold line delimits the domain of enablers that are both useful and feasible. (Adapted from Ikiz Kaya et al. 2021a)

mean values of 4.00 or higher. In sum, the local enablers scored higher in feasibility than the European and national level enablers. Ikiz Kaya et al. (2021a) provide more detailed explanations on the enabler analysis and the findings in their paper (Ikiz Kaya et al. 2021a).

4 Discussion and Conclusions

In this research, we investigated the barriers to adaptive reuse and examined the solutions, policy-related instruments and enablers that can tackle these challenges and support adaptive reuse practices towards circular economy transition. Through participatory HUL workshops, we engaged with stakeholders in the four CLIC pilots and identified the main barriers to adaptive reuse that predominantly focus on governance-related issued, followed closely by economic, social, and legislative barriers. The underlying parameters of the identified barriers also served as an initiative for formulating policy solutions and a complex set of toolkit recommendation to tackle these challenges. Followed by a semi-systematic review, we identified 18 policy enablers that can be adopted at three levels: European, national, and local. To investigate how a variety of local stakeholders from the CLIC cases evaluate the usefulness and feasibility of these enablers for their individual local contexts, an online survey was conducted. The findings showed that almost all the enablers were assessed to be useful and feasible, but their degree of adaptability changes significantly.

The stakeholder reflections and experiences deduced from the HUL workshops and survey results show that some barriers and policy enablers play a fundamental role to support cultural heritage adaptive reuse strategies and practices. To tackle the governance-related challenges emerging from lack of collaboration, communication and coordination between a wide range of relevant stakeholders, policies and regulations that support participatory and multi-level decision-making processes adopted at national and local levels are essential. In the last decade, citizen engagement, collaborative and participatory approaches for urban regeneration and adaptive reuse have gained more importance at the European level as well. Faro Convention and the Pact of Amsterdam are examples for this (European Commission n.d.; Council of Europe 2005). The survey results also demonstrated the importance of bottom-up approaches to decision-making and policy development, in terms of its usability and feasibility as policy enabler by local stakeholders. Best participatory practices, strategies and toolkits, such as the Faro Convention Action Plan, can provide guidelines for more participatory and collaborative processes (Faro Convention action plan handbook 2018).

Lack of funding and supporting financial and investment tools was also one of the main concerns raised by local stakeholders in all the pilot cities. According to the findings, EU funding was assessed as the most useful and feasible enabler of adaptive reuse at the European level. To provide financial support for the reuse of existing buildings and sites, it was agreed by the stakeholders that public financial support in the form of subsidies, tax and other market-based incentives, are necessary to facilitate the process. To support and facilitate these enablers at the national level, again the EU funds, i.e. European Structural and Investment Funds, and EU directives could be the catalysers for better feasibility and integrated actions (European Regional Development Fund n.d.; Veldpaus et al. 2019).

Another note is that the goal of embedding adaptive reuse of cultural heritage into the circular economy framework has been rather new, and it has been guided mainly by the EU policies and programs (Fusco Girard 2020). This integrated framework and its enablers are not yet fully grasped by the local stakeholders (Ikiz Kaya et al. 2021c). The findings of the survey indicate that Circular Economy Action Plan and governance facilitation are the least feasible policy-related tools. To raise awareness on this new framework and to increase the usability and feasibility of these enablers, better policy communication and educational tools are necessary. Awareness raising tools and activities, coupled with right educational tools can be instrumental for this objective.

This assessment on barriers, solutions, instruments, and policy enablers can stimulate innovative and collaborative projects on adaptive reuse of cultural heritage and can scale up actions across borders. The findings deduced from this multiple case study had been Europe-based, and more empirical data can be gathered from additional cases and adaptive reuse practices worldwide. A broadened investigation of the barriers, solutions and tools at diverse local contexts can contribute to the improvement and extension of multi-level policy enablers and their upscaling globally. The results of this study can further inform future policies at multiple levels that will leverage and scale up circular actions through adaptive reuse.

Acknowledgements This research was funded by the European Union Horizon 2020 Research & Innovation Program Project ID: 776758, CLIC—Circular models Leveraging Investments in Cultural heritage adaptive reuse. We thank all the stakeholders and respondents participating in the data collection. Thanks are due to the local CLIC partners for helping in organizing the HUL workshop; namely, Marco Acri, Suzanna Belosevic, Mirta Klaričić, Caroline Laudon, Raffaele Lupacchini and his co-workers of the Municipality of Salerno, Björn Ohlén, Carlijn Roovers, Charlot Schans, Joey van Loo, Dunja Zagorac, and the team of CNR-IRISS working on the CLIC project. We thank the facilitators of the workshops, i.e. Marco Acri, Martina Bosone, Gaia Daldanise, Gamze Dane, Cristina Garzillo, Antonia Gravagnuolo, Silvia Iodice, Lu Lu, Paloma Guzman Molina, Ana Pereira Roders, Julia Rey-Pérez, and Ruba Saleh. Thanks are also to Gamze Dane, Lu, Lu, Julia Rey Perez for their contribution in the literature search on the barriers to cultural heritage adaptive reuse. N.P thanks Ana Pereira Roders for her assistance in developing the HUL workshop methodology.

References

Abastante F, Lami IM, Mecca B (2020) How to revitalise a historic district: a stakeholders-oriented assessment framework of adaptive reuse. In: Mondini G, Oppio A, Stanghellini S, Bottero M, Abastante F (eds) Values and functions for future cities. Springer, Cham, pp 3–20. https://doi.org/10.1007/978-3-030-23786-8_1

Allegro I, Lupu A (2018) Models of public private partnership and financial tools for the cultural heritage valorisation. Urban Inf 278:1–5. https://doi.org/10.5281/zenodo.3333432

Architects' Council of Europe (2018) Leeuwarden declaration: adaptive re-use of the built heritage: preserving and enhancing the values of our built heritage for future generations. Leeuwarden

Australia ICOMOS (2013) The Burra Charter: The Australia ICOMOS charter for places of cultural significance

Ball R (1999) Developers, regeneration and sustainability issues in the reuse of vacant industrial buildings. Build Res Inf 27:140–148. https://doi.org/10.1080/096132199369480

Barber S (2003) Municipal tax incentives in Victoria, British Columbia – a case study. Plan Canada 43:20–22. https://doi.org/10.25316/IR-599

Blagojević MR, Tufegdžić A (2016) The new technology era requirements and sustainable approach to industrial heritage renewal. Energy Build 115:148–153. https://doi.org/10.1016/j.enbuild.2015.07.062

Bosone M, Micheletti S, Gravagnuolo A, Garzillo C, Wildman A (2019) Towards a ciruclar governance for the adaptive reuse of cultural heritage. BDC – Boll del Cent Calza Bini 19:279–305

Braun V, Clarke V, Hayfield N, Terry G (2019) Thematic analysis. In: Handbook of research methods in health social sciences. Springer, Singapore. https://doi.org/10.1007/978-981-10-5251-4_103
Bromley RDF, Tallon AR, Thomas CJ (2005) City centre regeneration through residential development: contributing to sustainability. Urban Stud 42:2407–2429. https://doi.org/10.1080/00420980500379537
Bruce T, Zuo J, Rameezdeen R, Pullen S (2015) Factors influencing the retrofitting of existing office buildings using Adelaide, South Australia as a case study. Struct Surv 33:150–166. https://doi.org/10.1108/SS-05-2014-0019
Bullen PA, Love PED (2010) The rhetoric of adaptive reuse or reality of demolition: views from the field. Cities 27:215–224. https://doi.org/10.1016/j.cities.2009.12.005
Bullen PA, Love PED (2011a) Adaptive reuse of heritage buildings. Struct Surv 29:411–421. https://doi.org/10.1108/02630801111182439
Bullen PA, Love PED (2011b) Factors influencing the adaptive re-use of buildings. J Eng Des Technol 9:32–46. https://doi.org/10.1108/17260531111121459
Bullen PA, Love PED (2011c) Adaptive reuse of heritage buildings: sustaining an icon or eyesore. In Proceedings of the COBRA 2011 – proceedings of RICS construction and property conference 2011, pp 1652–1662
Colantonio A (2010) Urban social sustainability themes and assessment methods. Proc Inst Civ Eng – Urban Des Plan 163:79–88. https://doi.org/10.1680/udap.2010.163.2.79
Conejos S, Langston C, Chan EHWW, Chew MYLL (2016) Governance of heritage buildings: Australian regulatory barriers to adaptive reuse. Build Res Inf 3218:507–519. https://doi.org/10.1080/09613218.2016.1156951
Cooper I (2001) Post-occupancy evaluation – where are you? Build Res Inf 29:158–163. https://doi.org/10.1080/09613210010016820
Council of Europe (2005) Council of Europe Framework Convention on the Value of Cultural Heritage for Society, 1–9
Darchen S, Tremblay D-G (2013) The local governance of culture-led regeneration projects: a comparison between Montreal and Toronto. Urban Res Pract 6:140–157. https://doi.org/10.1080/17535069.2013.808433
Della Spina L (2020) Adaptive sustainable reuse for cultural heritage: a multiple criteria decision aiding approach supporting urban development processes. Sustainability 12:1363. https://doi.org/10.3390/su12041363
Department of Environment and Heritage (2004) Adaptive reuse. Canberra, ISBN 0 642 55030 1
Douglas J (2006) Building adaptation, 2nd edn. Butterworth-Heinemann. ISBN 9780750666671
Ellison L, Sayce S, Smith J (2007) Socially responsible property investment: Quantifying the relationship between sustainability and investment property worth. J Prop Res 24:191–219. https://doi.org/10.1080/09599910701599266
Erik Bradley P, Kohler N (2007) Methodology for the survival analysis of urban building stocks. Build Res Inf 35:529–542. https://doi.org/10.1080/09613210701266939
European Commission (2019) European Framework for Action on Cultural Heritage. European Commission, Brussels
European Commission (n.d.) Pact of Amsterdam: Urban Agenda for the EU. Brussels
European Commission (2015) Circular economy action plan: for a cleaner and more competitive Europe. European Commission, Belgium
European Regional Development Fund (n.d.)
Faro Convention action plan handbook 2018–2019 Available online: https://www.coe.int/en/web/culture-and-heritage/faro-action-plan. Accessed on 5 Mar 2021
Fayad S, Buckley K (2019) The transformational power of the HUL approach: lessons from Ballarat, Australia, 2012–2017. In: Perieira Roders A, Bandarin F, (eds) Reshaping urban conservation: the historic urban landscape approach in action. Springer, Singapore, 2019; pp 123–148.

Foster G (2019) Planning the circular city: Focus on buildings' environmental impact. BDC – Boll del Cent Calza Bini 19:117–123. https://doi.org/10.6092/2284-4732/7063

Foster G (2020) Circular economy strategies for adaptive reuse of cultural heritage buildings to reduce environmental impacts. Resour Conserv Recycl 152:104507. https://doi.org/10.1016/j.resconrec.2019.104507

Foster G, Kreinin H (2020) A review of environmental impact indicators of cultural heritage buildings: a circular economy perspective. Environ Res Lett 15:043003. https://doi.org/10.1088/1748-9326/ab751e

Fox C (2003) The engineering and management of Retrofit projects in the process industries. European Construction Institute, Loughborough

Fusco Girard L (2019) Implementing the circular economy: the role of cultural heritage as the entry point.Which evaluation approaches? BDC – Bollettion del Cent Calza Bini 9:245–277. https://doi.org/10.6092/2284-4732/7269

Fusco Girard L (2020) The circular economy in transforming a died heritage site into a living ecosystem, to be managed as a complex adaptive organism. Aestimum. https://doi.org/10.13128/aestim-9428

Fusco Girard L, Gravagnuolo A (2017) Circular economy and cultural heritage/landscape regeneration. Circular business, financing and governance models for a competitive Europe. BDC – Bollettion del Cent Calza Bini 17

Gravagnuolo A, Fusco Girard L, Ost C, Saleh R (2017) Evaluation Criteria for a circular adaptive reuse of cultural heritage. BDC – Boll del Cent Calza Bini 17:185–216. https://doi.org/10.6092/2284-4732/6040

Gregory J (1997) Rehabilitation: new ways for older housing. Campbell J, ed; NSW Department of Housing, Liverpool

Gustafsson C (2019) Conservation 3.0 – cultural heritage as a driver for regional growth. SCIRES it 9:21–32. https://doi.org/10.2423/i22394303v9n1p21

Guzmán PCC, Roders ARP, Colenbrander BJFJF, Perieira Roders A, Colenbrander BJFJF (2017) Measuring links between cultural heritage management and sustainable urban development: an overview of global monitoring tools. Cities 60:192–201. https://doi.org/10.1016/j.cities.2016.09.005

Harnack M, Stollmann J (2016) Which second life? Adaptive reuse as a contested paradigm. In Proceedings of the 14th international Docomomo conference – adaptive reuse: the modern movement towards the future 2016; Tostões, A., Ferreira, Z., Eds.; Docomomo, 2016; pp 796–804

Hevey D (2018) Network analysis: a brief overview and tutorial. Heal Psychol Behav Med 6:301–328. https://doi.org/10.1080/21642850.2018.1521283

Ikiz Kaya D, Pintossi N, Dane G (2021a) An empirical analysis of driving factors and policy enablers of heritage adaptive reuse within the circular economy framework. Sustainability 13:2479. https://doi.org/10.3390/su13052479

Ikiz Kaya D, Pintossi N, Acri M, Debevec M (2021b) Deliverable 1.6: Report on project-long assessment

Ikiz Kaya D, Dane G, Pintossi N, Koot CAM (2021c) Subjective circularity performance analysis of adaptive heritage reuse practices in the Netherlands. Sustain Cities Soc 70:102869

Issa T, Chang V, Issa T (2010) Sustainable business strategies and PESTEL framework. GSTF Int J Comput 1:73–80

Kohler N, Yang W (2007) Long-term management of building stocks. Build Res Inf 35:351–362. https://doi.org/10.1080/09613210701308962

Kronenburg R (2007) Flexible: architecture that responds to change. Laurence King Publishing, London

Kurul E (2007) A qualitative approach to exploring adaptive re-use processes. Facilities 25:554–570. https://doi.org/10.1108/02632770710822634

Lami IM, Mecca B (2020) Assessing social sustainability for achieving sustainable architecture. Sustainability 13:142. https://doi.org/10.3390/su13010142

Langston C, Shen LY (2007) Application of the adaptive reuse potential model in Hong Kong: a case study of Lui Seng Chun. Int J Strateg Prop Manag 11:193–207. https://doi.org/10.1080/1648715X.2007.9637569
Langston C, Wong FKWW, Hui ECMM, Shen LYL-Y (2008) Strategic assessment of building adaptive reuse opportunities in Hong Kong. Build Environ 43:1709–1718. https://doi.org/10.1016/j.buildenv.2007.10.017
Leadbeter P (2013) Adaptive reuse of heritage buildings – do current planning and heritage controls support the concept? Environ Plan Law J 30:491–507
Lime Survey (n.d.) Available online: https://tueindhoven.limequery.com. Accessed on 18 Sept 2019
Lufkin PS, Desai A, Janke J (2005) Estimating the restoration and modernization costs of infrastructure and facilities. Public Works Manag Policy 10:40
Lützkendorf T, Lorenz D (2005) Sustainable property investment: valuing sustainable buildings through property performance assessment. Build Res Inf 33:212–234. https://doi.org/10.1080/09613210500070359
Mohamed R, Boyle R, Yang AY, Tangari J (2017) Adaptive reuse: a review and analysis of its relationship to the 3 Es of sustainability. Facilities 35:138–154. https://doi.org/10.1108/F-12-2014-01081
Morandotti M, Besana D, Cecchini C, Chiesa A (2019) A decision-making process for resilience assessment in adaptive reuse. IOP Conf Ser Earth Environ Sci 296:012031. https://doi.org/10.1088/1755-1315/296/1/012031
Muminović E, Radosavljević U, Beganović D (2020) Strategic planning and management model for the regeneration of historic urban landscapes: the case of historic center of Novi Pazar in Serbia. Sustainability 12:1323. https://doi.org/10.3390/su12041323
Murtagh WJ (2005) Keeping time: the history and theory of preservation in America. Wiley; ISBN 978-0-471-47377-0.
Othman AAE, Elsaay H (2018) Adaptive reuse: an innovative approach for generating sustainable values for historic buildings in developing countries. Organ Technol Manag Constr Int J 10:1704–1718. https://doi.org/10.2478/otmcj-2018-0002
Oxford College of Marketing (n.d.) What is a PESTEL analysis? Available online: https://blog.oxfordcollegeofmarketing.com/2016/06/30/pestel-analysis/. Accessed on 1 Mar 2021
Pendlebury J, Townshend T, Gilroy R (2004) The conservation of English cultural built heritage: a force for social inclusion? Int J Herit Stud 10:11–31. https://doi.org/10.1080/1352725032000194222
Pintossi N, Ikiz Kaya D, Pereira Roders A (2020) Cultural heritage adaptive reuse: Stakeholders identifying challenges and solutions in Salerno. In: Proceedings of the HERITAGE 2020 – 7th international conference on heritage and sustainable development. Green Lines Institute for Sustainable Development, Coimbra, Portugal
Polese M, Stren R (eds) (2000; ISBN 9781442682399) The social sustainability of cities. University of Toronto Press. https://doi.org/10.3138/9781442682399
Prada IF, Bungau C, Zsak I-G (2019) Regeneration of the industrial heritage in the central area of Oradea, Romania. IOP Conf Ser Mater Sci Eng 603. https://doi.org/10.1088/1757-899X/603/4/042005
Remøy H, Van Der Voordt T (2014) Adaptive reuse of office buildings into housing: opportunities and risks. Build Res Inf 42:381–390. https://doi.org/10.1080/09613218.2014.865922
Reyers J, Mansfield J (2001) The assessment of risk in conservation refurbishment projects. Struct Surv 19:238–244. https://doi.org/10.1108/02630800110412480
Rodwell D (2003) Sustainability and the holistic approach to the conservation of historic cities. J Archit Conserv 9:58–73. https://doi.org/10.1080/13556207.2003.10785335
Ross BE, Chen DA, Conejos S, Khademi A (2016) Enabling adaptable buildings: results of a preliminary expert survey. Procedia Eng 145:420–427. https://doi.org/10.1016/j.proeng.2016.04.009
Santi G, Leporelli E, Di Sivo M (2019) Improving sustainability in architectural research: biopsychosocial requirements in the design of urban spaces. Sustainability 11:1585. https://doi.org/10.3390/su11061585

Schöpfel J (2010) Towards a Prague definition of grey literature. In: Proceedings of the twelfth international conference on grey literature: Transparency in grey literature. Grey Tech approaches to high tech issues. Prague, 6–7 December 2010; Prague, pp 11–26
Shahi S, Esnaashary Esfahani M, Bachmann C, Haas C (2020) A definition framework for building adaptation projects. Sustain Cities Soc 63:102345. https://doi.org/10.1016/j.scs.2020.102345
Shipley R, Utz S, Parsons M (2006) Does adaptive reuse pay? A study of the business of building renovation in Ontario, Canada. Int J Herit Stud 12:505–520. https://doi.org/10.1080/13527250600940181
Snyder H (2019) Literature review as a research methodology: an overview and guidelines. J Bus Res 104:333–339. https://doi.org/10.1016/j.jbusres.2019.07.039
Steinberg F (1996) Conservation and rehabilitation of urban heritage in developing countries. Habitat Int 20:463–475. https://doi.org/10.1016/0197-3975(96)00012-4
Stubbs M (2004) Heritage-sustainability: developing a methodology for the sustainable appraisal of the historic environment. Plan Pract Res 19:285–305. https://doi.org/10.1080/0269745042000323229
Tam VWY, Hao JJL (2019) Adaptive reuse in sustainable development. Int J Constr Manag 19:509–521. https://doi.org/10.1080/15623599.2018.1459154
Tanrıkul A, Hoşkara Ş (2019) A new framework for the regeneration process of Mediterranean historic city centres. Sustainability 11:4483. https://doi.org/10.3390/su11164483
Throsby D (2016) Investment in urban heritage conservation in developing countries: concepts, methods and data. City Cult Soc 7:81–86. https://doi.org/10.1016/j.ccs.2015.11.002
Tongco DC (2007) Purposive sampling as a tool for informant selection. Ethnobot Res Appl 5:147–158
Torggler B, Murphy R, France C, Baltà Portolés J (2015) UNESCO's work on culture and sustainable development: an evaluation of a policy theme
Torrieri F, Fumo M, Sarnataro M, Ausiello G (2019) An integrated decision support system for the sustainable reuse of the Former Monastery of "Ritiro del Carmine" in Campania region. Sustainability 11:5244. https://doi.org/10.3390/su11195244
Tweed C, Sutherland M (2007) Built cultural heritage and sustainable urban development. Landsc Urban Plan 83:62–69. https://doi.org/10.1016/j.landurbplan.2007.05.008
UNESCO (2011) Recommendation on the Historic Urban Landscape
UNESCO (n.d.) International Bureau of Education Concept of Governance Available online: http://www.ibe.unesco.org/en/geqaf/technical-notes/concept-governance#:~:text=Governancehasbeendefinedto,%2Candbroad-basedparticipation. Accessed on 20 Jan 2021
United Nations General Assembly (2015) Transforming our world: The 2030 Agenda for sustainable development, p 41
Van Oers R (2013) Swahili historic urban landscapes: report on the historic urban landscape workshops and field activities on the Swahili coast in East Africa 2011–2012
Vecchio M, Arku G (2020) Promoting adaptive reuse in Ontario: a planning policy tool for making the best of manufacturing decline. Urban Plan 5:338–350. https://doi.org/10.17645/up.v5i3.3188
Veldpaus L, Fava F, Brodowicz D (2019) Mapping of current heritage re-use policies and regulations in Europe: complex policy overview of adaptive heritage re-use. OpenHeritage
Velthuis K, Spennemann DHR (2007) The future of defunct religious buildings: Dutch approaches to their adaptive re-use. Cult Trends 16:43–66. https://doi.org/10.1080/09548960601106979
Verhoef ET, Nijkamp P (2004) Spatial externalities and the urban economy. In: Capello R, Nijkamp P (eds) Urban dynamics and growth: advances in urban economics. Elsevier B.V, pp 87–120. https://doi.org/10.1016/S0573-8555(04)66004-4
Ward V, House A, Hamer S (2009) Developing a framework for transferring knowledge into action: a thematic analysis of the literature. J Health Serv Res Policy 14:156–164. https://doi.org/10.1258/jhsrp.2009.008120
WHITRAP; City of Ballarat (2016) The HUL Guidebook: managing heritage in dynamic and constantly changing urban environments. A practical guide to UNESCO's recommendation on the historic urban landscape, pp 1–59

Wilkinson SJ (2018) The context for building resilience through sustainable change of use adaptation. In: Wilkinson SJ, Remøy H (eds) Building urban resilience through change of use. Wiley, pp 1–20. https://doi.org/10.1002/9781119231455.ch1
Wilkinson S, James K, Reed R (2009) Delivering sustainability through the adaptive reuse of commercial buildings: the Melbourne CBD challenge. In: Proceedings of the PRRES 2008: proceedings of the Pacific Rim Real Estate Society 15th annual conference. Pacific Rim Real Estate Society (PPRES), Sydney, pp 1–19
Witcher BJ, Chau VS (2010) Strategic management: principles and practice; Cenage Learning; ISBN 978-1-84480-993-6
Yüksel I (2012) Developing a multi-criteria decision making model for PESTEL analysis. Int J Bus Manag 7. https://doi.org/10.5539/ijbm.v7n24p52
Yung EHK, Chan EHW (2012) Implementation challenges to the adaptive reuse of heritage buildings: towards the goals of sustainable, low carbon cities. Habitat Int 36:352–361. https://doi.org/10.1016/j.habitatint.2011.11.001
Yung EHK, Zhang Q, Chan EHW (2017) Underlying social factors for evaluating heritage conservation in urban renewal districts. Habitat Int 66:135–148. https://doi.org/10.1016/j.habitatint.2017.06.004
Zhang Y (2011) Boundaries of power: politics of urban preservation in two chicago neighborhoods. Urban Aff Rev 47:511–540. https://doi.org/10.1177/1078087411400376
Zhang Y, Kang S, Koo J-H (2019) What is the critical factor and relationship of urban regeneration in a historic district?: a case of the Nanluoguxiang Area in Beijing, China. Sustainability 11:6772. https://doi.org/10.3390/su11236772

Chapter 15
A New Approach for Cultural Heritage Adaptive Reuse: Circular Governance

Cristina Garzillo, Allison Wildman, Intza Balenciaga, and Ane Izulain

1 Introduction: New Pathways and Principles for Cultural Heritage Governance

The launch of the New European Bauhaus initiative has marked a milestone in the recognition and valorisation of cultural values and cultural heritage by European leaders as part of a sustainable thriving future. But cultural heritage assets—both tangible and intangible, protected and unprotected—are continuously under threat from a variety of economic, sociological, and environmental pressures (e.g., poor territorial planning, unconstrained tourism, disinvestment…).

Most existing cultural heritage governance models are binary with limited stakeholder diversity and require substantial economic resources—usually from a strained and shrinking tax base—to maintain. This traditional expert-based "single custodian" model of cultural heritage management and financing has long been the dominant governance model for cultural heritage assets, but it is vulnerable and faces significant challenges for its long-term sustainability and resiliency.

As the world's physical, economic, environmental and cultural contexts continue to evolve, an alternative approach to cultural heritage governance is needed to preserve and valorise cultural heritage sites in new and different ways to adapt to the new realities—and ultimately in a more inclusive and sustainable way. This alternative approach requires transparency, openness, and circular processes that engage a

C. Garzillo (✉) · A. Izulain
Built Environment, Culture and Heritage, ICLEI Local Governments for Sustainability, Freiburg, Germany
e-mail: cristina.garzillo@iclei.org

A. Wildman · I. Balenciaga
Governance Innovation, ICLEI Local Governments for Sustainability, Freiburg, Germany
e-mail: allison.wildman@iclei.org

L. Fusco Girard, A. Gravagnuolo (eds.), *Adaptive Reuse of Cultural Heritage*,
https://doi.org/10.1007/978-3-031-67628-4_15

broad range of stakeholders to foster inclusive decision-making and shared long-term responsibility for adaptively reusing cultural heritage assets—a principled process we call the *CLIC Circular Governance Approach*.

This approach builds on a foundation from the Five Principles of Good Governance (Graham et al. 2003) and UNESCO's governance of cultural heritage definition (UNESCO 2013), "ICOMOS's Quality Principles" (ICOMOS 2020), as well as the "Circular Economy principles of reuse/conservation and circularity" (Byström 2018). The governance approach is examined explicitly in the context of how cultural heritage adaptive reuse projects can be co-created, appropriately designed and developed, and sustained over time, and how they can engage and embed Heritage Communities in the process.

The following values and principles define the CLIC Circular Governance approach:

- **Participatory**: open the process to all members of society so that they can contribute a legitimate voice. Participation is not unidirectional. It should not simply be the practice of informing the public, but rather enabling the spaces (physical and virtual) and conditions for all interested community members to engage in open dialogues about community cultural heritage assets.
- **Inclusive**: the process engages a wide variety of public and private actors with diverse experiences and expertise, and not just those in the cultural heritage field. Diverse perspectives can offer new angles and potential solutions to problems hidden in groups with similar views and practices. By inviting and enabling a wide variety of participants to contribute in cultural heritage processes, the Heritage Communities concept is reinforced, which only strengthens the potential for collaborative, sustainable, community-managed cultural heritage adaptive reuse projects.
- **Transparent**: governance processes and decision-making processes should be transparent so that they are easier to understand from the outside and enable new actors to better engage and participate in the long term. Transparency is a cornerstone of good governance and co-functions with another Circular Governance principle, Accountability.
- **Accountable**: the process is accountable to the public (and future generations) and communicates clear, concise, and sufficient information about decisions, and accepts responsibility for its actions. Together with "Transparency", these principles provide a foundation for delivering high-quality authentic adaptive reuse projects, and fostering mutual trust and long-term organisational resiliency.
- **Collaborative**: the process encourages partnerships between different actors to share in the "ownership" of the processes, programs, and projects through collaborative ideation, development, execution, and management. Collaboration adds value to adaptive reuse processes by bringing together resources and talent from a variety of sources and reinforces the concept of Heritage Communities.
- **Circular (Focused and Iterative)**: focus on concrete objectives through a co-creation process that includes visioning, design development, long-term goal setting, shared commitments, and built-in feedback loops, such as 5-year plan

updates, quality control monitoring, or annual performance reporting. Communities and societies are dynamic. Needs and aspirations change, particularly as global influences, like rapidly evolving technologies and climate change, impact regions. The adaptive reuse of cultural heritage assets is one mechanism to adjust to this changing landscape, by both preserving historic cultural assets and adapting them for present needs. However, its governance processes need to balance long-term goals (e.g., physical preservation, cultural storytelling, safeguarding cultural values) with the evolving needs of a modern society in crisis.
- **Fair and Just**: strive to improve the well-being of society and provide a voice for the voiceless, particularly for intangible cultural heritage aspects and the environment. Many voices have been missing from cultural heritage discussions and decisions, which directly affect underrepresented populations. This principle intends to reset historical imbalances and provide an opportunity for underrepresented, marginalised, or voiceless entities, such as future generations, to be considered in the cultural heritage adaptive reuse process.

1.1 Exploring the CLIC Circular Governance Approach

The CLIC Circular Governance Approach is not government, but a value-based, principled approach for valorising, protecting, and sustaining cultural heritage assets as a **common good** for society. This approach specifically addresses the governance of cultural heritage adaptive reuse projects.

The CLIC project aims to operationalise cultural heritage conservation through change/adaptation—specifically through the adaptive reuse of cultural heritage assets. According to the Historic Urban Landscape approach (HUL), the intention is to preserve while managing change, seeking a balance between conservation and development (UNESCO 2011). Within this framework, the aim is to understand if a Circular Governance approach can help reframe the notion that adaptive reuse of cultural heritage is a community investment and a more broadly supported common good, and not just a cost.

Many studies have provided valuable insights and a wealth of information on local governance processes across Europe and beyond, but they have not investigated governance processes that specifically address adaptive reuse of cultural heritage. In particular, these studies did not scrutinise the relationship between adaptive reuse processes (which tend to be linear) and the process of circular governance. This was one of the central research questions for the CLIC project.

CLIC is interested in how circular business models, circular financial tools and a circular governance approach can be used to integrate cultural heritage adaptive reuse in the perspective of the circular economy model and circular city implementation. Adaptively reusing cultural heritage sites is a fundamental component of the circular economy and circular city model that the European Union is adopting to replace current linear models. Cultural heritage is the entry point for implementing the circular city.

The European Investment Bank (EIB) provides guidance on how adaptive reuse of cultural heritage assets can contribute to the circular city (Byström 2018). While cultural heritage is not explicitly addressed in the document, it can be inferred in "Step 5: Consider options for extending use and life of idle assets and products" and in "Step 6: Construct and procure circular buildings, energy and mobility systems." These steps emphasise repurposing and/or sharing idle and abandoned buildings, and ensuring that buildings are designed to be flexible, modular and as potential material banks for disassembly.

Applying a Circular Governance Approach to cultural heritage adaptive reuse projects not only reduces waste, raw material consumption and energy use, but it also reuses knowledge, preserves tangible and intangible heritage elements (like traditional construction methods, materials, and processes), engages a wider support community for long-term custodianship, and fosters new synergistic business, finance and governance partnership models. Hence, the research sought to find out if and how a Circular Governance approach to adaptive reuse of cultural heritage is being used in selected cities and regions, and which cooperation models and tools can best help communities continuously re-invent and revive the functions/use of cultural heritage sites.

1.1.1 Defining Custodianship

In an effort to better understand and analyse the diverse array of information from illustrative case study analysis of existing shared governance arrangements for cultural heritage adaptive reuse projects in 16 international cities, CLIC used a typology cluster analysis to map stakeholder roles and relationships, identify process patterns, and catalogue governance similarities between the cases.

The case studies analysis revealed a variety of ownership/management governance relationships between public, third-sector (namely civil society organisations) and private actors. The cases were clustered and organised by custodianship—that is, the ownership-management structure and relationship that defines the entities responsible for the heritage asset and its long-term physical, economic and cultural sustainability. Nearly all of our case study examples were publicly owned heritage assets, but many cases used a variety of multi-actor governance models to realise the project. As such, the conclusion was that the majority of the cases fell into one of three self-defined custodian governance models: Public Custodian, Community Custodian, or Private Custodian for the Common Good (Fig. 15.1).

Public Custodian

A Public Custodian governance model is one in which a public entity (local, regional or national) entirely owns, manages / programs, finances and governs the adaptive reuse of the heritage asset. It is important to note that although the public entity plays a central role, the public custodian model does not preclude the involvement of other stakeholders, particularly those in Heritage Communities.

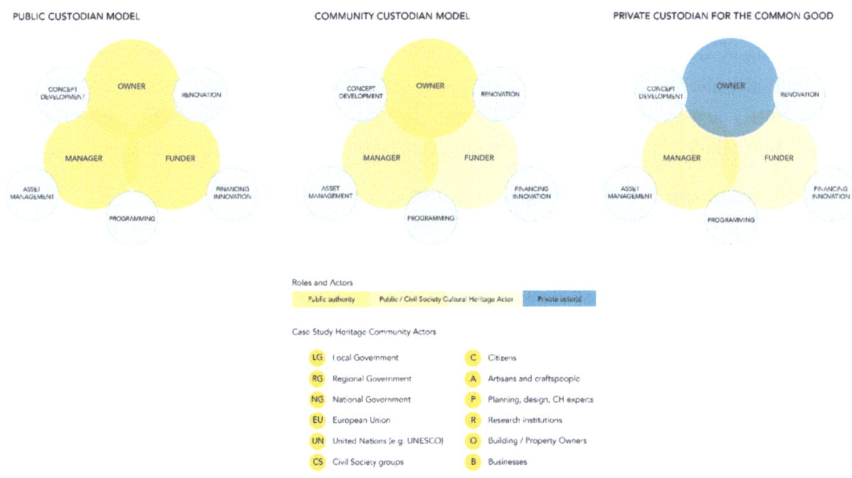

Fig. 15.1 Custodian governance models

In the case studies, the public authority often self-initiated and financed cultural heritage adaptive reuse projects as a catalyst for urban regeneration or to valorise marginalised socio-economic groups or cultures. In other cases, a Public Custodian governance model is a modernised version of traditional heritage governance. Instead of simply preserving and monumentalising the heritage asset, the public entity sought to adapt and actively use the resource for public purposes and the common good in a contemporary way.

Community Custodian

A Community Custodian governance model builds on the Public Custodian model, in as much that a public entity owns the heritage asset, but one or more Heritage Community actors are responsible for the management and long-term success of the asset. This multi-actor governance arrangement is largely defined by the owner-manager relationship and the degree of autonomy and support (financial and administrative) given to the Heritage Community actor(s) by the public entity. As such, the Community Custodian governance model is a spectrum, with many governance variations arrayed on its axis.

To illustrate, on one end of the spectrum, there are Community Custodian governance models in which the public entity plays a very prominent background role with strong financial, administrative and governance support, and the public-facing Heritage Community actor(s) have limited autonomy or decision-making power as individual organisations.

On the opposite end of the spectrum are Community Custodian governance models where the public entity is the "paper owner" of the asset and has almost no role in the governance arrangement; the Heritage Community actor(s) are entirely responsible for the asset through contractual agreements/pacts/partnerships, legal precedence, or other means.

Governance variations fall between these two rather extreme points on the Community Custodian spectrum and they can manifest in a variety of ways. However, the primary assumption of the Community Custodian model is that the public entity owns the asset and continues to play some role—no matter how small—in a shared multi-actor governance arrangement.

Private Custodian for the Common Good
A Private Custodian for the Common Good governance model is one in which a private entity collaborates with public or third-sector actors to preserve a heritage asset that has a common good. The end goal is to preserve and sustainably use the asset, not to make profit.

There were very few examples of privately-held adaptive reuse projects in our case study collection. Nevertheless, two cases uniquely illustrate where interventions targeted privately-held heritage assets through a multi-actor Heritage Community collaboration to preserve the asset for the common good. These examples show innovative multi-actor approaches to preserve both tangible and intangible cultural heritage, and it's interesting to observe how the privately-held assets are sustainably managed over time.

The character of ownership/management governance relationships, and the ways in which governments at different levels interact with civil societies and other relevant stakeholders is examined in the next section.

1.2 Connecting Beauty, Circularity and Human-Centred Approach: Heritage Innovation Partnerships and Local Action Plans for Adaptive Reuse of Cultural Heritage

The Heritage Innovation Partnerships or HIPs are experimental multi-actor partnerships convened to co-create and test adaptive reuse processes in selected regions/cities/sites across Europe. Each HIP was convened by a tandem of local partners: one academic and the other from the city-region (either the local authority in the cases of Salerno and Rijeka, the regional authority in that of Västra Götaland, and an NGO in Amsterdam). The work was conducted over a three-year period (2018–2021) and was facilitated by ICLEI, a city network that represents local governments in all relevant policy processes for sustainability in Europe.

The starting point for the experimentation built on the fundamental "Heritage Community" concept, which is deeply embedded in the Faro Convention on the Value of Cultural Heritage for the Society (2005) and namely consists of "people who value specific aspects of cultural heritage which they wish, within the framework of public action, to sustain and transmit to future generations". The HIPs are structured to recognize the inseparability of human and natural influences in the cultural environment and encourage policy-makers, experts and groups in civil society to be forward looking, inclusive, and circular in their approach to adaptive reuse of cultural heritage.

A second particular concept of specific relevance to the HIPs is the idea of cultural heritage as a 'common good', a hybrid between public and private. As common goods, heritage resources require an evolved framework of collective, multi-level, multi-stakeholder governance. The very nature of cultural heritage therefore implies the need for governance models that can adequately valorize and manage our heritage commons. This in turn calls for collaborative approaches that offer an active role for all types of users, including civil society organizations, social enterprise, civic foundations, and community hubs.

Considering these two concepts—Heritage Communities and heritage as a common good—the HIPs were structured to facilitate a bottom-up approach to collectively explore the cultural, economic, social and environmental potential of circular adaptive reuse practices for cultural heritage in European cities, particularly as they relate to efforts that improve community wellbeing and quality of life.

The questions that guide this section are as follows: (a) How did the HIP process influence the development of a Local Action Plan for adaptive reuse? (b) How did Circular Governance support sustainable adaptive reuse of cultural heritage? and; (c) How have the HIPs been an enabling factor for experimenting with CLIC models and tools?

1.2.1 HIPs and Local Action Plans for Adaptive Reuse

The HIP processes varied greatly according to the city, region and local NGO and evolved considerably both over place and time. Nevertheless, a relationship between the process and the conclusive output—the Local Action Plan for adaptive reuse (LAP)—was evident. The influence of the HIP process is notable both in individual actions and in general "way of doing things". In this sense, the HIPs can be an entrance point to prepare local heritage communities for the changes necessary to move to circular adaptive reuse processes.

Although local/regional governments still are the principal decision-makers for many cultural heritage assets, the attitude of creating alliances and unlocking possibilities for alternative cooperation models through the HIP process was of central importance to develop stronger relationships and co-create the Local Action Plan for adaptive reuse. For example, the HIP process in the city of Salerno resulted in two very concrete actions for the LAP. Firstly, a Regulation for the management of Cultural Heritage as a common good was created. The Regulation serves three different levels: small urban common goods (i.e., small squares, public gardens and spaces); medium-scale goods with high social impact and low market attractiveness; and large goods with high market attractiveness requiring advanced financial and technical skills. Complementary to this action, the municipality created a new Circular City/Urban Regeneration Office that follows the HIP process and ethos to build on various forms of expertise and perspectives and capitalize on the momentum generated during the project.

Salerno: A New Circular and Human-Centred City Model

Authors: Martina Bosone, Serena Micheletti, Antonia Gravagnuolo (CNR)

Context: The municipality of Salerno is located in southern Italy and covers an area of 59.22 km^2 with a population in 2020 of about 130,000 inhabitants. In Salerno was born the Salerno Medical School, the oldest and most illustrious medieval institution of Western Europe for the exercise and teaching of medicine. Most of Salerno cultural heritage is located in the historical centre but, unfortunately, some assets are in a state of abandonment/degradation/disuse.

The CLIC project supported the activation of participatory processes through HIPs (Heritage Innovation Partnerships): multi-actor partnerships, coordinated by local administrations and research bodies, involving and enabling local stakeholders in the experimentation of a "circular" and collaborative governance model. The HIPs improve local knowledge, ideas, skills and cooperation to provide all the tools for the co-creation of the Local Action Plan (LAP) of the city of Salerno and the implementation of strategies and projects for the adaptive re-use of cultural heritage (ARCH) towards culturally, socially and economically inclusive societies.

The participatory process revealed, also through the Historic Urban Landscape workshop, the major critical issues and opportunities in reuse processes at cultural, political, and regulatory level. Some identified weaknesses were:

- Lack of interest and participation of the local community;
- High level of degradation of the cultural heritage;
- Political uncertainty;
- Lack of communication;
- Long and complex bureaucratic process;
- Lack of interest of administrations;
- Gaps in regulations;
- Lack of funding.

Objectives: The Salerno LAP was developed by the Municipality with the support of CNR-IRISS through a participatory action carried out from October 2018, involving more than 50 local organizations (civic associations, entrepreneurs, startups, banks, foundations, public institutions, researchers and activist groups).

The LAP objectives are:

- Co-develop and plan concrete actions for the adaptive reuse of abandoned and underused cultural heritage;
- Build consensus on objectives and strategies;
- Identify priorities for action/intervention;
- Activate public-private-social synergistic relations for ARCH.

(continued)

The LAP also aims to achieve a circular city goal through circular solutions in the built environment (water recovery, renewable energy, material reuse and recycling, biomaterials, nature-based solutions, etc.), energy efficiency actions and positive energy balance of historic and modern building stock, green infrastructure (city renaturation, green mobility, etc.), reconnection of city-port and circular port area, innovative business and financing model and circular startups. Moreover, the purpose is to implement the circular and human-centred city model through the ARCH by transforming the abandoned/degraded/underused cultural heritage in a "vital place", attracting new tourists and businesses and to enhance the quality of life for residents.

Innovations: In Salerno, the elaboration of a Regulation for the shared management of cultural heritage as "common good" is being tested. The Regulation aims to establish a regulatory, transparent and impartial framework to define actions for care, shared management and re-generation of urban common goods implementable through "Collaboration Pacts" between citizens and public administration.

The Salerno Municipality has launched a public consultation to experiment with a possible governance tool aimed to collect implementable proposals for the re-use of "Edifici Mondo" in the perspective of circular economy, through the collaboration between public administration, professionals and active citizens. Fourteen proposals were submitted, ten of them were selected by Municipality for participation in the Business Model Workshop (held by ICHEC—Brussels Management School, CNR-IRISS and the City of Salerno) and for four of them a pre-feasibility study was elaborated.

Moreover, based on stakeholders' proposal, the CLIC Permanent Laboratory has been established since October 2019 to provide a permanent meeting place to discuss bottom-up proposals from different stakeholders for the ARCH and heritage-led urban regeneration, towards "Salerno circular city".

In the LAP a specific action on the Medical School of Salerno was included to enhance its value as cultural identity and "brand" of Salerno as city of health and wellbeing. This action favours both the development of the cultural and tourist offer based on the authenticity of "intrinsic values" and the promotion of heritage-led entrepreneurship for sustainable and circular development.

Furthermore, the platform CLIC Knowledge Information Hub was developed to facilitate opportunities in ARCH addressed to community members, funders, entrepreneurs and civil society organizations to co-develop new ideas and projects.

Finally, the participatory process is a cross-cutting element in LAP to build a shared vision for the definition of a strategic orientation plan for the transition towards "Salerno circular city".

Outcomes: The LAP aspires to be an "action-oriented" plan, developed assessing the main feasibility and sustainability conditions during the

(continued)

participatory planning phase to implement the proposed actions. The Plan contains synoptic sheets to monitor the contribution of planned activities to the achievement of established objectives. For each action are indicated the involved sectors of the Municipality, the link with existing plans and programs, the human and financial resources, the involved stakeholders, the timeline and the monitoring indicators.

The HIP process in Västra Götaland was more challenging due to its rural character and geographic scale. Many privately-owned individual heritage sites are separated by large geographical distances, and the HIP leaders found that it was difficult to engage with stakeholders in multiple, remote locations. This resulted in the partners feeling disconnected from the process and not "buying into" the process as strongly as in other HIPs. The regional authority recognized that new and unusual concepts need time to develop in the region. As a result, an innovative policy action was included in the LAP to disseminate knowledge about the value of adaptive reuse in the circular economy and to strengthen the link between building maintenance and circularity.

Another important action resulting from the HIP process was to create a digital platform/tool to connect heritage building property owners with potential tenants who share the same value system around cultural heritage and adaptive reuse.

Västra-Götaland: A Regional Approach to Circular Adaptive Reuse

Authors: Vera Telemo, Björn Ohlén, Jermina Stanojev, Crister Gustafsson (VGR, UU)

Context: In Västra Götaland, Sweden, the region's Cultural Development Department worked on the adaptive reuse of four industrial heritage sites located in four different rural areas: Fengersfors papermill, Gustavsfors workshop, Strömsfors workshop and Forsvik Bruk. The four sites were included due to their ability to attract or have the potential to attract new groups to the areas through innovative use of their cultural heritage, despite overall population decline, a characteristic of the region.

Objectives: The main objective of the Local Action Plan (LAP) was to identify and develop methods and governance tools for adaptive reuse of industrial environments, which could specifically address how they contribute to and strengthen local development and attract new users. The plan addresses place-transformation rather than reuse of specific buildings. The LAP was developed in close exchange with the local community and it starts from their own driving forces and initiatives. By building on existing local processes, the hope and aim is to continue the work initiated as part of the HIP process after the project is over. To ensure continuity after this point in time, the region has been using some of the same methods from CLIC in other local processes to support local development.

(continued)

Innovations: New working methods developed through the HIP process, combining knowledge in the cultural heritage sector with knowledge from the circular economy sector(s). One of the key outcomes from this process is the adaptation of the circular business model canvas that can be applied to an entire building/site, and not just a single business. It is possible to use this process to find synergies between different businesses operating at the same site, which creates a more robust and circular business model for adaptive reuse of cultural heritage. This new way of using the business model canvas will continue to be used to coach organizations interested in developing a cultural heritage site.

Outcomes: Within the CLIC project, Västra-Götaland worked at two levels: regional and local. As such, the resulting actions in the Action Plan will be implemented at two levels. The regional Action Plan focuses on governance and how regional working methods can be improved to support places where local actors want to build development based on their cultural heritage. This action plan is owned and managed by the Cultural Development Department in the region and aims to strengthen the link between cultural heritage, circular economy, and local development in the coming years. The plan contains specific goals and activities and—importantly—how these relate to other regional strategies and governing documents. The LAP has eight objectives, which vary in scope and timeframe. Developing a regional network of property owners who want to develop a cultural environment on their site is one example. Another objective is to spread the use and knowledge of circular business models to more actors to stimulate adaptive reuse of cultural heritage sites in the Västra Götaland region.

In two of the Västra Götaland pilot cases—Forsvik and Fengersfors—it was possible to connect the HIP process to existing local processes and, in these places, the work continues through those local Actions Plans. In Gustavsfors and Strömsfors, the HIP process has stopped due to lack of interest from the local community.

Reflections: There have been several challenges developing an Action Plan that affects areas (both geographical and subject-related) that the region itself does not have a mandate to pursue and that requires cross-sectoral and multi-level collaboration across different governance levels. At the same time, these challenges are some of the most important to work on and improve, while the cultural heritage sector is a cross-cutting theme that interacts with other policy areas.

Another challenge has been the rural context of the pilot cases in Västra Götaland. Stimulating development in cultural heritage buildings in urban areas is often market driven or subsidized through larger revitalization programs. Rural areas lack support systems and financing mechanisms to stimulate adaptive reuse of cultural heritage sites.

Pakhuis de Zwijger is a non-governmental cultural organization and cultural hub located in a cooling warehouse in the Eastern Docklands of Amsterdam that was—like many cultural institutions throughout Europe—significantly affected during the COVID-19 pandemic. Here the HIP process helped the organization initiate a learning process and provided a useful model to help restructure their governance model, diversify their funding sources, and develop a more resilient business model for long-term sustainability. The organisation also plans to further develop its portfolio of expertise in order to make circularity an integrated part of this recovery and resilience strategy.

Pakhuis de Zwijger: A Mid-Process Pivot to Recovery and Resilience
Authors: Deniz Ikiz Kaya, Thomas van de Sandt

Context: Pakhuis de Zwijger (PdZ) is a non-governmental organization located in the Eastern Docklands area of Amsterdam offering an independent platform for community gathering and public debates since 2006. The foundation is housed in a former cooling warehouse constructed in 1933–34, which is representative of the Dutch modernism style. The building became redundant in the 1980s, and its adaptive reuse process was initiated through grassroots initiatives. The reuse project was developed with the building owner, Stadsherstel Amsterdam—limited shareholder company for architectural restoration–, the project architectural group, the project developers, municipality and the involved cultural organisations as the future users. This public-private-people partnership continued in the governance and funding of the project and its future operation where the Stadsherstel, Monumentenfonds, and the De Zwijger Foundation (under which diverse cultural organisations and creative industries are organised as future users) collaborated. The building was inaugurated as PdZ, a cultural centre and public debate house.

A number of challenges limited the scope of influence for the adaptive reuse of the Pakhuis building, including restrictive regulatory and legislative frameworks (the current land use plan does not allow commercial exploitation of the building, and the building's monument status limits interventions to the exterior, etc.); limited influence within the existing governance structure (PdZ Foundation is not the building owner and has limited decision-making power); and financial barriers (no sustainable funding streams, impact of the COVID-19 pandemic on economic revenues, and rising land prices).

Objectives: The main aims of the HIP process were to place heritage values and cultural significance assigned to PdZ at the core of place branding, to create circular financial revenues and business models, to support sustainable building usage and operations, and to exchange knowledge with similar cultural institutions. To these aims, numerous meetings, workshops and programs were organized with a wide range of stakeholders, including the City of Amsterdam, the shareholders and supervisors of PdZ, private companies, experts, cultural institutions in Amsterdam, and building owners/users in the area.

(continued)

While the overall aims remained largely the same throughout the process, the COVID-19 pandemic put some objectives into tighter focus as the organisation was forced to innovate in real-time to remain viable. These included resilient business models, a Corporate Story and new governance arrangements, and strengthening collaboration networks with other institutions in the neighborhood.

Innovations: The following innovative instruments and strategies evolved through the HIP Dialogue process:

- Composing a PdZ Corporate Story and using it for an online petition to receive public subsidies;
- Hosting debates on how cultural institutions can recover from the impacts of the COVID-19 pandemic and future-proof their institutions with more resilient business and governance models;
- Starting a circular heritage initiative in the neighbourhood;
- Creating a program series on circular Cultural Heritage in the Netherlands.
- Outcomes: A number of cultural, administrative, financial and business-related findings and outputs were developed throughout the HIP process, including:
- Identification of challenges to adaptive reuse for the PdZ;
- Thirty-one measures to be implemented to make PdZ more sustainable, including establishing a sustainability team in the organisation;
- Local Impact Assessment of PdZ in its neighbourhood (pre- and post-COVID periods) and development of strategies to increase its local impact (i.e., neighbourhood meetings, open air cultural activities)
- a PdZ Corporate Story document and its wider dissemination
- Identification and assessment of PdZ's business model
- Restructuring the governance of PdZ in order for the PdZ foundation to be less dependent on the founders and owners, and to be eligible for more public funding;
- Sharing the story and experience of PdZ with public institutions, experts and local cultural institutions, and establishing a circular heritage initiative (along with livecasts and debates).

Reflections: One of the biggest challenges for PdZ during the three-year HIP process was the devastating financial impact of the COVID-19 pandemic. PdZ's pre-pandemic business model relied heavily on revenue generated from venue rental and its commercial spaces (restaurant/cafe). The public health restrictions taken in 2020 and 2021 (to date) have almost completely eliminated its primary economic revenues. In response, PdZ searched for alternative revenue sources and new ways to continue its cultural offer. The PdZ Corporate Story developed as part of the HIP process helped both inform its application for public pandemic relief aid from the Dutch government, and to brand and share its story/significance with a wider audience. In addition, the business model workshop focused on building resilience and creating alternative models, which helped PdZ develop new ideas and share them with similar cultural institutions in the neighbourhood.

The HIP process in Rijeka grew from the expert knowledge of the academic partner who was steering local HIP together with the local government. The conclusions deriving from the meetings were incorporated in the Local Action Plan through principles of the circular economy and focused on revitalising cultural heritage buildings within Rijeka's newly developed "Cultural Corridor". The "Cultural Corridor" is an integrated urban spatial system, designed to turn underused inner spaces into vibrant and comfortable cultural places through intense social and cultural capital preservation, creation and valorisation, in line with Historic Urban Landscape fundamental concepts. Adaptive reuse and heritage retrofitting (introducing green roofs, photovoltaic, solar panels, re-greening) is linked in Rijeka with other objectives, which include natural and cultural assets, the city's green system and its continuous interconnection to increase walkability, introducing nature-based solutions to environmental challenges, and cross-sectoral integration for a more inclusive society.

Rijeka: Developing the Cultural Corridor as a regenerative urban model

Authors: Maja Debevec, Marco Acri, Suzana Romac Belošević

Context: Rijeka is Croatia's third largest city, situated at the Rječina River's delta on the northern Adriatic coast. The city's strategic location enabled it to develop Croatia's largest seaport. Yet, like many post-industrial cities, Rijeka has suffered from decline as its core economies transition. The most prominent example of the shift is the number of vacant heritage buildings on the waterfront, but also [add more info here]. In 20XX, the European Commission selected Rijeka as the 2020 European Capital of Culture with the theme "Port of Diversity". Building on this achievement and drawing from a rich, multifaceted history, the city prepared for the year by investing in "New Cultural Infrastructure" for a regeneration process that paid particular attention to the social dimension. The Rijkea pilot, co-led by the City of Rijeka and the University of Nova Gorica, built on this framework for the participatory HIP process and implemented new forms of Historic Urban Landscape (HUL) valorisation inspired by the circular economy with stakeholders from [list a few].

Objectives: The HIP process initially focused on four primary cultural heritage assets that were dispersed through the city: the Galeb Ship, the modernist Energana power plant, the RiHub collaborative incubator of the creative city, and the Benčić Cultural Complex. However, the revitalisation objectives evolved after the early HIP Dialogues revealed the need for a holistic, integrated urban revitalisation approach. Inspired by the HUL approach, which emphasizes circularity and new economic opportunities, the objectives pivoted from the initial dispersed assets to new targets related to the Rječina River and the Delta. The four objectives were:

(continued)

1. Regenerate the Rječina River, its urban area and the sea waterfront;
2. Improve engagement with citizens and stakeholders to inspire new forms of circular businesses in heritage adaptive reuse;
3. Enhance the tangible and intangible cultural heritage protection, valorisation and preservation;
4. Improve the internal and external accessibility of the city to reduce pollution.

Innovations: The HIP process in Rijeka generated three innovations that help integrate the circular economy and HUL regeneration: the Cultural Corridor, Urban Seeding and (Third option)

The Cultural Corridor is a model for integrated urban planning co-created by the City of Rijeka and the University of Nova Gorica. It is designed as an incubator for heritage valorisation activities that are in line with the circular economy principles to promote cultural and natural heritage reuse, innovative circular businesses and uses, and transform deprived urban spaces into places.

It can be seen as applying the common-pool resources management concept by the Ostrom Theory, thus as a commonly managed resource characterised by sustainability and cultural creativity. In Rijeka, this has been defined by almost precise boundaries, mapping of cultural resources and underused assets, by sustainable objectives and cooperative governance principles.

Social capital and innovation can be used as a motor for transforming and co-creating the area through the participatory process, cultural and social activities with the overall goal of improving living conditions and well-being of the citizens through the enhancement of their landscape. In this light, HIP4 and HIP5 were organised to capitalise on the citizens' vision. The Urban Seeding (HIP4), intensive workshop of three months, was meant to produce inspirational, low-cost circular interventions for urban revitalisation, named "urban seeds", and the Circular Business Workshop (HIP5), to support the creation of a sustainable circular business model for the Children's House in Benčić Complex.

Outcomes: The HIP process in Rijeka generated three important results for possible urban strategies to integrate the circular economy and HUL regeneration: the Cultural Corridor, Urban Seeding and [third options].

The Cultural Corridor must be seen as a model for integrated urban planning, designed as an incubator of heritage valorisation activities in line with the circular economy principles. The cultural corridor promotes cultural and natural heritage reuse, innovative circular businesses and uses, transforming a deprived urban space into a place. It can be seen as applying the common-pool resources management concept by the Ostrom Theory, thus as a commonly managed resource characterised by sustainability and cultural creativity. In Rijeka, this has been defined by almost precise boundaries, mapping of cultural resources and underused assets, by sustainable objectives and cooperative governance principles.

(continued)

Social capital and innovation can be used as a motor for transforming and co-creating the area through the participatory process, cultural and social activities with the overall goal of improving living conditions and well-being of the citizens through the enhancement of their landscape.

The Cultural Corridor in the long-term can be achieved through regular urban planning and the use of various and flexible tools used interrelatedly and guided in an expert-led interdisciplinary process. Therefore, the multifaceted approach addresses the historic urban landscape through innovative initiatives and diverse approaches (workshops, education, labs), methods and tools (participatory activities and bottom-up initiatives, urban seeding, educational booklets).

The Urban Seeding was designed as a tool for regeneration adopting spatial based experimentation as a leading principle. It is designed as a collaborative and participatory, versatile and dynamic process acting as a catalyst for revitalisation by the systematic step-by-step implementation of the urban low-cost and small-scale interventions meant to enable the city's attention and capability users. The urban seeds must inspire citizens about the sustainable cultural potential of the city, as well as enable them to be aware of such potential.

In Rijeka, the Urban Seeding was tested as the workshop, thanks to a multidisciplinary mentored study-work for students and young professionals of different disciplines. Choosing well-educated youngsters was strategic for their learning capability, employment inspiration potential and commitment to the city as their future living environment. The proposals of the different groups, integrated on a multi-scale level (urban-scale, building scale and object scale) and across the groups-work, matched the unused cultural heritage in existing and new programs/activities and upgraded the urban spatial and green system in both city axes of the Cultural Corridor, including the river and sea waterfronts. The integrated proposals were self-evaluated and prioritised based on the circularity principles, the cultural valorisation impacts, the feasibility in terms of time and costs. As a follow-up, the Decision Support System designed with the University of Portsmouth was used to test the proposed projects/actions while considering criteria, objectives, constraints, and potential synergies.

Overall, the long-term vision for cultural heritage policies in the four HIPs was clear, easy to understand and even attractive. However, the pathway consisted of various small steps that could appear far away from the big picture and were not seen in a rigid chronological order, but rather in circular terms. Despite in all the pilots the level and type of engagement from the HIP participants was different, they have mostly internalized the vision and stand behind the long-term objectives while at the same time being responsible for implementing it step by step in a circular fashion.

1.2.2 Circular Governance as a Support Mechanism in HIPs

At the time of writing, in March 2021, it is clear that cultural heritage needs to be seen in a circular governance perspective, although just a few years ago there was little awareness that adaptive reuse could be a starting point for a deep understanding of circularity. At the start of the project, the HIP pilots still had a traditional expert-based, conservation perspective on cultural heritage. The management of adaptive reuse of cultural heritage was mainly thought as top-down, public and expert-based. While this by no means suggested that adaptive reuse should be restricted to its cultural dimension, all of the pilots strived for an "horizontal" approach, meaning that HIP participants looked at issues from different perspectives through ongoing internal training on cultural heritage and the circular economy, and cross-sectoral commitment to cultural heritage and circular economy (e.g., involving the business and environmental departments/stakeholders/local organisations in the process).

All of the pilots managed to make this connection to some extent (Circular Office in Salerno; Regional cultural-environment group as a forum for reconciliation between Västra Götaland and county administrative board; Resilient business board in Pakhuis de Zwijger; Cultural corridor in Rijeka).

In general, the HIPs in each pilot established a core group of participants and structure that facilitated constant input and support throughout the process, yet was flexible enough to enable specific stakeholders to participate on an ad-hoc basis. Stakeholders included those directly affected by adaptive reuse of cultural heritage, those attempting to help and also those who could help but were unaware of their role, including "difficult" or sceptical stakeholders. It has to be noted that despite the fundamental assumption that "Circular Governance is a necessary precondition for sustainable adaptive reuse of cultural heritage", at the start of the project the language and terminology regarding "circularity", "circular city", "circular governance", as well as "micro-community" was problematic in the pilots, particularly in cases where these terms are infrequently used or less known. This unfamiliarity with the vocabulary surrounding the main theme resulted in an initial lack of confidence of some stakeholders.

One of the challenges identified at the start of the HIP process was the "difficulty in keeping the stakeholders connected throughout the whole project". Clearly, due to the COVID-19 pandemic, patience and sensitivity were even more necessary to ensure that the planned steps continued to the greatest extent feasible and more effort was needed to translate complex concepts and sophisticated CLIC tools into a set of practical and action-oriented steps.

Discerning the role and influence of the partnership and its interaction with other partnerships, as well as challenges in promoting actions beyond conventional meetings, was difficult during this period. Nevertheless, some interested stakeholders took this opportunity to participate in additional informal meetings between the formal HIP "Dialogues" and official project meetings, which formed the groundwork for longer-term relationships (e.g., in Salerno).

The HIP "Dialogues" were complemented by four Peer Review Visits involving representatives from the four CLIC pilot territories (Salerno, Rijeka, Västra Götaland, Pakhuis de Zwijger in Amsterdam). These Peer Review Visits were facilitated by ICLEI with the aim to exchange local experiences, gather inspiration from others, and engage in a review of what worked, what did not work, and what could be done differently. Through the Peer Reviews, the pilots received feedback from their peers that focused on local experiences through the exchange of cases that worked and did not work, different mechanisms to overcome existing problems, building an understanding of the issues, situations and challenges, and exploring new ideas, options and solutions.

It is important to recognize the role of public support which continues to be essential for the safeguard and promotion of cultural heritage. The public government is as well the driver that brings cultural heritage in the circular perspective in the agenda, and facilitates the dialogue because it has the legitimate authority to incorporate the conclusions from the HIP processes within new policies.

HIPs as an Experimental Arena for New Circular Adaptive Reuse Tools

Six tools of the CLIC integrated multi-level toolkit developed as part of the project to facilitate circular adaptive reuse policies and processes were implemented within the framework of the HIP process. While tools are definitely needed, only the HIP stakeholders were in the position to understand and test the feasibility and "real-life" applicability/usefulness of such tools. The HIP process provided an experimental arena where the tools could be observed and assessed first, instead of implementing them all at once. The results achieved with the knowledge and planning tools, can then be mainstreamed into new mechanisms and structures. Specifically:

- The *Perceptions mapping* tool mapped feelings, perceptions and opinions about cultural heritage and its spatial interaction within the lived environment in Amsterdam, Rijeka, Salerno and the four locations in Västra Götaland Region (Fengersfors, Forsvik, Gustavsfors and Stromsfors).
- The *Economic Landscapes* identified and mapped cultural capital and spatial integration of cultural capital with urban economic functions in Amsterdam, Rijeka, Salerno and the four locations in Västra Götaland Region (Fengersfors, Forsvik, Gustavsfors and Stromsfors).
- The *circular business model workshops* took place in Amsterdam, Rijeka, Salerno and Fengersfors in Västra Götaland Region. The aim was to co-design ideas/solutions for adaptive reuse of cultural heritage that could make a sound and sustainable business model: starting from a unique value proposition and considering the available resources and identified needs; testing desirability; identifying partnerships, users, and beneficiaries; and ensuring sustainable impacts.

- The *Social Network Analysis* mapped interactions among people and organization in the adaptive reuse of cultural heritage to identify the existing relations and evaluate the networking initiatives in Salerno.
- The implementation of the *Decision Support System* is still ongoing in Salerno and Rijeka to aid the development of sustainable and circular strategies for the reuse of cultural heritage: identifying a set of actions and taking into consideration constraints and potential synergies among these actions.
- The *Cultural Corridor* was implemented in Rijeka as an integrated urban spatial system generating a new paradigm for sustainable use of inner cities and urban deprived areas, adopting the culture, cultural heritage and environment as a cross-leading principle for the adaptive reuse of urban heritage.

1.3 Enhancing Opportunities for Collaboration and Funding Through Data Management

Gabriella Monteleone, FacilityLive

Another very important tool for policy-makers is the *CLIC Knowledge Information Hub*, an innovative data management system enhancing opportunities for collaboration between diverse stakeholders and helping increase funding for cultural heritage. Indeed, the understanding and reuse of cultural heritage involves the collection, storage and processing of all forms of relevant information referring to it. The heterogeneous sets of data help understanding the heritage building or site and their management is an essential task for the dissemination and reuse of the assets. This is the reason behind the creation of the CLIC Knowledge Information Hub.

The platform has been designed to manage and visualise in an innovative way the information related to heritage assets to inform investors, policy-makers, entrepreneurs and civil society organisations on the opportunities and costs related to available heritage assets to start heritage regeneration actions.

It allows a city/region to showcase local reuse practices and present the abandoned heritage as assets/opportunities for adaptive reuse investments towards the implementation of a "circular city" model. Detailed data and information for the reuse of a building/site are provided as well as the communication of eventual calls for tender. The application, based on the requirements from the four CLIC pilot cities and region represents different geographic, historic, cultural and political contexts and could be transferable to other cities/regions.

The CLIC Knowledge Information Hub not only empowers city and regions providing an open platform for cataloguing, presenting and sharing reuse opportunities but it is also of interest to anyone who wishes to look into the adaptive reuse practices of European cities and regions and learn from their experiences as well as citizens interested in exploring the projects of their city, and investors searching for opportunities on a specific territory.

The pilot application is available on the CLIC Knowledge and Information Hub at www.clicplatform.eu, and its demo is available for all stakeholders across Europe.

Through the platform, the user can consult information on all opportunities across Europe (see here) and search for specific opportunities (available spaces, to build and reuse, to design and reuse) according to their characteristics (e.g. by typology of the heritage building/site/landscape -Civil, Residential, Religious, Military, Industrial, Commercial, Leisure, Agricultural-), or state of conservation and intervention, etc. The user can consult information on all opportunities in a specific city/region (see here).

Specifically, the platform provides:

- the specific data related to the involved heritage (see example here): its typology, protection, the typology of construction, the indoor and outdoor square metres, number of floors, the owner, the urban planning zone, the energy performance as well as specific documents as planimetries. In addition, images, contacts, news, links to websites as well as historic sites' opening times can be found.
- all the details related to the adaptive reuse process: the reuse status of the heritage building/site (to be renovated or already renovated) and the current reuse phase, its state of use, the allowed uses, the last intervention year and the intervention needed, the estimated or actual investment range and costs for the reuse project, its intervention timeframe timeline and managing aspects as the governance model, management type and manager.

Practically the CLIC Knowledge Information Hub has been developed by means of FacilityLive, an innovative technology for the organization and management of information that allows people to have a compelling user experience in information search. The end result is a new way of accessing and using cultural heritage information.

References

Byström J (2018) The 15 circular steps for cities. European Investment Bank

Graham J, Amos B, Plumptre T (2003) Principles for good governance in the 21st century, Policy brief no. 15. Institute on Governance, Ottawa

ICOMOS (2020) European quality principles for EU-funded interventions with potential impact upon cultural heritage – revised edition November 2020. Available at: https://www.icomos.org/en/about-icomos/committees/regional-activities-europe/58799-european-quality-principles-for-eu-funded-interventions-with-potential-impact-upon-cultural-heritage

UNESCO (2011) The UNESCO recommendation on the historic urban landscape. Available at: https://whc.unesco.org/uploads/activities/documents/activity-638-98.pdf

UNESCO (2013) Managing cultural world heritage: world heritage resource manual. Available at: https://whc.unesco.org/document/125839

Part IV
Circular Business Models and Financial Instruments

Introduction

The adaptive reuse of cultural heritage offers a unique opportunity to align preservation efforts with sustainable economic development through the implementation of circular business models. This section explores how such models can be applied to heritage projects. It also examines the role of innovative financial instruments in supporting these initiatives, from public-private partnerships to impact investing and green finance. Together, circular business models and tailored financial strategies are key to unlocking the economic potential of adaptive reuse while ensuring cultural and environmental sustainability.

Chapter 16
Circular Business Model for Cultural Heritage Adaptive Reuse: An Iterative Journey

Ruba Saleh and Christian Ost

1 Introduction: Urban Conservation as a Lever of Sustainable Development

After World War I, an architectural and urban movement for protecting heritage emerged. As a result, the Charter of Athens introduced in 1930 the need for legislation at national level for the preservation of historic sites. The main focus of rehabilitation and restoration works was initially addressed outstanding historical fabrics: monuments, sites or historic centers. The development of national legislative frameworks and planning tools was triggered by the massive destruction of European cities caused by World War II. The main objective was to safeguard and reconstruct the urban heritage. Urban conservation started evolving as a doctrine in the second half of the twentieth century. Indeed, the Charter of Venice in 1964, builds on the charter of Athens and expands the scope "The concept of a historic monument embraces not only the single architectural work but also the urban or rural setting in which is found the evidence of a particular civilization, a significant development or a historic event" (The charter of Venice 1964). The discussion on values was focused on the artistic and historic values and the main intention was to safeguard monuments or works of arts. In 1972, Meadows et al., published the famous report 'The Limits to Growth' which delivered at the time a clear message "Man can create a society in which he can live indefinitely on earth if he imposes limits on himself and his production of material goods to achieve a state of global equilibrium with population and production in carefully selected balance" (Meadows

R. Saleh (✉)
Economics, ICHEC Brussel Management School, Brussels, Belgium
e-mail: ruba.saleh@ichec.be

C. Ost
Raymond Lemaire International Centre for Conservation [RLICC], KU Leuven University, Leuven, Belgium

L. Fusco Girard, A. Gravagnuolo (eds.), *Adaptive Reuse of Cultural Heritage*,
https://doi.org/10.1007/978-3-031-67628-4_16

et al. 1972). As an indication of the fact that preservation of scarce resources was an overarching concern at the time, the World Heritage Convention was adopted a few weeks later, "The 1972 landmark for cultural conservation coincided with the awareness of excessive growth, which is itself nurtured by an environmental concern towards the protection of natural resources around the globe" (Ost 2021). Indeed, the World Heritage Convention was the first international convention to establish a neat link between nature conservation and the preservation of cultural properties. Moreover, it also recognized people's interactions with nature and the need to preserve the balance between the two.

The growing multidisciplinary literature and understanding of sustainability in urban context, made it clear that the city has to be analyzed as a dynamic ecosystem which needs to be conserved in its historical integrity. The Declaration of Amsterdam of 1975 was a turning point were the definition of a 'historic centre' as the main focus of the conservation process was extended to the entire built environment together with social and cultural context "While the emphasis was thus moving away from the conventional value judgements in vogue previously, such as identification of picturesque views and vistas, protection of particularly important architectural elements, and street elevations, a new interest was given to the protection and rehabilitation of the urban fabric as an overall support to the historical significance and the social structure of the city" (Zancheti and Jokilehto 1997:38). The urban mandate of conservation was reaffirmed in 1976 in the UNESCO Recommendation concerning the Safeguarding and Contemporary Role of Historic Areas and the Washington Charter for the conservation of historic towns and urban areas adopted by ICOMOS General Assembly in 1987 which clearly states: "This charter concerns historic urban areas, large and small, including cities, towns and historic centres or quarters, together with their natural and man-made environments. Beyond their role as historical documents, these areas embody the values of traditional urban cultures". The Nara document on Authenticity in 1994 was the fruit of intercultural dialogue that "underscores the importance of considering the cultural and social values of all societies, and emphasizes respect for other cultures, other values, and the tangible and intangible expressions that form part of the heritage of every culture" (Luxen 2004). An important turning point for urban conservation where the focus shifted from being focused on the process itself towards questioning why to conserve (what are the values being safeguarded) and for whom? (Luxen 2004; Ost 2021). And from artistic and historic values to multilayered values related to different perspectives (Van Balen 2008). This opening toward the diversity of cultures, social-life system and heritage led to the adoption in 2003 of the UNESCO Convention for the Safeguarding of the Intangible Cultural Heritage and the Faro Convention in 2005 by the Council of Europe. As described above, the drawing up of new charters and conventions demonstrate how the concept of heritage and its management evolved in time and space and the urban conservation planning process can be defined as "an activity that aims at the preservation of the creative change of values within the context of the continuity of the urban structure. Obviously, this is a circular or dialectical process, to put it in old-fashioned terms" (Zancheti and Jokilehto 1997:47).

In 2011, UNESCO adopted the Recommendation on the Historic Urban Landscape. By mapping the multilayers and interconnections between the cultural natural and human values, this new approach posits urban heritage as a resource for the entire city and its sustainable development. "The Historic Urban Landscape gives urban conservation a new perspective, an invitation to "'reconnect' the so-called 'historic' and 'modern' city "in order to enable a full-understanding of the significance of urban heritage, its changing dimensions and its power to inspire and guide contemporary urban rehabilitation processes and urban development" (Bandarin 2015:16).

The Hangzhou Declaration in 2013, placed cultural heritage (tangible and intangible) at the heart of sustainable development policies "The safeguarding of historic urban and rural areas and of their associated traditional knowledge and practices reduces the environmental footprints of societies, promoting more ecologically sustainable patterns of production and consumption and sustainable urban and architectural design solutions. Access to essential environmental goods and services for the livelihood of communities should be secured through the stronger protection and more sustainable use of biological and cultural diversity, as well as by the safeguarding of relevant traditional knowledge and skills, paying particular attention to those of indigenous peoples, in synergy with other forms of scientific knowledge" (The Hangzhou Declaration 2013). The 2030 Agenda for Sustainable Development adopted by the General Assembly in 2015, highlights under Goal n.11, Make cities and human settlements inclusive, safe, resilient and sustainable, the role of heritage specifically under Target 11.4, Strengthen efforts to protect and safeguard the world's cultural and natural heritage, and recognizes cultural heritage as a lever for development (United Nations 2015a). Based on this, the Habitat III Issue Papers highlights the need for new planning and governance models as "culturally sensitive urban development models" (United Nations 2015b). As tools aimed to "mitigate urban conflicts and reduce cities' ecological footprint, with a view to build more compact, inclusive and resilient cities" (ibidem). The UNESCO report, Culture Urban Future, strengthens the narrative, by presenting in 2016 a global portrait of the role of culture as a basis for a sustainable urban development through global seminal case studies. In the same year, the New Urban Agenda was adopted at the Habitat III conference in Quito. This agenda builds on Goal n.11 and declares that "We will support the leveraging of cultural heritage for sustainable urban development and recognize its role in stimulating participation and responsibility. We will promote innovative and sustainable use of architectural monuments and sites, with the intention of value creation, through respectful restoration and adaptation. We will engage indigenous peoples and local communities in the promotion and dissemination of knowledge of tangible and intangible cultural heritage and protection of traditional expressions and languages, including through the use of new technologies and techniques" (New Urban Agenda 2016:32). The role of cultural heritage in shaping a sustainable living environment was put into context by the Davos Declaration "Cultural heritage is a crucial component of high-quality Baukultur. The way we use, maintain and protect our cultural heritage today will be crucial for the future development of a high-quality built environment" (Davos Declaration

2018:3). In the same year and following the successful launch and implementation of the European Year of Cultural Heritage (EYCH) in 2018, the European Commission publishes the New European Agenda for Culture aimed at pursuing and scaling up the efforts of EYCH. The Agenda emphasis the strong connection to sustainability "restoration and upgrading of cultural and natural heritage contributes to growth potential and sustainability" (COM 2018:5) and presents cultural heritage as a shared resource.

In 2019, the Bucharest Declaration of the Ministers of Culture and their representatives on the role of culture in building Europe's future, stresses on the need to enable the connection between cultural heritage and sustainability through policy and capacity building: "We emphasize the importance of implementing cross-cutting policies that strengthen the intrinsic value of cultural heritage. Such policies need to create the conditions that allow cultural heritage to contribute to sustainable development. To this end, objectives need to be set in different policy areas, including smart, sustainable and inclusive growth, as well as the promotion of architectural policies" (Bucharest Declaration 2019). This declaration was followed in the same year by the Barcelona Declaration which elaborate on what tourism and cultural heritage sectors could achieve by working closely in line with the SDGs (Barcelona Declaration 2019). The European Quality Principles which was drawn by ICOMOS and funded by the EC defines what it means by quality in relation to cultural heritage and provides a wide range of adopted International and European charters/conventions/principles in this regards "Quality in relation to cultural heritage can in any case be seen as multi-dimensional, bearing environmental, cultural, social, and economic values. Notions of cultural diversity, inclusivity, and an understanding of intangible heritage contribute important perspectives to defining future actions and interventions" (ICOMOS 2019). It reiterates that cultural heritage is a dynamic common good which involves a variety of actors and stakeholders and to which contribute a wide array of disciplines. "Heritage-led regeneration that would increase the attractiveness and competitiveness of different historic areas is a cornerstone of regional economic policy. All this points to the need to balance heritage conservation and socio-economic development through integrated and innovative management strategies, taking into account the fact that cultural heritage is not renewable nor replaceable" (ICOMOS 2019:23). It also delineates how the quality of interventions on cultural heritage should be ensured and reflects on evaluation and monitoring "Cultural heritage projects should also be evaluated in relation to their contribution to the circular economy and to the circular territorial development model. New respectful and compatible uses of cultural heritage should always be clearly and explicitly connected to its "intrinsic value"" (ICOMOS 2019:39).

Recently, Ursula von der Leyen, President of the European Commission, launched the New European Bauhaus as part of the Green Deal. "I want NextGenerationEU to kickstart a European renovation wave and make our Union a leader in the circular economy. But this is not just an environmental or economic project: it needs to be a new cultural project for Europe"[1]

[1] https://europa.eu/new-european-bauhaus/index_en

1.1 Urban Conservation and Circular Economy

The gradual evolution of urban conservation led to significant progress in: expanding the scope from individual monuments to cultural landscapes (The Declaration of Amsterdam 1975; UNESCO Recommendation concerning the Safeguarding and Contemporary Role of Historic Areas 1976; ICOMOS Charter for the conservation of historic towns and urban areas 1987; The European landscape convention 2000; UNESCO Underwater Cultural Heritage Convention 2001; ICOMOS Charter on Cultural Routes 2008a; ICOMOS Charter on the Interpretation and Preservation of Cultural Heritage Sites 2008b); establishing a neat link between nature conservation and the preservation of cultural properties and thus embracing a sustainable development perspective (The World Heritage Convention 1972; the Convention on Biological Diversity 1992; UNESCO Recommendation on the Historic Urban Landscape 2001; ICOMOS 2019); incorporating multilayered values in addition to the originally established artistic and historic values in the Charter of Venice in 1964 (The charter of Venice 1964; the Nara document on Authenticity, 1994; the Burra Charter 2013; Charter on the Built Vernacular Heritage 1999); and becoming more inclusive by taking into consideration not only the tangible but also the intangible heritage and expanding the range of actors, stakeholders and disciplines involved in urban conservation (UNESCO Convention for the Safeguarding of the Intangible Cultural Heritage 2003; UNESCO Convention on the Protection and Promotion of the Diversity of Cultural Expressions 2005; the Faro Convention in 2005). An important turning point for urban conservation was when the focus shifted from the process itself towards questioning why to conserve (what are the values being safeguarded) and for whom? Adaptive Reuse of Cultural Heritage became a two faceted tool which tackles contemporary needs with high quality reuse projects and provides evidence of the multilayered positive impacts (Gravagnuolo et al. 2017, 2018; Ost 2019a; Foster 2020).

The Circular Economy is an economic philosophy which challenges the existing linear economy and calls for a new system for exchange and production. It aims at: reducing waste at source, increasing resource efficiency use at all stages of the value chain, preserving the natural capital and reducing environmental impacts, and increase people's well-being (Geissdoerfer et al. 2016; Reichel et al. 2016; Kirchherr et al. 2017; Marin and De Meulder 2018; Korhonen et al. 2018; Moragaa et al. 2019) It is designed to be regenerative, replacing the linear end-of-life concept with new circular flows of reuse in an integrated process. Inspired by and designed to mimic nature's cycle, the CE aims at meeting a sustainable future.

Cities are known to be hotbeds of culture, creativity, diversity, economic growth, experimentation and innovation (UNESCO 2016). Today they are part of the problem and solution at the same time. The growing pressure on urban resources, infrastructures, services and housing caused by half of the global population living in cities (OECD and European Commission 2020); almost two-thirds of global energy demand; 70% alarming level of production of greenhouse gas emissions, 50% global waste (OECD 2020) need urgent policy measures and actions (CSCP 2018;

EMF 2017; EMF and ARUP 2019a, b; EIB 2018, 2019; Eurocities 2017; Prendeville et al. 2018; World Economic Forum 2018). The Global Resources Outlook provided in 2019 key insights regarding consumption and production impact and found that "resource use has more than tripled since 1970 to reach 92 billion tons in 2017…Resource extraction and processing cause over 90 per cent of global biodiversity loss and water stress, and more than half of global climate change impacts… Environmental impacts of material consumption are 3 to 6 times greater in high-income countries than in low-income countries. Without action, resource use would more than double from current levels to 190 billion tons by 2060. Related impacts would exceed the planetary boundaries and endanger human well-being" (The Global Resources Outlook 2019).

A city that acknowledges that there are planetary boundaries understands the role urban heritage could play in contributing to a sustainable urban development. In 2019, Heritage Counts demonstrated that: huge amounts of carbon are locked up in existing historic buildings; that adaptive reuse can reduce the need for new carbon-generating construction activities and therefore, reducing the need for new material extraction and reducing waste production (Heritage Counts 2019).

In 2018, Circle Economy published its first circularity gap report which states that the global economy is only 9.1% circular. Its 2020 report found that the gap is actually widening and the global economy is only 8.6% circular. This is because the world is continuing business as usual: "high rates of extraction; ongoing stock build-up; plus, low levels of end-of-use processing and cycling" (Circle Economy 2020). This implies an enormous amount of wasted potential for reuse or recycling of existing materials. Indeed, while on the one hand, cities waste resources, energy and water and on the other underutilize space, products, and assets, especially heritage assets. In 2019, the circularity gap report states that "A 1.5 °C world is circular...The circular agenda and low-carbon agenda are complementary and mutually supportive…The pathway to a low-carbon future is circular". The 2021 circularity gap report, calls for extending the functional lifetime of buildings and infrastructure as a pathway towards satisfying one of the global needs and wants[2]: "more co-housing, increased renovation, refurbishment, modular design that can easily adapt to needs over time is needed, and re-occupancy of underused and disused buildings to meet housing demand within global stock limits. Optimising flexible and multi-functional space for non-residential purposes also frees up existing stock for (redesignation as) housing... Retrofitting existing housing stock will both slow flows by extending the lifespan and in some cases cycle flows through reuse of materials in renovation and refurbishment on-site" (Circle Economy 2021).

In 2015, the European Commission adopted its Circular Economy Action Plan, which includes measures aimed at stimulating Europe's transition towards a circular

[2]According to the circularity gap report 2021, 59.1 billion tonnes of Global Greenhouse Gas GGH emissions were emitted in 2019 linked to seven global societal needs & wants, as follows: Mobility (17,1 Gt), housing (13,5 Gt), nutrition (10,0 Gt), services (6,4 Gt), consumables (5,6 Gt), communication (3,5 Gt), healthcare (3,0 Gt) (Circle Economy 2021).

economy, boost global competitiveness, foster sustainable economic growth and generate new jobs (COM 2015).

The action plan established a programme of action, with measures covering the whole cycle: from production and consumption to waste management and the market for secondary raw materials and a revised legislative proposal on waste. It also sets out the timeline for the completion of actions. The proposed actions aim at contributing to "closing the loop" of product lifecycles through greater recycling and re-use, and to bring benefits for both the environment and the economy. A revised legislative framework on waste entered into force in July 2018 setting clear targets for reduction of waste and establishing a long-term path for waste management and recycling.

At the end of 2019, the EC launches the Green Deal which provides an action plan to "boost the efficient use of resources by moving to a clean, circular economy restore biodiversity and cut pollution" (COM 2019). A deal that paves the way towards becoming a global leader in CE and a climate neutral continent by 2050.

As part of the action plan (COM 2020a), the EC publishes its Renovation Wave Strategy (COM 2020b) which on the one hand presents the weight of the built environment on sustainability "Overall, buildings are responsible for about 40% of the EU's total energy consumption, and for 36% of its greenhouse gas emissions from energy. only 11% of the EU existing building stock undergoes some level of renovation each year". While on the other, it showcases the substantial potential offered by the building stock for greening the economy, creating jobs and improving the quality of life.

Back in 1997, Zancheti and Jokilehto argued that "Sustainable development is seen today as a powerful motivation for urban conservation planning. Basically, it would consist of a process of urban development based on the constant reuse of existing built and natural resources, associated with a low ingress of energy for their adaptation to new requirements conceived in society. It is also viewed as a process founded in the local culture, in an equitable distribution of urban services, the use of democratic principles of management, and the maintenance and regeneration of traditional social values and practices" (Zancheti and Jokilehto 1997:47). To pave the way towards smooth implementation of the Renovation Wave Strategy, urban conservation and circular city policies should harmonize (Foster and Saleh 2021).

1.2 Cultural Entrepreneurship Momentum

The current health and economic crisis has devastated the artistic and cultural sectors, caught between the impossible organization and production of activities, and the impossibility for the public to attend/interact. Scenarios to exit the crisis foresee economic and financial difficulties for the cultural offer, but expect also changes in behavior for the public and consumers.

It should be remembered that any solution to the crisis involves periods of transition during which new things are created, while others disappear. It is no mere coincidence that the year 2021 is the International Year of the Creative Economy for Sustainable Development. This concerns in particular the cultural and creative industries which have for many years a central place in European policies (KEA 2006; KEA & PPMI 2019; Kern 2020).

In this context of change, it is necessary both to learn how to better withstand crises that will inevitably return, and how to innovate in terms of products, processes and artistic organizations. Resilience means changing the modes of financing (Ost 2016, 2018, 2019b), the partnerships and the way of modulating cultural products and services to changes in behavior, and to the new needs from audiences (Ost and Saleh 2021a). It is a question of creativity, innovation, technology, but also of the behavior of artistic and cultural actors. These changes must be accompanied by additional resources to the cultural sector, such that cultural actors can be more resilient during transition. Although there is no commonly agreed upon definition of cultural entrepreneurship, we define it as a "set of activities aimed at harnessing a cultural business opportunity. The novelty stands in being innovative in transforming cultural values into economic values. The process of creating new cultural expressions could be also interpreted as the business of transforming intangible values (performing arts, artistic creation, traditions and knowledge, etc…) into tangible assets in the form of cultural capital. The process of creating new adaptive reuse of heritage buildings is about the business of transforming abandoned, underused or not in use cultural heritage into common goods which reflect needs and aspirations of the contemporary local community with respect to environment and social practices and interactions. By transforming the cultural asset, the cultural entrepreneur harnesses the existing cultural (tangible and intangible) and economic values and transform them into enhanced cultural, economic, social and environmental impacts, outcomes and benefits" (Ost and Saleh 2021a).

It is timely for cultural entrepreneurs to embrace circular and innovative organizational forms of business and finance for managing their cultural activities. Cultural entrepreneurship is auspicious because it endows innovative concepts on quantity, quality and manipulation of resources in tandem with novel ways of empowering consumers, prosumers[3] and the community at large in governing the common goods.

2 Sustainable Business Model Innovation

A literature review carried by Zott *et al.* covering the period between 1975 and 2009 demonstrated that the concept of business model "became prevalent with the advent of the Internet in the mid-1990s, and it has been gathering momentum since then" (Zott et al. 2011). Following the review, the scholars identify four important themes around the notion of business models as "a *new unit of analysis*, offering a *systemic*

[3] We refer to prosumers as active citizens willing to be part of the reflection, co-design, testing and customization of the products/services according to his/her needs.

perspective on how to "do business," encompassing *boundary-spanning activities* (performed by a focal firm or others), and focusing on *value creation* as well as value capture" (Zott et al. 2011). Moreover, the scholars' shed light not only on the usefulness of the business model as a means for planning and systemic organizational analysis but also as a powerful communication tool through which entrepreneurs can communicate complex business solutions to potential investors (Zott et al. 2011).

Although Osterwalder posits a business model ontology in 2004, the first disruptive business model canvas is visualized by Osterwalder and Pigneur in 2010. It is considered a viable systematic approach to structure a business model and since its launch it became a recognized model (Lewandowski 2016). The canvas come to be widely employed because it equips business model designers with a tool that "*efficiently design high quality (reliable, consistent and effective) monetarily profitable business models*" (Upward 2013:11). So, what is a business model? According to Osterwalder and Pigneur: "*A Business Model describes the rationale of how an organization creates, delivers and captures value*" (Osterwalder and Pigneur 2010:14).

From a heritage economics perspective, we envisage the business model as a tool that enables cultural entrepreneurs with a medium to analyze the rationale of their business, namely: the adaptive reuse of cultural heritage. This model helps the cultural entrepreneurs visualize the logic behind what values are being preserved and co-created? for whom, how (through which processes), and why (to achieve which goals/impacts)? (Saleh et al. 2020; Ost and Saleh 2021b).

Osterwalder and Pigneur describe their Business Model Canvas, through nine blocks covering four main areas namely; customers, supply, infrastructure, and financial viability. However, these articulated four areas of the Business Model Canvas provide tools for an organization to generate value in monetary terms only with no consideration of other possible values. For this specific reason, a scientific debate is blossoming in the last decade on whether the Business Model Ontology could inform the design and evaluation of Sustainable Business Models (SBM)? In 2014, based on their structured literature review, Bocken et al., classified SBM in eight different business model archetypes grouped in three innovation categories: technical, social and organizational (2014:48). In 2016, Schaltegger et al. define a Sustainable Business Model as follows:

> A business model for sustainability helps describing, analyzing, managing, and communicating (i) a company's sustainable value proposition to its customers, and all other stakeholders, (ii) how it creates and delivers this value, (iii) and how it captures economic value while maintaining or regenerating natural, social, and economic capital beyond its organizational boundaries (2016:6).

In 2018, Geissdoerfer et al., conduct a comprehensive review of the Sustainable Business Model innovation literature and conclude that there are four types of sustainable business model innovation (2018:408). The authors provide different sustainable business model types (four) and strategies (nine). The circular business model is considered one of the sustainable business model types. Lately, there is a thriving debate on the need to develop circular business models (Lewandowski 2016; Kirchherr et al. 2017) and the need for using them as tools for monitoring value co-creation and capture (Reichel et al. 2016) and more recently as enablers of the circular economy (Kirchherr et al. 2017; WBCSD 2017; WBCSD and Climate-KIC 2018; Thelen et al. 2018; EpE and INEC 2018; The EIB 2019).

3 Circular Business Model and Human Flourishing

Cities, corporates and cultural entrepreneurs are the forerunners in testing circular business models (Cheshire 2016; Acharya et al. 2018; Arup and Bam 2018). However, the discussion is bourgeoning and it is pretty much focused on urban metabolisms, specific loops, pilot projects/case-studies, or strategies (Van Renswoude et al. 2015; DSGC 2015; Lewandowski 2016; Reichel et al. 2016; EIT RawMaterials circulator[4], Geissdoerfer et al. 2018).

Osterwalder and Pigneur's Business Model Canvas was adapted by Ellen MacArthur Foundation as a Circular Business Model (EMF 2016). By adding some questions related to the circular economy context, the EMF aims to help companies use a circular lens for their business. However, the four articulated areas of the Business Model Canvas namely: customers, supply, infrastructure, and financial viability, provide tools for an organization to generate value in monetary terms only with no consideration of other possible values. The main dilemma is how can a society and its related organizations' attain the sustainable development goals without addressing the cultural, social and environmental values in an integrated manner?

- Under the framework of H2020 project CLIC, the authors designed a circular business model for the adaptive reuse of cultural heritage and implemented it CLIC's four pilots: Pakhuis de Zwijger in Amsterdam, The Netherlands; Fengersfors, Vastra Götaland Region, Sweden; Rijeka, Croatia; and Salerno, Italy. The authors aim was to develop a circular business model for adaptive reuse of cultural heritage in line with the guidelines of the Historic Urban Landscape approach. An innovative business model which can cope with specific issues of governance and financing processes of adaptive reuse. Adaptive reuse of cultural heritage is considered today key in urban conservation and sustainable development. After reviewing a number of sustainable business model canvases, the authors found inspiring the speculation of Upward and Jones (2015) about a tri-impactful and tri-profitable Business Model.

Upward and Jones build on the Business Model Ontology and develop a Strongly Sustainable Business Model (SSBM). They base their speculation on questioning whether a successful business can be defined as such based on monetary returns

[4]The circulator is a web tool providing an overview of relevant Circular Business Models and applied case-studies: http://www.circulator.eu/about

only? More importantly, they query the aspired outcome in terms of sustainable performance. Their thorough research leads to the development of an ontology for a Strongly Sustainable Business Model (SSBM) composed of Four Formative Propositions (definition of a strongly sustainable firm; definition of value; definition of a business mode; and definition of Tri-profit) and five Instrumental Principals (conception of an SSBM; boundaries of an SSBM; validation of a SSBM; necessary financial viability of a sustainable model; and; modelling social benefits and environmental regeneration (Upward and Jones 2015:11–14).

According to the two scholars the four Formative Propositions highlight the lacking concepts in the classical Business Model Ontology BMO (Osterwalder 2004) and set the scene for the proposed new ontology. While the five Instrumental Principals portray the concepts and relationships within the new proposed ontology.

- One of the initiatives of the SSBM is the Flourishing Business Canvas. It is based on Upward and Jones's (2015) vision of a world where enterprises commit not only to do less harm but also to sustaining human flourishing under the framework of the Sustainable Development Goals. According to the scholars, this model aims at "not only do no harm, but do well by only doing good". Thus, it is calculated on a tri-profit metric, namely; economic viability, social benefits and environmental regeneration.

The authors embraced the flourishing business model because it is based on a tri-profit metric and therefore, it fulfills CLIC's vision and multidisciplinary human-centered framework. By building on this model, the authors analyzed the different sections (How, What, For Whom, Why) in line with the Historic Urban Landscape approach and under the framework of the circular economy and developed the following prototype (Fig. 16.1):

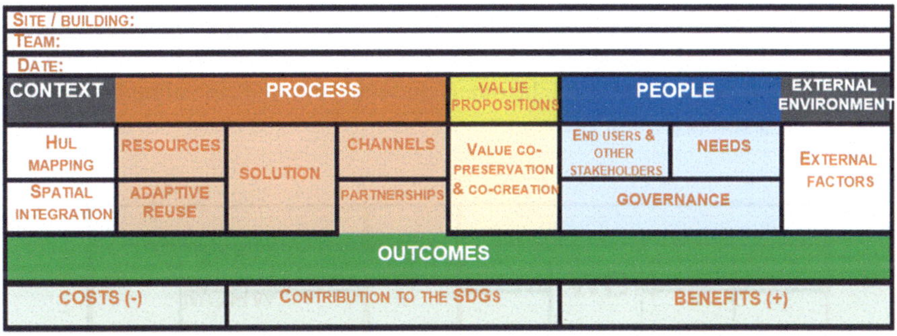

Fig. 16.1 Adapted circular business model for cultural heritage adaptive reuse. (Ost and Saleh 2019)

4 Circular Business Model for Cultural Heritage Adaptive Reuse: Structure and Tools

The structure of this circular business model is based on four blocks, namely:

- **What (value): What** are the values being co-created and **What** is the vision behind co-creation or co-destruction values? through a supply-driven reflection.
- **How (process): How** to process? And **How** can we co-accomplish what we want? Using which resources? and a demand-driven reflection.
- **For whom (people): For whom** building the model? Who are the customers and users? What are their needs? The community at large? What governance model? and
- **Why (outcomes): Why** are we building this model? With which means shall we evaluate how tri-impactful it is? (Fig. 16.2)

Moreover, the business model was developed and assessed at every stage in regards to:

- Desirability (is anyone interested in the proposed feature?)
- Feasibility (is it feasible? Do we have the technology, resources and capabilities to fulfil it?)
- Viability: (is it economically sound? What possible revenue streams and BMs to adopt?) and
- Impact: Does it respond to societal challenges? Does it generate positive economic, environments, social and cultural impacts?

For the sake of clarity, we will explain the issues to be tackled under each of the four blocks (value, process, people, outcomes) and encompassing fifteen elements. Specific questions were designed for every block in order to keep in mind both the Historic Urban Landscape Approach and the Circular Economy perspective. However, it is worth mentioning that the blocks were neither developed separately nor followed a chronological order. On the contrary, they were intertwined and the co-design of one block impacted the others and thus, implied amending and/or pivoting the idea/solution.

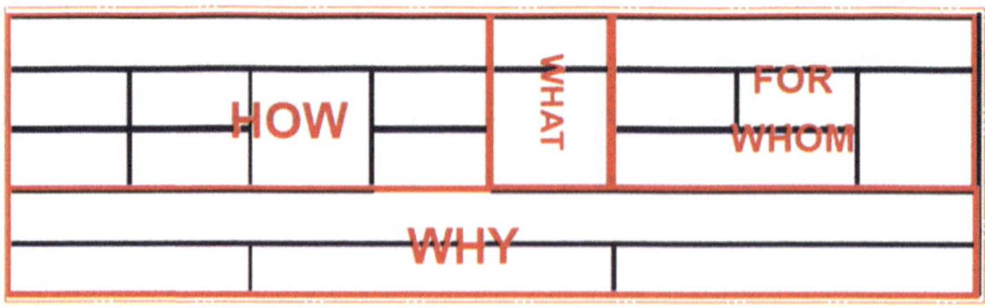

Fig. 16.2 Structure of the adapted circular business model for cultural heritage adaptive reuse. (Ost and Saleh 2019)

The Value Propositions

Value Co-preservation & Co-creation

- What are the different value propositions brought to the different stakeholders?
- How does each value proposition relate to meeting a stakeholder need?
- What is so unique in what you are proposing?
- What are the value propositions from the functional, emotional, economic, environmental & social perspectives?

The Process

Context: The Historic Urban Landscape

- What is the broader urban context / landscape that embeds the project?
- How to describe the context's human, natural and cultural layers (cultural capital)?

Context: Spatial Integration

- Describe the spatial integration of the project (transportation, mobility, businesses, public services, points of attractiveness in the macro area, walkability, utilities, infrastructures, regulations, land uses) within its context?

Solution

- What are the main features/components of the idea/solution
- What do users, customers and other stakeholders have access to (products, services, activities, etc…)?

Adaptive Reuse

- What reuse is embedded in the value proposition?
- How to monitor and design out any negative externalities?
- What circular flows of materials are enabled (construction/demolition/waste)?
- What are the assets that are further valorized / less wasted (embedded energy, local skills, techniques and knowledge, environmental / social capital)?
- What is in place regarding heritage community network/s and feedback mechanisms (disseminate and exchange)?

Resources

- Which resources are required for your project:
    - building (bricks & mortars);
    - land;
    - skills, expertise & know how;
    - materials;
    - assets
- Which materials are moved or transformed during the process of creating, delivering and maintaining value propositions?

Channels

- What channels are used to communicate, give access and deliver value propositions?
- How do services/products/ solutions reach end users?
- What are the channels (physical and virtual) used to make the offer (a) known (b) available?
- What are the channels put in place to stay in touch with end-users? to get their feedbacks re their satisfaction & needs / expectations?
- How far does the supply chain integrate the "reduce, recycle, reuse" principle?

Partnerships

- What are the formal stakeholder relationships required to deliver your value propositions i.e multi-actor agreements?
- Who are the organization's partners and what agreements for resources and activities should be made with them?
- Which partnerships with circular organizations (flows of materials, information and capital) are put in place so that the circular / closed loop dimension is feasible?

People

End Users, Customers & Other Stakeholders

Cultural heritage is a common good embedded in a broader urban context:

- Who are the end-users of the project? The customers?
- Who are the other stakeholders of the project (producers, users, consumers, decision makers, policy makers, investors, supporters, etc...) besides users and customers?
- Who are the 5–8 most important stakeholders?
- Are the stakeholders influencial (power) and/or committed (interest)?

Needs

The adaptive reuse project aims to benefit to the various stakeholders and more specifically to the local community. Independently from the project:

- What are the key needs and pain points of your main stakeholders?
- What are their expectations towards the project?
- Which needs do the project intend to address?

Governance

- Which stakeholders get to make decisions about what the organization does, where and how activities are undertaken and how resources are transformed? In other words, who makes decisions about:
  - What to conserve and reuse?
  - Managing the common good (public, private and community)?

- Setting priorities: where (which area first) and when (high priority, medium priority, low priority)?
- How are local communities actively involved, mobilized, involved in the governance?

External Environment: External Factors

Many external factors can be either opportunities i.e. levers for the success of the project or risks / threats i.e. items that can adversely impact the project. Undertake a PESTEL analysis by taking into consideration the political, economic, social, environmental, technological and legal factors.

The Outcomes

Benefits (+)

- How is the financial sustainability of the project guaranteed?
- What are the business models that can generate enough revenue streams to increase the economic viability of the project and ensure its resilience thanks to revenue diversification?
- What are the financial vehicles used for the project?
- How are ethical and local investments stimulated?
- How is civil economy stimulated?
- What are the non-financial benefits generated by the project: social (i.e. happiness, wellbeing), environmental (i.e. reuse and regeneration of resources)?
- What are the short term and long term sustainable benefits?
- How can the positive societal impacts of your project be amplified 100 times?

Costs (−)

- What are the financial costs of the project in terms of the conservation work and adaptive reuse?
- What are the capital expenditures and operational costs associated to the solution?
- What are the negative externalities of the project in non-financial terms:
  - social (i.e. gentrification, mass tourism, loss of intangible assets; social conflicts, decreased happiness, illness)
  - environmental (i.e. harm/depletion of natural resources)? What are the short term and long term sustainable costs?
  - opportunity costs of the decision of adaptive reuse (loss of local jobs and commerce, loss of biodiversity, dismantlement of the buildings and new development projects, etc.).
  - loss of authenticity and integrity coming from the adaptive reuse decision
- What can be done to mitigate/neutralize these adverse impacts? Which costs could be shared or lowered through other users and partners? How can negative externalities be 100 times less impactful?

Contribution to the Whole—Sustainable Development Goals

Adaptive reuse aims to enhancing a heritage building and its values and to significantly contribute to Sustainable Development. What are the Sustainable Development Goals (SDGs) that are addressed by this adaptive reuse project?

4.1 Implementation

Between 18 February and 18 June 2020, ICHEC's team led a circular business model workshop, in every CLIC pilot. The workshop was either conducted physically or virtually, and dedicated to the adaptive reuse of a representative heritage asset proposed by the local partner. Every workshop was prepared and discussed thoroughly with each local partner and its related academic partner in order to agree on the challenge to tackle; the storyboard; the modality of identifying and involving the stakeholders (open call; or by invitation); workshop format (number of days/sessions; tools); participants' profiles; experts to be invited/involved in different capacities and logistics. The workshops were designed and facilitated by Philippe Drouillon, ICHEC's team member and catalyst of positive impact territories and organizations.

Vastra Götaland Region in Sweden choose to focus on the adaptive reuse of the paper mill in Fengersfors. Thirty-nine people participated proactively in the co-design workshop and five complementary business models were co-designed by the workshop participants. This was the only face to face workshop.

The original plan for Pakhuis de Zwijger, in Amsterdam (NL) was to revisit their business model under the circular economy perspective but due to the outbreak of COVID19, Pakhuis de Zwijger, asked ICHEC's team to change the topic and focus of the workshop in order to synchronize with the current situation and reverberate the repercussions of COVID19 on public places closing their doors. Only one session took place and 6 people participated in the co-design workshop. To this end, very limited aspects of the business model were tackled with Pakhuis De Zwijger senior and middle management due to the time restraints.

The municipality of Rijeka in Croatia asked ICHEC's team to develop a workshop aimed at addressing the adaptive reuse of the Brick Building within the former Rikard Benčić industrial complex. 23 people registered to Rijeka's circular business workshop. However, only 10 people completed all the sessions. At the end of the six sessions, three complementary business models were co-designed by the workshop participants.

The Municipality of Salerno in Italy asked ICHEC's team to develop a workshop aimed at addressing the adaptive reuse of the complex of Edifici Mondo (convents of San Francesco and San Giacomo and San Pietro a Maiella and Palazzo San Massimo). Fifty-two people participated in session 1 where the local administration participated as well, while thirty people participated proactively in all the sessions. By the end of the fifth session, four business models were co-designed by the workshop participants.

The four workshops tackled public and private cultural heritage assets and were structured in a way to cover the four blocks (value propositions, process, people and outcomes). The fifteen elements of the circular business models were tackled during the workshop except for the context which was provided beforehand thanks to the economic landscape mapping conducted previously by ICHEC in CLIC pilots (see D3.2)[5]. As per the external environment, although it was discussed during the workshops, the authors filled it in after the conclusion of the workshops.

During the implementation of the workshops, the main challenges faced were related to:

- Technical barriers: The WIFI connection was not always ideal and Mural is a heavy program. Therefore, participants were invited to use their phones to connect to Zoom and computers to work on Mural;
- Language: In order to make people feel at ease and express themselves to the maximum, groups were invited to speak/work in the local language in the break-out rooms. Moreover, simultaneous translation in Italian was provided for Salerno's workshop; and
- Attendance: The problem of attendance was faced in Rijeka only. Although participants were consulted via Facebook survey beforehand on the most suitable timetable and dates, very few people committed to the workshop until the end.

During the testing and putting into force of the business model itself, the process went smoother for the private heritage asset in Fengersfors, Sweden. Indeed, the implementation of the co-designed circular business model for the adaptive reuse of the paper mill in Fengersfors is on-going. Following the workshop, ICHEC's team analyzed the workshop outcomes and a roadmap was set and agreed with Not Quite collective in order to put the Business Model into practice. From their end, Not Quite collective listed and tested their Riskiest Assumptions; confirmed their value streams; and updated Minimal Viable Solutions (MVS) for each revenue stream. The group of four committed persons who are developing the revenue stream for the next three years to come are starting to reap the benefits of their hard and coordinated work. The Phytoremediation solution was granted a funding and Not Quite is about to start the project with the help of a group of researchers. In the meantime, they keep pushing forward the other revenue streams.

In the cases of the publicly owned assets in Rijeka and Salerno, a few implementation challenges arose due to:

- Lack of human and financial resources.
- Need for lifelong learning platforms to update authorities and citizens on new governance approaches, in this case, the circular governance and the circular city perspective.
- Lack of commitment/official endorsement by the local administration due to administrative regulations and procedures (the need to launch a public bid/PPP) and political will;

[5] https://www.clicproject.eu/files/D3-2.pdf

- Need to create a congruency between research projects, in this case H2020 project CLIC, and previous national and European programs/projects approved by the Municipal administration;
- Lengthy and bureaucratic process of approving/deliberating a regulation for the shared management of cultural heritage as a common good.

Finally, the management of Pakhuis De Zwijger was dealing with tough decisions related to governance and staff management caused by COVID19 crisis and thus, BM testing and implementation was not a priority.

5 Conclusions

Urban conservation doctrine evolved in the past fifty decades to become sustainable, inclusive and responsive to shifting social and economic values. Building on its existing intrinsic values, the circular economy is the new allay to demonstrate how adaptive reuse of cultural heritage is crucial to reverse climate change trends, improve the quality of the urban environment and people's wellbeing.

The author's main objective was to develop a circular business model for adaptive reuse of cultural heritage in line with the guidelines of the Historic Urban Landscape approach. Adaptive reuse of cultural heritage is considered today key in urban conservation and sustainable development. After reviewing a number of sustainable business model canvases, the flourishing business model was selected and adapted by the authors to the context of adaptive reuse. The reason behind embracing the flourishing business model is that it is based on a tri-profit metric, namely; economic viability, social benefits and environmental regeneration. Therefore, the authors found that the flourishing business model was the most adequate model in order to fulfill CLIC's vision and its multidisciplinary human-centered framework.

This circular business model for cultural heritage adaptive reuse builds on an iterative process already started by the authors under the framework of H2020 project CLIC. This process started with a documentation and analysis of the supply (economic landscape) (Ost and Saleh 2019) and demand (perceptions mapping) (Saleh and Ost 2020; Ost and Saleh 2021c) and ends up with the business model for a specific heritage asset (Saleh et al. 2020; Ost and Saleh 2021b). The circular business model workshop was structured in a way to cover the process; the value propositions; the people and the outcomes. Specific questions were designed for every block in order to keep in mind both the Historic Urban Landscape Approach and the Circular Economy perspective. It was conceived as a co-design process during which participants proposed reuse ideas/solutions in relation to their territorial needs. Every workshop departed from the unique value proposition prioritised and agreed upon by the participants. By tapping into the collective intelligence, participants were divided into groups and every group delineated a list of possible solutions and activities based on the resources and identified needs in the territory. After identifying the different stakeholders, the desirability of these activities was tested and the Minimum Viable solution was updated. Revenue streams were developed with particular attention to environmental, social and economic impacts, both negative and positive. Moreover, groups were asked to select the sustainable development goals that they aim to fulfil in their project. Once this final phase was fine-tuned and the Minimum Viable Solution was updated, every group pitched its circular, desirable, feasible, and economically viable solution(s).

This co-design workshop aimed at democratizing the design process and enabling participation and pitching of creative proposals (Gudiksen et al. 2014; Mitchell and Buur 2010; Sani et al. 2018). It encouraged participants to partake in the decision making, project planning and evaluation of the needs and adaptive reuse projects. Thus, to develop their own sense of community as heritage communities (Council of Europe 2005). The ultimate objective was to co-design the new tri-impactful reuse with the people, instead of for the people. Moreover, the workshop embraced the circular economy as a regenerative economy applied to the economic, social, and environmental aspects. Thus, the circular economy was incorporated as a culture of regeneration. The adapted flourishing business model was instrumental in putting forward the heritage cultural asset as a connective infrastructure where solidarity and cooperation at the local level excels. Hence, the heritage asset was perceived as a common good for collectivity to preserve and regenerate.

This is why, the authors believe that this Circular Business model is a useful tool aimed at informing the decision-making process and thus, should be part of the action plan of public and private owners aspiring at evaluating and putting forward coherent investment projects with sustainable urban conservation processes (Boxes 16.1, 16.2, 16.3, and 16.4).

Box 16.1: Business Model Workshop at Pakhuis de Zwijger, Amsterdam, the Netherlands
Partner: Pakhuis De Zwijger

Asset: Pakhuis De Zwijger building (industrial heritage)

Owner of the asset: Stadsherstel

Challenge: How to make Pakhuis de Zwijger more resilient to shocks? i.e. COVID19

Workshop participants: 5 from Pakhuis De Zwijger + 1 from TU/e, Eindhoven University of Technology + 2 ICHEC team

Workshop format: Virtual workshop. One session only on Friday April 24, 2020 from 09:30 to 13:00 CET

Workshop tools
ICHEC's team provided:

- Zoom plenary and break-out rooms
- Pre-set Mural walls

(continued)

Box 16.1 (continued)

Topics addressed during the workshop:

The process, value proposition, people and the outcome; the circular business model for cultural heritage adaptive reuse; Revenue streams cards; Business model cards (circular, inclusive and local collaboration); the 9Rs strategies, and the twelve vital functions.

Results:

General framework of the current business model.

Box 16.2: Business Model Workshop in Rijeka, Croatia
Partner: Municipality of Rijeka

Asset: The Brick Building within the former Rikard Benčić industrial complex

Owner of the asset: Municipality of Rijeka, Croatia

Challenge: How to make the management of the children's house sustainable from an economic viewpoint?
What type of circular child-friendly and child oriented activities should take place?

Workshop participants: 6 stakeholders + 2 Rijeka municipality + 2 University of Nova Gorica + 2 ICHEC team

Workshop format: Virtual workshop
Remote session 1: Idea generation. Tuesday 28 April, 2020, 13:00–17:30 CET
Remote session 2: Solution prototyping. Wednesday 29 April 2020, 13:00–17:00 CET
Remote session 3: Desirability testing. Tuesday 5 May 2020, 13:00–15:00 CET
Remote session 4: Feasibility assessment. Friday 8 May 2020, 13:00–17:00 CET
Remote session 5: Viability assessment. Tuesday 12 May 2020, 13:00–16:00 CET
Remote session 6: Solution fine-tuning and pitching. Tuesday 19 May 2020, 11:00–13:00 CET
Remote session 7: Road-map and way forward. Tuesday 2 June 2020, 11:00–12:00 CET

Workshop tools
ICHEC's team provided:

- Zoom plenary and break-out rooms
- Pre-set Mural walls

(continued)

Box 16.2 (continued)

Topics addressed during the workshop:

Context, objectives and broader landscape of related topics; the process, value proposition, people and the outcome; the 9Rs strategies, the twelve vital functions, value proposition definition and example; Test action plan interview guide and questions; the circular business model for cultural heritage adaptive reuse; Steps for defining RAT; MVS; VP, prototyping and MVS and VP updates; Revenue streams cards; Business model cards (circular, inclusive and local collaboration); Grids for quantifying revenue streams and costs; Sustainable Development Goals; Methodology for defining desirability, feasibility, viability and impacts; Reasons behind start-up failure; and pulse surveys.

Results: A business model encompassing three revenue streams:

1. Health and food Hub
2. Creativity Hub
3. STEM Hub

Box 16.3: Business Model Workshop in Salerno, Italy
Partner: Municipality of Salerno

Asset: The complex of Edifici Mondo (convents of San Francesco and San Giacomo and San Pietro a Maiella and Palazzo San Massimo).

Owner of the asset: Municipality of Salerno, Italy

Challenge: How to combine and increase the strength and robustness of the selected proposals from the open call.

Workshop participants: 31 stakeholders + 2 Municipality of Salerno + 7 IRISS-CNR + 2 ICHEC team + 1 ICHEC expert

Workshop format: Virtual workshop
Remote session 1: Idea pitching and prioritization. Tuesday 14 May, 2020, 13:30–17:50 CET
Remote session 2: Solution prototyping and desirability testing. Monday 18 May, 2020, 09:00–13:05 CET
Remote session 3: Feasibility assessment. Monday 25 May 2020, 13:30–16:50 CET
Remote session 4: Viability assessment. Thursday 4 June 2020, 13:30–17:00 CET

(continued)

Box 16.3 (continued)

Remote session 5: Solution fine-tuning and pitching. Monday 8 June 2020, 09:30–12:45 CET

Remote session 6: Roadmap from visionary design to action. Thursday 18 June 2020, 10:45–12:15 CET

Workshop tools

- Zoom plenary and break-out rooms
- Pre-set Mural walls

Topics addressed during the workshop:

Context, objectives and broader landscape of related topics; the process, value proposition, people and the outcome; the 9Rs strategies, the twelve vital functions, value proposition definition and example; Test action plan interview guide and questions; the circular business model for cultural heritage adaptive reuse; Steps for defining RAT; MVS; VP, prototyping and MVS and VP updates; Revenue streams cards; Business model cards (circular, inclusive and local collaboration); Grids for quantifying revenue streams and costs; Sustainable Development Goals; Methodology for defining desirability, feasibility, viability and impacts; Reasons behind start-up failure; the governance circles model; and Pulse surveys.

Results: Four business models which could be consolidated as a cultural ecosystem for the city of Salerno as follow:

1. Hippocratica Hills Health Heritage Hub and water paths BM
2. House of music BM
3. The identity between tradition and innovation BM
4. Solidarity condominium BM

Box 16.4: Business Model Workshop Västra Götaland Region, Sweden
Partner: Västra Götaland Region

Asset: Fengersfors paper mill

Owner of the asset: Private owner

Beneficiaries: The Not Quite collective (https://www.notquite.se/sv-SE/english/about-not-quite-27707600)

Challenge: What solutions could be co-designed in order to make the business model of the new paper mill town ecologically and socially sound and financially sustainable?

(continued)

Box 16.4 (continued)

Workshop participants: 39 Stakeholders + Västra Götaland Region + 2 Uppsala University + 2 ICHEC team

Workshop format: Face to face workshop.
Day 1: Idea generation. Tuesday 18 February 2020, 18:00–21:00 CET.
Day 2: Solution prototyping. Wednesday 19 February, 09:00–17:30 CET
Day 3: Solution fine-tuning and pitching. Thursday 20 February 2020, 10:00–19:30 CET

Workshop tools
The Not Quite collective provided the following:

- A workshop room
- The vision of the New Mill project
- Workshop materials: Flipchart, pens, colors, post-it, and DIY material for mock ups.
- Food and beverage: Delicious meals and coffee breaks were prepared by the coffee-shop on-site specifically for the workshop. This coffee shop opens usually only during summer time.

ICHEC's team provided:

- The circular business model for cultural heritage adaptive reuse in A1 format;
- The economic landscape maps of Fengersfors
- Revenue streams cards
- Business model cards (circular, inclusive and local collaboration)
- Presentations including definitions (circularity, RAT; MVS; VP, etc…); process; methodology relevant examples and the way forward

Topics addressed during the workshop:

Context, objectives and broader landscape of related topics; the process, value proposition, people and the outcome; the 9Rs strategies, the twelve vital functions, value proposition definition and example; Test action plan interview guide and questions; the circular business model for cultural heritage adaptive reuse; Steps for defining RAT; MVS; VP, prototyping and MVS and VP updates; Revenue streams cards; Business model cards (circular, inclusive and local collaboration); Grids for quantifying revenue streams and costs; Sustainable Development Goals; Methodology for defining desirability, feasibility, viability and impacts; Reasons behind start-up failure; the governance model.

(continued)

Box 16.4 (continued)

Results: A business model encompassing five revenue streams:

1. Accommodation, retreats, events
2. Rural innovations Center
3. Phytoremediation services
4. Hub/cluster/network of creative industries
5. Ecosystem of sustainable businesses (food, arts, agriculture, construction, education & training).

A group of four committed persons pursue developing the revenue stream for the next three years to come. The Phytoremediation solution was granted a funding and Not Quite is about to start the project with the help of a group of researchers. In the meantime, they keep pushing forward the other revenue streams.

References

Acharya D, Boyd R, Finch O (2018) From principles to practices: first steps towards a circular built environment. ARUP and Ellen Macarthur Foundation, report

Arup GC, Bam NM (2018) Circular business models for the built environment. Arup BAM:1–44

Bandarin F (2015) Urban conservation and the end of planning. In: Bandarin F, Van Oers R (eds) Reconnecting the city: the historic urban landscape approach and the future of urban heritage. Wiley, Chichester, pp 1–16. https://doi.org/10.1002/9781118383940

Barcelona Declaration (2019) Barcelona declaration of tourism and cultural heritage: better places to live, better places to visit. Available at: https://onedrive.live.com/?authkey=%21ALGmRQscySOLV5Q&cid=19E1928B8C6B7F5A&id=19E1928B8C6B7F5A%21157090&parId=19E1928B8C6B7F5A%21105860&o=OneUp

Bocken NMP, Short SW, Rana P, Evans S (2014) A literature and practice review to develop sustainable business model archetypes. J Clean Prod 65(2014):42–56. https://doi.org/10.1016/j.jclepro.2013.11.039

Bucharest Declaration (2019) Bucharest Declaration of the Ministers of culture and their representatives on the role of culture in building Europe's future. Available at: https://www.consilium.europa.eu/media/39209/190416-bucharest-declaration-on-the-role-of-culture-in-building-europes-future.pdf

Cheshire D (2016) Building revolutions: applying the circular economy to the built environment. RIBA Publishing, London

Circle Economy (2018) The circularity gap report. An analysis of the circular state of the global economy. Report. Available at: https://www.circle-economy.com/resources/the-circularity-gap-report-our-world-is-only-9-circular

Circle Economy (2020) The circularity gap report. Report. Available at: https://www.circle-economy.com/resources/circularity-gap-report-2020#:~:text=The%20Circularity%20Gap%20Report%202020,was%20first%20launched%20in%202018

Circle Economy (2021) The circularity gap report. Report. Available at: https://www.circle-economy.com/resources/circularity-gap-report-2021

Communication from the Commission to the European Parliament, the Council, The European Economic and Social Committee and the Committee of the Regions COM (2015) Closing the

loop – an EU action plan for the Circular Economy. COM (2015) 614 final. Available at: https://eur-lex.europa.eu/legal-content/EN/TXT/?uri=CELEX%3A52015DC0614
Communication from the Commission to the European Parliament, the Council, The European Economic and Social Committee and the Committee of the Regions COM (2018) A New European Agenda for Culture. COM (2018) 267 final. Available at: https://eur-lex.europa.eu/legal-content/EN/TXT/?uri=COM%3A2018%3A267%3AFIN
Communication from the Commission to the European Parliament, the Council, The European Economic and Social Committee and the Committee of the Regions COM (2019) The European Green Deal. COM/2019/640 final. Available at: https://eur-lex.europa.eu/resource.html?uri=cellar:b828d165-1c22-11ea-8c1f-01aa75ed71a1.0002.02/DOC_1&format=PDF
Communication from the Commission to the European Parliament, the Council, The European Economic and Social Committee and the Committee of the Regions COM (2020a) A new Circular Economy Action Plan For a cleaner and more competitive Europe. COM/2020/98 final. Available at: https://eur-lex.europa.eu/legal-content/EN/TXT/?uri=CELEX%3A52020DC0098
Communication from the Commission to the European Parliament, the Council, The European Economic and Social Committee and the Committee of the Regions COM (2020b) A Renovation Wave for Europe – greening our buildings, creating jobs, improving lives. COM/2020/662 final. Available at: https://eur-lex.europa.eu/legal-content/EN/TXT/?uri=CELEX%3A52020DC0662
Council of Europe (2000) European Landscape Convention. Florence 20. 10. 2000. European Treaty Series. No. 176. 7 p. Available at: https://rm.coe.int/CoERMPublicCommonSearchServices/DisplayDCTMContent?documentId=0900001680080621
Council of Europe (2005) Framework convention on the value of cultural heritage for society (Faro Convention). www.conventions.coe.int
Davos Declaration (2018) Towards a high-quality Baukultur for Europe, Davos Declaration. Available at: https://davosdeclaration2018.ch/
Dutch Sustainable Growth Coalition DSGC (2015) Circular economy: DSGC companies on their journey of implementing circular business models. Available at: https://www.dsgc.nl/publications/dsgc-circular-economy.pdf. Accessed on 22 Nov 2018
Ellen Macarthur Foundation (2016) Business Model Canvas. Available at: https://www.ellenmacarthurfoundation.org/assets/design/Business_Model_Canvas_Final.pdf
Ellen Macarthur Foundation (2017) Cities in the circular economy: an initial exploration. Report. Available at: https://www.ellenmacarthurfoundation.org/assets/downloads/publications/Cities-in-the-CE_An-Initial-Exploration.pdf
Ellen Macarthur Foundation and ARUP (2019a) City governments and their role in enabling a circular economy transition. Report. Available at: https://www.ellenmacarthurfoundation.org/assets/downloads/CE-in-Cities_Policy-Levers_Mar19.pdf
Ellen MacArthur Foundation and ARUP (2019b) circular economy in cities. Report. Available at: https://www.ellenmacarthurfoundation.org/assets/downloads/CE-in-Cities-Project-Guide_Mar19.pdf
EpE and INEC (2018) Circular economy indicators for businesses. Report. Available at: https://institut-economie-circulaire.fr/wp-content/uploads/2018/10/2018_indicateurs_economie_circulaire_entreprises.pdf
Eurocities (2017). Full circle. Cities and the circular economy. Report. Available at: http://nws.eurocities.eu/MediaShell/media/2017cities_and_circular_economy-web-spreads.pdf
European Investment Bank (2018) The 15 circular steps for cities. Report. Available at: https://www.eib.org/attachments/thematic/circular_economy_15_steps_for_cities_en.pdf
European Investment Bank (2019) The EIB Circular Economy Guide. Supporting the circular transition. Report. Available at: https://www.eib.org/attachments/thematic/circular_economy_guide_en.pdf

Foster G (2020) Circular economy strategies for adaptive reuse of cultural heritage buildings to reduce environmental impacts. Resour Conserv Recycl 152(2020):104507. https://doi.org/10.1016/j.resconrec.2019.104507

Foster G, Saleh R (2021) The adaptive reuse of cultural heritage in European circular 2 city plans: a systematic review. Sustainability, Special Issue: The Valorization of the Cultural Heritage and Landscape as the Entrance Point for the Circular City Strategy. Forthcoming

Geissdoerfer M, Savaget P, Bocken NMP, Jan Hultink E (2016) The circular economy – a new sustainability paradigm? J Clean Prod 143:757–768. https://doi.org/10.1016/j.jclepro.2016.12.048

Geissdoerfer M, Vladimirova D, Evans S (2018) Sustainable business model innovation: a review. J Clean Prod 198:401–416. https://doi.org/10.1016/j.jclepro.2018.06.240

Gravagnuolo A, Fusco Girard L, Ost C, Saleh R (2017) Evaluation criteria for a circular adaptive reuse of cultural heritage. *BDC – Bollettino Del Centro Calza Bini* 17:2

Gravagnuolo A, Saleh R, Ost C, Fusco GL (2018) Towards an evaluation framework to assess cultural heritage adaptive reuse impacts in the perspective of the circular economy. Urbanistica informazioni 278 s.i:28–31

Gudiksen S, Poulsen SB, Buur J (2014) Making business models. Int J CoCreat Design Arts CoDesign Through Mak 10(1):15–30

Heritage Counts (2019) There's no place like old homes. Re-use and recycle to reduce carbon. Report. Available at: https://historicengland.org.uk/content/heritage-counts/pub/2019/hc2019-re-use-recycle-to-reduce-carbon/

ICOMOS (1964). International Charter for the Conservation and Restoration of Monuments and Sites (Venice Charter). Available at: https://www.icomos.org/charters/venice_e.pdf

ICOMOS (1987) Charter for the conservation of historic towns and urban areas (Washington charter 1987). Available at: https://www.icomos.org/images/DOCUMENTS/Charters/towns_e.pdf

ICOMOS (1994) The Nara document on Authenticity. Available at: https://www.icomos.org/charters/nara-e.pdf

ICOMOS (1999) Charter on the Built Vernacular Heritage. Available at: https://www.icomos.org/images/DOCUMENTS/Charters/vernacular_e.pdf

ICOMOS (2008a) Charter on cultural routes. Available at: https://www.icomos.org/images/DOCUMENTS/Charters/culturalroutes_e.pdf

ICOMOS (2008b) Charter on the Interpretation and Presentation of Cultural Heritage Sites. Ratified by the 16th General Assembly of ICOMOS, Quebec (Canada) in 2008. Available at: https://www.icomos.org/images/DOCUMENTS/Charters/interpretation_e.pdf

ICOMOS (2019) The future of our pasts: engaging cultural heritage in climate action. Available at: https://indd.adobe.com/view/a9a551e3-3b23-4127-99fd-a7a80d91a29e

KEA & PPMI (2019) Research for CULT Committee – culture and creative sectors in the European Union – key future developments, challenges and opportunities. European Parliament, Policy Department for Structural and Cohesion Policies, Brussels

KEA European Affairs (2006) The economy of culture in Europe. Study prepared for the European Commission. In: Culture

Kern P (2020) The future of cultural policies, 1st edn. Les editions kea, Brussels

Kirchherr J, Reike D, Hekkert M (2017) Conceptualizing the circular economy: an analysis of 114 definitions. Resour Conserv Recycl 127:221–232. https://doi.org/10.1016/j.resconrec.2017.09.005

Korhonen J, Honkasalo A, Seppälä J (2018) Circular economy: the concept and its limitations. Ecol Econ 143:37–46

Lewandowski M (2016) Designing the business models for circular economy—towards the conceptual framework. Sustainability 8(1):43. Available at: https://www.mdpi.com/2071-1050/8/1/43

Luxen J-L (2004) Reflections on the use of heritage charters and conventions. Getty Conserv Instit Newslett 19(2):4

Marin J, De Meulder B (2018) Interpreting circularity. Circular city representations concealing transition drivers. Sustainability 10:1310. https://doi.org/10.3390/su10051310

Meadows DH, Meadows DL, Randers J, Behrens WW III (1972) The limits to growth. A report for the club of Rome's project on the predicament of mankind. Universe Books, New York
Mitchell R, Buur J (2010) Tangible business model sketches to support participatory innovation. DESIRE conference
Moragaa G, Huysvelda S, Mathieuxc F, Blenginic GA, Alaertsd L, Ackerd KV, de Meesterb S, Dewulfa J (2019) Circular economy indicators: what do they measure? Resour Conserv Recycl 146(2019):452–461
OECD (2020) The circular economy in cities and regions, OECD Urban Studies. OECD Publishing, Paris. https://doi.org/10.1787/10ac6ae4-en
OECD/European Commission (2020) Cities in the world: a new perspective on urbanisation, OECD Urban Studies. OECD Publishing, Paris. https://doi.org/10.1787/d0efcbda-en
Ost C (2016) Innovative financial approaches for culture in urban development. In: Culture urban future, global report on urban sustainable development. UNESCO, Paris
Ost C (2018) Inclusive economic development in the urban heritage context. In: Larsen P, Logan W (eds) World heritage and sustainable development. New directions in World Heritage Management. Routledge, Oxon/New York, pp 53–67
Ost C (2019a) Towards multi-criteria analysis for circular economy in adaptive reuse of cultural heritage. In: Fusco Girard L, Trillo C, Bosone M (eds) Matera, città del sistema ecologico uomo/società/natura. Il ruolo della cultura per la rigenerazione del sistema urbano/territoriale. Giannini Editore, Naples, pp 345–381
Ost C (2019b) Urban economics. In: Cody JW, Siravo F (eds) Historic cities – issues in urban conservation. Getty Publications, Los Angeles
Ost C (2021) Revisiting heritage conservation in its social and economic background. In: Pottgiesser U, Fatoric S, Hein C, de Maaker E, Roders AP (eds) LDE heritage conference on heritage and the sustainable development goals: proceedings. BK Books, pp 282–289. Retrieved from https://books.bk.tudelft.nl/index.php/press/catalog/book/781
Ost C, Saleh R (2019) D3.2 Economic landscapes maps of pilot cities. European Commission Research and Innovation portal. CLIC Consortium
Ost C, Saleh R (2021a) Cultural and creative sectors at a crossroad: post COVID-19 challenges and opportunities, Built Heritage. Special Issue Innovative Heritage-based Post-Crisis Urban Recovery Strategies. Forthcoming
Ost C, Saleh R (2021b) Innovative business model for adaptive reuse of cultural heritage in a circular economy perspective. Int J Entrep Small Bus. Special Issue Sustainable Entrepreneurship in the Cultural and Creative Sector: Contextualisation, Challenges and Prospects. Forthcoming
Ost C, Saleh R (2021c) Perceptions mapping: a participatory tool for urban conservation planning. In: Reflections on Cultural Heritage Theories and Practices: a series by the Raymond Lemaire International Centre for Conservation. KU Leuven, Belgium. Forthcoming
Osterwalder, A. (2004). The business model ontology: a proposition in a design Science Approach. PhD thesis, l'Ecole des Hautes Etudes Commerciales de l'Université de Lausanne, Lausanne, Switzerland
Osterwalder O, Pigneur Y (2010) Business model generation: a handbook for visionaries, game changers, and challengers. Wiley
Prendeville S, Cherim E, Bocken N (2018) Circular cities: mapping six cities in transition. Environ Innov Soc Trans 26(2018):171–194
Reichel A, De Schoenmakere M, Gillabel J (2016) Circular economy in Europe, Developing the knowledge base. European Environmental Agency (EEA), Report No 2/2016. Available at: https://www.eea.europa.eu/publications/circular-economy-in-europe
Saleh R, Ost C (2020) Introduction to perceptions mapping: the case of Salerno, Italy. TRIA – Territorio Della Ricerca su insediamenti e ambiente. Italian Scientific Publishing, Naples. https://doi.org/10.6092/2281-4574/6639
Saleh R, Drouillon P, Ost C (2020) D4.5 circular business model workshops for cultural heritage adaptive reuse. Deliverable. CLIC Consortium

Sani M, Lynch B, Visser J, Gariboldi A (2018) Mapping of practices in the EU Member States on Participatory governance of cultural heritage to support the OMC working group under the same name (Work Plan for Culture 2015–2018). European Expert Network on Culture EENC
Schaltegger S, Hansen EG, Lüdeke-Freund F (2016) Business models for sustainability: origins, present research, and future avenues. Organ Environ 29(1):3–10. https://doi.org/10.1177/1086026615599806
The Burra Charter (2013) The Australia ICOMOS Charter for places of cultural significance. Available at: https://australia.icomos.org/wp-content/uploads/The-Burra-Charter-2013-Adopted-31.10.2013.pdf
The Collaborating Centre on Sustainable Consumption and Production (CSCP) (2018) Circular economy guidebook for cities. Report. Available at: https://www.scp-centre.org/wp-content/uploads/2019/03/Circular_Cities_Publication.pdf
The Declaration of Amsterdam (1975) Congress on the European architectural heritage. 21–25 October 1975. Available at: https://www.icomos.org/en/resources/charters-and-texts/179-articles-en-francais/ressources/charters-and-standards/169-the-declaration-of-amsterdam
Thelen D, van Acoleyen M, Huurman W, Thomaes T, van Brunschot C, Edgerton B, Kubbinga B (2018) Scaling the circular built environment. Pathways for business and government, Arcadis, World Business Council for Sustainable Development (WBCSD), and Circle Economy. Report. Available at: https://www.wbcsd.org/contentwbc/download/6173/85923/1
UN Environment (2019) Global resources outlook. Natural resources for the future we want. Implications for business leaders. Report. Available at: https://www.resourcepanel.org/reports/global-resources-outlook
UNESCO (1972) The world heritage convention. Available at: https://whc.unesco.org/en/conventiontext/
UNESCO (1976) Recommendation concerning the safeguarding and contemporary role of historic areas. Available at: http://portal.unesco.org/en/ev.php-URL_ID=13133&URL_DO=DO_TOPIC&URL_SECTION=201.html
UNESCO (2001) Underwater Cultural Heritage Convention. Available at: http://www.unesco.org/new/en/culture/themes/underwater-cultural-heritage/2001-convention/official-text/
UNESCO (2003) Convention for the safeguarding of the intangible cultural heritage. Available at: http://portal.unesco.org/en/ev.php-URL_ID=17716&URL_DO=DO_TOPIC&URL_SECTION=201.html
UNESCO (2005) Convention on the Protection and Promotion of the Diversity of Cultural Expressions. Available at: https://en.unesco.org/creativity/convention
UNESCO (2011) Recommendation on the Historic Urban Landscape. In: Records of the general conference – 31st session
UNESCO (2013) The Hangzhou Declaration Placing Culture at the Heart of Sustainable Development Policies. www.unesco.org/new/fileadmin/MULTIMEDIA/HQ/CLT/images/FinalHangzhouDeclaration20130517.pdf
UNESCO (2016) Culture urban future: global report on culture for sustainable urban development. Summary. www.unesdoc.unesco.org/images/0024/002462/246291E.pdf
United Nations (1992) The Convention on Biological Diversity. Available at: http://www.pngcepa.com/wp-content/uploads/2018/07/CBD-Convention.pdf
United Nations (2015a) Transforming our World: The 2030 Agenda for Sustainable Development. www.sustainabledevelopment.un.org/post2015/transformingourworld
United Nations (2015b) Habitat III issue papers 4 – urban culture and heritage. https://habitat3.org/documents-and-archive/preparatory-documents/issue-papers/
United Nations (2016) HABITAT III Draft New Urban Agenda. www.habitat3.org/the-new-urban-agenda/
Upward A (2013) Towards an ontology and canvas for strongly sustainable business models: a systemic design science exploration. Masters thesis, York University, Toronto, Canada
Upward A, Jones PH (2015) An ontology for strongly sustainable business models: defining an enterprise framework compatible with natural and social science. Organ Environ, Special

Issue: Business Models for Sustainability: Entrepreneurship, Innovation, and Transformation (On-Line First), 1–27. https://doi.org/10.1177/1086026615592933

Van Balen K (2008) The Nara Grid: an evaluation scheme based on the Nara document on authenticity. APT Bull J Preserv Technol 39:2–3

Van Renswoude K, ten Wolde A, Joustra JD (2015) Circular business models—Part 1: an introduction to IMSA's circular business model scan. IMSA, Amsterdam

WBCSD and Climate-KIC (2018) Circular metrics. Landscape analysis. Report. Available at: https://docs.wbcsd.org/2018/06/Circular_Metrics-Landscape_analysis.pdf

World Business Council for Sustainable Development WBCSD (2017) 8 Business cases for the circular economy. Report. Available at: https://www.wbcsd.org/contentwbc/download/3598/47010/1

World Economic Forum (2018) Circular economy in cities evolving the model for a sustainable urban future. While paper. Available at: https://www.weforum.org/whitepapers/circular-economy-in-cities-evolving-the-model-for-a-sustainable-urban-future

Zancheti SM, Jokilehto J (1997) Values and urban conservation planning: some reflections on principles and definitions. J Archit Conserv 3(1):37–51. https://doi.org/10.1080/13556207.1997.10785179

Zott C, Amit RH, Massa L (2011) The business model: recent developments and future research. J Manag. https://doi.org/10.1177/0149206311406265

Chapter 17
Business Models for Cultural Heritage Adaptive Reuse

Immacolata Vellecco and Assunta Martone

1 Introduction

Literature on business models is a pillar of management research and its recent contributions have broadened traditional frameworks to include sustainability and circular economy issues. However, contributions focused on business models of cultural heritage have always been scarce; moreover, most of the studies on adaptive reuse of cultural heritage are based on the analysis of individual assets. A few studies make in-depth analysis of the business model and they hardly produce generalizable results, due to the idiosyncratic nature of cultural heritage and its link with the cultural, social, environmental and economic context, that makes it difficult to replicate the process and decisions of reuse (in structural and managerial terms).

A wide literature in business and management sciences applied the business model conceptual framework to different industries companies. Recent research tried to apply the concepts developed in business model studies to circular economy, mainly focusing manufacturing industries and new waste management strategies but Business Model perspective and circular economic approaches are completely absent in Cultural Heritage adaptive reuse, while on this issue the perspective of the public economy appears very relevant.

Cultural heritage recovery and maintenance has traditionally been in charge to of the public sector, which used public resources to maintain heritage "public goods" in optimal state of conservation and accessibility. Moreover, the public sector (Institutions) has the role of setting the "rules" for heritage conservation, maintenance and reuse, especially to avoid destructive interventions, especially on heritage goods owned or managed by private actors.

I. Vellecco · A. Martone (✉)
IRISS Institute for Innovation and Services for Development, CNR National Research Council, Naples, Italy
e-mail: i.vellecco@iriss.cnr.it; a.martone@iriss.cnr.it

L. Fusco Girard, A. Gravagnuolo (eds.), *Adaptive Reuse of Cultural Heritage*,
https://doi.org/10.1007/978-3-031-67628-4_17

Public investments have mainly focused on the evaluation of cultural heritage adaptive reuse projects highlighting social and environmental costs and benefits. However, traditional public funding sources and in general public financial resources are decreasing. In this context of fragile public finance, business model and economic sustainability of reuse projects are increasing their importance in order to leverage private investment and to ensure economic self-sustainability, with the main goal to avoid waste of public resources.

An increasing overlapping of perspectives has recently attained to Cultural Heritage, some of them within the theories of New Public Management, merging business service approach and governance issues, as cultural heritage management has to comply legal, economic and organizational specificities. Furthermore, the value generated by reuse initiatives has ambiguous definitions and a plurality of stakeholders and the presence of network economies make the decision-making processes very complex.

The study of Misirlisoy and Günçe (2016) offers holistic approach and unified factors for the successful implementation of Cultural Heritage adaptive reuse projects, paying attention to the decision making process and on the role and involvement of different stakeholders. The study also recommends deep analysis of the existing fabric, which includes original function, physical characteristics, adaptive reuse potentials and needs of the district. The decision of the new function according to the needs of the region is important in terms of the life of the adaptive reuse project.

Conservation actions should be decided and adaptive reuse potentials of the architectural for the new use should be evaluated. The main aim should be preserving the values and originality of the building and its context; however, the economic sustainability of the building is important for the future of the built heritage.

A relevant stream of research focus on the role of Private Companies in the whole Project Cycle, as private partners may contribute at different stages (i.e. Project design, Finance, Build, Operate) and with different roles (Promoter, Sponsor, In-kind contributor, Evaluator). Furthermore, their role may vary from delivering non-core/outsourced services to a full project cycle involvement. Neverthless, institutional PPP differ from public procurement as well as from privatizations (Macdonald 2011; Macdonald and Cheong 2014), and imply risks and responsibility sharing, often strictly regulated by national (country specific) laws.

The majority of research efforts have focused on case studies. Although this research design prevents findings generalizability to different industries, contexts, or countries, it is still appropriate to study the early state of art of circular business models in Cultural Heritage Adaptive Reuse (CHAR).

The following part of the work offers a literature review of the evolving concept of business model toward sustainability and circularity paradigms, which caused new element addition to the old frameworks in order to enlarge the analysis to determinants and impacts of the business model, which can be studied at enterprise and/or at network level.

A further paragraph describes a set of European reuse projects, starting from the typology of building and location and exploring the new possible use, mainly focusing on the new use function given to heritage buildings of different typologies and set in rural or urban area of different countries. New uses are the base to pitch some typologies of business models and for some generalization that are proposed in a separate further paragraph.

Conclusions remind some limitation of the analysis, suggesting further avenues of research.

2 Literature Review

In business disciplines, business model (BM) is a unit of analysis to explain how value is created and delivered. BM is seen as an antecedent of heterogeneity in firm performance; specifically, BMs are argued to be an important factor contributing to firm performance. As some types of BMs are found to outperform others, successful BMs are seen as examples to be imitated or replicated.

The most well-known and widely used framework is the Business Model Canvas (Osterwalder and Pigneur 2010), based on nine elements: key partners, key activities, key resources, value proposition, customer relationships, channels, customer segments, cost structure, and revenue streams.

Fielt (2014) also notes that it is hard to comprehend a definition of business model without a better understanding of the value concept. He explicitly includes the customer value (or use value) where other definitions are less clear by referring to value in general or include business value (or exchange value). The focus is on the value creation from the customer perspective and linking value creation to value capture. Moreover, while the focus is on the organization, the business network needs to be included as well, when it plays a critical role in creating and capturing customer value. However, a more strategic perspective is required to fully understand value capture and business value.

The emphasis given to organizational, market and societal transformations distinguishes the discourse on business models for sustainability from their conventional antecedents, which focus on organizational value appropriation, that is, one-dimensional profit maximization, without considering the consequences for the wider social and ecological contexts. In the emerging field of sustainable business model (SBM), an increasing number of scholars and practitioners go beyond value creation in economic or financial terms and explore the potential of business models to solve ecological and social problems. (Bocken et al. 2014; Lüdeke-Freund and Dembek 2017; Schaltegger et al. 2016). Several authors describe iconic cases of companies that aim at reducing the pressure on ecological and social systems through their business models (e.g., Boons et al. 2013; Stubbs and Cocklin 2008). Some of these cases and business model types bear the potential to provide inspiration or even useful solutions for established companies and start-ups facing similar challenges.

Taking a "strongly sustainable" perspective, Upward and Jones (2016) formulate four propositions:

1. A strongly sustainable business model creates ecological, social and economic value and takes its embedding value network into account, which implies an extended understanding of the value that is proposed, delivered and finally created.
2. The concept of value itself broadens to forms of value that meet the needs of actors in aesthetic, psychological, physiological, utilitarian and/or monetary terms.
3. This extended perspective on a business model's value network and extended understanding of value requires a systemic conception of business models as being embedded within wider ecological, societal and economic contexts.
4. A new kind of metric, "tri-profit", is required to integrate all forms of value creation into one single measure, instead of measuring these in parallel, as with traditional triple-bottom-line approaches.

Joyce and Paquin (2016) suggest the Triple Layered Business Model Canvas as a tool for exploring sustainability-oriented business model innovation. A holistic impact approach is proposed, linking economic impacts to social and environmental ones. It also creates two new dynamics for analysis: horizontal coherence and vertical coherence. This tool fits a decision-making approach, both in a business and in a policy maker perspective.

Furthermore, the business model perspective is particularly interesting in the context of sustainability because it highlights that new value creation logic of an organization potentially allows (and calls) for new internal governance forms such as cooperatives, public private partnerships, or social businesses, thus helping transcend narrow for-profit and profit-maximizing models (Schaltegger et al. 2016).

Circular Economy oriented BM also add uncertainties and complexity to conventional BM. Firm activities play an important role in the various conceptualizations of business models that have been proposed. New variables have to be considered, for instance, reverse on top of forward logistics; quality, quantity and timing of returns of resources; customers perceptions and preferences for 'as new' (Bocken et al. 2018). This requires a systemic and transdisciplinary view, which has been reflected in recent publications exploring the interfaces of CE-oriented Business Model Innovation (BMI) with other innovation perspectives, such as product design, value chain and digital technologies (Bocken et al. 2016; Foss and Saebi 2017; Geissdoerfer et al. 2018).

Pieroni et al. (2019) provide a review of approaches for business model innovation for circular economy and/or sustainability, finding opportunities to seize synergies from the intersection of both streams. They acknowledge resource efficiency, resources longevity and economic growth at the intersection of Circular Economy and Sustainable Development approaches.

Urbinati et al. (2017)'s taxonomy suggests three available modes of integrating CE principles in BMs: *downstream circular* (altering value capture and delivery, through new revenue schemes and customer interface e e.g. pay-per-use models), *upstream circular* (changing value creation systems, e.g. reverse logistics), or *fully circular* (combining upstream and downstream principles).

Circular business models may be also studied taking into account:

- Business ecosystem level (Antikainen and Valkokari 2016): (i.e. Trends and Drivers, Regulation; Financing or technology opportunities; consumers consciousness as well as stakeholders involvement and policy commitment);
- The Adoption factors (Lewandowski 2016), as transition towards circular business model must be supported by various organizational capabilities and external factors.

According to this view, circular business models in Cultural Heritage adaptive reuse require a wider perspective, overcoming the focus on the micro-business perspective and enlarging the view to an extended stakeholder's network, as the value proposition is the result of a negotiation process among different stakeholders both in decision making and in financing, building and operating.

The impact of the circular economy models and sustainability should measure value creation for all stakeholders and, then, the challenge of re-designing business ecosystems is to find the "win-win-win" setting that balances the self-interests of involved actors and sustainability impacts.

3 Methodology

This work investigates 34 case studies of cultural heritage adaptive reuse.

As the typologies of building were considered relevant in the decision and success of the reuse projects, the analysis focuses on coastal buildings (lighthouses), religious buildings (monasteries and churches), forts and castles, industrial buildings, and minor buildings. Furthermore, some rural and urban cases highlight connections between reuse initiatives, new functions of the buildings and local (contextual) needs.

Building theory from case studies is a research strategy that involves using one or more cases to create theoretical constructs, propositions and/or midrange theory from case-based, empirical evidence (Eisenhardt 1989). The central notion is to use cases as the basis from which to develop theory inductively. The theory is emergent in the sense that it is situated in and developed by recognizing patterns of relationships among constructs within and across cases and their underlying logical arguments. Central to building theory from case studies is replication logic (Eisenhardt 1989). That is, each case serves as a distinct experiment that stands on its own as an analytic unit. Like a series of related laboratory experiments, multiple cases are discrete experiments that serve as replications, contrasts, and extensions to the emerging theory (Yin 1994).

According to Lambert (2015), it is widely recognized that classification is a necessary step in understanding a research area, however throughout history there has been continuous debate about the best way to classify objects, what criteria to use, and what purpose the classification can serve.

Each of the many classifications is conceived to meet the specific needs of the researcher, and they vary considerably in terms of purpose and the scientific rigor used in their development. Some classifications are constructed using a large number of business model characteristics and potentially serve a relatively wide range of purposes and others are based on a small number of business model characteristics, serve specific purposes and, consequently, facilitate only a limited range of generalizations.

Furthermore, according to Ritter and Lettl (2018), business model can be seen as a theoretical mechanism for combining different literature streams. As such, business-model research is positioned as a central connecting component in the further development of strategic management field of Cultural Heritage Reuse.

Then, the following analysis mainly focus the pattern of "original use"/"new use", studying connections between typologies of buildings (as resources) and new functions, chosen in the view of market need, users and forecasted demand, which includes the respect of the four pillars of sustainable conservation (economic, social, environmental, cultural).

4 Old Buildings and New Uses: An Overview

4.1 Coastal Buildings—Lighthouses

The traditional economic analysis considered the lighthouse as a classic example of public good: those who don't pay cannot be excluded from consuming it, and one's consumption does not reduce the consumption of others. So, general taxation funds public production (Stuart Mill, 1848). This old vision was questioned by Ronald H. Coase (1991 "Nobel prize") in "The lighthouse in economics" (1974), he outlined that in the lighthouses system, in England and Wales, sixteenth to nineteenth century, private individuals embarked on financing, building, and maintaining numerous lighthouses.

Nowadays lighthouses are often decommissioned. They become obsolete due to changes activated by new technologies like GPS and sonar. Therefore, lighthouses are at risk of deterioration while rehabilitating them is a difficult challenge also because represent an ancient heritage.

There are several experiments of adaptive re-use of buildings and maintenance projects. The lighthouses positioned close to urban agglomerations sometimes turn in clock towers (Old Colombo Lighthouse in Colombo, Sri Lanka).

Some turn in art studio—as an "artist in residence" program operating in Port Bickerton lighthouse in Nova Scotia—or art space as Lighthouse in Maryport in the northwest of England, which in 2009 was used as a gallery space by an arts collective. In other cases, natural scientists develop lighthouses into animal observatories or wildlife refuges (Low Light Lighthouse on the Isle of May in Scotland; Seahorse Key lighthouse in Florida).

The Northern Lighthouse Board (NLB) responsible for Scotland and the Isle of Man and the General Lighthouse Authorities (GLAs) of the United Kingdom and Ireland developed a new initiative for the Scotland's Outstanding Lighthouses. The initiative aims to promote and drive tourism to Scotland's coastal communities and increase awareness of the role and history of NLB's unique heritage.

Exploitation of lighthouses through tourist activity contributes to their protection and revitalization. Their involvement in tourist offer would not obstruct their primary role in assuring the safety of sailing. The Croatian lighthouses are an example on how they can contribute to Croatian tourist offer as a unique tourist category.

Palagruža is the most attractive site in this group. Due to equipment expenses, regular maintenance, transport and staff, tourism rent fee is usually high and increases due to the transfers to and from the site. Maintenance expenses are very high, as lighthouse inventory decays rapidly because of exposure to weather conditions and salt and this affects profit gain. Distance from the mainland and nearby inhabited villages affects operating costs. Connection to the mainland and to local water supply network allows having water supply costs significantly reduced. Water supply expenses for Palagruža island lighthouse take up to 50% of the income, while these expenses for lighthouses on the mainland take up only 15–20% of the income. Then, lighthouses may be a fruitful field for development of circular economy as eco-innovation, applying water saving application and green energy technologies.

Lighthouses enhancement attempts to prevent their falling into a state of decay, regenerating the places and helping the activation of local economies in favor of citizenship, enriching the public heritage of refurbished structures for the community.

In recent years, the Italian Agenzia del Demanio (State Property Agency) together with Ministero della Difesa, through Agenzia Difesa Servizi Spa, has activated the "Valore Paese Fari" initiative, currently being implemented, with the aim of increasing the economic and social value of the assets and territories in which the lighthouses are sited, thus contributing the competitiveness of the entire Country. The aim is to recover public assets, owned by the State and local authorities, so that it is no longer just a cost for the community, but also a lever for territorial and social development, based on public-private partnership in management and/or financing reuse initiatives.

The Genova lighthouse is the only one in the world to be a symbol of a city and repository of a large part of its history. The construction of the base seems to date back to 1128, an age in which Genova was a maritime independent republic and one of the powerful cities in the Mediterranean. The historic value of the lighthouse and its symbolic value made it a cornerstone of the old waterfront reuse project. The result is a leisure and meeting area for citizens and tourist, where the ancient maritime identity of the city finds new ways to increase the social capital. The case suggests a light stress on direct economic returns on the single building reuse, adequately considering also indirect economic effects, due to the social and economic regeneration of the surrounding area.

As point of interest in environment education and research, lighthouses similar to Capo d'Orso may catalyze environmental cultural change and sharing. More generally, lighthouses are well suited to a request for tourism that is attentive to the environment and culture, connected to unspoiled places and places of landscape-environmental interest.

4.2 Religious Heritage

The religious real estate patrimony presents countless properties with high historical and artistic value which, also due to the current period of vocation crisis, have not received the attention necessary for optimal conservation, and which, in many cases, have been closed and abandoned.

The religious structures are embedded in the collective memories of the members of a society and are a source of identity; therefore, the public becomes concerned with maintenance the integrity of these edifices and conserving their cultural heritage symbols. Therefore, over time the protection of the religious architectural heritage has been increasingly recognized as a cultural obligation.

The proper reuse of buildings is one of the best ways to ensure their survival and a change of the original function has to keep the intangible meaning in the suitable reuse. Then, it is difficult to manage the limits and opportunities in the adaptive reuse of this type of heritage, while preserving its social and cultural significance.

In recent years, the decline of religious practices and the economic crisis have led to the abandonment of countless structures, often sold and privatized.

Cases show different uses of religious building, not only because they may be different in their structures (monasteries, churches) or in their state of conservation, but also because different communities may need different services and may feel different sensitivity to the aesthetic and/or authentic features.

Monasteries structures usually fit to hospitality function and the reuse as hotel has to question to what kind of customer targets address the services, choosing appropriate marketing channels, adequate and sustainable price for value, also profiting from other contextual attractive. Nature-based hospitality may be more adequate for rural areas or coastal zones, while high class hospitality may occur in urban historic areas, but alternatives are also offered if the religious owner also manages the site as a social business, hosting students or supporting social tourism.

Churches offer a larger number of reuse opportunities; some of these are very unusual, others are foregone, as the use as concert hall, museum or as art/handicraft showroom.

The Dominicanenkerk (Dominican church) is a Gothic monastery church situated in the city centre of Maastricht. The church was built in the thirteenth century while in 1796 the church's ecclesiastical function ended, and in 2006 got a new destiny. A bookshop was established inside the church. The major renovation went smoothly, thanks to a successful public-private partnership between the municipality, monument care, project developer and the new user Selexyz. The architect Merkx + Girod from Amsterdam won the Lensvelt Architectuur Prijs in Holland in

2007. The sacral elements such as the stained-glass windows, fresco's, vaults and the incidence of light have been saved.

Partnerships with important cultural players (as a University) can keep the mission of the building fitting to the original goals, enhancing human cultural and spiritual wellbeing, and educational scope.

The Charterhouse of Avigliana was founded as a Franciscan monastery in 1515. Over time, had several destinations and the community of Carthusian nuns who lived there decided to leave it. The Abele Group took action for the purchase and restoration, which ended in September 2011. The Certosa di Avigliana is currently owned by the Social & Human Purpose Fund of REAM sgr of Turin and managed by the social cooperative Binaria 1515 scs. Today it is a place of hospitality, education and location for events.

A very important question is limitation to new use that religious owner can impose both in the property transfer contract and in the rent or free use agreement. Uses in open contrast with the original and sacred function of the building are usually not admitted. Although limitations bear only the users or the first buyer, they reduce the typologies of entrepreneurial ventures which can settle in the site, and social enterprises seem the most adequate initiatives to be hosted, mainly in temporary use agreement.

Nevertheless social enterprises, even able to ensure the economic self-sustainability of their service delivery system, rarely can arrange a large amount of economic resource for the large investment required to restore or renovate a church. So, when the latter is in a very bad state, further partners might be necessary in order to provide financial support, preferably as grants (crowdsourcing community, social responsible corporation, private foundation), in order to prevent the burden of a loan repayment on social enterprises venture, usually having a fragile economic balance.

4.3 Forts and Castles

Forts were usually military buildings, aiming at defending the territory against enemies. They are robust and very large structures which have been surviving many centuries. They usually have high historical value as their storytelling is the narrative of wars changing people history and culture. This is the reason why they are usually state owned, although charging public finance with high restoring and maintenance costs.

Reuse projects are a great challenge as they need to keep historical and cultural meaning furthermore producing direct and/or indirect economic impacts.

In the following cases show how forts can be turned into cultural tourism attractions, saving historic heritage buildings for next generations.

The Hotwalls Studios are an enviable location with access to the waterfront and harbour views located in Old Portsmouth, UK; the reuse project as studios combined social goals with tourism attraction, reusing the site as a creative art district.

Fort Vechten and Fort Resort Beemster alongside the 85 km long "New Dutch Waterline defence system", in Netherlands. These forts are included in the same cultural route and base their attractive on same natural resource: water. Fort Vechten seems to target families and children as visitors, offering didactic and virtual experiences; this business model—high revenue oriented- may produce better economic results but may also conduct toward a Disneytization of the cultural heritage resources. Fort Resort Beemster uses water as a cultural and natural resource, adding wellness services to cultural ones.

Fort Monostor, that is part of the system of historic forts situated on opposite banks of the Danube River. It is a shared heritage resource across state borders, and its reuse project, still ongoing, is under the management of a Special (Project) Company, which may (or may not) manage the running phase when the building will be full readapted.

Suomenlinna, a former naval military fortress system (eighteenth century) near to Helsinki, Finland, and it is now a tourism attraction for over one million of visitors per year, and the success has been reached due to a Governance body very attentive to save the beautiful landscape and the precious natural environment from overexploitation.

Forts and Castels reuse have to manage a system of goals, saving cultural value and producing direct and indirect economic impacts. Most of them are turned into tourism attraction, and the package of services can include accommodation, entertainment, and wellness services, as well as museum areas, live performance. Large spaces can be also rent for conferences, meeting and other social and cultural events. Most of them are listed buildings, others are going to be.

All these cases show a certain awareness of circular economy fundaments, and social and economic impacts seem more to guide reuse projects, also taking into account cultural values, landscape quality protection and natural resources saving.

4.4 Industrial Heritage

Buildings for industrial production were usually located near provisional areas of raw material, near ports or near densely populated settlements where it was easy to hire cheap workers. Then factories and their surrounding areas (with workers houses, schools, hospital, churches, municipal building, open spaces and other facilities) gained a visual identity according to building materials and technologies of the time.

The economic development of western economies, the decline of industrial production and the contemporary transition toward a service economic system often left industrial quarters or cities largely disused, as a consequence of a decreasing population and of people migration to more vibrant and modern cities, searching new working opportunities.

Large disused building or whole city quarters have become a great resource and opportunity to reuse and regenerate, strengthening new social and economic place identities. The success of the reuse initiatives also depends on appropriate decisions regarding the core and ancillary services the building has to provide, according to urban planning, based on visionary forecasting of the emerging demand.

Industrial buildings (and mainly ex manufactures), although considered historical heritage, are usually not listed and provide wide room to be adapted, at affordable cost and without constraints to restore as it was. Thus, architects can choose to mix original structures, materials and style with contemporary ones, also adding new modern spaces to old structures and profiting of advanced technologies in the project and operation phase, as well as new materials and solutions, that increasingly boost resource saving and green technologies.

New uses can aim at tourist attraction, as the cases of Le Grand Hornu and C Mine, large complex in ex coal districts, in Boussu and Genk (Belgique), small towns in need for economic regeneration and a new image.

Le Grand Hornu is a neoclassical former mining complex, recently reused as a cultural centre for innovation and design. C Mine was a mining complex too, reused as a recreational and educative site, also boosting artistic entrepreneurship. Both the cases are well-known practices of cultural and economic regeneration in heritage tourism (at C Mine largely based on the tourism experience paradigm), creative economy and art development, with indirect economic impacts in terms of new business ventures and jobs. It should be noted that both the regeneration projects were managed by a Special (Project) Company, able to fully integrate the regeneration phases and the network of partners in the project.

Other reuse projects of ex industrial buildings aim at creating innovation centers, fostering social networks, innovation partnerships and new entrepreneurial ventures. This business model frequently named as Hub, is supportive of the local development, offering services addressed to main local industries. Some examples are:

- Simonsland, former textile industry in Boras (Sverige) reused as a multifunctional fashion center;
- Inredia, and old shirt fabric whose new mission is linking furniture industry with interior design;
- The Lichttoren (Light tower) a factory for Philips, a light bulb manufacturing company in Eindhoven which has been adapted as a living, working and leisure complex;
- Brew House, a former brew fabric in Goteborg, recently reused as a "culture fabric", as a venue for musical events and as business support for creative start-ups.

Simonsland and Inredia address the needs for design and innovation of local manufacture (furniture and textile). Brew House is a culture oriented hub; it couples the mission of talent incubator for cultural and creative industries with commercial high rewarding activities in hosting and organizing event and concerts; this is the typical hybrid model of non-profit organization as it is. The site owes the success to the ability in managing a wide network of service providers and customers.

The Lichttoren has more conventional goals of residential, working and leisure complex, although the concept of "loft" as spatial module to rent or buy was very innovative at the time of its reuse project.

Further reflections can also be proposed on all the cases: alliances and strong partnership are fundamental drivers of the success both in the project concept and operation phase and in the delivery of the new services the building was re-functionalized to.

4.5 Minor Heritage Sites

The lack of financial resources together with the constraints on interventions on the built cultural heritage are threatening the survival of the buildings with the risk of losing the benefits that they can potentially offer. When the built heritage is no longer able to fulfil its initial functions because they are no longer useful or because it is too expensive to provide them, the risk of the structures being abandoned becomes high. Nevertheless, demand for new spaces and for new services may result into unexpected solutions, both in the choice of spaces to reuse and in the services they offer.

Some interesting cases need to be cited:

- The case of Cavallerizza Reale used as stables at the Royal Palace in Turin, stands out as a valuable example of civic commitment towards cultural heritage, as a community anticipated a privatization decision limiting the use of heritage, taking a bottom-up action to revitalize the building through innovative financing (crowdfunding).
- Dynamo was an historic abandoned space underneath near the central train station in Bologna; it is a valuable example of circular economy in reusing a neglected space to promote sustainable mobility by bike sharing.
- The Bourbon Pheasant standing inside the Royal Park of Capodimonte in Naples was reused during Covid-19 emergency as vaccine hub, also offering users cultural information about the Museum collections and a coupon for free entrance to the next temporary exhibition. In this way, the use during the Covid-19 phase was turned into a chance to renew the attachment of the citizens to one of the most precious museum and park of the city. The initiative is also a lesson about the use value of minor heritage buildings, as their lower artistic importance enables more flexible and quick reuse solutions.
- The reuse of the historic gardens and greenhouse of Regina Margherita in Bologna show how gardens reuse can accelerate social innovation and creative community's start-ups, also highlighting the role social enterprises (in this case, a social cooperative) as partners and managers of reuse projects aiming at creating social value.

4.6 Rural Heritage

Rural buildings are often set in wider rural areas, and both may be object of reuse projects, with the aim of increase the value produced by these resources. Reusing local buildings and areas for recreation and leisure can activate economic dynamics owed to the attraction of tourism, generating new jobs and profits, as tourists demand for goods and services generates new jobs and profits.

Building reuse can also benefit social ties, as cultural initiatives can catalyze local inhabitants, providing space for young and old people, increasing young people place attachment and encouraging an active aging of the old ones.

Place attachment and social ties are a key resource in rural environment, moving people to cooperate, and acting bottom-up, bypassing bureaucracy and procedure to obtain financial resource. Some reuse initiatives described below have been carried out by local communities only relying on its own work. This cultural mind-set can also avoid external speculative incoming actors which in the medium or long term can totally change local identity and landscape.

Different reuse initiatives in the following examples highlight different options and drivers of rural building and landscapes. Some of them are based on local community cooperation and work and they clearly aim at improving people wellbeing while further economic goals can be reached without being stressing priorities. Small, personal and in-kind investments are sufficient resources to reach the projects goals. The following example can be indicated:

- Lanckorona Ecomuseum, creating a museum network spreading on four rural municipalities near Cracow;
- the project "Adopt a terrace" in the Brenta River Valley, reusing a former system of intensive tobacco cultivation, by assigning free use of small pieces of lend to local applicants for cultivation and gardening, in order to maintain and enhance the landscape;
- Škratelj Homestead, a ruined stable reused as cinema and social hub;

More ambitious projects are ReDock and H-Farm projects, applying leading edge green technologies to minimize the impact on the landscape, albeit the extension of the complex and the large and knowledge intensive community which has settled in the area.

ReDock reused a medieval village as tourism attraction, fruitfully applying the best eco-innovation technologies, in the view of creating a sustainable eco-friendly community. The village aims to be a blueprint for a sustainable future in the countryside. As circular economy strongly promotes natural resource saving and reuse, this project could be more in depth analyzed to learn more about problems and solutions in applying leading edge technological eco-innovation in a medieval village. Furthermore, an interesting field of research is the evaluation of middle-long term social impacts of the project on the local community, with a special attention to migrations and demographic trends.

H-Farm is a case of fast-growing, fast-changing area, reusing an ex rural building surrounded by the agrarian land near Venice and Treviso. Cooperation between Municipality, University and High tech industry transformed the place into a Hub for digital technologies where innovation, education and entrepreneurship are designing a new era.

4.7 Urban Sites

Heritage building reuse needs to consider the surrounding framework and, especially in urban context, heritage building reuse projects need to connect past and future identities that cities have experienced. Sustainability, as a function of increasing quality of life of urban residents is a driving force of social and ecological innovation inspiring cities planning and urban development at national and international level. So, the reuse project for a single building must set multiple contextual goals but standing alone rarely induces a leap forward for the entire City, toward ecological and social values. On the other hand, although multiple reuse and conservation initiatives can result as sustainable at a micro-level, a change in the view is ongoing at urban level, linking sustainability goals of the area and their monitoring to massive use of ICTs. Neverthless, an adaptive reuse project needs to take into account previous historical artefacts when they are of exceptional value, and repurpose ideas can be changed if communities acknowledge the importance to keep memory of the past. Some metropolitan initiatives described in the following can give examples or inspiration.

The House of Vans, a former underground tunnels, near London Waterloo station, has fruitful reused a neglected space which can be precious in high dense urban context where any square meter of soil can reach skyrocketing prices. Keeping the old visual identity, the project has reinforced the uniqueness of the new space and functions.

Pianofabrik is the reuse project of an ex industrial building (an old laboratory of piano makers) now complementing with the existing cultural offer of the city of Bruxelles. The project aimed at creating a space playing the role of multifunctional hub for the intercultural and international community of Bruxelles, fostering social networks, innovation partnerships and new entrepreneurial ventures. A leading edge cultural offer and events couples with support to arts, entrepreneurship and bottom-up initiatives. The site owes the success to the ability in managing a wide network of service providers and customers, also promoting diversity and inclusiveness as social values.

The Kultur-Token (KT) is one of the Vienna's Smart City initiatives to enhance city life whilst addressing the global challenge of climate change with local greenhouse gas/air pollution reductions. By targeting culture and cultural heritage as rewards, the Kultur-Token business model recognizes and celebrates culture and cultural heritage's role as a multi-faceted and valued "commodity" in the future sustainable city.

El Mercat del Born, an historical market in Barcelona, now hosting a cultural center, is a case of a very long "stop and go" rehabilitation process, offering important elements to comment. Different ages of the city history have found connection in the site due to the last reuse project, which is able to show different historical layers from XII century to the siege of the city of Barcelona in 1714, to contemporary age. Furthermore, the meaning of the site strictly connects city history and past events to current civic proud, and this connection is supported to last by the current reuse functions of the site as a cultural and memorial centre. The case also offers a fruitful field for social research to study the link of community action in cultural heritage reuse to community awareness and knowledge of its history.

5 Patterns of Reuse as Business Models

Case studies analysis shows the importance of the building typology in enabling new reuse functions, supporting the study of Misirlisoy and Günçe (2016) and highlighting some typical (and replicable) patterns and business models.

Reuse models seem strictly linked to the context needs and local development trends. Tourism destinations often reuse heritage building to offer new attractions able to increase tourist flows or to increase the staying of tourists. Projects of reuse may be highly costly and often require specialized designer to make into practice new ideas, while high level marketing have to promote the attraction through different tourism channels. Investment costs couple with high running cost, making public private partnerships riskily and heavy.

Forts and castles, like the lighthouses, are often reused as tourism attractions, although they offer a larger mix of services; virtual experience and historic narratives can couple with different leisure and entertainment services, fostering more frequent access both by residents and nearby inhabitants. The goal of economic self-sustainability can be easily reached when a flexible renting of spaces for single events increases revenues. Monasteries can be easily reused as hospitality structures, targeting social goals and accessible tourism and/or rural tourism, profiting of their remote and wild location. Otherwise, building reuse projects can target high class experience enhancing the original features of the built environment coupled with the excellence of the service delivery system.

Churches structures offer a larger set of choices in the activities they can host, albeit limits can be imposed to initiatives in open contrast with the original sense of the place.

The Multifunctional hubs, located in industrial heritage or other sites (rural, urban, etc.), provide spaces for people networking and entertainment. This general trend is usually conceived according to the history and the evolving needs of the local community, delivering business services complementing historic industrial specialization of the local "milieu", otherwise fostering new business start-up in cultural and creative industries, or technological ones. An efficient planning of use and renting of the space is the mean to reach (and overcome) economic self-sustainability, hosting high revenue initiatives alternated with low profit or social ones.

Projects involving clusters of buildings need an area based strategy: residential use needs to be supported by services and infrastructures, while office uses need to take into account different facilities and the evolving occupational trends. Clusters of building may also support tourism and cultural industries, creating networks of museums or branded hospitality, profiting of shared distribution channels.

Nevertheless, reuse supporting tourism development needs to consider modern success factors of tourism destinations (branding, themes, experience, online booking) as well as a life-cycle approach to the destination itself. These uses also require an adequate supply of hospitality services and infrastructure and a high-quality context (architectural and relating to landscape).

Therefore, although the building typology plays a very important role enabling, or sometimes hindering, adaptive reuse and new functions, local context matters too. Rural buildings can give local communities space to create and manage new ventures, also fostering social cohesion through entertainment, hobbies, cultural initiatives as cinema and museums. Heritage building reuse in rural areas near large towns or metropoles enhances rural areas attraction, also equipping them with residential or leisure services, aiming at people living, enjoying and wellbeing in a natural and green environment. Some initiatives can also bring about a rapid turnaround in development, as new players can provide ideas and financial resources to change the local milieu, provided they find some consensus and support at the local community level.

Some reuse cases within urban contexts are generating economic and social value by neglected buildings or spaces, in high dense populated and built context; a strong demand for space and soil makes more spaces and building attractive to renovate and reuse, also allowing limited economic and financial risks, as foreseen demand for new uses is high.

Renovation and reuse projects also support urban identity and urban strategy, targeting one or more goals as tourism development, social networks and innovation partnership, social inclusion, leisure and entertainment, quality of life, sometimes addressing the global challenge of climate change with local greenhouse gas / air pollution reductions.

6 Conclusions

Although structural rehabilitation and new services are the backbone of the reuse projects, they do not fully describe the business models. The core of projects is the value proposition, a multiple-level player field ranking and balancing economic, social, cultural, environmental goals. Local communities and different stakeholders co-create a valuable project, not only taking into account present needs but also sharing a long-term vision of local development. This phase is the very critical black box of the reuse project, as legitimacy, leadership, competencies, and financial resources may fall into a never-ending negotiation, missing the main goal of governing the commons. This result is achieved only if local stakeholders are able to find an effective self-organization.

High skilled facilitators of local decision could speed the process, avoiding unworkable or unfeasible initiatives, and raising awareness of constraints and opportunities. Therefore, while enabling local communities to build strongly sustainable business models, constraints and incentives should also be re-defined as new priorities emerge for a common future of people and the planet.

The cases herein described do not allow to validate circular business models as detailed by theory and research (Bocken et al. 2018 Upward and Jones 2016; Joice and Paquin 2016). Nevertheless, most of the cases fit circular economy as a general perspective (see Pieroni et al. 2019; Urbinati et al. 2017) of resource efficiency, shared use and economic growth.

The model of Urbinati et al. seems the best responsive to frame the results of case studies, highlighting upstream, downstream and fully circular models. Upstream circular models mainly involve innovation in building technologies (i.e. energy saving or green energy). Indeed, some cases of reuse aim at eco-innovation, and fit the concept of circular economy as efficient use of natural resources. Downstream circular models mainly refer to social or multifunctional uses, while fully circular models involve both of the perspectives including, of course, efficient economic management.

It is unquestionable that circular economy principles, mainly in the sense of eco-innovation and natural resources saving, must scale up both at urban planning and at citizens' level. This leapfrog requires updated knowledge and the use of leading-edge technologies, including ICT-based solutions for smart mobility and smart environment supported by municipally based partnerships.

Economic sustainability is now receiving much more attention than in the past, challenging other sustainability goals and paying more attention to management abilities both in project design and operation phases. Therefore, multifunctional reuse is common to many projects, as core services or attractions and ancillary services or facilities enable longer staying of users as well as customer satisfaction and repeated visits.

Nevertheless, cultural heritage is not a mere space to use and profit. Frequently projects lay on a wider concept of circularity, based on the sustainable development paradigm and targeting social, economic, cultural goals, in the view of use cultural heritage for people and next generations, not too burdening public finance for renovation and maintenance.

Cultural heritage produces many intangible and long-term benefits: the education of young people, the strengthening of the identity processes, the inclusion of disadvantaged social groups or minorities and immigrants, the development of tolerance and human dignity based on the knowledge and protection of cultural diversity. It also enables network and learning economies. Therefore, cultural heritage is the main input and enabling infrastructure of culture.

Culture nourishes the human personality and it is the basis of educational processes. It enriches the endowment—concepts, images, information, emotions—available to individual and community, thus facilitating reasoning, logic and semantic associations, analogies and contamination.

Therefore, culture provides people with more opportunities and a general ability to find solutions to problems as well as a flexible attitude in dealing with the "new". That is the reason why culture is assuming an increasingly strategic role as a synergistic agent that provides other sectors of the production system with contents, tools, creative practices, increasing value added.

These patterns induce many local systems to invest more heavily in allowing a deeper integration between culture and the various aspects of social everyday life. By this way, cultural heritage adaptive reuse turns a stock of historical resources into an engine able to mobilize the best energies of the community, and to leverage human and social capital.

This perspective renews the role of financial resources and financial policies, as both of which may be a serious constraint toward a complex value enhancement which grounds on the sustainability paradigm.

A similar critical role has the choice of the management at the end of the structure rehabilitation: at this stage, social entrepreneurial ventures could provide a level player field of new business model toward social innovation and sustainability.

References

Antikainen M, Valkokari K (2016) A framework for sustainable circular business model innovation. Technol Innov Manag Rev 6(7)

Bocken NM, Short SW, Rana P, Evans S (2014) A literature and practice review to develop sustainable business model archetypes. J Clean Prod 65:42–56

Bocken N, Miller K, Evans S (2016) Assessing the environmental impact of new Circular business models. In: Conference "New Business Models" – exploring a changing view on organizing value creation – Toulouse, France, 16–17 June 2016, Toulouse, France

Bocken N, Boons F, Baldassarre B (2018), Sustainable business model experimentation by understanding ecologies of business models. J Clean Prod 208:1498–1512. https://doi.org/10.1016/j.jclepro.2018.10.159

Boons F et al (2013) Sustainable innovation, business models and economic performance: an overview. J Clean Prod 45:1–8

Eisenhardt KM (1989) Building theories from case study research. Acad Manag Rev 14(4):532–550

Fielt E (2014) Conceptualising business models: definitions, frameworks and classifications. J Bus Model 1(1):85–105

Foss NJ, Saebi T (2017) Fifteen years of research on business model innovation: how far have we come, and where should we go? J Manag 43(1):200–227

Geissdoerfer M, Morioka SN, de Carvalho MM, Evans S (2018) Business models and supply chains for the circular economy. J Clean Prod 190:712–721

Joyce A, Paquin RL (2016) The triple layered business model canvas: a tool to design more sustainable business models. J Clean Prod 135:1474–1486. https://doi.org/10.1016/j.jclepro.2016.06.067

Lambert SC (2015) The importance of classification to business model research. J Bus 3(1):49–61

Lewandowski M (2016) Designing the business models for circular economy—towards the conceptual framework. Sustainability 8(1):43

Lüdeke-Freund F, Dembek K (2017) Sustainable business model research and practice: emerging field or passing fancy? J Clean Prod 168:1668–1678
Macdonald S (2011) Leveraging heritage: public-private, and third-sector partnerships for the conservation of the historic urban environment. In: ICOMOS 17th General Assembly, 2011-11-27/2011-12-02, Paris, pp 893–904
Macdonald S, Cheong C (2014) The role of public-private partnerships and the third sector in conserving heritage buildings, sites, and historic urban areas. Getty Conservation Institute, Los Angeles
Misirlisoy D, Günçe K (2016) Adaptive reuse strategies for heritage buildings: a holistic approach. Sustain Cities Soc 26:91–98
Osterwalder A, Pigneur Y (2010) Business model generation: a handbook for visionaries, game changers, and challengers. Wiley, Hoboken
Pieroni MPP, McAloone TC, Pigosso DCA (2019) Business model innovation for circular economy and sustainability: a review of approaches. J Clean Prod 215:198–216
Ritter T, Lettl C (2018) The wider implications of business-model research. Long Range Plan 51(1):1–8
Schaltegger S, Hansen EG, Lüdeke-Freund F (2016) Business models for sustainability: origins, present research, and future avenues. Organ Environ 29(1):3–10
Stubbs W, Cocklin C (2008) Conceptualizing a "sustainability business model". Organ Environ 21(2):103–127
Upward A, Jones P (2016) An ontology for strongly sustainable business models: defining an enterprise framework compatible with natural and social science. Organ Environ 29(1):97–123
Urbinati A, Chiaroni D, Chiesa V (2017) Towards a new taxonomy of circular economy business models. J Clean Prod 168:487–498
Yin RK (1994) Discovering the future of the case study. Method in evaluation research. Eval Pract 15(3):283–290

Chapter 18
Key Learnings from the "Kultur-Token" Sustainable Business Modelling Case Study

Gillian Foster

1 Introduction

In 2020, the Vienna University of Economics and Business' Institute for Ecological Economics and Institute for Cryptoeconomics (WU) conducted a sustainable business modeling case study for a municipal initiative, the City of Vienna's "Kultur-Token" (KT). The KT is one of the City of Vienna's pilot initiatives to enhance city life with culture—while addressing the global challenge of climate change with local greenhouse gas/air pollution reductions.

The WU Institutes published the open-access working paper, "Kultur-Token" Sustainable Business Model: Visualizing, Tokenizing, and Rewarding Mobility Behavior in Vienna, Austria. The study, hereinafter "the KT Report" was the result of a collaboration between the Institute for Ecological Economics CLIC staff and the Research Institute for Cryptoeconomics, and the City of Vienna (Foster et al. 2020). The KT Report is also a case study of the European Union Horizon 2020 research project "Circular models Leveraging Investments in Cultural heritage adaptive reuse" (CLIC). The KT Report is one of the CLIC case studies that expand on the CLIC concept, conceptually and practically. It is about an app rather than a building renovation, monument, or adaptive reuse of cultural heritage (ARCH).

The KT is part of the Vienna Smart City concept because it uses digital technologies like block chain, geo-tracking, and cellular communication. Simply put, KT is a downloadable app for mobile phone users that rewards Viennese for low-carbon mobility choices. The KT Report describes the KT as follows:

> The [pilot] KT app users create a personal account on users' phones to track their own mobility behaviors and be rewarded for low-carbon choices. Four different transport modes are measured: car; bicycle/scooter; walking; and public transport (buses, trams, and trains).

G. Foster (✉)
Institute for Ecological Economics, Vienna University of Economics and Business, Vienna, Austria
e-mail: gillianjfoster@wu.ac.at

L. Fusco Girard, A. Gravagnuolo (eds.), *Adaptive Reuse of Cultural Heritage*,
https://doi.org/10.1007/978-3-031-67628-4_18

> When a user travels by bicycle or scooter, walking or public transport, (s)he is rewarded with "Kultur-Token" (KT). One full KT, equivalent to 20 kg CO_2 avoided, can be exchanged in the app's marketplace for one voucher for a ticket to various cultural events and venues in Vienna. Users choose from tickets available in the marketplace on a one-token-equals-one-ticket basis.
>
> The individual mobility behavior data are provided with sensors in the users' smartphones that gather information on GPS locations, speed of travel, acceleration, etc. The data collected allows the app to calculate: the route traveled; the time it took to travel a certain distance; and the distinctive start-stop patterns that define the most likely transport mode. KT is able to achieve an average accuracy of 90% for the detection of travel mode and distance. As users travel by any mode other than by car, the algorithms calculate the CO_2 emission equivalent of an average car trip in comparison to the low-carbon mode and award this amount as avoided CO_2.
>
> KT recognizes citizens transitioning to sustainable transport modes and at the same time provides additional access to cultural experiences for citizens. An increase in sustainable mobility is encouraged by providing individuals with an incentive to avoid travel by car and by visualization of their personal travel-related CO_2 profile. Further, the app provides citizens with a new way of exploring and gaining access to culture. As these offers are a bonus for the KT, they do not require entrance fees, thereby lowering financial barriers to access cultural institutions for the Viennese. (Foster et al. 2020)

The current article highlights the key learnings from the KT Report for the CLIC community of practice audience (CLIC research partners and pilot projects, cultural heritage scholars, and others considering business modeling for cultural heritage-related enterprises). The purposes of this article are to recap and reflect on the KT Report's findings from the perspective of CLIC and present recommendations. Hence, the current work investigates two questions, "What parallels can be drawn between the KT Report and other CLIC case studies?" and "What are the implications of the KT Report for CLIC?" This research is unique in the scientific literature because it is the first reflection on the new KT Report. The KT was not studied extensively before the KT Report and no other business modelling of it exists. The current article focuses on the KT Report's implications relevant to CLIC for the first time; therefore, the discussion herein is novel in the literature.

The remaining sections of the article proceed as follows. Section 2 answers the first question to bridge the KT Report and CLIC by reviewing the research philosophy, objectives, and WU business modelling research methods, that link ARCH enterprises and the KT. This section includes an empirical comparison between key features of the KT and CLIC Best Practices for ARCH. Section 3 summarizes the implications with key lessons learned and recommendations for sustainable business modelling in the context of circular ARCH enterprises. Section 4 concludes with broad observations and future research directions resulting from this study.

2 What Parallels Can be Drawn Between the KT Report and Other CLIC Case Studies?

This section explores the KT Report's role as a CLIC business model case study. The task of exploring and documenting new sustainable business models relevant to ARCH is one of the main objectives of CLIC. What parallels can be drawn between the KT, the KT Report, and more typical ARCH case studies that are anchored to a reuse of a building or site? In retrospect, the research philosophy, objectives, and WU's perspective on business modelling for the KT Report and CLIC are in alignment.

The KT Report's research philosophy is an expression of "relationality" in urbanism as defined by Fusco Girard (2013). According to Fusco Girard, "relationality is reflected in the ability to explore/understand complex systems recognizing interdependences, links, connections, also when they are implicit; assuming a multidimensional point of view; it is a way to interpret reality in a comprehensive/holistic perspective that does not exclude, but integrates economic, aesthetic/visual, fairness aspects and values, etc." (Fusco Girard 2013). Relational research is needed because the KT is complex. It bridges macro- and micro- socioeconomic and geographic scales and integrates the shared social values of culture and creativity, fairness/inclusiveness and the shared social value of reducing carbon emissions from urban mobility in order to address climate change. To the authors' knowledge, the KT Report represents the first time that the Flourishing Business Canvas (Elkington and Upward 2016; Jones and Upward 2014; Upward 2014) is used for a sustainable behavioral changeapp. The KT Report provided an opportunity to test the Flourishing Business Canvas[1] as a research, evaluation, and strategy tool for business models that connect citizens, governments, and the cultural and creative industry. The KT Report is scientific research for understanding a complex initiative in Vienna, Austria. The study's research method, business modeling, is guided by the theoretical perspective of relationality with the objective of naming and integrating disparate aspects of the KT's socioeconomic network.

2.1 *The Business Modeling Research Method of the KT Report*

Business models provide a common understanding about the general dynamics of a simple or complex process or enterprise, provide a common vocabulary and most important can coalesce thinking around a set of ideas embedded in the process they describe. Business modelling is used for research and strategy. Historically, strategic business modelling is applied to profitmaking entities. The main goal is to establish the company's value proposition(s) in a way that meets customers segment's needs, to generate sales revenue and profit. The 2005 Business Model Canvas

[1] Upward, A. (2014). "Flourishing Business Canvas v2.0." http://www.flourishingbusiness.org/.

(Osterwalder et al. 2005) is indicative of this approach, which is widely applied in research for various enterprises. In the view of the authors of the Kultur-Token Report, a profit-centred approach is not suitable for an enterprise that seeks to achieve sustainability. Foremost, because the approach does not adequately consider the biophysical resources provided by the earth upon which the functioning of the company depends. In addition, it does not adequately consider the waste streams that result from the company's activities that must be absorbed (if possible) by the earth's natural systems upon which life on earth is sustained. Further, the underlying profit ethic focuses the exercise on paying customers. Therefore, the more common forms of business modelling for research do not align with relationality or ecological economics, hence were not suitable for analyzing the Kultur-Token initiative and likewise many circular adaptive reuse of cultural heritage (ARCH) enterprises.

According to WU's desk research in preparation for the KT Report in 2020, business modelling for research and strategy has evolved to include non-profits, social enterprises such as b-corporations, and other forms of economic and non-economic activity. Recent research confirms (Lüdeke-Freund and Dembek 2017) that Sustainable Business Modeling is rapidly becoming established as an academic field of academia, industry, and government (Lüdeke-Freund and Dembek 2017). The key element is that trade, commerce, or business activity is a means to a community development end rather than ignored or counted as an externality. For example, Burkett (2013) states that, social enterprises, "Have an economic, social, cultural or environmental mission aligned to public or community benefit" (Burkett 2013). According to Breuer et al. (2018) better understanding of business models is "needed to empower sustainability-oriented entrepreneurs, to support experimentation with business model design patterns and to facilitate comparative research" (Breuer et al. 2018). For example, the Burkett (2013) framework sets out questions for business modelling for social enterprises that differ from or expand on the common concepts of value, impact, and customer. Gauthier and Gilomen (2015) compared business model in energy efficiency case studies in France. Their research concluded that organizations attempting sustainable business models could achieve four levels of innovation with various degrees of sustainability as a result. These authors identified agency and inclusion of diverse agents as a key factors of new and transformative value propositions that can change not only the enterprises business model but also change the whole market thereby challenging existing rubrics and reflecting transition towards sustainability (Gauthier and Gilomen 2015). Bocken et al. (2014), identified archetypes of sustainable business models grouped as Technological, Social, and Organizational (Bocken et al. 2014). Each of the groups are relevant to circular economy initiatives and ARCH, but none more so than the technological business models, described as, "maximise material and energy efficiency; substitute with renewable and natural processes; and create value from waste" (Bocken et al. 2014). In particular, the latter archetype, which targets "underutilised assets and capabilities as a form of wasted value that could be recaptured" (Bocken et al. 2014) is relevant exemplifies ARCH and CLIC's typical case studies.

For most ARCH case studies, the social and organizational benefits of the business model flow from giving new purpose to an under-utilized physical asset. In contast, the KT rewards low-carbon mobility with vouchers for tickets to cultural venues and events. From the perspective of Bocken et al.'s archetypes, the KT's capacity to offer free tickets to cultural venues and events captures and repurposes waste. In fact, it assigns new values to the cultural venues and events beyond the price of admission. The KT valorizes the cultural and cultural heritage assets from the individual (micro) to the national (macro) scales. The WU's business modeling must explain and capture the multiscale environmental and social values produced and consumed by the KT enterprise, not only the financial ones.

The Flourishing Business Canvas is based on a "strong sustainability" ontology, combining "financial rewards, social benefits, and environmental regeneration" (Upward and Jones 2015). The structure of the Flourishing Business Canvas reflects the ecological economics ontology of embeddedness. Human economic activity is embedded within social systems, which is in turn embedded within ecological systems. The authors found the Flourishing Business Canvas fit for the purpose of understanding circular economy and sustainability initiatives like the KT and ARCH. Based on the review of existing business models, WU decided to adopt the Flourishing Business Canvas for the KT to best capture relationality, an ecological economics perspective and the environmental and social goals of the KT enterprise. A full description of ecological economics is beyond the scope of the current article; therefore, we refer readers to the following (Anderson and M'Gonigle 2012; Bruel et al. 2019; Common and Stagl 2005; Martínez-Alier and Muradian 2015; van den Bergh 2001).

The Flourishing Business Canvas was also adapted for the new business model for ARCH developed by the CLIC partner university ICHEC Brussels Management School. This new CLIC business model was tested in the circular business model workshops with CLIC Partners. Detailed information about the workshops can be found on the CLIC website (Saleh et al. 2020).[2] Therefore, CLIC has applied and tested the Flourishing Business Canvas as both a research method with the KT report and as a practical strategy development tool with the partner organizations managing adaptive reuse projects.[3] In each case, the Flourishing Business Canvas was adapted/expanded to better fit the case study.

In general, the KT Report follows the Flourishing Business Canvas guidance of (Elkington and Upward 2016). However, our study changed two elements of the Flourishing Business Canvas method to explain better the KT. First, the environment dimensions' discussion of ecosystem services is expanded to include the concept of cultural ecosystem services. This allows for deeper consideration of the cultural and cultural heritage goals of the KT. Second, monetary valuation of nature is contested within ecological economics; therefore, we chose not to apply monetization in the analysis. WU explicitly chose not to monetize the value of the ecosystem for the KT's activities. These transparent methodological choices are relevant to

[2] Saleh, R., Drouillon, P., & Ost, C., (2020), D4.5 Circular Business Model Workshops for Cultural heritage adaptive reuse. https://www.clicproject.eu/files/D4-5.pdf

[3] Ibid.

ARCH projects and enterprises because the value of cultural heritage is often contested, unknown, or intangible. For example, the CLIC WP4 Deliverable D4.5 documents how the circular business model workshop participants found it difficult to assign monetary valuation to the cultural heritage assets.[4]

An additional element that WU implemented in the KT Report to identify the ecosystem context, which is not often used in conjunction with most Flourishing Business Canvas research, is Bronfenbrenner's Ecological Theory of Human Development. The WU authors adopted the concept from Jones (2017b) who applied it first (Jones 2017b). Another article by Jones, "Social Ecologies of Flourishing: Designing Conditions that Sustain Culture", details his application of Bronfenbrenner (Jones 2017a). See Fig. 18.1. The WU desk research of Flourishing Business Canvas applications found only one other explicit use of the Ecological Theory of Human Development with the Flourishing Business Canvas in the literature. This example is the 2017 business modelling in the health care sector by Sonowal, Jyotish. "Hospital as a Business for Flourishing." (Sonowal 2017). The Ecological Theory of Human Development is a practical framework for envisioning

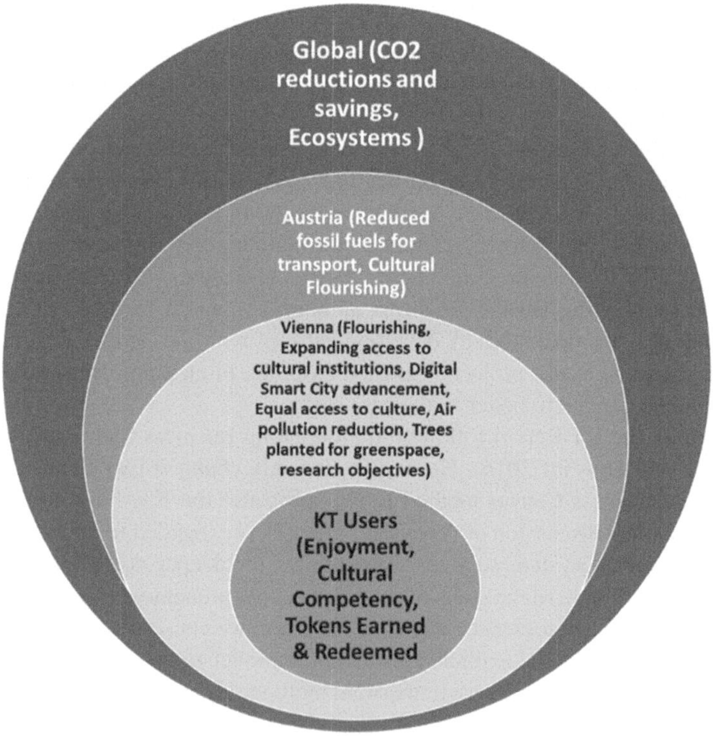

Fig 18.1 Ecosystem Context for the KT Sustainable Business Model. The socioecological systems model is adapted from Jones (2017a, b)

[4] Ibid.

the values of the enterprise from the ecosystem to microsystem levels. The WU authors further specified the ecosystem context from macro to microsystem levels according to an ecological economics framing and the KT case and setting in Vienna. "The ecosystem context is the combined environmental, social, and economic or financial constraints and opportunities within which all aspects of the enterprise function. Clarity about the ecosystem context is critical to identify who pays the costs and who accrues the benefits of achieving the KT's goals." (Foster et al. 2020) The WU found that defining the ecosystem as shown in Fig. 18.1 was a central task of the research. The final version was reached through an iterative process that evolved as more became known about perspectives at different scales, for example the KT User or the City of Vienna. Implementing the ecosystem context concept as described is not time consuming and creates tangibility and local focus to otherwise general or amorphous concepts such as "climate change".

2.2 Comparison of KT to CLIC ARCH Best Practice

The business modelling results of the KT Report are used here to examine empirically parallels between the KT and ARCH projects. Six key features of the KT highlighted in the KT report are compared to the 96 CLIC Best Practice case studies found on the online CLIC Knowledge Information Hub (Hub), which is a state-of-the-art encyclopedia of ARCH projects. The best practices collated by the Hub are freely accessible and searchable by several pertinent criteria such as public and or private funding, social impact, and economic spillovers. The footnotes below note the search term and website links that are used to retrieve the relevant data on the best practices from the Hub.

1. **The KT is a publicly run enterprise**. Many ARCH examples are owned and/or operated by governments of cities and other municipalities, non-governmental organizations, and public and private organizations. Of the 96 CLIC Best Practice case studies found on the CLIC Knowledge Information Hub, 82 or 85% involve public financing.[5]
2. **The KT enterprise does not have a profit ethic**. Only 34 or 35% of the ARCH best practices are considered as having "very high economic spillovers".[6]
3. **Sustainability, as realizing low-carbon mobility and greenhouse gas (GHG) reductions is driving the KT's actions**. This is comparable to the Hub's "circular metabolism" assessment defined as follows. "It reflects the degree of reduction in the consumption of resources in terms of renewable energy sources and low energy consumption systems, water recycling systems, reduction of GHG emissions and construction wastes, recycling of materials, use of local traditional

[5] https://clicplatform.eu/search?q=Public Accessed on October 2, 2021.

[6] https://clicplatform.eu/search?q=very%20high%20economic%20spillovers Accessed on October 2, 2021.

materials, bio-materials and/or reuse materials as well as recovery/increase of biomass/green spaces and adoption of nature-based solutions." (https://clicplatform.eu) Roughly 22% or 21 best practices are assessed as having "very high circular metabolism".[7]

4. **Sustainability, as a function of increasing quality of life of residents is driving the KT's actions**. It is impossible to tell from the information on the Hub if social impacts are the driving force of the best practice as is noted in the KT report. However, the Hub lists 54, or 56% of the best practices as having "very high social impacts".[8]
5. **The KT concept of key stakeholders is broadened beyond customer of the enterprise to society**. This concept is reflected in the CLIC best practices as well. The Hub notes that the participation in the ARCH project included public consultation for 22 or 23% of best practices.
6. **By valorizing access to cultural venues and events, the KT expresses a desire to embed cultural heritage and culture in modern urban life**. In principle, all of the CLIC best practice case studies of ARCH included on the Hub share this goal. The Hub states its purpose as follows, "to showcase local reuse practices and highlight the abandoned and underused cultural heritage as asset/opportunity for adaptive reuse initiatives, stimulating sustainable investments towards the implementation of a "circular city" model."[9]

In summary, the analysis shows that despite the obvious differences, the KT and ARCH share several key features as shown by the above comparison to ARCH best practice data accessible on the Hub. As a result, the learnings of the KT business modelling research is applicable to CLIC and ARCH.

3 What are the Implications of the KT Report for CLIC?

This section presents the implications of the KT report for CLIC as three key lessons learned followed by recommendations for business modelling in the context of circular adaptive reuse of cultural heritage projects like those of the CLIC partners and beyond. The discussion highlights relevant CLIC partner projects.

First, consider if the financial goal of the enterprise is for-profit, non-profit, or a combination. This is the defining element of many ARCH projects. If the ownership of the cultural heritage asset is governmental, public-private partnership, non-profit, or even private, the goal of its operation may be to recover the cost of rehabilitation, cost or operation and maintenance, or simply to collect a nominal

[7] https://clicplatform.eu/search?q=circular%20metabolism%20very%20high Accessed on October 2, 2021.

[8] https://clicplatform.eu/search?q=social%20impact%20very%20high Accessed on October 2, 2021.

[9] https://clicplatform.eu/ Website landing page accessed on October 2, 2021.

charge for access and entry. Two CLIC pilot project examples described on the Hub are the Giardino della Minerva in Salerno, Italy and the Benčić Complex ex-factory to become Culture Park in Rijeka, Croatia. For many ARCH projects, a profit-focused business model is not appropriate. Using a sustainable business model that is not profit-centric, for example for non-profits and social enterprises, is a better fit for ARCH.

Second, the positive and negative environmental impacts of the ARCH project are better captured and explained using the Flourishing Business Canvas's ecological economics ontology based on the concept of strong sustainability with the addition of cultural ecosystem services. Another aspect of value relevant for ARCH included in the Flourishing Business Canvas is the notion of co-creation of value and destruction of value. Value is co-created be diverse actors and agents, not only the cultural heritage owner. In addition, values are co-created when the project meets the needs of those who engage with the project, including, but not limited to customers. For example, individual enjoyment of the ARCH project is a value proposition. The value of community level enjoyment can be diminished by increasing exclusive access to individual members of the public; therefore, the value of enjoyment is destroyed. In addition, ARCH projects are encouraged to reject the mantra of "monetize or ignore" for cultural heritage values. Intangible values can be described and understood using a cultural ecosystem services model that does not rely solely on monetization of all values, for example monetized enjoyment.

Third, the complex social context of ARCH as recognized by urban conservation theories calls for a clear alliteration of the social context to underpin all aspects of business modelling for ARCH. This assertion is underpinned by the CLIC research findings. According to Ikiz Kaya et al. (2021) one implication of surveying CLIC partners is that "the challenges that emerge in participatory administrative structures and processes (i.e., lack of collaboration, communication and coordination between stakeholders) should be addressed, especially at the local and national levels" (Ikiz Kaya et al. 2021). The KT report also noted the complex social context and the need for better integrating the actors and agents present. Therefore, WU recommends the Flourishing Business Canvas-based CLIC Circular Business Model for Adaptive Reuse projects per (Saleh et al. 2020).

In addition to the CLIC Circular Business Model for Adaptive Reuse projects, and as a new contribution to CLIC business modeling, WU recommends that ARCH projects apply the Bronfenbrenner's Ecological Theory of Human Development as shown in Fig. 18.1. The experience of the KT Report shows the usefulness of a systems perspective for ARCH. This framework allows the researchers and business-modelling participants to better define and understand the purpose of an ARCH project as a socially and ecologically embedded enterprise with multiple interconnected outcomes.

4 Conclusion

This article derives outcomes for CLIC through a retrospective examination of commonalties between ARCH and the Kultur-Token. The research perspective of relationality and the research objectives and methods of the KT Report are explained herein. The article notes the six key features of the Kultur-Token that resulted from the business modelling research and compare these to research data on ARCH best practices. The key lessons learned are summarized with recommendations.

In conclusion, the WU sustainable business model research and application of the Flourishing Business Canvas to the Kultur-Token was an a-typical CLIC case study; however, this non-adaptive reuse project provides practical recommendations to improve future business modelling for adaptive reuse of cultural heritage projects. The deliberation on the research experience of the KT report published in 2020 and an empirical ex-post assessment using the data of the Hub are now possible in 2021. An interesting line of research in future is to examine the long-term impacts of CLIC's business modeling research on practice. For example, will the recommendations of a Flourishing Business Canvas-based business model with a strong sustainability approach take hold for ARCH in years to come?

Funding Sources and Acknowledgments This research was funded under the framework of Horizon 2020 research project CLIC: Circular models Leveraging Investments in Cultural heritage adaptive reuse. This project has received funding from the European Union's Horizon 2020 research and innovation program under Grant Agreement No 776758. The research is conducted under Deliverable D4.4/5—Report: Circular business models for cultural heritage adaptive reuse. CLIC is a project in which the Institute for Ecological Economics, Vienna University of Economics and Business is Partner. The Research Institute for Cryptoeconomics' contribution to this research cited herein is funded by Vienna's digital innovation budget (Digitalisierungsbudget der Stadt Wien).

References

Anderson B, M'Gonigle M (2012) Does ecological economics have a future?: Contradiction and reinvention in the age of climate change. Ecol Econ 84:37–48

Bocken NM, Short SW, Rana P, Evans S (2014) A literature and practice review to develop sustainable business model archetypes. J Clean Prod 65:42–56

Breuer H, Fichter K, Lüdeke-Freund F, Tiemann I (2018) Sustainability-oriented business model development: Principles, criteria, and tools. Int J Entrep Ventur 10(2):256–286

Bruel A, Kronenberg J, Troussier N, Guillaume BJJOIE (2019) Linking industrial ecology and ecological economics: a theoretical and empirical foundation for the circular economy. J Industrial Ecol 23(1):12–21

Burkett I (2013) Using the business model canvas for social enterprise design. Recuperado de. http://knode.com.au/wp-content/uploads/Knode_BusModCanv4SocEntDesign_E1LR_30p.pdf

Common M, Stagl S (2005) Ecological economics: an introduction. Cambridge University Press

Elkington R, Upward A (2016) Leadership as enabling function for flourishing by design. J Glob Responsib 7(1):126–144

Foster GJ, Lamura MJ, Hackel J (2020) "Kultur-Token" sustainable business model: visualizing, tokenizing, and rewarding mobility behavior in Vienna, Austria, Working Paper Series,

WU Vienna. WU Vienna University of Economics and Business, Vienna. https://epub.wu.ac.at/7777/1/Kultur-Token%20FBMC%20WorkingPaper.pdf
Fusco Girard L (2013) Toward a smart sustainable development of port cities/areas: the role of the Historic Urban Landscape approach. Sustainability 5(10):4329–4348 https://doi.org/10.3390/su5104329
Gauthier C, Gilomen B (2015) Business models for sustainability: energy efficiency in urban districts. Organ Environ 29(1):124–144
Ikiz Kaya D, Pintossi N, Dane G (2021) An empirical analysis of driving factors and policy enablers of heritage adaptive reuse within the circular economy framework. Sustainability 13(5):2479
Jones P (2017a) Social ecologies of flourishing: designing conditions that sustain culture. In: Design for a sustainable culture. Routledge, pp 38–54
Jones P (2017b) Soft service design outside the envelope of healthcare. Routledge, pp 39–57
Jones P, Upward A (2014) Caring for the future: the systemic design of flourishing enterprises
Lüdeke-Freund F, Dembek K (2017) Sustainable business model research and practice: emerging field or passing fancy? J Clean Prod 168:1668–1678
Martínez-Alier J, Muradian R (2015) Handbook of ecological economics. Edward Elgar Publishing
Osterwalder A, Pigneur Y, Tucci CL (2005) Clarifying business models: origins, present, and future of the concept. Commun Assoc Infor Syst 16(1):1
Saleh R, Drouillon P, Ost C (2020) Deliverable D4. 5 Circular business model workshops for cultural heritage adaptive reuse. 258
Sonowal J (2017) Hospital as a business for flourishing
Upward A (2014) Flourishing Business Canvas v2.0. http://www.flourishingbusiness.org/. Accessed 29 May 2020
Upward A, Jones P (2015) An ontology for strongly sustainable business models: defining an enterprise framework compatible with natural and social science. Organ Environ 29(1):97–123
van den Bergh JC (2001) Ecological economics: themes, approaches, and differences with environmental economics. Reg Environ Change 2(1):13–23

Chapter 19
Investment Leverage for Adaptive Reuse of Cultural Heritage

Tracy Pickerill

1 Introduction

The investment needs of urban, peri urban and rural areas to achieve sustainable growth varies depending on the size of urban conurbations and rural clusters, demographic trends, geographic location, political and economic stability and the quality of existing assets, including cultural heritage assets. The reticence of traditional capital markets to quantifying intangible socio-cultural, environmental, aesthetic, and spiritual values in the cultural heritage investment decision making process, can lead to market failure, investment barriers and a consequent trade-off between value judgments relating to the value preferences of diverse stakeholders. Altered financial markets stemming from political and economic turbulence over the last decade has exacerbated this situation and negatively impacted on the financial viability of many cultural heritage activities. Market failure coupled with pure profit orientation, from some sectors of the marketplace, can potentially result in inappropriate adaptation or demolition of built heritage assets unless gap finance can be sourced from limited public finances or sympathetic philanthropic benefactors, leaving the fate of many cultural assets to profit driven capital markets and political whim.

Four bodies of knowledge are critical to the choice of hybrid financial instruments (combinations of complementary financial and non-financial tools) designed to leverage investment flows for cultural heritage adaptive reuse, while remaining mindful of the physical, humanistic and ecological synergies that exist within complex cultural landscapes:

T. Pickerill (✉)
Surveying and Construction Innovation, TU Dublin, Dublin, Ireland
e-mail: tracy.pickerill@tudublin.ie

L. Fusco Girard, A. Gravagnuolo (eds.), *Adaptive Reuse of Cultural Heritage*,
https://doi.org/10.1007/978-3-031-67628-4_19

(i) Tool Knowledge: to identify type and operational characteristics of financial tools;
(ii) Design Knowledge: to facilitate targeted and tailored blending of financial tools to leverage circular investment streams, mitigate risk and avoid displacement;
(iii) Stakeholder knowledge: to understand the motivations and behaviour traits of investors to participate in collaborative investment ventures;
(iv) Impact knowledge: to pre-define strategic impact goals at the design stage.

Inter-related economic demand-side and supply-side policy tools are the fundamental building blocks with which governments traditionally implement heritage policy. Keynesian economic theory promotes direct government intervention (demand side policies), via government spending such as grants and direct government loans, funded by taxation and borrowing. Classical economic theory promotes indirect government intervention (supply side policies), such as fiscal incentives to create a positive economic environment to entice private investment. The concept of a regenerative, generative and symbiotic 'Triple Circular' adaptive reuse strategy developed by Fusco Girard (2019) provides a valuable guiding principle for decision makers to inform the choice and design of holistic regenerative funding mechanisms that align with the ultimate goal of achieving value creation where societal, environmental and financial benefits are in equilibrium.

1.1 Redefining Capital Leverage Thinking

The capital markets are designed to prioritize capital preservation and maximize profit for wealth creation, with little regard to regenerative social, cultural and environmental value creation. The ongoing work of traditional philanthropic endowment foundations and evolving global corporate metrics, of ESG compliance and intentional measured and managed impact (IMM), provide a point of mediation between the harsh realities of the marketplace and the need for inclusive people-led and place-led gap funding mechanisms. In 1963, Martin Luther King Jr. referred to the inescapable network of mutuality and challenged the motives of charitable endeavor writing "Philanthropy is commendable, but it must not cause the philanthropist to overlook the circumstances of economic injustice which make philanthropy necessary" (Luther King 1963). Over sixty years later, these words still resonate in the quest for intentional and regenerative investment leverage strategies to engender long-term sustainable renewal of the ecosystem, including safeguarding cultural heritage assets. This ties in with the philosophy of Zeleny (2021) that in order to solve societal disruption and disequilibrium, one has to identify and understand the causes of human suffering and natural resource extraction, not simply mitigate the symptoms. In periods of rapid transition, solutions to degenerative ecosystems, must remain multidisciplinary and evolutionary to create resilient investment environments. With regard to cultural philanthropic impact investment markets, Fram (2018) suggests that those who influence the allocation of capital have an obligation to consider whether current practices solve or perpetuate the challenges impact investments aim to address and poses the conundrum: 'How can capital serve people and not the other way around?

1.2 Hybrid Financial and Non-financial Instruments

Financial instruments leverage private investment by channelling funds to final recipients via multi-layer structures, involving public &/or private co-financing modalities of beneficiaries, with the expectation of both social and financial returns (WEF and OECD 2015). Gianoncelli and Boiardi (2017) make a distinction between 'hybrid financial instruments' and 'tailored financing' in the context of venture philanthropy. Hybrid Financial Instruments are monetary contracts that combine features of traditional financial instruments (grant, debt and equity) to achieve the best possible alignment of risk, return and impact for investments. Tailored financing is the process whereby investors choose from a range of available financial instruments dependent on the risk, return and impact profile of the investor. Blended Finance Principles, developed by the OECD, state that blended finance should be anchored to a development rationale, designed to increase commercial finance, tailored to a local context, designed to ensure efficient partnering and monitored for transparency and results (OECD 2020). Pooled financial instruments entail collaborative partnerships to solve the parallel needs of local governments and local communities by raising combined funding on the capital markets via various financial instruments. Apart from knowledge sharing, the pooling of assets to raise finance allows for mitigation and diversification of risk for investors, creates economies of scale and enables less experienced partners to improve their capacity and credit rating. Key characteristics of hybrid 'blended' and 'pooled' finance include:

- Leverage: use of development finance and philanthropic funds to mobilize private capital investment;
- Impact Return: investments that drive measured social, environmental and economic progress;
- Financial Return: financial returns for private investors in line with market expectations, based on real and perceived risks;
- Revolving Funds: recycled flows of money via loan repayments (debt) or the realisation of investments (equity);
- Collaborative Fund Structures: access to private sector financial tools and expertise to support public policy goals (WEF and OECD 2015).

1.3 Combining Return, Risk and Impact Metrics

Risk adjusted market return is defined as financial returns for private investors in line with market expectations, based on real and perceived risk (WEF and OECD 2015). 'Return on Capital' focuses on the return investors earn while their invested capital remains at risk. 'Return of Capital' focuses on when investors receive their actual invested capital back. The use of financial instruments designed to achieve socio-cultural and environmental impacts, in addition to a risk adjusted financial return, carry an additional duty of care or 'impact liability', as failure of a financial instrument could negatively impact vulnerable people or communities. This creates a two-tiered risk spectrum of:

- Risk adjusted Market (financial) Return or Below Market Return
- Risk adjusted Impact Return

Gianoncelli and Boiardi (2018) make a distinction between the two extremes of 'investing for impact' and 'investing with impact', along a spectrum of impact capital investment strategies. Investors for impact (Impact first) prioritize the achievement of intentional long term social impact with or without financial return, ranging from financial loss to capital preservation. Investors with impact (Finance first) prioritize the achievement of positive financial returns, with the achievement of social impact as a secondary goal. At the extreme, investors with impact may limit their risk mitigation to minimum screening criteria of 'Do No Significant Harm' (Gianoncelli and Boiardi 2018). Within these two extremes, a range of social impact, financial return and risk mitigation investment strategies exist to support social innovations at different stages of development (ex-ante, seed, start-up, scale-up and ex-post evaluation) within different levels of market sector maturity. These strategies may overlap and complement each other in the central scale-up stage, where some investees begin to build their capacity and achieve financial independence and sustainability. By improving the regenerative operational and financial capacity of investees, the generative capacity (and by association the symbiotic capacity) is also scaled up.

1.4 Sustainable and Circular Financial Reporting

ECA (2019) defines sustainability reporting as the practice of measuring, disclosing and being accountable to internal and external stakeholders for organizational performance towards the goal of sustainable development. Investment strategies, including asset allocation, portfolio construction and risk management processes, effect the pricing of financial assets and the long-term risk and return opportunities of investments. In order to maintain their market share, financial entities must demonstrate that investments are financially sustainable, equitable and resilient, via transparent Environmental, Social & Governance (ESG) compliance, impact measurement and climate resilience building processes. Independent rating agencies[1] evaluate ESG analytics to inform investment markets. In recent years, the European Securities and Markets Authority (ESMA) raised concerns that unregulated and

[1] such as MSCI, KLD, Sustainalytics, FTSE, RobecoSAM, Asset4, Vigeo Eiris.

unsupervised ESG ratings increase the risk of capital misallocation, greenwashing and product mis-selling (ESMA 2021).[2] The EU Sustainable Finance Strategy including the Taxonomy Regulation (EU 2020) and Complementary Delegated Acts (Disclosures Delegated Act EU 2021/2178, Climate Delegated Act EU2021/2139 & amended Delegated Regulation EU2023/2485, Complementary Climate Delegated Act EU2022/1214 and Environmental Delegated Act EU2023/2486) has started to address this issue by providing detailed technical screening criteria for environmentally sustainable economic activities, including legal obligations for financial market participants. The Complementary Climate and Environmental Delegated Acts provide a suite of additional criteria, with evolving metrics and thresholds relating to different economic activities. The EU Taxonomy uses the European Industry classification system (NACE) as an indicative orientation of eligible economic activities. It is envisaged that future reviews of the Delegated Acts will reflect technological progress and evolution in climate and environmental policy. The Regulation on the transparency and integrity of Environmental, Social and Governance rating activities (ESGR) had been adopted by the European Parliament pending formal approval (EU 2023). In essence, the 'living' Taxonomy is a green finance criteria and performance tool to encourage investors to allocate capital to assist the transition to a low carbon, resilient and resource-efficient economy, by measuring sustainable, and circular transitionary investment flows (including targeting EU funding priorities). The Taxonomy sets performance thresholds,[3] including (i) Climate change mitigation (ii) Climate change adaptation (iii) Sustainable use and protection of water & marine resources (iv) Transition to Circular Economy (v) Pollution prevention and control (vi) Protection and restoration of biodiversity & ecosystems (EU TEG 2020a). Reinforcing the social dimension, EU Taxonomy principles for recovery and resilience identify the need to build social, economic and ecological ecosystem resilience in preparation for future climate related disruption (EU TEG 2020b). In parallel with protecting vulnerable communities, the principles highlight the importance of creating robust collaborative financial investment frameworks to attract private investment to support and reinforce both member state and collaborative transnational investment towards reaching zero emission targets.

2 Developing a Capital Leverage Toolkit for Adaptive Reuse of Cultural Heritage

The toolkit incorporates umbrella clusters of financial (grant and endowment, tax, debt, equity) and non-financial (regulation, real estate, risk mitigation, risk performance, capacity building, impact metric and digital network) instruments to

[2] ESMA also cited the lack of a legally binding ESG definition and comparability among providers of ESG ratings.

[3] requiring economic activities to (i) Substantially contribute to at least 1 objective; (ii) Do No Significant Harm (DNSH) to the other 5; & (iii) comply with minimum safeguards.

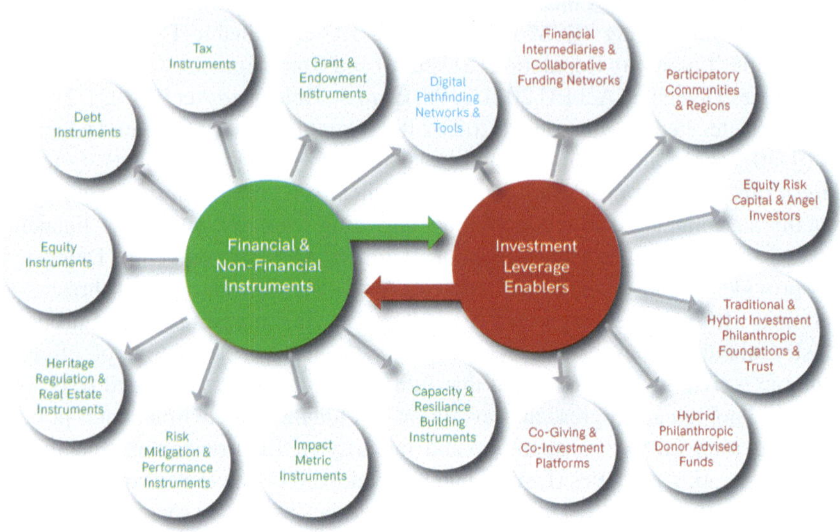

Fig. 19.1 Panoptic visualisation of CLIC investment leverage toolkit. (Source: Author)

encourage the flow of private investment funds for cultural heritage adaptive reuse activities. Creating value, within the cultural heritage financial landscape, is as much about understanding existing and opportunistic relationships between collaborative capital leverage enablers, as about blending complementary financial instruments. The clusters are based on analysis of European and International finance literature and Pan European financial directives (EC 2015a, b; WEF and OECD 2015; EIB 2015, 2020a, b; ECA 2020; OECD 2020). Flexible financial & non-financial instruments can be:

- Blended with complementary instruments within tailored place-led or typology-led[4] initiatives.
- Pooled within multilateral partnership collaborations initiated by financial intermediaries;

Figure 19.1 provides a panoptic visualization of the CLIC Toolkit highlighting the relationships between 'blended' complementary Financial and Non-financial Instruments and 'pooled' Capital Leverage Enablers.

The ultimate choice and design of funding mechanisms is not just a technical decision, based on the most efficient way to solve a particular funding gap, it is influenced by political and economic context, pre-existing institutional structures, the inclusion of diverse communities and finally ideological pre-dispositions regarding socio-cultural and environmental goals.

[4] 'Typology based' can refer to building type (such as Civic, Spiritual, Industrial, Residential, Commercial) or vintage architectural style (such as, inter alia, Ancient, Islamic, Classical, Byzantine, Romanesque, Gothic, Renaissance, Baroque, Rococo, Neo-classical, Art Nouveau, Neo-Gothic, Art Deco, Modernist, Bauhaus…..).

2.1 Investment Motivations of Market Spheres

Understanding diverse investment motivations between the different economic market spheres is key to designing hybrid funding instruments tailored to the specific needs of investment beneficiaries and final recipients. Collaborative funding strategies, involving formal or informal structures, both within and between market spheres, requires the design of complex funding mechanisms that align the financial and impact goals of collaborating stakeholders, bearing in mind their return expectations, investment timeframe and appetite for risk. While the importance of public action to leverage private investment should never be underestimated, it is also important to highlight the significant role of grassroots community finance strategies and actions to initiate social impact activities. The four main economic market spheres include:

Government 1st Market Sphere Direct & indirect public action for the common good, funded by tax collection and borrowing. Many local governments face multiple financial constraints including insufficient and unpredictable monetary support from central government, weak fiscal structures, poor revenue generation from civic assets and legal constraints. Public financial and non-financial instruments initiated at local government level are backed by national and regional government administrative, regulatory and financial resources, which in turn are backed by European Union financial and technical support structures.

For-Profit 2nd Market Sphere Profit motivated corporate entities with tax liability. Corporate structures can vary across a broad spectrum of risk adjusted profit motivation ranging from socially agnostic corporations that prioritize financial returns to Environmental, Social and Governance (ESG) and EU Taxonomy compliant corporations.

Non-Profit 3rd Market Sphere Mission motivated community organizations, philanthropic cultural foundations and charitable trusts with tax exempt status. Many informal local community groups and formal non-profit grassroots community organizations, in deprived urban neighborhoods and isolated rural communities, often lack the financial track record, entrepreneurial capacity and network infrastructure to overcome exclusion from traditional capital markets. Community-led cultural heritage activates traditionally rely heavily on philanthropic cultural foundations and charitable trusts to bridge the financial and entrepreneurial gap between local community action and access to capital market finance.

Hybrid 4th Market Sphere Emerging philanthropic and market savvy 'hybrid' entities motivated by patient profit generation (or at least capital preservation) in addition to intentional Impact Measurement and Management (IMM). Complex legal and tax structures are necessary due to combined hybrid for-profit, reduced-profit and non-profit structures. This hybrid structure is an innovative way to address the issue of access to finance as the non-profit stream can attract grant aid, while the for-profit stream can attract social investment, thereby increasing the potential to pool resources (Gianoncelli and Boiardi 2017). Achleitner and SpeissKnafl (2012) identify emerging investment leverage enablers in social capital markets as development banks, venture philanthropy funds, social Investment funds and advisors, social stock exchanges & funding platforms. Salamon (2014) adds capital aggregators, philanthropic banks, corporate originated philanthropic foundations, enterprise brokers and climate exchanges to this evolving list.

Cross-sector partnership models including local communities and citizens allows the risks and returns of cultural heritage strategies to be shared. The diversification offered by partnership structures, ranging from high-budget to a simple exchange of information and expertise, permits a wide range of resources and skills to be brought together.

3 Investment Leverage Enablers

The clusters of investment leverage enablers, for adaptive reuse of cultural heritage, within the CLIC toolkit, are categorized as:

(i) Financial Intermediaries & Collaborative Funding Networks
(ii) Participatory Communities and Regions
(iii) Equity Risk Capital & Angel (Impact) Investors
(iv) Traditional & Hybrid Investment Philanthropic Foundations & Trusts
(v) Hybrid Philanthropic Donor Advised Funds
(vi) Co-Giving & Co-Investment Platforms: Lottery Funds & Crowd Funding
(vii) Digital Pathfinding Networks and Tools (Fig. 19.2)

3.1 Financial Intermediaries and Collaborative Funding Networks

'Financial Intermediaries',[5] are financial entities that initiate the leverage of 'pooled' capital from a collaborative funding network of investment partners. The financial intermediary then channels the pooled funds, via selected financial instruments, to underfunded target recipients within a pre-defined geographic location, sector or disadvantaged community, that cannot access capital in traditional investment markets. To hedge risk, financial intermediaries typically make investments in

[5] WEF and OECD (2015) defines financial intermediaries as institutions that facilitate the channeling of funds between investor and investee company or project and between lenders and borrowers.

Fig. 19.2 CLIC toolkit investment leverage enablers. (Source: Author)

stages—Inception & Design / Adaptation / Use & Operate—allowing them to only provide additional capital infusions when certain milestones have been achieved. Financial intermediaries can operate within for-profit, non-profit or hybrid fund structures including the provision of venture philanthropy capital, impact investment capital, and impact foundation capital (Richter 2014). World Bank (2013) categorises financial Intermediaries as:

- Bank Intermediaries including commercial, credit union & savings banks; and
- Non-bank Intermediaries including government agencies, NGOs, foundations, social equity fund providers, community development & microfinance entities.

Although funding collaboratives take a variety of forms, a common distinguishing feature is that they scale up their collective impact by networking to make coordinated decisions about the strategic targeting, leveraging and allocation of their pooled resources (Eikenberry and Bearman 2014).

3.2 Participatory Communities and Regions

Cultural heritage adaptive reuse activities are reinforced by the participation and commitment of regional and community-led actions to enhance living and working conditions in deprived urban neighborhoods and isolated rural communities. Throughout Europe, significant number of community-led social enterprise organizations, stemming from collective awareness, are evolving within the social enterprise arena to address unmet societal challenges such as the need to promote social justice (equity, diversity & inclusion) and environmental protection. This dynamic is more prevalent in member states where there is a strong social economy and 3rd market sphere tradition paving the way for transformative social enterprise entities. Academic institutions have a key role to play in this sphere through societal engagement activities and knowledge sharing.

3.3 Equity Risk Capital and Angel Investors

Direct equity risk capital, also 'termed 'venture capital', is provided in tandem with human and institutional capacity building consultancy support for target recipients, to achieve intentional and measurable social, cultural and environmental impact (Tuan 2014). Equity venture philanthropy fund activities increasingly incorporate people and community led placemaking goals and the need to achieve intentional impact measurement and management (IMM) goals in addition to ESG and EU Taxonomy compliance (mandatory or voluntary) and alignment with SDGs mapping in their decision-making process.

Indirect equity investment opportunities in the venture capital arena take place via public and private stock and bond exchanges. The evolution of social stock exchange platforms, since 2013, provides a transparent market mechanism for enterprises with a socio-cultural and environmental mission to raise finance on the capital markets while protecting their triple bottom line corporate structure.[6] The distinction here is that social and environmental exchanges encourage efficient allocation of capital by transparently pricing financial instruments against financial and social returns. Exchanges broadly take the form of regulated environmental (trading carbon credits) and social stock exchanges (trading social enterprise bonds) and private self-regulated platforms syndicating social enterprise partnerships

[6] Chicago Climate Exchange [CCX]; Euronext ESG funds: Amsterdam Brussels Dublin Lisbon London Oslo Paris; European Climate Exchange [ECX]; European Union Emissions Trading Scheme [EU ETS]; Impact Exchange of Mauritius [IIX]; Impact Investment Exchange [IIX] Singapore; Mission Markets trading platform supporting social and environmental capital markets; Social Stock Exchange Limited [SSE] UK.

(Robeco and Booz and Co 2009; Shahnaz et al. 2014). Social enterprise brokers source and connect social and environmental investors to mission focused investment opportunities with the potential to deliver risk adjusted market returns in addition to socio-cultural and environmental impacts (Hagerman and Wood 2014). The choice of financial instruments and funding collaboratives entail different financial risk and return profiles, leverage opportunities and investment scale requirement giving enterprise brokers an important role in structuring impact investment strategies.

3.4 Traditional and Hybrid Investment Philanthropic Foundations and Trusts

Philanthropic communities are uniquely placed to play an essential role in future rebuilding and recovery efforts for Europe (McKinsey 2020). Philanthropic foundations fulfil a diversity of policy driven non-profit functions across Europe (circa 800 Foundations) with combined budgets amounting to €60 billion per annum (ECF and AKS 2020). The charitable non-profit status of heritage trusts is also an important asset in the process of persuading the public to donate funds, legacies and property, due to the inalienable status of donated assets. While traditional foundations have the potential to provide a cohesive and inclusive link between European & national policy makers and civil society, cross border investment synergies are severely hindered by political lethargy in terms of unsolved legislative and taxation collaboration barriers (EFT and TGE 2017). Hybrid investment philanthropic foundations involve the restructuring of traditional non-profit grant-giving endowment functions to enable parallel for-profit investment functions. Foundation's core assets are used as collateral to leverage additional investment funds in the quest for financial return in addition to intentional measured impact. Salamon (2014) uses the term 'Philanthropic Bank' to describe this evolution, pointing out that not all foundations have the capacity to follow this route due to legal and administrative challenges.

3.5 Hybrid Philanthropic Donor Advised Funds

Hybrid philanthropic donor advised funds involve the restructuring of for-profit financial institutions, offering mutual investment funds, to create a parallel non-profit tax-exempt Donor Advised Fund (DAF). This facilitates donors to create an individual tax-deductible charitable account within the ring-fenced philanthropic arm of the financial entity and choose an investment strategy based on their specific mission and risk profile (Cohen 2014).

3.6 *Co-giving and Co-investment Platforms*

Within the cultural heritage arena, an array of digital platforms have emerged in recent years including cultural heritage matching crowd funding and lottery funding ventures.

3.6.1 Heritage Crowd Funding Ventures

In the last decade, crowdfunding models across Europe have evolved from a fringe fund raising tool to a recognized financial instrument providing capital leverage for ventures, ranging from local community initiatives to large urban regeneration initiatives. With advancement in digital technology, crowd funding ventures open up inclusive donating, investing opportunities in addition to community engagement, skills development and marketing research. Baeck et al. (2017) identify the emergence of public & private partnerships, including matching funding schemes, within crowdfunding platforms. While donation-based crowdfunding requires no return, equity-based and loan-based investment crowdfunding initiatives require a revenue generation activity, which either pays back the loan with interest or generates net profit for investors. Regulation for donation-based and reward-based crowdfunding does not exist at the EU level, although investment-based equity and debt models will be subject to both EU and member state regulation relating to the underlying use of equity and debt financial instruments (Mazur 2017) (EC 2017).

3.6.2 Heritage Lottery Funding

In 2018, €18 billion generated by European Lotteries (EL) members, was returned to society in their respective countries and channeled towards national tax revenues, research, sports, culture and social projects (EL 2018). Cultural heritage projects and activities benefit within the culture category, although it is difficult to establish contribution to heritage as categories vary by country. Lotto Baden-Wurttemberg (Germany), the UK Heritage Lottery Fund (HLF) and the French 'Loto du Patrimoine' have developed 'heritage specific' lottery fund models (Baeck et al. 2017).

3.7 *Digital Pathfinding Networks*

Pathways for the creation of liaisons between impact driven entrepreneurial investors and resource strapped cultural heritage projects and communities can be hindered where investors and investees simply cannot find each other. Evolving technological advancements, such as digital portals, within the cultural philanthropic investment landscape, has made this knowledge sharing and matchmaking process more transparent and achievable, although still complicated to decipher due to the myriad of emerging online giving and investment platforms.

4 Financial and Non-financial Instruments

In parallel with the categorisation of emerging impact investment leverage enablers, the CLIC toolkit highlights both traditional and evolving financial, and non-financial instruments within nine 'umbrella categories' as follows:

(i) Grant & Endowment Instruments
(ii) Tax Instruments
(iii) Debt Instruments (Bonds & Loans)
(iv) Equity Instruments
(v) Heritage Regulation & Real Estate Instruments
(vi) Risk Mitigation & Performance Instruments
(vii) Impact Metric Instruments
(viii) Capacity Building Instruments
(ix) Digital Pathfinding Networks & Tools (Fig. 19.3)

4.1 Philanthropic Grant and Endowment Instruments

The traditional grant instrument of providing a cash transfer from donor to target recipient, with no expected repayment, remains unchanged although some enhancements have emerged in the marketplace such as targeting grant aid to achieve intentional measured impacts and innovations in the area of idea generation allied to funded competitions, prizes and crowdfunding that engender public participation. Some grant instruments require matching contributions from grant recipients in the form of financial contributions or 'in kind' benefits such as voluntary services (Pickerill and Armitage 2009).

4.2 Tax Instruments

Tax incentives offer an effective mechanism to encourage private investment in cultural heritage activities. Indirect, tax based incentives involve no direct transfer of money although foregone taxes represent a cost to the state. Traditional tax incentives include Income / corporation tax credits, sponsorship, property, VAT, transfer, inheritance and capital gains tax (Pickerill and Pickard 2007). To achieve SDGs, and by association cultural heritage policy goals, countries need to ensure equitable

Fig. 19.3 CLIC toolkit financial and non-financial instruments. (Source: Author)

and inclusive tax policy (UN Habitat 2020) including cross border tax policies (EFT and TGE 2017). Land Value Capture (LVC) is a process where local governments can increase their tax revenue based on the increased value of privately held land derived from rezoning, infrastructure and other public investments (Peterson 2009) (Noring 2019).

4.3 Debt Instruments

Debt is a means of achieving leverage when the debt allows the borrower to increase its assets which in turn generates more net revenue (McVeigh and Sass 2014). Debt instruments include loans and bonds.

Both senior debt (commercial bank loans) and mezzanine equity (venture risk capital loans) interest rates may be significantly increased to reflect real, or perceived, additional investment risk associated with start-up enterprise and adaptive reuse or landscapes enhancement initiatives in non-prime locations or disadvantaged communities. This has triggered a rise in social purpose venture risk capital loans at subsidized interest rates. Micro finance investment vehicles work on a similar basis to social purpose venture capital, where an investment fund or financial intermediary, serves as a conduit for capital flowing into microfinance investment institutions for distribution to micro enterprise entrepreneurs (Salamon 2014).

Bond issues, also termed fixed term securities, are a common traditional debt instrument used to finance long term capital investment (CapEx) projects or shorter term ongoing operating expenditures (OpEx). Bond securities with shorter maturities are termed 'Notes'. While private non-profit social purpose entities can initiate a bond issue on a taxable basis, most bonds are issued by government agencies, either solo or within a public private partnership, to ensure tax-exempt investment status for bond purchasers (Balboni and Berenbach 2014).

4.4 Equity Instruments

An equity instrument is an investment tool which enables the investor to purchase an ownership share in a socio-cultural venture with a claim on future returns in the form of value creation (capital gains) and/or profitable revenue streams (dividends). Traditional equity investors forsake the less risky collateral backed returns of debt instruments by speculating for risk-adjusted higher returns. Critically, if the investment venture fails, equity holders are at risk of losing their initial capital investment as their share of any liquidated assets is subordinate to debt instruments. Private equity investors (investing directly in enterprises not floated on a formal stock exchange) face additional liquidity and return risk in relation to early-stage adaptive reuse projects and socio-cultural enterprises with limited cash flow expectations (Brand and Kohler 2014). Equity investors can invest in various stages of a project or enterprise with phased finance, although it is most often used at an early inception, design and development stage at a point where activities are unable to secure debt financing. Social equity instruments play a vital role in absorbing early-stage losses until a cultural venture begins to generate enough revenue and create enough value to access traditional debt and equity tools.

4.5 Heritage Regulation and Real Estate Instruments

Within the Alternative Investment Fund (AIF) regulatory regime, Real Estate Investment Funds (REIFs) can invest in underlying real estate assets via whooly owned Special Purpose Vehicles (SPVs) in order to ringfence the liability relating to each real asset. The SPV can also facilitate tax efficiency by avoiding or reducing withholding tax on dividend and interest payments. Real estate fund managers have the option to set up a Real Estate Investment Trust (REIT) within the remit of AIF, by listing the REIT structure on a recognized Stock Exchange (Fox and Rooney 2015).

REITs are designed to deliver tax efficient dividends payments to investors with modest long-term share price appreciation. European member states with REIT enabling legislation in place include Belgium, Finland, France, Germany, Greece, Hungry, Ireland, Italy, Luxembourg, Spain, The Netherlands (PCW 2020).[7] The main operating restrictions are that the REIT company must primarily be in the business of owning and operating real estate and must payout a pre-determined percentage of taxable income (varies by country) on dividends to shareholders, some of which may be paid in stock. Shareholders are then taxed on their dividend income. REITS can trade at significant discount to the value of underlying real estate assets as they are subject to the whims of stock market sentiment (Linneman and Kirsch 2020).

A revolving fund structure provides a pro-active tool to purchase and adapt endangered heritage structures and onward sell (or let) them to sympathetic new owners (or tenants) with protective covenants. A variety of public or private entities can operate a revolving fund, but the majority of funds that buy and sell properties are managed by private, non-profit entities in order to attract capital from donors. The most common sources of start-up funds for revolving funds are grants from government agencies, cultural foundations or corporations. Lending institutions can provide funding in the form of debt finance secured by the fund's assets or personal guarantees. As funds gradually become depleted, fund-raising is an ongoing activity of most revolving funds. Non-profit cultural heritage revolving funds rely on techniques such as renovation lease agreements to ensure long-term protection of the properties they sell on.

Syndication is a method of raising capital for large scale real estate investments in which a group of investors pool their capital to invest in a single property. The main benefit of a syndicate structure is that is allows investors to control investment decisions, compared to real estate investment companies (REIFs and REITs) where investors lose control over the investment decision making process (Cook 2016).

4.5.1 Regulatory Property Rights

Creating a balance between regulatory tools and financial instruments forms the basis with which many state agencies implement cultural heritage policy. Legislative measures are of limited use unless political commitment exists to monitor compliance and use sanctions for non-compliance, such as, fines &/or expropriation. A range of regulatory real estate mechanisms, each with their own benefits, risks and implications for funding leverage, include Land Value Capture (LVC), developer contributions, planning bonus agreements, co-operative housing (land, equity &

[7] International REIT enabling legislation is in place in Australia, Brazil, Bulgaria, Canada, Hong Kong, India, Japan, Malaysia, Mexico, New Zealand, Singapore, South Africa, South Korea, Taiwan, Thailand, Turkey, United Kingdom & USA (PCW 2020).

debt sharing), social value leases (linked to impact metrics), renovation leases, district heating & cooling partnerships, vacancy & derelict land tax, transfer development rights, split landlord & tenant incentives, digital real time energy efficiency monitoring and leveraging government land assets. The main sources of Land Value Capture are the recouping of unearned land value uplifts benefiting private property owners due to beneficial planning rights (land re-zoning) and close proximity to publicly funded infrastructure and urban regeneration initiatives, causing spiking land values (linked to tax Instruments).

4.6 Risk Mitigation and Performance Instruments

As in all capital markets, debt default carries the risk of bankruptcy, damaged reputation and poor credit rating negating future borrowing ability. Lenders may be induced to issue loans, reduce interest rates, or extend the loan term for cultural heritage projects upon receipt of a loan guarantee, or reserve cash deposit, from credit worthy public or private third-party partners to the borrower. This shifts the risk from the lender to the credit enhancer in order to induce the lender to make a loan. Cultural heritage finance initiatives can mitigate risk for commercial bank lenders by providing collateral assets, such as real estate or business assets, to mitigate the risk of loan default. Governments and private foundations may also provide guarantees for mission focused bond issues to enhance their credit rating.

A complexity arises in the case of service-based cultural initiatives which lack tangible assets as collateral to mitigate risk. Performance Risk Financing (PRF) is a relatively new concept (not yet piloted in the cultural heritage arena) that combines public-led performance-based procurement (service contracting) with private investors and non-profit or for-profit service providers to achieve targeted financial return with measured social impact (Von Glahn and Whistler 2014). When a service provider achieves the agreed upon social outcomes or impacts, the state pays the project, and by extension re-pays the investors, the principle plus a risk premium. Measurable outcomes can relate to human wellbeing and could be structured to address additional societal concerns such as environmental and circular sustainable practices.

4.7 Impact Metric Instruments

Measurement metrics relating to cultural heritage span a number of market sectors including economic (financial), socio-cultural and environmental impact. Internal Rate of Return (IRR) is one of the most common metrics used in financial analysis to estimate the performance and profitability of potential investments, in parallel with the interpretation of risk relating to specific projects or activities.[8] IRR analysis provides a tool to benchmark a potential investment across the spectrum of asset classes (such as Equity, Debt & Real Assets). However, the IRR measurement metric is limited in its analytic insight, as it does not explicitly measure the risk elements of investments, such as: the size and length of the investment, timing and stability of expected cash flows, investment management capacity, liquidity/exit risk. In the specific case of comparative investment decisions, relating to adaptation or retrofit of built heritage property assets, IRR does not distinguish between location, quality of adaptation, cost overrun risk, tenant credit risk, leasing risk, or exit liquidity. The final risk return decision ultimately lies with the investor.

A cornerstone of impact investing is the ability to translate impact intention into impact results. Impact Measurement and Management (IMM) is the process of identifying the impact achieved and assessing the projected and realized impact of financial instruments to communities and the planet. The four key elements are intentionality, financial returns, range of asset classes and impact measurement. GIIN (2020a, 2021) acknowledge that although impact measurement and management (IMM) practices have matured in the last decade (such as, IRIS+ Core Metric sets and Impact Management Projects (IMP) Financial Markets), opportunities for further refinement remain. The impact measurement process may entail a combination of both internal institutional analysis and external independent analysis. The European Venture Philanthropy Association (EPVA) provide a step-by-step toolkit of strategies and best practice to assist investors for impact (Gianoncelli and Picon Martinez 2020). GIINs (2020b) analysis on 'The State of Impact Measurement and Management Practice' highlights interpretation, comparison and validation of impact results, to negate 'impact washing', as the most significant challenge facing impact investors.

GIIN (2021) make an observation that SDG alignment & mapping and Environmental, Social and Governance (ESG) Integration strategies do not fulfil the criteria for intentional measured impact. Real estate and construction sustainability metrics are evolving to include tenant and community engagement initiatives (ARUP and EMF 2020). Investors and developers voluntarily engage with the independent sustainability certification benchmarking processes to establish sustainability and circularity performance and gain market share from investors seeking responsible investment opportunities. Global real asset benchmark certification includes, inter alia, GRESB Score, LEED, BREEAM, EU Levels, WELL, WiredScore, RESET Air, nZEB, Net Zero Carbon Standard, DGNB. The social dimensions of safety, labour and human rights within ESG ratings are expanding to include indicators such as tenant wellness, satisfaction, justice (equal and affordable access to energy efficient buildings and housing) and tenant behavioural changes (Kempeneer et al. 2021).

[8] IRR is defined as the annual rate of return (annual growth expectation) that generates a Net Present Value (NPV) of zero for a stream of expected (or actual) cashflows (Linneman and Kirsch 2020).

4.8 Capacity Building Instruments

UN Habitat (2020) states that well resourced human and institutional capacity, such as coordinated multi-level governance structures, robust policy and legislative frameworks, strong leadership and collaborative skills is a necessary prerequisite to combat urban and rural landscape challenges. The provision of non-financial capacity building instruments, such as knowledge sharing, training, technical support, mobilization and matchmaking networks, is fundamental to strengthening financial sustainability and institutional resilience of recipients, in addition to reducing grant dependency (Gianoncelli and Boiardi 2017) (GIIN 2017).

4.9 Digital Pathfinding Tools

From a communication perspective, in the context of matchmaking, grant-giving, co-investing, knowledge sharing and volunteering, digital portals provide visible pathways for stakeholders to find each other in an emerging cultural financial landscape. In addition to providing secure electronic systems for the transfer of funds, digital technology innovations in recent years have revolutionized knowledge transfer and networking within the cultural heritage sphere (Stehle 2014). Advances in R&D for Proptech and Fintech are currently in an early evolutionary phase creating speculation about digital solutions for future technical, regulatory, security and data protection scenarios.

5 Illustrative Pooled Funds and Blended Financial and Non-financial Instruments

The potential combinations of financial and non-financial instruments with the above umbrella clusters are limitless. The following examples provide a brief illustrative contextual insight into the benefits of blending and pooling investment strategies.

5.1 Pooled Hybrid Private Investment Leverage

Financial intermediaries engage in collaborative investment strategies to enable mission focused place-led and heritage-led adaptive reuse, urban regeneration and community initiatives. The following examples illustrate for-profit and non-profit private investment funds acting as financial intermediaries to enable hybrid combinations of financial and non-financial instruments.

5.1.1 Fondation du Patrimoine (France)

Private non-profit foundation created with corporate endowment funding, working in partnership with local governments, Loto du Patrimoine, corporations and citizens to provide grants and capacity building assistance to regional and local heritage projects using tool combinations:

- Escheated inheritance
- Tax deductions
- Lottery funds
- Donations in-kind and crowd funding
- Donations from corporations and SMEs.[9]

5.1.2 Calvert Impact Capital (USA)

Private non-profit impact fund, launched with support from Ford, McArthur and Mott foundations, that leverages funds to provide micro finance to enterprise and adaptive reuse initiatives in disadvantaged communities. Funds are leveraged via:

- Community Investment Note (bond debt instrument)
- Collaborative co-lending syndication services (CIC 2020).

Calvert Impact Capital assesses three layers of impact measurement (Investor, Portfolio and Community) using both internal and external (IRIS+ & Impact Management Projects 5) data sources (CIC 2020).

5.1.3 IPUT plc. (Ireland)

Private for-profit Real Estate Investment Fund (REIF) that leverages capital in the marketplace using debt (revolving loans and bonds) and equity Instruments (institutional investors), via

- Institutional private equity shareholder investments
- US Private Placement Market investments (US Private Bond Market);
- Revolving Credit facility from Wells Fargo Bank including Green Finance (IPUT 2019, 2020; PFW 2020).

IPUT plc. engages with real estate industry energy, wellness, sustainability and ESG financial reporting benchmarks. The company also invests in public realm enhancements in addition to cultural, community, tenant and employee initiatives and events as part of their sustainability strategy. IPUT is a signatory to the World Green Buildings Council (WGBC) (IPUT 2019, 2020).

[9] Presentation by Celia Verot, Director General, EC Workshop on Complementary Funding for Cultural Heritage, 25–26th January 2021.

5.1.4 BlackRock (Luxembourg)

Circular Economy Fund, sub fund of the Blackrock Global Funds (BGF), launched on the stock exchange in partnership with the Ellen MacArthur Foundation, to drive investment in companies transitioning to circular economy business models, driven by climate mitigation, resource scarcity, technological disruption and rapid urbanization. The fund is rated 'AA' on the Morgan Stanley Capital International (MSCI) ESG Index (BGF 2021).

5.1.5 Detroit Home Mortgage Programme (USA)

Non-profit grassroots community collaborative funding structure, involving partnership with government agencies, national and local banks, cultural foundations (Kresge, Ford and Community Foundation of South Michigan) and various non-profit neighborhood-based community groups. The programme is designed is rebuild the city's failed housing market by using a mix of grants, loans and loan guarantees that enable buyers to purchase and adapt abandoned housing (OCC 2018).

5.1.6 Irish Landmark Trust and Dublin Civic Trust (Ireland)

Irish landmark Trust is a private non-profit trust that leases abandoned heritage assets (such as lighthouses, castles and gate lodges) and carries out restoration and adaptive reuse works, for letting as self-catering holiday accommodation. The negotiated lease term is usually circa 50 years and once a lease expires, the heritage asset reverts to the original owner (https://www.irishlandmark.com/). By contrast, Dublin Civic Trust, also a private non-profit trust, purchases and undertakes restoration and adaptive reuse works to heritage buildings at risk of demolition, using a revolving fund structure. Completed projects are sold to recapitalise the revolving fund, in addition to charitable donations. The Trust is a member of Europa Nostra, ICOMOS, Building Limes Forum Ireland, INTBAU (traditional building, architecture and Urbanism network) and European Environmental Bureau (EEB) (http://www.dublincivictrust.ie/).

5.1.7 Stadsherstel Amstel Foundation (The Netherlands)

Stadshertstel Foundation is a hybrid for-profit limited-liability company and a non-profit public housing corporation. The hybrid structure is justified by the fact that the annual dividend to shareholders is fixed at 5%. The dividend is exempt from tax as long as the upgraded value of a building does not exceed the cost of works. In the case of dissolution, shareholders only receive their initial capital outlay plus the dividend. The company purchases endangered historic buildings in very poor structural condition and buildings whose rehabilitation will contribute to city regeneration, such as corner buildings to consolidate streetscapes (https://stadsherstel.com/).

5.2 Blended Hybrid Financial Instrument Combinations in North America

Politically, a strong pro-market bias underlies the choice of financial instruments in North America, hence the selection of illustrative examples from the USA, where private and community owners are permitted to 'piggyback' different financial instruments on top of each other until adaptive reuse projects become financially viable. All works to historic structures are carried out under the supervision of the State Historic Preservation Officer (NPS 2017). These examples show that every funding solution is different relating to the tailored needs of recipients. Flexible policy allows private owners, who are experiencing cashflow issues during the development phase, to syndicate their future tax credits to third party investors (for a percentage fee) and receive the cash payments upfront. In some states, where owners do not have a high enough income to claim a tax credit, the benefit can be converted into an alternative mortgage credit. Illustrative case studies include:

5.2.1 Oliver Neighborhood, East Baltimore (USA)

Low-income housing initiative, enabled via partnership structure, to adapt a street of abandoned historic row houses to low-income community housing. A partnership was formed in 2016 between TD Bank (99% ownership), East Baltimore Historic Group (1% ownership), ReBUILD Metro (faith-based community non-profit), Community Reinvestment Funds USA, City of Baltimore and State of Maryland. The East Baltimore Historic group act as superior housing landlord and sublet to low-income tenants.

Oliver Neighborhood Historic Low-income Housing, East Baltimore, Maryland, USA	
Construction loan TD Bank	$12,175,500
Historic Tax Credit TD Bank as equity investor	$6,438,534
ReBUILD Metro loan financed by 10 year unsecured promissory Notes to 26 investors	$6,648,364
City of Baltimore grant	$3,171,000
State of Maryland grant	$3,100,000
Deferred Developer fee	$2,340,782
US Treasury Magnet Fund	$600,000
East Baltimore Historic Group	$10,000
Total project Cost	$34,484,190

Source: (OCC 2018)

5.2.2 Germania Print Works, Milwaukee (USA)

Adaptive reuse of a disused industrial complex in 2017, (built 1896), involved conversion from an abandoned printworks to a mixed retail and low-income housing development.

Germania Print Works, Milwauke, Wisconsin, USA	
Federal Historic Tax Credit (via investor equity)	$3,276,892
Federal Low Income Housing Tax Credit (via investor equity)	$1,909,856
Wisconsin State enabled Bond finance	$10,213,000
Wisconsin State enabled low interest loan	$1,400,000
Wisconsin State Historic Tax Credit (via investor equity)	$2,619,871
City of Milwaukee Tax Increment Financing Loan	$1,500,000
Accrued Interest	$51,418
Deferred Developer Fee	$1,206,782
Total project Cost	$22,177,819 (100%)

Source: Rutgers and NPS (2019)

6 Conclusions

Cultural heritage adaptive reuse investment activities involve long-term time horizons and investment strategies. Strategic blending of both financial and non-financial instruments, combined with mutually beneficial partnerships between public, private and people-led investment leverage enablers, has the potential to activate positive cash flows for cultural heritage activities. The positive impact that combined state and philanthropic finance contribute to both human and cultural capital can never be overstated. However, a significant unmet demand for funding to halt the ongoing depletion of Europe's finite heritage assets still remains, as there is a misalignment between the investment needs of the owners and curators of cultural assets and what mainstream investors are willing to finance.

A number of emerging investment leverage enablers, with a mission to 'create value and give back', signal an appetite for 'real' long-term sustainable investment opportunities in the marketplace. Market oriented capital leverage enablers, such as pension funds, insurance companies and real estate funds, have an obligation to their co-investors and shareholders to mitigate risk and optimize investment returns, in parallel with ESG compliance and intentional impact (IMM) returns. This indicates that the investment mindset is increasingly recognizing the supply-side investment perspective (investors desire to make risk-adjusted market and impact investments) to compliment traditional demand-side perspective (direct state funding mechanisms for owners, curators and local communities).

The spectrum of capital leverage enablers, ranging from philanthropic grant-givers (impact first) to purely profit motivated investors (finance first), captures a broad knowledge base and skillset. Partnership arrangements between capital leverage enablers, within these two extremes, will facilitate the cross fertilization of knowledge and skillsets.

The design of collaborative funding networks, to support grassroots enterprise and reuse of obsolete heritage resources, must remain flexible to respond to local regional socio-cultural and economic contexts and allied political priorities. Many investment leverage instruments can only be materialized with political support, via legislative and revenue enactment. This highlights the importance of evolving intentional impact metrics (IMM), Taxonomy and ESG reporting standards and regulations, providing transparent and robust evidence-based analytics on the positive impacts of cultural heritage activities, to bolster investor confidence and reduce the risk of capital misallocation, greenwashing and product mis-selling.

A critical investment bottleneck still exists in creating sustainable investment leverage for small scale local community cultural heritage activities, particularly in disadvantaged urban and depopulated rural locations. Work still remains to cultivate 'connective networking Infrastructure' to foster tactical relationships between 'grassroot' cultural heritage initiatives and cash-rich investment markets. The use of risk mitigation and capacity building instruments, aligned with the development of digital network pathfinding tools, is vital to bridging the connective infrastructure gap to open up micro investment leverage opportunities for local communities. Sustainable hybrid funding strategies should also seek to strengthen the financial sustainability and resilience of recipients.

Capital market invest strategies rely heavily on economic market sentiment in addition to increasing reliance on intentional impact metrics. To encourage private investment capital flows to urban and rural, built and landscape heritage, cultural heritage must be recognized as a *'Market Asset Class'*. This would facilitate both traditional capital market financial metrics and evolving socio-cultural and environmental impact metrics to generate transparent evidence-based comparable market data, relating to the costs and benefits of cultural heritage activities. Only then, can informed and transparent market and impact risk assessment take place, to inspire greater investment confidence.

The potential contribution of a Pan-European cultural heritage philanthropic foundation to engender greater financial synergy between citizens, local communities, capital markets and local government agencies remains untapped. The rise of combined market driven and impact focused venture capital investment initiatives, combined with the rise in citizen-led co-giving and co-investment crowd funding platforms, indicates an eagerness among proactive people and communities to coalesce in order to safeguard tangible and intangible cultural heritage resources.

Indicators, such as the arrival of the BlackRock Circular Economy Fund supporting adopters, enablers and beneficiaries of circular economy, the launch of the EU Taxonomy 'green finance' criteria tool imply that cultural heritage has the potential to become a vector for circular economy transitionary ambitions with the capacity to make a real contribution to the SDGs, the Renovation Wave and the European Green Deal. Evidenced by the fact that some financial and non-financial instrument combinations and investment leverage partnership arrangements, highlighted in this chapter, did not exist 20 years ago, it is not unreasonable to predict that the financial landscape will further evolve, in line with expansion in human and institutional capacity and the ongoing need for human centric solutions to dysfunctional eco-systems. Future manifestations of hybrid circular financial solutions will not be effective unless the selection and design process remains empathetic to the needs of local communities and ecosystems, in parallel with vulnerable heritage resources. Only then will capital investment strategies work for people, and not the other way around.

Acknowledgements This research was funded by the European Union Horizon 2020 Project ID: 776758, CLIC—Circular models Leveraging Investments in Cultural heritage adaptive reuse. Figure 19.1 Infographic designed by Cathal Power, TU Dublin.

References

Achleitner, SpeissKnafl (2012) Financing social entrepreneurship. In: Tokarski V, Ernst (eds) Social entrepreneurship and social business. Springer Gabler

ARUP, EMF (2020) From principles to practice: realizing the value of circular economy in real estate. https://www.arup.com/perspectives/publications/research/section/realising-the-value-of-circular-economy-in-real-estate

Baeck P, Bone J, Mitchell S (2017) Matching the crowd, Nesta. Dept. of Digital, on behalf of Culture & Media Sport, Arts Council England, Lottery Heritage Fund

Balboni E, Berenbach S (2014) Fixed Income Securities. In: Salamon LM (ed) New Frontiers of Philanthropy. Oxford University Press, ISBN:798-0-19-935754-3

BGF (2021) BlackRock Global Fund: Circular Economy Class A2 Factsheet January 2021 https://www.blackrock.com/ch/individual/en/literature/fact-sheet/bgf-circular-economy-class-a2-usd-factsheet-lu2041044095-ch-en-individual.pdf

Brand M, Kohler J (2014) Private equity investments. In: Salamon LM (ed) New Frontiers of Philanthropy. Oxford University Press, ISBN:798-0-19-935754-3

CIC (2020) Calvert Impact Capital Impact Report 2020: Investing for a more equitable world, USA

Cohen R (2014) Corporate originated charitable funds. In: Salamon LM (ed) New Frontiers of Philanthropy. Oxford University Press, ISBN:798-0-19-935754-3

Cook, S. (2016) Investing in real estate private equity, Poland

EC (2015a) Guidance for Member States on Financial Instruments–Glossary https://www.fi-compass.eu/publication/ec-regulatory-guidance-guidance-member-states-financial-instruments-glossar

EC (2015b) Towards more efficient financial ecosystems: Innovation instruments to facilitate access to finance for cultural and creative sectors (CCS), European Agenda for Culture. European Commission http://ec.europa.eu/culture/library/index_en.htm

EC (2017) Crowd funding reshaping the crowd's engagement in culture, prepared by IDEA Consult, Ecorys & European Crowdfunding Network, authored by De Voldere, I., & Zeqo, K., https://doi.org/10.2766/011282

ECA (2019) European Court of Auditors, rapid case review: reporting on sustainability: a stocktake of EU Institutions and agencies, June 2019. https://www.eca.europa.eu/Lists/ECADocuments/RCR_Reporting_on_sustainability/RCR_Reporting_on_sustainability_EN.pdf

ECA (2020) European Court of Auditors, EU Investments in cultural sites: a topic that deserves more focus and coordination, Special Report 08/2020. https://op.europa.eu/webpub/eca/special-reports/cultural-investments-08-2020/en/

ECF, AKS (2020) Imagine Philanthropy for Europe, Ed. Wider Sense, European Cultural Foundation and Allianz Kulturstiftung

EFT, TGE (2017) Boosting cross border philanthropy in Europe: towards a tax efficient environment. http://www.efc.be/uploads/2019/03/Boosting-Cross-Border-Philanthropy-in-Europe-Towards-a-Tax-Efficient-Environment

EIB (2015) Financial instruments products – loans, guarantees, equity & quasi equity. Fi-Compass, ESF
EIB (2020a) Circular economy guide: supporting circular transition. https://doi.org/10.2867/917400
EIB (2020b) Stocktaking study on financial instruments by sector. Fi-Compass ERDF
Eikenberry AM, Bearman J (2014) Funding collaboratives. In: Salamon LM (ed) New frontiers of philanthropy. Oxford University Press
EL (2018) The European lotteries for the benefit of society: report on the lottery sector in Europe. The European Lotteries (EL)
ESMA (2021) Letter to commissioner in charge of financial services, financial stability and capital markets union of the European Commission, from European securities and markets authority, ESMA 30-379-423 28th January 2021, signed by Steven Maijoor
EU (2020) Regulation 2020/852 on the establishment of a framework to facilitate sustainable investment, and amending Regulation (EU) 2019/2008. 18th June 2020
EU (2023) Proposal for a regulation of the European Parliament and of the council on the transparency and integrity of Environmental, Social and Governance (ESG) rating activities. COM/2023/314 final. Available at: https://eur-lex.europa.eu/legal-content/EN/TXT/?uri=CELEX%3A52023PC0314
EU TEG (2020a) Taxonomy: final report of the Technical Expert Group on sustainable finance, March 2020
EU TEG (2020b) EU Technical Expert Group on sustainable finance, Statement of the EU TEG on Sustainable Finance. Five high level principles for Recovery and Resilience, July 2020
Fox R (2015) A guide to establishing alternative investment funds. Irish Funds Ireland
Fram S (2018) The next step in impact investing: breaking the shackles of extractive thinking. Stanf Soc Innov Rev, July 23rd 2018
Fusco Girard L (2019) Implementing the circular economy: the role of cultural heritage as the entry point. Which evaluation approaches? BDC Boll Del Cent Calza Bini 9:245–277
Gianoncelli A, Boiardi P (2017) Financing for social impact – the key role of tailored financing and hybrid finance. European Venture Philanthropy Association (EVPA), ISBN: 9789082494037
Gianoncelli A, Boiardi P (2018) Impact strategies – how investors drive social impact. European Venture Philanthropy Association (EVPA), ISBN: 9789082494082
Gianoncelli A, Picon Martinez A (2020) The Investing for Impact Toolkit. European Venture Philanthropy Association (EVPA), ISBN: 9789464202120
GIIN (2017) Beyond Investment: The Power of Capacity Building Support Global Impact Investing Network
GIIN (2020a) The state of impact measurement and practice, 2nd edn. Global Impact Investing Network
GIIN (2020b) Annual Impact Survey 2020, 10th edn. Network, Global Impact Investing
GIIN (2021) Impact investing decision making insights on financial performance. Global Impact Investing Network
Hagerman L, Wood D (2014) Enterprise Brokers. In: Salamon LM (ed) New Frontiers of Philanthropy. Oxford University Press
Hibernia REIT (2020a) Annual report 2020: transforming Dublin. Hibernia REIT plc
Hibernia REIT (2020b) Sustainability report 2020: delivering long-term value to our stakeholders. Hibernia REIT plc
IPUT (2019) Greening our portfolio: responsibility report. IPUT Real Estate Dublin
IPUT (2020) IPUT plc. Real Estate Quarterly Report, Q3 2020
Kempeneer S, Peeters M, Compernolle T (2021) Bringing the user back in the building: Analysis of ESG in real estate and a behavioral framework to guide future research. Sustainability 13:3239. https://doi.org/10.3390/su13063239
Linneman P, Kirsch B (2020) Real estate finance & investments risks and opportunities 5.1. Linneman Assoc, Philadelphia, ISBN 978-1-7923-3191-6

Luther King M (1963) Strength of love. Beacon Press, Boston, Republished Oct.2019, ISBN-10-080705190x

Mazur J (2017) Regulating Crowdfunding and Crowdinvesting, international resources for local communities, in Funding the Cooperative City by Patti, D., & Polyak, L., Cooperative City Books, Vienna

McKinsey (2020) Reimagining European philanthropy. McKinsey Global Institute

McVeigh N, Sass J (2014) Loans, loan guarantees and other credit enhancements. In: Salamon LM (ed) New frontiers of philanthropy. Oxford University Press, ISBN:798-0-19-935754-3

Noring L (2019) Public asset corporation: a new vehicle for urban regeneration and infrastructure finance. Cities 88:125–135, Elsevier. https://doi.org/10.1016/j.cities.2019.01.002

NPS (2017) The Secretary of the Interior's Standards for the Treatment of Historic Properties with guidelines for preservation, rehabilitation, restoring and reconstructing historic buildings. US Department of the Interior National Park Service Technical Preservation Services, Washington DC. https://www.nps.gov/tps/standards/treatment-guidelines-2017.htm

OCC (2018) Community Development Investments, Expanding Housing Opportunities, Single-Family rehabilitation financing programs. Office of the Controller of Currency, USA https://www.occ.gov/publications-and-resources/publications/community-affairs/community-developments-investments/feb-2018/pub-cdi-feb-2018.pdf

OECD (2020) Blended finance principles for sustainable development. http://www.oecd.org/officialdocuments/publicdisplaydocumentpdf/?cote=DCD/DAC(2020)42/FINAL&docLanguage=En

PCW (2020) Worldwide Real Estate Investment Trust (REIT) Regimes – Compare and Contrast

Peterson GE (2009) Unlocking land values to finance urban infrastructure. The World bank, Washington, DC

PFW (2020) IPUT agrees largest green revolving credit facility in Irish real estate market, Property Funds World, 29th May 2020. Accessed 13/1/2021

Pickerill T, Armitage L (2009) The management of built heritage, a comparative review of policies & practice in Western Europe, North America & Australia, Pacific Rim Real Estate Society Conference, Sydney. https://arrow.tudublin.ie/beschreccon/36/

Pickerill T, Pickard R (2007) A review of fiscal incentives to benefit heritage conservation, RICS Research Paper Series Vol. 7 No. 6

Richter L (2014) Capital aggregators. In: Salamon LM (ed) New frontiers of philanthropy. Oxford University Press, ISBN:798-0-19-935754-3

Robeco and Booz & Co (2009) Responsible investing: a Paradigm Shift from Niche to Mainstream

Rutgers, NPS (2019) Annual Report on the Economic Impact of the Federal Historic Tax Credit for Fiscal Year 2018, prepared by the Edward J. Bloustein School of Planning and Public Policy & National Park Service, US Dept. of the Interior technical Preservation Services, October 2019

Salamon LM (ed) (2014) New frontiers of philanthropy. Oxford University Press, ISBN: 978-0-19-935754-3

Shahnaz D, Kraybill R, Salamon LM (2014) Social and environmental exchanges. In: Salamon LM (ed) New frontiers of philanthropy. Oxford University Press, ISBN:798-0-19-935754-3

Stehle V (2014) Online portals and exchanges. In: Salamon LM (ed) New frontiers of philanthropy. Oxford University Press, ISBN:798-0-19-935754-3

Tuan M (2014) Capacity builders. In: Salamon LM (ed) New frontiers of philanthropy. Oxford University Press, ISBN:798-0-19-935754-3

UN Habitat (2020) World cities report: the value of sustainable urbanisation https://unhabitat.org/sites/default/files/2020/11/world_cities_report_2020_abridged_version.pdf

Von Glahn D, Whistler C (2014) Social impact bonds and pay for success financing. In: Salamon LM (ed) New frontiers of philanthropy. Oxford University Press, ISBN:798-0-19-935754-3

WEF, OECD (2015) Blended finance vol. 1 a primer for development finance and philanthropic funders. http://www3.weforum.org/docs/WEF_Blended_Finance_A_Primer_Development_Finance_Philanthropic_Funders.pdf
World Bank (2013) The new microfinance handbook
Zeleny M (2021) Entering the age of accelerated change. In: Search of equilibrium, human systems management, vol 40. IOS Press, pp 3–14. https://doi.org/10.3233/HMS-209002, ISBN 0167-2533/21

Chapter 20
Circular Financing Mechanisms for Adaptive Reuse of Cultural Heritage

Aliona Lupu and Ivo Allegro

1 Introduction

Cultural heritage represents a typical example of "common good" belonging to each citizen leaving in a certain place and the identity character of a community. It usually receives funding from the public sector for the purpose of its preservation.

The recent pandemic crisis has affected in an unprecedented way both our lives as well as cultural sector fruition. Indeed, the societal challenges have become more acute and this may bring to turn off the spotlights on already limited dedicated public budget to cultural heritage. The pandemic has affected every dimension of the cultural heritage value chain: loss of revenue, stop of maintenance and restoration works, the restriction of access and participation in cultural events etc. According to a recent report by the Joint Research Centre of the EC, over seven million of cultural and creative jobs are at risk due to the crisis (Montalto et al. 2020). The Council conclusions on risk management of cultural heritage emphasises the relevance of sustainability and resilience for cultural heritage management and mobilisation of financial resources to safeguard the endangered heritage (CoE 2020).

In this context, top-down public policies become more and more inefficient, because of their excessive concentration on cultural heritage preservation and less focus towards the exploitation of cultural heritage potential. In a period of increasing pressure on public budgets, this activates the perverse spiral of increasingly inadequate investments because of scarce available resources in the hands of the public decision-maker.

Moreover, in the absence of clear allocation criteria, spreading equally the available resources among the existing cultural heritage initiatives can even worsen the inefficient distribution of funds, because bad investments withdraw resources from good investments. This translates into the impossibility to reach a minimum

A. Lupu (✉) · I. Allegro
Iniziativa Cube, Naples, Italy
e-mail: aliona.lupu@iniziativa.cc; ivo.allegro@iniziativa.cc

L. Fusco Girard, A. Gravagnuolo (eds.), *Adaptive Reuse of Cultural Heritage*,
https://doi.org/10.1007/978-3-031-67628-4_20

efficient investment scale capable of triggering positive processes of circular and cumulative causation (Myrdal 1958) that can fuel self-propulsive developmental dynamics and, more generally, to exploit the full business potential. In the medium term, this situation can lead to the depletion of a non-trivial part of the cultural heritage, including its immaterial values. The consequence is the call for the "private hand" at the last stage with a wrong approach of "decommissioning", which often brings cultural heritage "on sale"[1] that brings to the denial of its fruition or the deformation of the cultural identity of a place with significant negative externalities against short-term economic benefits.

Opposite to the exclusive public intervention, the choice of privatization of cultural heritage often generates both immediate cash flows as well as side effects in medium/long term. The private sector can adopt a "cherry picking" approach by leaving less valuable cultural heritage in the public hands. This increases the problems of ordinary management and insufficient resources, activating, therefore, the perverse spiral of progressive decay of the residual cultural heritage. From a social point of view, in the medium term, the privatization can contribute to the community impoverishment of powerful instruments of collective memory and identity due to the inaccessibility of privatised goods or to the high costs of use.

In CLIC project, the cultural heritage adaptive reuse is suggested "*as the entry point for implementing the circular city, that is the specific spatial/territorial aspect of the circular economy*" (Fusco Girard 2019).

In the transition from a linear economy to a circular economy, financing represents one of the main critical barriers. The most affirmed circular business models are still perceived as highly risky by the investors: new revenues and ownership structures, cash flows spread out over time, longer payback periods, increased risk of default and demand for working capital, decrease in short-term margins etc. (Achterberg and Tilburg 2016).

At the same time, the long-term strength and robustness of circular business models are not accounted for in current financial decision-making models. To finance circular businesses investors needs to change their mind-sets and innovate the existing investment evaluation techniques with new ones.

In this perspective, the financial mechanisms described in this chapter will regard both physical adaptive reuse of cultural heritage (ARCH) initiatives (i.e. restoration, revitalisation and rehabilitation) as well as the "*new functions in the re-use* linked to the *innovative/creative* functions" (Fusco Girard 2019), interpreted, in particular, through the adoption of circular business models. The final aim of the financing mechanisms proposed is to lower the financing barriers that prevent the human-centric and favourable exploitation of cultural heritage.

[1] The documentary "Europe for sale" by Andreas Pichler.

2 Taxonomy and Definitions of Circular, Sustainable, Green, Social, Impact and ESG Finance for ARCH, Blended Finance and Leverage Concept: An Overview

One of the key objectives of Horizon2020 CLIC project were to develop circular financing mechanisms for the adaptive reuse of cultural heritage, also considering the impact generated by investments. The starting point was the focus on the "circularity" concept for finance. Indeed, the first step of the financial mechanisms design phase was the analysis of the existing taxonomy for **circular, sustainable, green, social, impact and ESG finance.** Policy makers, regulators and other financial sector stakeholders have recently advanced several relevant proposals for the mentioned definitions and taxonomy, aimed to become financial sector's standards. At global and European level, there is still a need for convergence on terminology among markets participants, stakeholders, as well as policy makers and regulators.

The first definition of **Circular Economy Finance** has been provided by the members of the FinanceCE working group,[2] founded by PGGM[3] and supported by the Ellen MacArthur Foundation. The members' goal is to create and stimulate a common understanding of circular economy finance:

> Circular Economy Finance is any type of instrument where the investments will be exclusively applied to finance or re-finance, in part or in full, new and/or existing eligible companies or projects in circular economy (ABN Amro 2018).

In 2018, the Sustainable Finance Study Group under G20, has extended the definition of **Sustainable Finance** as:

> Sustainable finance can be broadly understood as financing as well as related institutional and market arrangements that contribute to the achievement of strong, sustainable, balanced and inclusive growth, through supporting directly and indirectly the framework of the Sustainable Development Goals (SDGs). A proper framework for sustainable finance development may also improve the stability and efficiency of the financial markets by adequately addressing risks as well as market failures such as externalities (G20 Sustainable Finance Study Group 2018).

In the same period, the European Commission has also developed a rich policy agenda on sustainable finance, which plays a key role in mobilising the necessary financial resources to deliver on the policy objectives under the European Green Deal (European Commission 2019). In the EU's policy context, "*sustainable finance is understood as finance to support economic growth while reducing pressures on the environment and taking into account social and governance aspects. Sustainable finance also encompasses transparency on risks related to ESG factors that may impact the financial system, and the mitigation of such risks through the appropriate governance of financial and corporate actors*" (European Commission 2020).

---

[2] ABN AMRO, ING and RABOBANK.

[3] PGGM is a cooperative Dutch pension fund service provider.

Green Finance can be defined as *"financing of investments that provide environmental benefits in the broader context of environmentally sustainable development"*, addressing also the "adjustment of risk perception" "related to environment factors" (G20 Green Finance Study Group 2016).

The EU's Action Plan on Financing Sustainable Growth (March 2018) called for the creation of a classification system for sustainable activities or *Taxonomy*. As of 2020, **the design of financial instruments and investment strategies must take the EU Taxonomy performance thresholds into consideration**, as set out by the EU Technical Expert Group (TEG) on Sustainable Finance. The EU Taxonomy is a "green finance" criteria tool to assist investors, companies, issuers and project promoters navigate the "transition to a low carbon, resilient and resource-efficient economy". The Taxonomy sets technical screening criteria for six environmental objectives, requiring economic activities to substantially contribute to at least one objective; and Do No Significant Harm (DNSH) to the other five (EU TEG 2020).

The EU Taxonomy Objectives are:

(i) Climate change mitigation;
(ii) Climate change adaptation;
(iii) Sustainable protection of water & marine resources;
(iv) Transition to Circular Economy;
(v) Pollution prevention and control;
(vi) Protection and restoration of biodiversity & ecosystems (EU TEG 2020).

Interestingly, the development and use of the Taxonomy is rather appropriate with the development phase of circular financing under CLIC project. Indeed, CLIC final goal is perfectly aligned with the **Taxonomy's forth objective on the transition to a circular economy.** At the same time, based on the adaptive reuse of cultural heritage initiative and the circular business model to be adopted, one or more of the other five objectives can be addressed. Under the CLIC project framework, the Taxonomy is relevant for both businesses/non-profit organisation as well as for financial participants (private or public) interested in the implementation of CLIC circular financing mechanisms.

The main features of the Taxonomy have been considered during the design stage of new financial mechanisms for ARCH initiatives. Moreover, the Taxonomy already provides the technical screening criteria for the sector of building renovation—the most relevant one when addressing ARCH initiatives, if not considering also the economic activities of the deployed business models inside the adapted cultural heritage asset. Indeed, **the sector of building renovation is considered by the TEG providing environmental contributions to all the six objectives** underlying the Taxonomy.

The definition of **Social Finance** refers to and is consistent with the Social Bond Principles supported by ICMA (ICMA 2020; International Capital Market Association 2020). **Social Finance** is financing that supports actions mitigating or addressing a specific social issue and/or seeking to achieve positive social outcomes especially but not exclusively for a target population(s). Social finance project categories include but are not limited to, providing and/or promoting affordable basic infrastructure, access to essential services (such as health and healthcare), affordable housing, employment generation including through the potential effect on SME financing and microfinance, food security, and socioeconomic advancement and empowerment (ICMA 2020).

According to the definition of ICMA, **Impact Finance** is the financing of businesses or economic activities which produces verifiable and direct positive impact on the society and/or environment, based on agreed metrics and benchmarking while also seeking market aligned or better financial return (ICMA 2020).

ESG investing places itself within the investment spectrum between financial and social returns. One extreme of the spectrum, based only on pure social investing such as philanthropy, looks for social returns, related to environmental or social benefits, including with regard to human and worker rights, gender equality etc. The other extreme of the spectrum is characterised by the maximisation of investors' value through financial returns based on absolute or risk-adjusted measures of financial value (Boffo and Patalano 2020).

Within this spectrum, ESG investing aims at financial returns maximisation and application of ESG factors for the risks and opportunities assessment in a mid to long-term perspective.

The definitions for circular, sustainable, green, social, impact and ESG investment are characterised by rather **vague boundaries** depending on different factors. The interaction between the provided definitions is quite high and it can be assumed that **wider definitions incorporate narrower ones**. For example, Sustainable Finance can be considered as a wider definition incorporating impact investing, ESG investing, green finance and social finance etc. (Boffo and Patalano 2020).

Alongside with the different definitions of finance, the adoption of **blended finance** as an innovative approach to financing sustainable development is also important for the adaptive reuse of cultural heritage.

In the "Blended Finance Principles Guidance", the OECD defines blended finance as the *strategic use of development finance for the mobilisation of additional finance towards sustainable development.* The perspective addressed is that of developing countries and (multilateral) development banks and finance institutions. The OECD also identifies other market players to get engaged in blended finance such as foundations, philanthropic investors, institutional investors, commercial banks, private equity and venture capital funds, hedge funds, as well as corporations and SMEs (OECD 2020).

The **five core principles for blended finance** developed by OECD at a glance are the following:

- Anchor blended finance use to a development rationale;
- Design blended finance to increase the mobilisation of commercial finance;
- Tailor blended finance to local context;
- Focus on effective partnering for blended finance;
- Monitor blended finance for transparency and results (OECD 2020).

The "Blended Finance Principles" developed by the OECD, are used as guidance when structuring circular blended finance mechanisms under CLIC project.

Moreover, the "**leverage**" terminology behind the CLIC financial mechanisms is based on the definition provided by the World Bank, as "the ability of a public financial commitment to mobilise some larger multiple of private capital for investment in a specific project or undertaking" (WorldBank 2011; Griffiths 2012).

2.1 CLIC Project Approach Towards Circular Financing of ARCH Initiatives

Building on the definitions above, when addressing the financing of adaptive reuse of cultural heritage initiatives, it is important to consider the main steps of a circular investment pattern in ARCH, such as:

- **design**—this phase covers the planning of the CH transformation, ARCH design, circular business model viability, project readiness for investment preparation and sourcing of finance;
- **build**—this phase represents implementation-related investment, covering operational costs for construction, rehabilitation and adaptation of the project. It includes the building material sourcing in the perspective of the circular economy;
- **use and operate**—this phase refers to the new use of the adapted cultural heritage, the deployment of the viable circular business model and self-sustaining financing of the project's long-term running costs (Gravagnuolo et al. 2021).

Different investment phases may require different types of sustainable finance: e.g. for the design stage, social finance or grants can be used; for the building stage, circular and/or green financing can be used; for the use and operate phase—social, impact or ESG finance can be used, also based on the adopted business model.

Furthermore, circular ARCH initiatives can represent commercially viable investments or may not generate sufficient cash flow to cover, completely or partially, investment costs with a sufficient return on capital or the running costs after investments (particularly relevant for the use and operate phase).

In the CLIC project, three scenarios have been identified:

- **Cold ARCH**: a project that does not generate any cash flow for the investment disbursement and coverage of long-term running costs or that generates low level cash flows, insufficient to cover running costs;
- **Lukewarm ARCH**: a project that generate sufficient cash flow to cover long-term running costs but insufficient to counterbalance the investment disbursement or vice versa;
- **Hot ARCH**: a project that generate sufficient cash flow for the coverage of both the investment disbursement as well as long-term running costs (financially free standing).

Based on the three scenarios, different types of sustainable finance can be used, e.g. for cold ARCH, social finance can be chosen; for lukewarm and hot ARCH—ESG

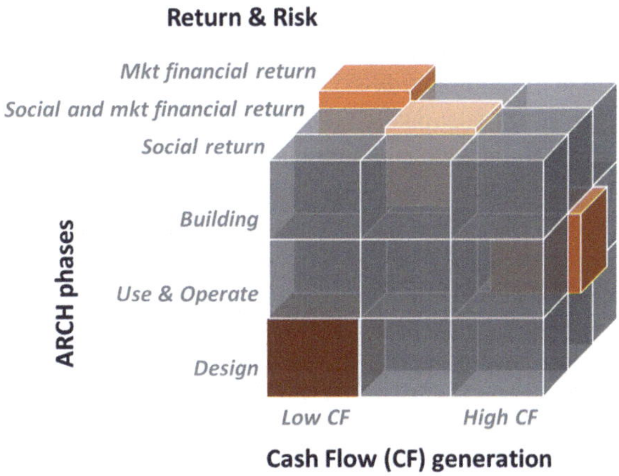

Fig. 20.1 CLIC project approach towards circular financing of ARCH initiatives. (Source: Authors)

investing can be the option. The ambiguity in the adoption of different types of sustainable finance can even increase because ARCH initiatives often evidence greater financial complexity due to the size of the investments necessary to allow adaptive reuse. Many actions to be implemented are connected to the possibility of "heating" "cold" or "lukewarm" operations by not acting at single asset/building level but in **a systemic and aggregate return approach**, valorizing area-based initiatives.

The CLIC overarching approach towards circular financing of ARCH initiatives shall be focused on a flexible vision in the use of (blended) financing instruments, based on different expected risk-adjusted returns, in function of the ARCH investment phases and the capacity of the adopted circular business models to generate cash flow and impacts. This approach can further support the private leverage investments in circular adaptive reuse of cultural heritage (Fig. 20.1).

3 CLIC Circular Financing Instruments for Adaptive Reuse of Cultural Heritage

3.1 ARCH Investment Readiness Facility

Many barriers prevent investment in adaptive reuse of cultural heritage in the perspective of circular economy. In the framework of CLIC Project, different barriers have been identified such as: regulation, lack of knowledge, decision-making, lack of incentives, limited community engagement, balancing cultural significance and economic viability, commercial risk and uncertainty, technical difficulties etc. (Ikiz Kaya et al. 2021).

More specifically, the lack of capacity for a systemic and aggregate return approach with respect to ARCH initiatives represents a relevant weakness in the field. This makes it difficult to identify an integrated mix of initiatives in order to get an attractive ARCH operation with acceptable social and financial returns. These aspects usually are not addressed because of processes complexity and risks embedded in the administrative and authorization procedures for the implementation activities. If not adequately addressed in the initiating phase, these issues may change into commissioning and procurement risks (i.e. the ARCH initiative will not achieve the approvals by the public stakeholders and/or the engaged community, with consequent delays or contentious relationships). Thus, to diminish this complexity, project promoters avoid to work on the bundling of ARCH initiatives but, in this way, they condemn the investment initiatives towards failure or non-feasibility.

To achieve the desired investment targets for ARCH and drive sustainable development, it is fundamental to build an **enabling environment for ARCH investments** at EU, national and/or local levels.

The design of "financeable" ARCH initiatives that will attract both traditional and impact investors must ensure sustainable profitability that matches investors' return expectations (social, social and market financial, market financial returns) as well as the expressed interests and needs of local stakeholders. The financing considerations must include cultural, social and environmental aspects as well.

Preparing the field for ARCH privately leveraged investments requires specific funding, in some cases with limited direct return expectations. Such financing requirements may be covered by public funds, foundations, technical cooperation agencies or others.

The findings of the empirical analysis under CLIC project (Ikiz Kaya et al. 2021) highlight several driving-factors and enablers that can accelerate adaptive reuse practices. The authors advise the inclusion of EU funding, as "the most useful and feasible enabler of heritage adaptive reuse" in the circular economy framework.

Public funds can be used to support the implementation of feasibility studies, to evaluate the economic viability and expected impacts of ARCH investments, to have legal support etc. The employment of public funds may contribute to enhance the investment maturity/readiness of the ARCH initiatives and mobilise private investments. This form of technical assistance, already used on a smaller scale in CH sector by several revolving funds (Pickard 2009), developed in CLIC project under the form of a financial supporting mechanism called **Investment Readiness Facility (IRF)** for ARCH, will allow to address the barriers and challenges mentioned above and improve the financial landscape for ARCH initiatives.

The idea behind the set-up of this mechanism has been inspired at least by two sources:

- The relevance of the "built environment" for ARCH initiatives and circular economy deployment. The built environment affects many sectors of the economy with relevant impacts on the environment and resources use. The expected "Strategy for a sustainable built environment" by the European Commission in 2021, will guarantee the integration of all relevant policies such as climate, energy and resource efficiency, management of construction and demolition waste, accessibility, digitalisation and skills. At the same time, it will promote circularity principles throughout the lifecycle of buildings (European Commission 2020);

- Linked with the previous point, there are several resemblances between ARCH and energy efficiency retrofitting initiatives. Both typologies, partly overlapping, can present similar characteristics regarding the economic and financial viability (with respect to "cold", "lukewarm" and "hot" initiatives), longer pay-back periods, and similar market failures.

The IRF has been designed upon the recent EC facilities[4] to support public and private bodies in developing bankable sustainable urban and energy projects and preparing and mobilise private investments in the sector.

All these facilities can support partially ARCH initiatives, addressing only specific aspects of the projects involving cultural heritage buildings and only through a limited number of eligible activities. Circular ARCH initiatives need stronger dedicated support.

The **Investment Readiness Facility for ARCH** will have the aim to bridge the financing gap for circular ARCH initiatives through supporting all activities necessary to prepare viable projects and mobilise the necessary investments.

The IRF can be implemented at EU level and/or in local context (national or regional level) and be complementary to the existing EC facilities, based on the investment strategy of each ARCH initiative. At EU level, the EC and its partners (e.g. EIB, Agencies etc.) can manage the facility. While, at national and regional level, the IRF can be managed by the respective public authorities and funded through dedicated European Structural Investment Funds (ESIFs).

The facility can be provided to the beneficiaries under the form of binding grant, which will be reimbursed in case of missing the contractual milestones and key performance indicators. The facility can be also managed through the CLIC Hybrid Fund. The features of the described mechanism should take into account both the **physical interventions** for the ARCH as well as the deployed **circular business models** in the specific adapted cultural asset (in the case of project-base initiative) or cultural area (in the case of area-based initiative). The facility may envisage two levels of financing: higher amount binding grants for big projects (physical interventions and business models) and lower amount grants for the feasibility studies demonstrating circular business models viability.

The instrument aims at supporting the following goals:

- Design circular business models and organisational innovation;
- Build technical, economic, financial, impact measurement and legal expertise;
- Ensure high degree of replicability of similar initiatives;
- Remove existing barriers (administrative, financial, market failures etc.);
- Mobilize private investments;

[4] https://ec.europa.eu/easme/en/project-development-assistance-pda, https://eiah.eib.org/about/initiative-urbis

- Bundle projects and mix interventions to reach critical size and "warm up" the initiative, exploit also financially and economically unsustainable projects and achieve the expected returns and impacts.

The **beneficiaries** of the IRF may be both public authorities (regions, municipalities, other local authorities and public entities) as well as private for-profit (large enterprises, SMEs) and non-profit organisation (social enterprises, B-corporations etc.). The overall budget dedicated to the facility can be shared between the two categories according to a predefined allocation.

The **targeted sector** will be represented by the circular adaptive reuse of cultural heritage initiatives.[5] All other complementary sectors that contribute to integrate the ARCH initiative may be considered eligible, e.g. energy efficiency, urban transport (in case of an area-base initiative) etc. The complementary sectors may be assessed case by case and, the list of these sectors may be integrated in the future. The overall eligibility of the initiative will be assessed if it responds fully to the circular economy framework (both for the built heritage as well as for the adopted business model or "use&operate" of the building).

A potential first list of **activities supported by the IRF** may be:

- Feasibility studies;
- Executive design of the works, including also the aspects relating to the energy efficiency measures and renewable energy;
- Financial and economic analysis;
- Cost-benefit and impact analysis;
- Relevance to the EU Taxonomy analysis;
- Environmental assessment;
- Ex-ante socio-cultural impacts assessment;
- Energy audits;
- Risk analysis;
- Legal support, also for drafting Energy Performance Contracts (EPC);
- Procurement phase designing;
- Financial structuring, considering also drafting Result-based financing and Pay-per-results schemes, third parties financing as ESCOs for the energy efficiency aspects;
- Capacity building;
- Evaluation of the addressed value-chain and their coherence with the circular economy approach;
- Co-design in participatory logic in order to contain the risk of commissioning and the onset of syndromes related to "not invented here" and to "not in my backyard" etc.

[5]...groups of buildings or sites which are legally protected or of prestige value, but also all groups of buildings in urban or rural settings which form a coherent whole by virtue of the homogeneity of their style or the imprint of the history of groups of people who have lived there (CoE 1991).

Moreover, the project proposals shall demonstrate organisational innovation, in particular through:

- innovation in the mobilisation of the investment programme—bundling, pooling, stakeholder engagement, community finance etc.;
- innovation in the financial engineering: alternatives to traditional financing etc.

The IRF instrument will support circular ARCH initiatives that generate impacts according to a set of criteria and indicators (Bosone et al. 2021), in coherence with the EU Taxonomy and the main definitions under the umbrella of sustainable finance mentioned above. At the end of the technical assistance programme based on the IRF, the investment contracts will be enhanced with the impacts scoreboard and the baseline, to be than monitored during the whole investment period at least on annual basis.

The IRF can be managed as an instrument on two levels: for small project (<= €5 million) and large projects (> €5 million). Moreover, similar to the other technical assistance facilities, it is recommended to have the instrument bind to the successful implementation of the supported investments. For this reason, the IRF can be enhanced by the application of a **leverage factor**—for example, of at least 15 for small projects and of at least 20 for large projects. This mean that for each euro invested in the technical assistance for ARCH initiatives, there should be mobilize respectively at least 15 and 20 euros in terms of investments. As evidence, it can be necessary to require the beneficiaries to demonstrate by the end of the technical assistance support that the framework contracts with the investors and other stakeholders have been signed.

The **process** for the submission of the project proposals can be open on a first-come-first –serve basis, subject to the availability of funds and the synergies with the already existing ARCH initiatives portfolios, at the discretion of the management team.

The IRF management team will also have to define the application format for the proposals, the general conditions for awarding the facility support (evaluation and selection criteria) and the awarding process.

After awarding the IRF support to the beneficiary and signing the contract with the managing authority, each IRF beneficiary has to complete the agreed activities, including the identification and selection of investors, within a maximum two-year timeframe.

3.2 *ARCH Hybrid Circular Impact Fund*

Non-profit organization, entrepreneurs and SMEs are essential for economic growth, job creation, boosting innovation and promoting social cohesion in particular when addressing the adaptive reuse of cultural heritage. In the context of the EU New Green Deal towards climate neutrality and the achievement of the Sustainable Development Goals, as well as the post-pandemic COVID-19 recovery, the role of entrepreneurship and of the European SMEs in contributing to the sustainable economic development, in improving the quality of life and producing social and cultural impacts, in particular in the cultural sector, is fundamental. Hence, the importance of financial support under the form of (social) impact investments given to new initiatives, also through the exploitation of ARCH, for their development and scale-up has become more and more visible, especially at regional levels.

Different new impact investment funds have appeared across Europe under the form of incubators, accelerators and programmes to support start-ups, social entrepreneurship, seed investors and many investment vehicles have emerged. None of these instruments focus specifically on the circular adaptive reuse of cultural heritage initiative and the respective circular business models.

The proposal of an **ARCH Hybrid Circular Impact Fund** (also "the Fund") aims at fostering the valorization of cultural heritage through sustainable adaptive reuse and circular business models with cultural and social impacts and financing valuable initiatives for people and environment. In accordance with CLIC project objectives, the main impact areas of the Fund investments are cultural and natural heritage protection and safeguard, social inclusion and integration, culture and education, health and well-being, resilient and sustainable cities and human settlements, sustainable use of terrestrial ecosystems, directly linked to the SDGs: No Poverty (SDG 1), Quality Education (SDG 4), Decent Work and Economic Growth (SDG 8), Sustainable Cities and Communities (SDG 11), Life on Land (SDG 15) etc.

The ARCH Hybrid Circular Fund has been designed on the following main features:

- Publicly owned with private co-investment requirement or only privately owned;
- Revolving approach;
- Blended finance;
- Use of investments/proceeds for the circular economy initiatives;
- Impact and/or ESG invest;
- Results-based financing;
- Project bundling.

As envisaged for the Investment Readiness Facility above, **the Fund can be set up at European, national or regional level.** At EU and national levels, it can be also implemented through the Fund of Funds—i.e. a pooled fund that invests in other funds. In this specific case, the Fund will be implemented at **regional level** with potential market replication in different European geographical areas.

From the governance point of view, the Fund can be **publicly owned with private co-investment requirement** for each project to be financed **or fully privately owned**. In the Italian context, for example, in the last twenty years many public funds have been launched with the private co-investment approach for the financing of innovative start-ups: funds such as Innova Venture of Lazio Region, Start-up, Start-hope of Abruzzo Region, Venture Capital Fund of Basilicata Region with the newest experience at national level of Fondo Nazionale Innovazione of CDP Venture Capital Sgr. The final goal of these funds were to increase the offer of risk capital for innovative start-ups and SMEs leveraging private investments and enhancing regional impacts. Many of these funds have been partially funded through ESIFs. Under this perspective, the co-investment approach can represent an effective modality to involve private investors in the financing of circular ARCH initiatives and businesses. In average, the private co-investment in the Fund can be settled at a level of about thirty percent.

The Fund will also adopt a **revolving approach** with the meaning of a pool of "patient" capital to be dedicated, under different forms of financial instruments, to specific circular ARCH initiatives, with the restriction that the monies are returned to the fund to be reused for similar activities. The most common source of up-front capital for the revolving funds is represented by grant aid from regional or local authorities but, also from local foundations and corporations. It is important to evidence once more time here the relevance of the ESIFs use to start-up the Fund.

Interestingly, with the expansion of impact invest funds and the effects of the low market rates, volatile financial markets and post-pandemic expectations, the phenomenon of **"permanent capital"** is taken pace on capital markets. The "permanent capital" is characterized by a type of investment where the available capital is managed for an unlimited period of time that can go past 15–20 years. It does not focus on the short term returns of investments, but use the funds for initiatives that create long-term value. This type of "permanent capital" is particularly well-suited for the revolving approach of the Fund and the long-term funding needs of ARCH initiatives that can present longer pay-back periods with respect to traditional businesses.

The Fund can include different circular financial instruments that can be also used under the **blended finance** formula. For example, the Fund can be set up with the following instruments:

- Equity and quasi equity;
- Low-rates loans;
- Minibonds or Basket of Bonds/Minibonds;
- Guarantees.

The Investment Readiness Facility can be managed also under the Fund umbrella with specific focus on private beneficiaries that can support through this instrument the project design phase and advance towards the financing stage.

All the financial instruments under the Fund management will be based on circular and sustainable finance principles and the EU Taxonomy. The use of different financial instruments on a case-by-case base and their blending, when necessary, can contribute towards further mobilization of private investments and long-term value creation.

The financial instruments under the Fund umbrella shall be used to finance only circular ARCH initiatives/projects and business models/organisations that generate long-term positive impacts. Based on the recent best practices at global and European level in issuing green, social and sustainable bonds, a "**use-of-proceeds**" **approach** has been chosen for the design of the financial instruments under the Fund umbrella (TEG 2020). Bond market used to raise capital for general corporate purposes, based on the risk profile of the issuer, which is explicated then in its credit rating and the interest paid. A "use-of-proceeds" approach provides transparency for investors, facilitate impact reporting, and allows the financing of any organization/company regardless their main business activity—thus, fostering the transition towards sustainable business models etc.

The proceeds of the envisaged financial instruments should finance circular ARCH initiatives that are also aligned with the requirements of the EU Taxonomy.

Adopting this approach, the **financial instruments can be used to finance circular organisations or companies or projects/area-based projects supporting circular ARCH**.

As example, the circular business models in ARCH shall address features as:

- Circular inputs: organisations/projects that substitute virgin raw materials with recycled materials originating from materials and resources recovery;
- Energy efficiency: retrofitting of the built environment for energy savings and less pollution, adoption of renewable energy etc.;
- Circular design: organisations/projects that adopt innovative design techniques and approaches to facilitate the recycling, reuse and life time extension;
- Spatial integration of the ARCH initiative: in particular for area-based projects, it is important to address the context in which the initiative is developed, considering also the integration of the whole supply chain from a circular perspective (mobility, infrastructures, regulations etc.) (Saleh et al. 2020);
- Create positive impacts: organisations/projects shall address SDGs and seek to achieve positive social, cultural and environmental outcomes. All designated circular ARCH initiatives should provide clear the mentioned benefits, which will be assessed and, where feasible, quantified by the players involved in the financing process;
- Sharing business models: organisations/projects that increase the capacity utilization of an ARCH asset during its useful life;
- Life time extension: organisations/projects that increase reuse/refurbishment/remanufacturing to extend the useful life of products and assets;
- Product-as-a-service: improve the circularity of the whole supply chain through product-as-a-service offerings etc.

In the case of social oriented circular business models involving ARCH, social project categories may include:

- the provision of affordable housing;
- employment generation, and programs designed to prevent unemployment stemming from socioeconomic crises, including through the potential effect of SME financing and microfinance;
- socioeconomic advancement and empowerment (e.g. equitable access to and control over assets, services, resources, and opportunities; equitable participation and integration into the market and society, including reduction of income inequality).

Many ARCH initiatives aim at addressing or mitigating specific cultural or social issues and achieving the well-being of society with positive impacts for a target population. The latter can include, but are not limited to those that are: unemployed, women, undereducated, migrants or displaced persons, living below the poverty line, excluded or marginalized populations or communities etc.

A crucial step in the circular financing is the **impact assessment of the ARCH investments**. The process shall regard the qualitative and quantitative assessment of all the generated impacts by the organization or by the project. Beyond the environmental impact assessment, an overarching importance has the evaluation of working conditions, human rights, gender equality, health and other determinants of wellbeing in the ARCH initiatives. All these impacts issues can be addressed through adjusted ESG[6] methodologies.

A common framework of indicators for the assessment of multidimensional impacts of ARCH initiatives adopting the circular economy principles have been developed under CLIC project with respect to the three ARCH circularity dimensions: regenerative (auto-poietic), symbiotic and generative capacity (Fusco Girard 2019; Iodice et al. 2021). This framework has been structured for ex-post evaluation. Similar circularity criteria, where possible, are suggested to be adopted for ex-ante evaluation of the ARCH investments in order to build the impact indicators baseline to allow the decision-making process for the ARCH initiative evaluation and selection, monitoring and reporting of the investments.

Additionally, in coherence with the impact assessment framework, investors could increasingly adopt **results-based financing (RBF) structures** in their financial instruments in a core evolutionary step to foster impact-driven investments and contribute to the sustainable development. RBF solutions may include *performance-based contractual agreements* and *pay by results schemes*. These formulae can be included in the agreements underpinning the Fund umbrella of financing instruments mentioned at the beginning of this section. This type of approach of RBF schemes can generate cost-savings for investors by ensuring that funds are spent only if the results and/or outcomes are achieved. Amongst other benefits, such vehicles can help promote stronger performance management, enabling constant improvement of investment programs, especially when funded partially by public resources.

3.2.1 ARCH Initiatives Bundling

As already mentioned in the introduction part of this chapter, ARCH initiatives often evidence greater financial complexity due to the size of the investments necessary to allow adaptive reuse. It is possible to have financially free standing ARCH initiatives and non-viable ones. Many actions to be implemented are connected to

[6] Environmental, Social and Governance.

the possibility of "heating" "cold" or "lukewarm" operations by not acting at single asset/building level but in a systemic and aggregate return approach.

This means that the umbrella of ARCH financing instruments can be adopted in a more efficient way from the financial point of view through bundling the investments, considering not only project-based initiatives but also area-based interventions, which allow the creation of critical mass and synergies with both higher financial returns as well as more relevant social-cultural and environmental impacts.

Area-based initiatives can be addressed under the framework of "integrated conservation systems", first advocated in Europe by the "Amsterdam Declaration" at the Congress on the European Architectural Heritage in 1975 (CoE 1975). Among the considerations made by the Congress, cultural heritage is considered not only individual buildings of exceptional quality and their surroundings, but also all areas of towns or villages of historic or cultural interest. The conservation issues must be addressed as major objective of town and country planning: "*The conservation of these architectural complexes can only be conceived in a wide perspective, embracing all buildings of cultural value, from the greatest to the humblest – not forgetting those of our own day together with their surroundings. This overall protection will complement the piecemeal protection of individual and isolated monuments and sites*" (CoE 1975).

An area-based scheme works upon a **balanced portfolio of integrated projects** that **bundled together** could give a boost to cultural heritage (including also industrial heritage) areas in need of investment, also thanks to the "heating" effects of the financial returns generated by the synergies of different assets commonly considered not sufficiently attractive if stand-alone.

The bundling of ARCH initiatives should aim to stop and reverse the decline of cultural and historic landscapes and townscapes (often affected by phenomena similar to those of "circular cumulative causation" outlined by Myrdal (1957) in the "*Economic Theory and Underdeveloped Regions*") by capitalising on their unique character to create attractive, vibrant, and interesting places where people want to live, work, visit and invest.

3.2.2 Process for Circular ARCH Initiatives Evaluation and Selection, Management of Investments and Reporting

The process for ARCH initiatives evaluation and selection is detailed in the Fund regulation where each financial instrument goal is described with respect to the Fund overall strategy and rationale. The Fund regulation details all expected socio-cultural and environmental objectives and economic-financial performance of the investments. Furthermore, the Fund regulation outlines the eligibility and exclusion criteria or any other process applied to identify and manage material risks.

The investments or proceeds from the Fund should be transparently tracked in the accounts of the receiver of these funds, in order to verify that investments keep contributing to ARCH initiatives and to the shift towards a circular economy during the lending/investment period.

The reporting process can be set up at project/portfolio level and at the Fund level. The funds receivers should report on the allocation of funds and the impact of the financed activities. The Fund managers should report on their investments and monitor the impact generated at project/portfolio levels. The reporting shall include reference to the alignment with the EU Taxonomy and, if necessary, with other similar guidelines (e.g. EU Green Bonds Standards). The reporting shall be done at least annually.

Finally, it is also recommended by many recent guidelines in sustainable finance, the use of an external review to confirm the alignment of circular ARCH portfolio with the key features of the circular and sustainable finance and EU Taxonomy, for example by "second party opinions" (e.g. auditors, circular economy and similar experts). At present, no globally recognised circular certifications exist for circular organisations, projects or products. This is even truer for the specificity of circular ARCH field. The voluntary external review market is in an early stage but its development is very relevant for the promotion of integrity, credibility, harmonisation and efficiency of the circular and sustainable finance, as well as investor confidence (ABN Amro 2018; TEG 2020).

3.3 Hybrid Public Private Partnership Approach

The increasing adoption of Public Private Partnerships (PPPs) for ARCH may contribute to increase private investments for the maintenance and valorisation of public cultural assets with positive effects on the efficiency of cultural heritage management.

Public administrations can implement ARCH initiatives through PPPs, with integral or partial but prevalent (to be compliant with Eurostat rules on the accounting of PPPs in public budgets) private financing, thus satisfying the twofold imposed function of preservation and valorisation. At the same time, public administrations can benefit from the expertise and management competences of the private sector in different phases of the process: design, implementation and management of the cultural public good.

PPPs, if implemented with correct logics and approaches, ensure a better risk sharing among different actors with higher efficiency in project implementation, greater mobilisation of private funds without worsening public finances, higher probability of success of the project etc. Moreover, in PPPs it is expected a perfect alignment of public and private interests in deploying the best value for money (VfM) (Allegro and Lupu 2018).

Opposite to the undoubted advantages of PPPs linked to the convergence of interests between public and private sectors that should lead to the implementation of a higher quality project, there are several weaknesses to be considered when executing PPPs. Information asymmetries between public and private may enable phenomena of moral hazard and adverse selection in perfect coherence with the "contract theory" of George Akerlof (1970) and "incomplete contract theory" pioneered by Sanford J. Grossman (1986), John H. Moore (1990) and Oliver D. Hart[7] (1995).

In certain periods, it is possible to assist at short term distorted perspective of the public administration derived from the activation of "cherry picking" selection processes where private sector select the projects with higher returns on investments. The result of these phenomena will reflect in an impoverishment of the public sector and in potential public administration financial tensions in near future due to the management of the remaining "poor" cultural heritage projects portfolio.

Knowledge gaps are also some of the main weaknesses when implementing PPPs: programming misaligned with respect to the effective needs of the public administration and the societal ones, shortcomings in the governance capacity of the public authority, partnership misaligned regarding risks sharing between public and private, difficulty in selecting the best projects, specific knowledge and competences lack.

From the private perspective, because of the "public" features of cultural heritage, the private sector may overestimate the returns of the investments and underestimate the related costs.

Thus, PPPs are not necessarily a magic bullet cure for the problems of scarce resources, mismanaged cultural heritage and the unique solution to the addressed challenge. However, at the same time, PPPs, if managed with more competences, innovation and transparency can be even more productive and sustainable for the valorisation of cultural heritage and industry especially in an era of scarce public resources and significant skills and management gaps on the part of the public sector. If properly designed, PPPs can provide considerable operating flexibility: for the public sector to be compliant with its own regulations and resources, for the private sector, to bring to the project different management models, know-how, financial and technical/technological inputs.

The use of PPPs in cultural sectors is relatively recent and rather limited. Many lessons learnt from the use of PPPs in more "traditional" fields (i.e. health, infrastructures, education etc.) can be transferred to the cultural industry with innovative sector-specific adjustments in a holistic perspective, that may regard active involvement of citizenship, creation of shared value for all actors, use of innovative circular business models and impact financing etc. The hybridisation through the adjustment of a strategic instrument such as PPP will contribute to achieve the final goal of cultural sustainability with relevant impact not only for the economic development of a territory but also for the social inclusiveness and enhancement of local communities. This effect can be obtained, for instance, by linking a certain part of the remuneration of the private party to the achievement of specific social results (according to the logic of "pay for results") or a certain social impact (according to the logic of "social impact investments"). The explicit consideration of these "social

---

[7] O. Hart and B. Holmström have been awarded the Nobel Prize in 2016 "for their contributions to contract theory".

effects" in the PPP contract can also help to overcome those possible "side effects" associated with the information asymmetries mentioned above.

The involvement of citizens as individuals (the 4th P for "People") in the definition of priorities of PPP for the implementation of ARCH is of particular relevance for the role of cultural heritage as "common good".

The decisions on an ARCH initiative, which probably is rooted in an urban or rural context/landscape, may have an impact on daily life of people that live in the surroundings of the cultural heritage. If citizens' involvement in the decision process is weak, this can bring to negative perceptions of the ARCH with potential unsuccessful implementation of the initiative also linked to the exponential growth in the commissioning risk (as mentioned in connection with the Investment Readiness Facility for ARCH). Moreover, it is important to give evidence that citizens can provide benefits to ARCH initiatives through the deployment of creative and innovative ideas and solutions.

Today, especially in the European context, there is a continued need for innovations for the cultural heritage valorisation, in particular with respect to business models innovation, for building a solid and transparent pipeline of economically and socially sustainable related investment projects in order to demonstrate the attractiveness of this sector to private investors.

Public administration can introduce in the tendering processes and contractual/service agreements elements for the deployment of innovative solutions for the cultural heritage valorisation. The tender proposals, also under PPPs projects, can be evaluated according innovation criteria such as organisational innovation in financial engineering (also, including impact investing, results-based financing and performance contracts with revenue-share formulae etc.), mobilisation of investments (bundling of different projects, different stakeholders engagement etc.). At the same time, similar features as for the Fund above regarding the circular economy ARCH initiatives aligned with the EU Taxonomy, use of blended finance etc. shall be common elements also of the PPP schemes.

The measurement of cultural shared value that integrate economic and financial analysis with cultural, social, governance and environmental analysis creates value for all the actors involved i.e. public, private for-profit, non-profit and, citizens.

Several key characteristics of impact investing can be mutualised to the cultural heritage projects and investments under a PPP scheme:

- the intention of the investor to generate cultural, social and/or environmental impacts;
- the expected return (social, impact of financial) on investment by the entrepreneur.

These characteristics evidence the need for transparency in the ex-ante definition of the cultural, social and environmental expected impacts from the investments, Key Performance Indicators (KPIs), the measurement methodologies and the period of measurement, the reporting of the cultural investments impact.

As already evidenced for the Fund, the measurement of impacts is becoming more and more relevant in investments with social spill overs, but at the same time it is really complex because of lack of track record and historical data. Different stakeholders may have different interests in the measurement of the impacts:

- public administrations have interest because of lack of resources and identification of the best Value for Money;
- non-profit organisations need to demonstrate the generated impacts for further funding, for creating changes and for transparency and responsibility;
- for-profit enterprises measure impacts to improve their investments, transparency, responsibility, innovation and reputation;
- investors are looking for impact investments.

In the ARCH domain, as already mentioned above, a framework under CLIC project with cultural, social, environmental and economic criteria have been developed with particular focus on the adaptive reuse of cultural heritage investment projects. Thus, the use of PPP in the ARCH field seems to require the use of a hybrid approach very focused on the impacts generated by the activity of the private partner.

4 Conclusions

Cultural heritage, intended as a "common good" belonging to each citizen, traditionally receives funding from public sector, primarily for the purpose of its preservation and without the aim to exploit it economically. In a period of increasing pressure on public budgets, this approach activates the perverse spiral of increasingly inadequate investments because of scarce available resources in the hands of the public decision-maker. Different scales of investment, adaptability of assets for new uses, attractiveness of urban areas, as well as ownership and governance structures, determine the adoption of different financing mechanisms for the cultural heritage valorisation. This chapter described the three financing mechanisms developed under the CLIC project specifically relevant for the adaptive reuse of cultural heritage (ARCH).

The proposal of the three financial mechanism under the CLIC framework project goes in this direction and yearn for an intense impulse towards sustainable financing of circular ARCH to preserve the cultural value and achieve the SDGs locally. An opportunity in the deployment of the mentioned financial mechanisms can come from the synergies between private patient capitals and European, national and regional public funds and, in particular, through new exploitation ways of ESIFs.

References

ABN Amro (2018) Circular Economy Finance. 2018 (Economy Circular Guidelines Finance), pp 1–4

Achterberg E, van Tilburg R (2016) Circular Economy Finance: clear analysis and innovative approaches to financing the circular economy. Circle Economy.

Akerlof GA (1970) The market for 'Lemons': quality uncertainty and the market mechanism. Q J Econ 84(3):488–500. https://doi.org/10.2307/1879431
Allegro I, Lupu A (2018) Models of public private partnership and financial tools for the cultural heritage valorisation. Urbanistica Informazioni – INU (278si):1–6
Bosone M et al (2021) Indicators for ex-post evaluation of cultural heritage adaptive reuse impacts in the perspective of the circular economy. Sustainability 13:4759. Multidisciplinary Digital Publishing Institute, 13(9), p. 4759. https://doi.org/10.3390/SU13094759
Boffo R, Patalano R (2020) ESG investing: practices, progress and challenges. OECD Publishing.
CoE (1975) Amsterdam declaration. Congress on the European Architectural Heritage
CoE (1991) Measures likely to promote the funding of the conservation of the architectural heritage
Council of Europe (2020) Risk management of cultural heritage. Council of Europe Publishing.
European Commission (2019) The European Green Deal
European Commission (2020) A new circular economy action plan: for a cleaner and more competitive Europe. https://doi.org/10.7312/columbia/9780231167352.003.0015
Fusco Girard L (2019) Implementing the circular economy: the role of cultural heritage as the entry point. Which evaluation approaches? BDC – Bollettino Del Centro Calza Bini 19(2):245–277. https://doi.org/10.6092/2284-4732/7269
G20 Green Finance Study Group (2016) G20 Green Finance Synthesis Report 2016, G20 Green Finance Synthesis Report (August), pp 1–11
G20 Sustainable Finance Study Group (2018) G20 Sustainable Finance Synthesis Report 2018
Gravagnuolo A, Angrisano M, Fusco Girard L (2021) Circular economy strategies in eight historic port cities. BDC – Bollettino Del Centro Calza Bini 21(1):245–277. https://doi.org/10.6092/2284-4732/7269
Griffiths J (2012) "Leveraging" private sector finance. Bretton Woods Project
Grossman SJ (1986) The informational role of warranties and private disclosure about product quality. J Law Econ 24(3):461–483
Hart O (1995) Firms, contracts, and financial structure. Clarendon Press.
ICMA (2020) Sustainable finance—high-level definitions (May), p 8
Ikiz Kaya D, Pintossi N, Dane G (2021) An empirical analysis of driving factors and policy enablers of heritage adaptive reuse within the circular economy framework. Sustainability (Switzerland) 13(5):1–25. https://doi.org/10.3390/SU13052479
International Capital Market Association (2020) Social bond principles. The Social Bond Principles (June), p 1
Iodice P, Senatore A, Buonomano A (2021) A novel multi-criteria optimization methodology for energy and economic savings in historical buildings. Energy Build 231:110595. https://doi.org/10.1016/j.enbuild.2020.110595
Montalto V, Sacco PL, Alberti V, Panella F, Saisana M (2020) European cultural and creative cities in COVID-19 times: Jobs at risk and the policy response. Publications Office of the European Union. https://doi.org/10.2760/120639
Moore JH (1990) Implementation in environments with complete information. In: Game theory in economics, vol. 1. Academic Press, pp 187–201
Myrdal G (1957) Economic theory and underdeveloped regions. Duckworth
Myrdal G (1958) Rich lands and poor. Harper & Row
OECD (2020) Blended finance principles guidance. OECD Publishing
Pickard R (2009) Funding the architectural heritage: a guide to policies and examples. Council of Europe Publishing, Strasbourg
Saleh R, Drouillon P, Ost C (2020) Circular business model workshops for cultural heritage adaptive reuse
TEG (2020) USABILITY GUIDE context and background information, (March)
WorldBank (2011) Mobilizing climate finance

Part V
Is it Possible to Incentivise Creativity in the Cultural Heritage Entrepreneurial Sector?

Introduction

Incentivizing creativity within the cultural heritage entrepreneurial sector is critical to fostering innovation and sustainable development, particularly in the context of the circular economy. This section explores the potential for stimulating creative solutions that not only preserve heritage but also reimagine it as a driver for economic growth. By examining strategies such as financial incentives, policy frameworks, and collaborative networks, this chapter delves into how the entrepreneurial sector can be empowered to integrate creativity and sustainability into adaptive reuse projects. The goal is to uncover pathways that enable heritage to thrive as a living, evolving resource within a circular economy model.

Chapter 21
Heritage-Led Entrepreneurial Ecosystems: Skills and Role of Startups and Innovation

Antonia Gravagnuolo, Aliona Lupu, Jermina Stanojev, and Valeria Catanese

1 Introduction

Cultural heritage is recognized as a powerful driver of sustainable development in cities and regions. The adaptive reuse and regeneration of cultural and natural heritage can become a driver of jobs creation, as demonstrated by many experiences (Gravagnuolo et al. 2021). At the same time, cultural heritage adaptive reuse can involve communities in innovative and unexpected ways, generating multiple positive impacts in the territories and boosting a more sustainable, resilient, and equitable development model to finally create cities of opportunities connecting culture and innovation[1] (Gravagnuolo et al. 2020).

[1] Tenth Session of the World Urban Forum, "Cities of opportunities: connecting culture and innovation" (8–13 February 2020, Abu Dhabi, UAE), https://wuf.unhabitat.org/cities-opportunities-connecting-culture-and-innovation

A. Gravagnuolo
CNR ISPC Institute of Heritage Science, National Research Council, Naples, Italy
e-mail: antonia.gravagnuolo@cnr.it

A. Lupu
Iniziativa Cube, Naples, Italy
e-mail: aliona.lupu@iniziativa.cc

J. Stanojev
Department of Art History, Uppsala University, Visby, Sweden
e-mail: jermina.stanojev@konstvet.uu.se

V. Catanese (✉)
IRISS Institute for Innovation and Services for Development, CNR National Research Council, Naples, Italy
e-mail: valeria.catanese@cnr.it

L. Fusco Girard, A. Gravagnuolo (eds.), *Adaptive Reuse of Cultural Heritage*,
https://doi.org/10.1007/978-3-031-67628-4_21

The EU is committed to promoting culture in its international relations, and particularly the diversity of cultures in the EU [2] (European Union 2012). Promoting culture as a vital element in EU international relations has been one of the three main objectives of the *European Agenda for Culture* since 2007[3] (European Commission 2007). Here the culture sector is regarded as an increasingly source of job creation, that is a key goal in European Union, contributing to economic wealth. The culture sector is also an excellent conduit for promoting social inclusion and supporting cultural diversity.

The European Commission's 2014 Communication *Towards an integrated approach to cultural heritage for Europe* underlined the importance of maximising the intrinsic, economic, and societal value of cultural heritage, in order to promote inter-cultural dialogue. In the agenda for cultural heritage research and innovation *Getting Cultural Heritage to Work for Europe*[4] (Thurley et al. 2015), cultural heritage is understood as a production factor and hereby an important resource for innovation, social inclusion, and sustainability.

One of the first collective reports within the EU policy framework, under the Structured Dialogue of the European Commission, *Towards an integrated approach to cultural heritage for Europe* – Prospectus on "Skills, Training and Knowledge Transfer for Traditional and Emerging Heritage Professions" (Ateca Amestoy et al. 2017)[5] it has recognised that "the cultural heritage sector must also be responsive and ready to grow, reacting to changing conditions, including new demands coming from diverse European societies, new technologies, new uses for cultural heritage and emerging markets. The sector needs to be encouraging both cultural entrepreneurship and a personal/professional expansion into new roles, professions and competencies, enriching and complementing the existing field."

These issues have been defined in the context of a more overarching approach understanding that "the creation of new job profiles commonly takes place at the margins where different fields or disciplines interface, this should be a point of particular attention, and cultural entrepreneurship encouraged in". Similarly to some scientific publications developed some years later, the group of the Structured Dialogue noted that "the field of cultural heritage research and innovation is quite fragmented. Subsectors are scattered among different approaches, and ideas for cultural heritage innovation are mainly linked to innovative technological processes for conservation of cultural heritage, rather than to innovative governance, networks, reuse, entrepreneurship etc." Therefore, an urgent need for "creating a dialogue with the educational sector about innovation, entrepreneurship, and two-way influence between principle and practice in order to develop relevant programmes" has been identified.

---

[2] http://eur-lex.europa.eu/legal-content/EN/TXT/PDF/?uri=CELEX:12012E/TXT&from=en#page=75&zoom=100&view=FitB

[3] https://eur-lex.europa.eu/legal-content/EN/TXT/HTML/?uri=CELEX:52007DC0242&from=EN

[4] European Commission. *Getting Cultural Heritage to Work for Europe Report of the Horizon 2020 Expert Group on Cultural Heritage*; European Commission: Brussels, Belgium, 2015.

[5] www.voicesofculture.eu/skills

The European Parliament resolution of 8 September 2015 "Towards an integrated approach to cultural heritage for Europe" (2014/2149(INI)) under point 41 "The economic and strategic potential of cultural heritage" "*Invites the Commission to consider the possibility of the European Institute of Innovation and Technology (EIT) establishing, under its next Strategic Innovation Agenda, a Knowledge and Innovation Communities (KIC) in the field of cultural heritage and creative industries, thus directly supporting a holistic vision of research and innovation*". The Structured Dialogue report was among the first ones from the sector to clearly call for setting up a KIC on cultural heritage (and creative industries).

These needs were further clearly reflected in the work of the OMC (Open Method of Coordination) Working Group of Member States' Experts, the report "Fostering cooperation in the European Union on skills, training and knowledge transfer in cultural heritage professions" (2019) where "entrepreneurship" has been listed under a skill needed for cultural heritage transversal competences.

These developments have been consequently followed by a call in 2020 for cultural heritage under the Blueprint for sectoral cooperation on skills that was introduced by the Skills Agenda for Europe 2016 to build on previous work by the European Commission and sectoral partners to address sector skills mismatches, in particular the European sector skills alliances. The European Skills Agenda 2020 has confirmed the Blueprint as a key initiative to create new strategic approaches and cooperation for concrete skills development solutions in the industrial ecosystems as introduced by the updated EU industrial policy. Blueprint Alliances gather key stakeholders from industrial ecosystems that can include for example business, trade unions, research institutions, education and training institutions, public authorities etc. For cultural heritage sector, the project CHARTER—Cultural Heritage Actions to Refine Training, Education and Roles has started the implementation with an aim to create a lasting, comprehensive sectoral skills strategy to ensure Europe has the necessary cultural heritage skills to support sustainable societies and economies. The project shall use strategic collaboration and innovative methodologies to bridge the gaps between educational and occupational systems, employer and job market needs and gaps, to reduce skills shortages, gaps and mismatches.

Furthermore, recent academic research has shown that innovation and entrepreneurship in the field of reuse and enhancement of cultural heritage and landscape could be underestimated (Saleh and Ost 2023). Entrepreneurial attitude and capacity are not yet significantly developed in the heritage sector (Ost and Saleh 2021; Gravagnuolo et al. 2017). The reasons can be summarized as follows: on one hand, the lack/scarcity of private investors disposed to funding businesses in the field of cultural heritage, considered as a potentially risky investment, not remunerative enough; on the other hand, people working in the heritage field tend not to recognize themselves as possible entrepreneurial subjects, revealing almost a kind of reluctance of the sector to "move away" from economic and financial issues, without intentionally pursuing economic-financial self-sustainability in heritage-related activities. Basically, there are many valuable ideas but there is often a lack of practical experience and funding to turn them into innovative services and businesses (Catanese and Gravagnuolo 2021).

Another reason for that, is that the knowledge triangle model has not been integrated in cultural heritage transforming the knowledge triangle model in a way that meets societal renewal needs of innovation in cultural heritage (Stanojev and Gustafsson 2022). The results of the study (Stanojev and Gustafsson 2022) indicate that some of knowledge triangle stakeholders are present in the field of cultural heritage, however they are not engaged in a right way or not interested to be involved. The findings suggest that the progressive (re)combinations of their engagement, with community participation and other instruments, tools and stakeholders could create visionary foundations for advanced visions, building up a common basis for open innovation activities. So far, the knowledge triangle model has not been adjusted for assessing innovation in the cultural heritage and landscape from the perspective of research, education and business innovation.

Examples of circular business models that can be implemented in the heritage sector are related to social and cooperative enterprise models, with the simultaneous reduction of costs through circularization of processes and creation of social, cultural, and economic value (Dalberg Global Development Advisors 2014). Social entrepreneurship is particularly linked to circular economy business models as they share "mission-driven" organisational forms (Porter and Kramer 2011). As highlighted by Sacco and Teti (2017), 8 dimensions can be identified in which cultural value is produced through cultural, creative and heritage-related business activities: (1) *Innovation*—The investment in cultural heritage and cultural activities stimulates cooperative behaviours, innovation-oriented mind-sets in the community and innovative entrepreneurship; (2) *Wellbeing*—Enhancement of emotional, perceived and physical wellbeing (thus minor costs for public health systems); (3) *Sustainability*—Socially beneficial behaviours (minor costs for city management), long lasting stakeholders' value; (4) *Social cohesion*—reduction of crime rates, social conflict solving, cultural diversity and identity; (5) *New entrepreneurship models*—Stimulation of divergent thinking, new business models, enhanced employment opportunity; (6) *Lifelong learning*—Enhanced adaptive capacity/resilience, increased economic efficiency over time; (7) *Soft power*—Soft skills increasing productivity, visibility, city/country brand value; investment attractiveness; (8) *Local identity*—Sense of belonging to a community, involvement of civil society and private owners in financing and managing heritage (cost-sharing), enhancement of city/territory as a location for living, travelling, working and making business.

In the heritage field, the pro-active role of civil society organizations, social enterprise, civic foundations, community hubs was explored in the Horizon 2020 CLIC project and stressed as a resource for cultural, economic, social and environmental growth in Europe and for improving the "good life" in the city/territory (MacDonalds 2011; Guntram et al. 2014; McKeever et al. 2015; Ragozino 2016). The role of "Heritage Commons" adaptive reuse in stimulating cooperation, mutuality, subsidiarity, local identity and sense of belonging to EU values was explored, evaluating the multidimensional beneficial impacts (Michiara 2016).

Adaptive reuse of cultural heritage can be seen as a process that substantially includes cooperation and conflict solving in the contemporary complex society. It is not only producing a technical design project, a viable business model for its economic-financial management, public quality control procedures and regulations. It is related to the construction of common values, community, sense of public benefit/interest (Fusco Girard 2020, 2021a, b). In transition times, the traditional models of knowledge and value creation become obsolete. The innovation comes out of a "creative hybridization" process that crosses all sectors (cultural, social, economic, environmental, technological) to produce new ideas/processes/values (Fusco Girard and Gravagnuolo 2017).

Despite the economic crises, the third sector (also addressed as "third sphere" or social economy, i.e., next to private and public sectors) is growing. This sector plays an important role in terms of social cohesion and inclusiveness, and also of economic value generation and for improving the living conditions. Here we assume a pro-active role of this sector for promoting the reuse of historic assets, both creating economic value and social—environmental value.

Social economy organisations traditionally refer to the set of associations, cooperatives, mutual organisations, and foundations whose activity is driven by values of solidarity, the primacy of people over capital, and democratic and participative governance (OECD 2018). The social economy includes a broad range of entities with different business and organisational models, operating in a number of economic sectors: from agriculture to construction, from information and communication to energy and climate, from financial and insurance activities to human health, social works activities, culture, arts and media etc. Despite this diversity, their main common principles and features can be summarized as follows: *the primacy of people as well as social and/or environmental purpose over profit; the reinvestment of most of the profits and surpluses to carry out activities in the interest of members/users ("collective interest") or society at large ("general interest") and democratic and/or participatory governance* (European Commission 2021a, b).

A further characteristic of social economy organisations is that they typically operate developing their economic activities at the local level. The local tie reinforces their ability to activate synergies across value chains with local stakeholders (including policy makers, SMEs, researchers, and citizens) and experiment with new and cooperative ways of working to provide place-based solutions able to satisfy societal needs (OECD and European Commission 2022). According to the European Economic and Social Committee study on the "Recent Evolutions of the Social Economy in the European Union" in 2017 social economy enterprises and organisations in the European Union accounted to 2.8 million, employed 13.6 million people and contributed to 8% of the EU's GDP (Rafael Chaves Ávila and José Luis Monzón Campos 2019).

Social economy can significantly support the implementation of the circular development model and the achievement of SDGs in different ways. The Circular Economy Action Plan recognizes the potential of the social economy as a "pioneer in job creation linked to the circular economy" (European Commission 2020), starting from reusing and recycling activities and gradually expanding to diverse sectors paying special attention to inclusion and social cohesion. Indeed, social enterprises have opened up job and training opportunities for more vulnerable groups, including people with disabilities (European Commission 2021a, b) and marginalized individuals (migrants, people who have experienced drug/crime problems etc). In this perspective, they promote more equitable and fairer societies.

Moreover, social economy organizations can promote more resilient societies, as recently demonstrated by the COVID-19 crisis, by providing themselves a range of crucial services related to health and social services and collaborate with local authorities to integrate services and mitigate the social impact of the crisis especially for the most vulnerable groups. Going beyond the concrete contribution to the most critical phases of the COVID emergency, putting social and environmental concerns at the hearth of their operations and business models, social enterprises can help transform society and the economy by experimenting alternative business models, inspiring sustainable and inclusive practices, identifying new sectors and providing services able to start processes of urban regeneration and territory revitalization, in particular in remote areas or in disadvantaged contexts (OECD 2020).

Social economy organisations are credited with showing the economic potential of sectors strongly related with circular economy that had previously been ignored by investors because they could not guarantee the same returns on investment, even though the resulting economic and social benefits were substantial (OECD and European Commission 2022), which is the case of cultural heritage adaptive reuse and valorization (Ragozino 2016).

Against this backdrop, the present contribution aims to share the experience of the Horizon 2020 CLIC project in developing innovative circular business models for cultural heritage adaptive reuse, through the selection and training of start-ups during the entrepreneurial event of the CLIC Startup Competition and subsequent mentoring programme.

The following sections present a virtuous case study as an example of successful social entrepreneurship and circular business model of heritage reuse and regeneration, the Catacombs of San Gennaro in Naples, Italy. Then, the rationale of the CLIC Startup Competition initiative is presented, as well as the evaluation criteria for the selection and development of innovative business models. The results and training programme are presented, drafting first conclusions and proposing a critical reflection on the role of social entrepreneurship and innovative business models for circular cultural heritage adaptive reuse.

2 The Case Study of the Catacombs of San Gennaro in Naples, Italy

A virtuous example of heritage-led social entrepreneurship which has been selected by the CLIC project as one of the best practices of urban regeneration through the recovery and valorisation of religious heritage is represented by the Catacombs of

San Gennaro[6], Naples, Italy. The development model of the Catacombs, based on a bottom-up valorisation of an abandoned cultural heritage, which has generated so many multiple positive impacts on the local community, has been investigated also at the international level to understand its replicability in different contexts.

The Catacombs are an archaeological site/religious heritage located in one of the most densely populated and characteristic neighbourhoods of Naples, the "Rione Sanità", and are an important part of the city's history strongly intertwined with that of its patron saint, San Gennaro. The original nucleus of the Catacombs of San Gennaro dates back to the second century AD. One of the largest catacombs in Southern Italy, the Catacombs remained for a long time in a state of neglect until a small group of local young people began to take care of this heritage, recognizing its immense cultural value and extraordinary potential, founding in 2006 the cooperative enterprise "La Paranza", which is currently responsible for the management of the Catacombs. Through the recovery and valorisation action conducted by the cooperative La Paranza and the Community Foundation "San Gennaro" engaging the local community, the ancient heritage site returned to live, starting a broad requalification process that has gradually interested the entire district. Inspired to the principles of circular and social economy, the recovery of the Catacombs has produced significant positive impacts for the community: creation of new jobs, enhanced wellbeing, social cohesion, urban regeneration, education and local empowerment, solidarity relationships, thus turning in an engine of economic and cultural growth in the deprived area of Naples historic centre called "Sanità" neighbourhood (Europa Nostra, ICOMOS and Climate Heritage Network 2021). The Catacombs have entered into a symbiotic relationship with the neighbourhood, generating a virtuous process that has encouraged the creation of other businesses (economic growth) but also of educational and leisure activities that take place in buldings that have been reused that contribute to social cohesion. Just to make an example, the Sanitansamble association,[7] which is inspired by the experience of 'El Sistema', an educational model with free access for children and young people in Venezuela promoting collective musical practice as a means of community organisation and development in difficult areas and social contexts. The Sanitansamble has more than 80 young people, including children and adolescents aged 7–24, in two orchestral ensembles which have performed at many prestigious events and concerts in local theatres, including the Nuovo Teatro Sanità located in the district, and national theatres. Another example of adaptive reuse of buildings in the district is represented by the Casa del Monacone, an ancient convent transformed in a hospitality structure.

By reversing the bad "reputation" of the Sanità district that blocked tourism, thanks to the support of the local community that has realized the opportunities stemming from the recovery project and to the strategic position respect the historical centre of Naples, the district has turned into one of the most attractive city tourism destinations.

---

[6] https://catacombedinapoli.it/en/

[7] http://sanitansamble.it/la-nostra-storia/

Data referred to 2021 report significant milestones: 13,500 sqm of the Catacombs have been recovered, 44 jobs for young people created and 87.000 people visited the sites (in 2019, before the pandemic period, visitors reached 160.000 per year) (Buglione et al. 2021). In June 2022, the Catacombs of San Gennaro were selected among the 30 most outstanding heritage achievements, being honoured with the European Heritage Award, promoted by Europa Nostra, for the following motivation: "*This cooperative of young people has undertaken an extraordinary task for the benefit of heritage and the people of Naples, creating a sense of ownership among the community for their local heritage. They have helped transforming a previously underdeveloped area into an attractive destination for tourists, while recovering a fascinating, hidden element of the city's heritage. With great determination and ingenuity, La Paranza Cooperative has demonstrated how to care for heritage, in spite of limited resources*[8]".

The regeneration process of the Sanità District is still ongoing. The internationally renowned architect and senator for life Renzo Piano has chosen Naples, together with Bari and Rovigo, for the 2022 edition of the G124 project,[9] aimed at carrying out requalification projects in peripheral urban areas that the senator defines as "suburban mending" actions.

Something we in G124 have done: small mending interventions that can trigger regeneration also through new trades, micro-businesses, start-ups, light and widespread construction sites, thus creating new jobs. These are just sparks, but they stimulate the pride of those who live there. Because as Italo Calvino wrote, 'there are fragments of happy cities that continually take shape and vanish, hidden in unhappy cities'. These fragments must be discovered and valued. It takes love, even if it is in the form of anger, it takes identity, it takes pride of being a periphery[10].

Inspired by the experience of entrepreneurial best practices of cultural heritage adaptive reuse and regeneration, the CLIC Startup Competition called for innovative ideas to turn cultural heritage from a cost into an investment, generating new jobs and enhancing cohesion and community collaboration.

[8] https://www.europeanheritageawards.eu/winners/la-paranza-cooperative/

[9] G124 is the code that identifies a room in Palazzo Giustiniani in the Senate assigned to the architect and senator for life Renzo Piano as well as the name of his group working on the periphery and the city to be. The G124 group employs young architects on an annual contract who are paid from Renzo Piano's parliamentary salary, which has been entirely earmarked for this project. Each year the architects will be replaced by others selected through a special call for applications. For further information: https://renzopianog124.com

[10] PERIFERIE 1. Diario del rammendo delle nostre città, 2014, https://renzopianog124.com/pubblicazioni/diario-delle-periferie-2-marghera/

3 CLIC International Startup Competition: Why and How?

The CLIC international Startup Competition was designed as a capacity-building programme addressed to early stage and pre-seed startups, informal teams and already active entrepreneurs working on a business idea for the adaptive reuse and enhancement of cultural heritage inspired to the circular economy principles for enhanced sustainability, social impact, and environmental regeneration. Consistently with the objectives of the CLIC project and the work carried out within the Task Force on "Circular models for cultural heritage adaptive reuse in cities and regions",[11] the initiative was conceived to support the development of sustainable business ideas coherent with economic and social models to place cultural heritage and historic urban landscapes adaptive reuse at the forefront for the implementation of a European model of circular economy and circular city-region centred on the regeneration of cultural and natural capital.

The CLIC Startup Competition intended to increase the entrepreneurial capacity in the heritage sector by investing in capacity building. Start-ups usually need not only financial support, but also—and above all in the early stage when the project is still in a very embryonic phase—non-financial support as advice, knowledge and assistance from experts who can help to develop a consistent, solid business model, to get in contact with a network of relationships and investors and plan the commercialization of their ideas, in order to have rapid development prospects and reduce the failure risk which is particularly high in the startup phase.

The topics of the startup competition were identified in line with the challenges of circular adaptive reuse of cultural heritage as emerged in the CLIC research. The call emphasized the role of social impacts and ethical finance for a long-term and equitable recovery from the pandemic crisis. Therefore, the initiative was oriented to projects and ideas able to reduce the already existing social inequalities further accentuated by the COVID-19 emergency especially in the cultural sector, to mitigate the negative consequences of the crisis and to design solutions for cultural heritage adaptive reuse able to contribute to the health and wellbeing of people in the post-covid recovery phase.

The call was opened to startups, as well as to individual innovators and informal teams, including foundations, civil society organizations, social enterprises and cooperatives, creative collective of artists and innovators, focusing on the following heritage-related fields, as per CLIC Startup Competition Call (Catanese and Gravagnuolo 2021).

Cultural, Safe and Sustainable Tourism
E.G. Startups/ideas addressed to innovative solutions to develop new products and services in cultural and sustainable tourism sector, taking into account the new and future scenarios for cultural tourism in Europe and beyond. Their focus could include adaptive reuse of abandoned heritage sites for cultural tourism functions; cultural tourism development in less-known cultural heritage and landscape areas

---
[11] https://www.clicproject.eu/taskforce/

and post-tourism experiences; "inclusive" and human-centred solutions designed for people with special needs (including also senior tourists, pregnant women, parents with small children, people with dietary problems, etc.); reduction of pressure on overcrowded heritage tourism destinations. All these solutions, by broadening the range of choices for the final users increasing the attractiveness of areas marginal compared to the common tourist hotspots and fulfilling the needs of specific groups, were expected to guarantee tourists a safer and more enjoyable experience. Startups/ideas should have taken into account also the negative impacts of tourism on the environment and local communities as well as ways to reduce and mitigate them.

Creative, Cultural, Education and Entertainment Industries
E.G. Startups/ideas in the creative and cultural industry including education and entertainment able to conceive innovative products and services to stimulate cultural heritage adaptive reuse projects. Their focus could include designing creative circular economy solutions for a virtual/immersive cultural heritage experience (for example, Augmented Reality, virtual guides, etc.); raising awareness, knowledge and engagement of local communities in cultural heritage adaptive reuse projects; engaging young audiences through educational and entertainment services (for example, through gamification).

Heritage Community, Social Innovation and Ethical Finance
E.G. Startups/ideas oriented to social innovation for cultural heritage adaptive reuse, able to build up and strengthen the "heritage community", bridging it with relevant stakeholders and possible funders. Their focus could include innovative fundraising activities/products (for example, crowdfunding, blockchain, tokenization); cooperative and co-created solutions/projects; bottom-up, participatory, collaborative, and cooperative approaches for the conservation, valorisation, maintenance and adaptive reuse of cultural heritage.

Technologies and Materials for the Circular City and Building
E.G. Startups/ideas developing technological solutions that contribute to the realization of a circular city and circular building. Their focus could include green energy solutions and technologies and responsible use of resources in a lifecycle perspective; management & treatment solutions for wastes reduction, reuse and recycle; low-tech and nature-based solutions for the conservation and adaptation of different types of cultural heritage; virtualization and Industry 4.0 technologies for restoration such as 3D scanning and 3D printing; data management tools (open digital platforms) for cultural heritage towards circular city and circular building implementation. Technological solutions should demonstrate how they ensure compatibility and respect of cultural heritage values.

Abandoned Cultural Landscapes Regeneration
E.G. Startups/ideas dealing with innovative products and services for the development, reuse, and regeneration of abandoned cultural landscape areas, paying special attention to rural, internal and mountain areas, villages and less accessible natural and cultural heritage.

The energy node was considered central, in line with the European Green Deal. As an example of the interest in this sector, the Energy Efficient Mortgages Initiative[12] aims to mobilise capital markets and implement ESG best practices in the financial sector in support of the objectives of the EU Green Deal and Renovation Wave Strategy. Investment in energy efficient buildings and energy saving renovations need particular attention. Innovation in the energy sector, focusing on heritage buildings and sites, is particularly relevant. Therefore, the CLIC Startup Competition paid particular attention to startups providing energy efficiency systems and renewable energy systems for historic buildings and landscapes. One of the categories of the competition call was dedicated to "circular cities and buildings", explicitly calling for startups/ideas "developing technological solutions aimed at circular cities and buildings, whose focus can include [...] green energy solutions and technologies".

3.1 The Evaluation Process

A total of 73 innovative ideas were collected through the CLIC Startup Competition, coming from Europe, USA, Asia, and Africa. As is clear from Fig. 21.1, half of the participants were pre-seeded start-ups, but the other half were evenly distributed among informal teams and individual innovators who, in most cases, had not yet formed themselves into startups, cooperatives, associations etc. The type of participants showed that, at the general level, very often those who are concretely engaged in the valorisation of both tangible and intangible heritage, despite having a valid business idea—albeit at an embryonic stage—do not perceive themselves as potential entrepreneurs (Figs. 21.1 and 21.2).

As described in the CLIC Startup Competition Report, the preselection of the most promising startups/teams was based on the following criteria:

Coherence with CLIC Approach

Criterion 1—Cultural heritage regeneration and environmental, human, social regeneration

Startups/projects should have a clear focus on the conservation, adaptive reuse, and valorisation of cultural heritage, tangible or intangible, adopting a circular economy approach based on environmental, human, and social capital regeneration. Attention to the environmental impacts, social and cultural impacts of the business activity. Attention to turn marginal, abandoned, and underused cultural resources into drivers of territorial and community regeneration. Startups/projects avoid wastes and over-exploitation of natural, cultural, and social resources, generate/use renewable energy, generate/use nature-based solutions to regenerate ecosystems. Startups/projects enhance multi-stakeholder and multi-level synergies and cooperation at urban/territorial scale and enhance human capital and knowledge opportunities especially in marginalized urban and rural areas.

[12] https://energyefficientmortgages.eu/

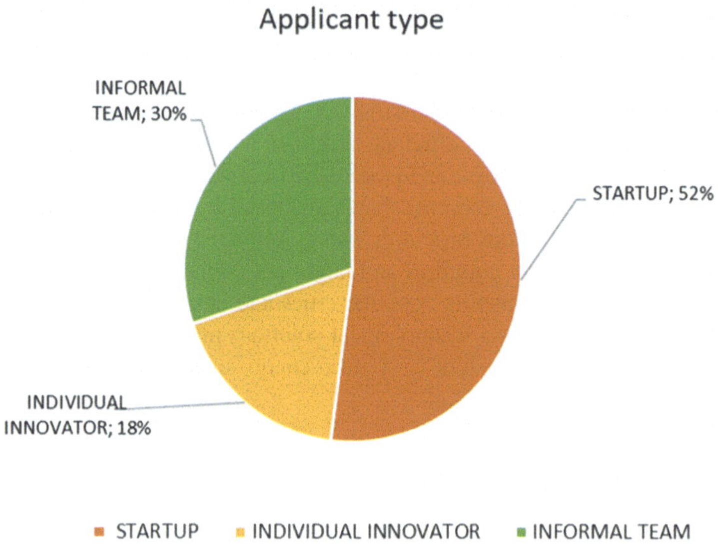

Fig. 21.1 Participants distribution per typology

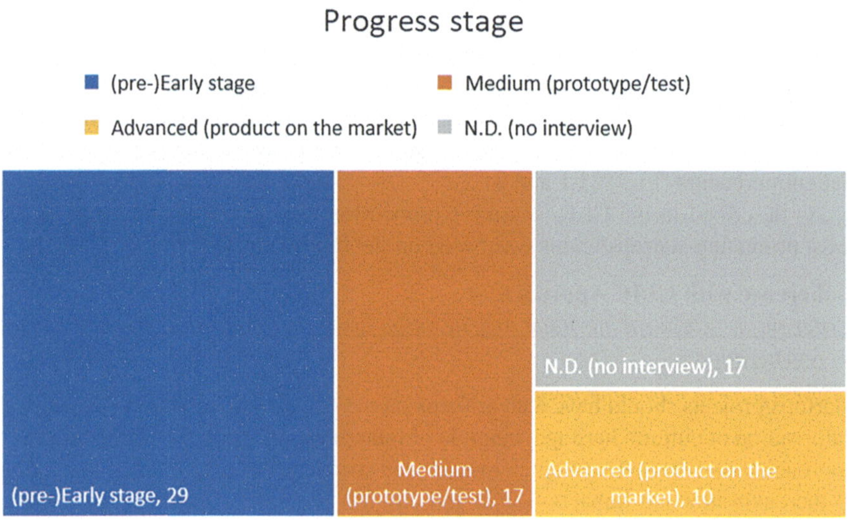

Fig. 21.2 Participants distribution per progress stage

Market Potential
Criterion 2—Financial self-sustainability

Startups/projects should have a clear financial self-sustainability, ensuring adequate revenue streams compared to investments and management costs. This criterion includes the evaluation of the target market and the peculiar value proposition in relation to the target problem or need and competitors. Here it is also evaluated the capacity of the startup/project to develop a Minimum Viable Product, or, in case of advanced startups, the success achieved with the MVP. Financial self-sustainability is evaluated as "potential" in relation to early-stage projects, based on their target market, competitors, revenue model, strategy. For advanced startups, financial self-sustainability is evaluated according to their financial metrics.

Criterion 3—Team and skills

The team carrying out the startup/project should have a clear entrepreneurial approach and include all necessary and complementary skills to realize the innovative product/service and sell it in the market. This criterion includes the evaluation of industry knowledge, unique skills, leadership, key relationships, prior successes and/or failures.

The applicant teams were divided into 3 main groups:
Circular tourism;
Circular creative industries and social innovation;
Circular creative cities and regions (Fig. 21.3).

The awarded teams were identified by the Jury based on the three above-mentioned criteria, the effectiveness of their pitch and the ability to provide adequate and convincing answers to the remarks made by the invited experts. The Jury members did evaluate not only the robustness and self-sustainability of the development model of the projects, or the added value compared to competitors in the market but paid special attention to the capacity of the project to generate positive environmental and social impacts—especially in terms of inclusiveness (gender equality, minority cultures integration) and local community empowerment.

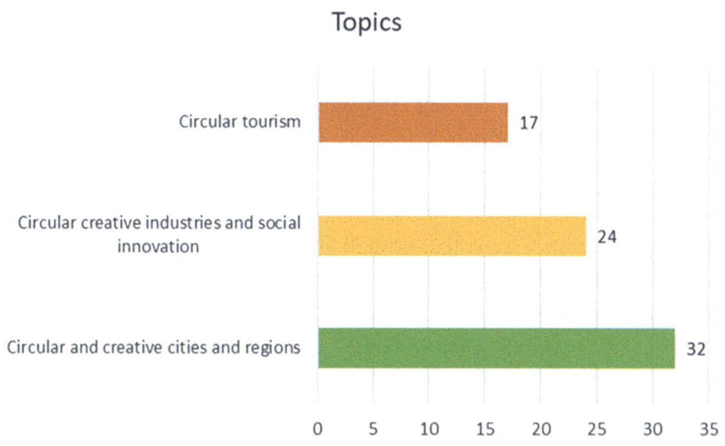

Fig. 21.3 Participants distribution per topic

Evaluation Criteria at a Glance

Value proposition in relation to the circular solution: description of the challenge and circular solution proposed—"how your product/service contributes to circular economy and tangible/intangible cultural heritage valorization, reuse, conservation? Please focus also on environmental and social sustainability, inclusiveness and local communities' empowerment"

- does the startup clearly focus on cultural heritage reuse and valorization?
- does it adopt a clear circular economy approach?
- does it generate positive environmental and social impacts?
- does it pay attention to inclusiveness, gender equality and minority cultures integration?

Market potential: "description of the market you are entering, is it a growing market? Who are your competitors (direct and indirect)? What can you do better than your competitors? Description of your business model – with basic metrics: annual revenues in € actual/expected costs"

- does the startup addresses a large and/or enlarging market?
- is the problem or need addressed real?
- is the solution proposed efficient and of high quality?
- is the solution proposed sufficiently developed?
- is there a clear and robust revenue model?
- is there a clear competitive advantage with respect to other solutions available in the market?

Team/Skills

- is the team complete?
- has the team sufficient expertise to develop the proposed solution?
- are there sufficient or potential networks to be engaged?

Expertise—"What do you need to start/scale-up your business? What do you expect from this competition?"

The best 15 teams were invited to present their business idea at the final event of the CLIC Startup Competition, selected through an evaluation process structured in different phases which involved an international Jury including European Commission representatives, researchers in the cultural heritage field, investors and innovation specialists, experts in startups, representatives of cultural and creative foundations, social designers, and experts from the social investment sector. During the training phase, the teams had the opportunity to better define their business idea, exploring potential solutions to overcome the weaknesses of their projects, emphasizing their strengths and—most importantly—embedding more and more the circular model in their business plan.

The main award consisted of a four-month acceleration programme provided by experts in finance and entrepreneurship which once again demonstrates the urgent need felt by the organizers of filling the gap of entrepreneurial skills in the field of cultural heritage as a key factor to support on the ground a new generation of circular startups with a "vocation" for social innovation, thus maximising the impact of the capacity building action and of the entire research project.

3.2 CLIC Mentoring Programme: Towards a new Generation of Circular Startups

The approach of the mentoring programme (Catanese and Gravagnuolo 2021) was based on the use of a mix of practical methodologies including business innovation roadmap tool, investment readiness level, business model canvas. This approach was aimed at facilitating the involved managers to generate a holistic and systemic picture of their business and to elaborate a strategic master plan for implementation measures (Fig. 21.4).

The investment readiness level of the following aspects was evaluated:

- Leadership team capability
- Product readiness
- Market readiness
- Financial readiness

The programme was structured in a series of training sessions, interspersed with practical sessions and in follow-up sessions with each team. Theoretical explanations on Business Modelling and Business Planning were provided. Furthermore, during the plenary sessions, thanks to breakout rooms, participants were divided into smaller groups, both to carry out the activities/exercises proposed by the experts and to share ideas, get tips, hint, and exchange information among themselves.

Specifically, the first session was focused on the study of the Business Model, the learning objectives were:

- Understand the key concepts of the business model, through the definition of "Business Model"
- Acquire knowledge of the constitutive elements of the Business Model Canvas for value creation (Customer Segments, Value Propositions, Channels, Customer

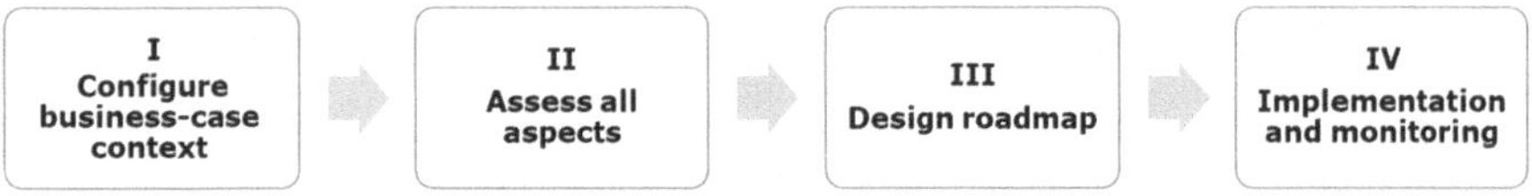

Fig. 21.4 Mentoring programme scheme for circular heritage startups

Relationships, Revenue Streams, Key Resources, Key Activities, Key Partnerships and Cost Structure)
- Find out how to apply the Business Model Canvas to map their Business Model
- Acquire a basic understanding of the business model design process and related frameworks.

The teams identifyed how to create and deliver value to customers and analysed the different components of a business model canvas.

The second session focused on Business Planning. Learning objectives consisted of:

- Understand the key concepts of business planning
- Acquire awareness of one's goals, timing of action and the potential of one's business
- Find out how to transform the circular Business Model Canvas into their Business Plan
- Become aware of what really needs to be done to secure funding.

The activity performed consisted of a SWOT analysis with the aim of evaluating the strengths, weaknesses, opportunities and threats of the project in order to identify the main advantages to offer to customers and therefore identify the key innovative elements that differentiate their business idea from competitors.

At the end of the second session and the next one-to-one sessions, the teams:

- Defined their clear value proposition
- Identified their main product/service to develop a Minimum Viable Service (MVS) or Product (MVP)
- Set business goals and established measurable objectives
- Analyzed target markets/clients and competition
- Defined the "go to market strategy"
- Identified the main costs/needs to achieve their goals.

The participants were able to prepare, with the help of experts, an Initial Business Plan, identifying the strengths and weaknesses of the start-up and the chances of success, to understand where to make corrections and adjustments to their idea.

The third session was focused on the explanation of the Business Model and the Business Plan. The purpose was to complete the drafting of the business plan started during the previous sessions and identify an action plan in order to:

- Set the strategic goal
- Set up the activities/steps that must be performed to achieve the strategic goal
- Identify the resources responsible for each activity
- Define deadlines and goals
- Identify the final need
- Identify indicators and metrics to monitor progress.

The expected output was a Final Business Plan and an Action Plan.

The last session was dedicated to innovation, entrepreneurship, and financing strategies with a focus on the scouting of funding sources. The following topics were explored:

- The concept of innovation
- EU support for research and development programs
- Impact investment and financing tools
- Reward-based crowdfunding.

The training programme developed aimed at enhancing the skills levels in heritage-led businesses, contributing to the development of the sector. The most promising startups resulted from the CLIC initiative are presented below.

4 Startups Business Models Contributing to Circular Cultural Heritage Adaptive Reuse

4.1 Nice Visions

Nice Visions[13] is an innovative company based in Slovakia which has developed Solar Tiles, a solar system for facades and roofs specially designed for areas where appearance is as important as energy production such as urban centers and historic buildings. By combining renewable energy sources (the energy production of photovoltaic panels) with the use of ceramic tiles, Nice Visions ensures increased energy efficiency of historic buildings, without ruining cultural heritage. In addition, the product is designed to comply with the PV Cycle recycling scheme, bringing most materials back into the material cycle. This creative company focuses on creative strategies on how to fulfill basic human needs within the urban context, including transition to a zero-carbon future among their top priorities. The team is committed to develop innovative solutions to integrate renewable energy sources in urban areas (Fig. 21.5).

Fig. 21.5 Nice Visions: a solar system for facades and roofs

[13] https://hello.nicevisions.com/

4.2 *ExtrArtis—Artists in Residence for Creative Tourism Economy*

ExtrArtis[14] is a cultural project that aims to involve emerging artists through the experimentation of the practice of Artist Residency, letting them have a salary for their work and materials during the experience, without facing the costs of accommodation in residences with a high historical and artistic value. The project was created to enhance the less known cultural heritage, interpreting art as an engine of creativity for tangible and intangible transformation of the territory in which it intervenes. The Italian team of innovators promotes a cultural heritage enhancement project while co-creating and testing sustainable human-centered innovations for circular cultural tourism through collaborative innovation networks/methodologies and improved investment strategies (Fig. 21.6).

4.3 *Taste of Terraces*

Together with local farmers of the area of the Amalfi Coast, Taste of Terraces promotes sustainable development and environmental protection, dedicating special attention to the recovery of local biodiversity and agricultural productions at risks of extinction. The team has incentivized the re-cultivation of the historical Re Urberto (or so-called "Fiascone") tomato to preserve the rich cultural landscapes terraces of the area. Taste of Terraces started in 2009 to search and established partnerships via local crowdfunding to re-cultivate and preserve this old heritage variety of tomatoes, bringing back many native species (fruits, vegetables, wheat, etc.) that would have been lost otherwise, thus securing biodiversity and preserving the terraces which are part of the UNESCO World Heritage (Fig. 21.7).

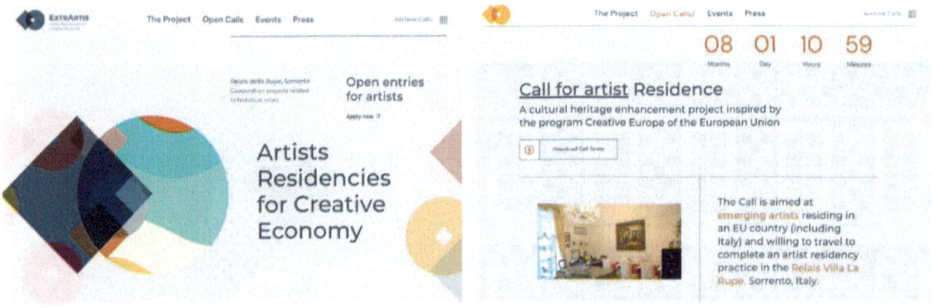

Fig. 21.6 ExtrArtis, Sorrento (Italy)

[14] https://www.extrartis.com/extrartis-artists-residencies-for-creative-economy/

Fig. 21.7 Taste of terraces, Amalfi Coast, Italy

Through the Startup Competition, the circular business model was developed in detail, the online shop was launched and a new promotional product with a local artist was created. The start-up born on the terraces now aims to strengthen and expand its partnerships, generating new jobs in the field of landscape recovery and regeneration.

Currently, more than 10 hectares of neglected terraces have been recovered thanks to the ancient tomato recovery project, with more than 15 family farmers involved in the process, which manage to integrate their yearly income through the activity. It is estimated that around 40% of the terraces once present on the Amalfi Coast are now abandoned, causing, among other things, hydrogeological instability problems linked to the poor maintenance of sloping land. The Re Umberto (Fiascone) tomato recovery project is seeking to recover tangible and intangible cultural heritage and landscape, for the benefit of the territory, also generating potential new job opportunities and a small economic return for the families and family farmers involved, concretely implementing a regenerative and sustainable economic model that could become viable for the future.

4.4 A Terraced Soundscape

The Italian association "A Terraced Soundscape[15]" addresses the issues of sustainable tourism and regeneration of abandoned cultural landscape in Canal di Brenta. Born in 2010, the association has participated to the competition presenting the

[15] https://www.adottaunterrazzamento.org

Fig. 21.8 A Terraced Soundscape, Casarette, Valbrenta (Italy)

project of requalification of an historical and rural contrada in the province of Vicenza, named Casarette. By signing an agreement with the Municipality, the association has taken on the responsibility of the management of this important natural heritage to recover terraced landscape of the valley. The long-term goal of the association is to conclude the requalification of the housing unit in Casarette, investing in agricultural experimentation, entertainment of various kinds leveraging on both natural and cultural resources (hiking, artistic works, firstly the project of reproducing the soundscape of the valley hence the name of the business idea) and training, thanks to the construction on site of an educational farm, aimed at rediscovering the importance of the ancient techniques of agriculture and dry-stone craftsmanship. Furthermore, the collaboration with ITLA Italia—the International Terraced Landscapes Alliance Italia—will be oriented to the creation of a branch of the Italian School of Stone dry (Fig. 21.8).

The recovery of Casarette has the merit of rebuilding a cultural context where people worked in contact with the nature by collaborating. The team wants to challenge the noise of the mine and the highway running into the valley by reproducing the soundscape of the valley into an acoustic installation.

4.5 Marte

MARTE is a startup in the field of art and technology founded in 2019. It aims to connect artists, institutions, and small spaces to develop a cultural economic model to support emerging, independent, and experimental artists, while promoting culture and the regeneration of spaces in rural areas and suburbs. The company promotes a creative, open-distribution model to improve opportunities for artists to disseminate their projects, encourage innovative collaboration and regenerate the identity of local communities. The startup combines research, innovation, and digital assets with environmental sustainability. An added value of this project is represented by its replicability.

5 Conclusions

The CLIC Startup Competition built a large community of innovators that included international, national, and regional organizations, more than 70 startups and teams of innovators operating in all continents, and researchers. The initiative was able to reach a large audience, mobilizing the interest of high-level experts and organizations in Europe and beyond.

The interest in this competition showed a potential in entrepreneurship linked to the cultural heritage sector, adopting a circular economy perspective. The startups participating were mostly in early stage of development, which shows the need of entrepreneurial skills development in the heritage sector. The mentoring programme provided, as well as the additional special prizes, are expected to benefit the startups awarded as they were hungry of this kind of non-financial support, to reach a level of development suitable to seek for more direct financial support and become self-sustainable over the medium-long term (Catanese et al. 2021).

In general, the technological startups resulted in a more advanced stage of development, while social innovation-oriented startups were generally less advanced with regard to market potential criteria but ranked higher in circularity coherence criteria. This aspect is particularly relevant because it sheds light on the ecosystem of actors engaged in cultural heritage adaptive reuse from a social and cultural perspective, which is made mostly by third sector actors, civil society organizations, NGOs, or even informal teams of innovators not related to any legal person. These kind of "potential" startups find difficulties in seeing themselves as "entrepreneurial actors", and cases their original audience reflects a quite defensive perception towards the idea of "making money" or generating "private profit" through cultural heritage "common good" resources. However, this perception can be substantially subverted if a "human-centred" circular economy approach is adopted, which can encourage innovators to develop projects able to become self-sustainable over time, generate jobs for youths and marginalized social groups through the diverse revenue streams, and adding a wide range of positive social, cultural and environmental impacts linked to the success of the startup itself—fostering the positive approach of 'not only doing business that does not harm people and the planet, but also doing business that makes good' (Saleh and Ost 2023). The initiative and business projects collected highlighted through practical examples how the business and financial sector can contribute to address the challenges of cultural heritage adaptive reuse (Ost 2016; Pickerill 2021; Fusco Girard and Gravagnuolo 2017).

The CLIC Startup Competition was a learning experience for participants. The circular business models proposed and co-developed through the selection process, including the training session, provided new skills to existing and potential entrepreneurs in the heritage sector, stimulating innovation towards circular adaptive reuse of cultural heritage (European Commission, Directorate-General for Education, Youth, Sport and Culture 2019a, b). Further initiatives could be advocated to enhance skills in the heritage sector, particularly in line with the objectives of the European Year of Skills and those that are already in place, CHARTER project, the new EIT KIC Culture & Creativity initiative of the European Institute of Innovation and Technology (EIT) the New European Bauhaus initiative, the C-Ship Cultural Entrepreneurship programme developed at ICHEC Brussels Management School, and other programmes, projects and initiatives oriented to fundraising and entrepreneurship for cultural and heritage sector, such as the guide for cultural entrepreneurs "So you need Money" of the Creative FLIP project (IDEA Consult 2021).

In conclusion, the results of the CLIC project were promising in terms of generating new business ideas and projects, showing a potential for research and education institutions to become active in providing entrepreneurial and innovation skills in the cultural and heritage sectors. Clearly, some projects could not reach all objectives, however it can be expected that the enhancement of skills and entrepreneurial mindset will benefit the sector indirectly and contribute to the European innovation and entrepreneurial ecosystem for culture and cultural heritage.

References

Ateca Amestoy V, Aguilella Cueco D, Baatz W, Karatza M, Konstantinidis K, Marçal E, Macfarlane A, McKeon S, Stanojev J (2017) Brainstorming report. Towards an integrated approach to cultural heritage for Europe – prospectus on 'Skills training and knowledge transfer for traditional and emerging heritage professions', Voices of Culture, October 2017. Internet resource: http://www.voicesofculture.eu/wp-content/uploads/2017/12/VoC-Skills-and-training-Final-report-with-Appendix1.pdf

Buglione B, Angrisano I, Gravagnuolo FG (2021) H2020 CLIC project Deliverable 1.4, Database of best practices

Catanese V, Gravagnuolo A (2021) H2020 CLIC project Deliverable 6.12, Report on Start-up Competition

Catanese V, Gravagnuolo A, Bosone M, Micheletti S, Saleh R, Pintossi N, Kaya DI, Monteleone G, Lupu A, Palomba C, Telemo V, Dell'Aversana I, Gerardi L, Ammirati F, Gaggiotti G, Alessia Gianoncelli A, Piergiovanni LH (2021) H2020 CLIC project Deliverable 6.14, Plan of after-project exploitaition activities

Dalberg Global Development Advisors (2014) Innovative financing for development: scalable business models that produce economic, social, and environmental outcomes. Innovative Financing Initiative An initiative of the Global Development Incubator. September 2014.

Europa Nostra, ICOMOS, Climate Heritage Network (2021) European Cultural Heritage Green Paper "Putting Europe's shared heritage at the heart of the European Green Deal"

European Commission (2007) Communication from the Commission to the European Parliament, the Council, the European Economic and Social Committee and the Committee of the Regions on a European agenda for culture in a globalizing world {SEC(2007) 570}, https://eur-lex.europa.eu/legal-content/EN/TXT/HTML/?uri=CELEX:52007DC0242&from=EN

European Commission (2020) Circular Economy action plan. For a cleaner and more competitive Europe, European Union

European Commission (2021a) Communication from the Commission to the European Parliament, the Council, the European Economic and Social Committee and the Committee of the Regions. Building an economy that works for people: an action plan for the social economy.

European Commission (2021b) Union of equality: strategy for the rights of persons with disabilities 2021–2030. European Union

European Commission, Directorate-General for Education, Youth, Sport and Culture (2019a) European framework for action on cultural heritage. Publications Office. https://data.europa.eu/doi/10.2766/949707

European Commission, Directorate-General for Education, Youth, Sport and Culture (2019b) Fostering cooperation in the European Union on skills, training and knowledge transfer in cultural heritage professions. Publications Office. https://data.europa.eu/doi/10.2766/92718

European Union (2012) Consolidated version of the treaty on the functioning of the European Union, 26 October 2012, OJ L. 326/47-326/390; 26.10.2012, https://eur-lex.europa.eu/legal-content/EN/TXT/PDF/?uri=CELEX:12012E/TXT&from=en#page=75&zoom=100&view=FitB

Fusco Girard L (2020) Metropolitan cities: which development strategies? Which new perspectives for improving productivity? Which governance tools?, Urban empires cities as global rulers in the new urban world, 1st edn. Routledge, New York

Fusco Girard L (2021a) The circular economy in transforming a died heritage site into a living ecosystem, to be managed as a complex adaptive organism. Aestimum 77:145–180. https://doi.org/10.13128/aestim-9788

Fusco Girard L (2021b) The evolutionary circular and human centered city: towards an ecological and humanistic "re-generation" of the current city governance. Human System Management. https://doi.org/10.3233/HSM-211218

Fusco Girard L, Gravagnuolo A (2017) Circular economy and cultural heritage/landscape regeneration. Circular business, financing and governance models for a competitive Europe. BDC. Bollettino Del Centro Calza Bini (1):35–52

Gravagnuolo A, Fusco Girard L, Ost C, Saleh R (2017) Evaluation criteria for a circular adaptive reuse of cultural heritage. BDC Bollettino Del Centro Calza Bini 17(2):185–216

Gravagnuolo A, Angrisano M, Nativo M (2020) Evaluation of environmental impacts of historic buildings conservation through life cycle assessment in a circular economy perspective, Aestimum, Special Issue "Sustainable development and circular economy", https://doi.org/10.13128/aestim-10004

Gravagnuolo A, Micheletti S, Bosone M (2021) A participatory approach for "circular" adaptive reuse of cultural heritage. Building a heritage community in Salerno, Italy. Sustainability (Switzerland) 13(9). https://doi.org/10.3390/SU13094812

Guntram G, Hornung-Prähauser V, Strasser A (2014) Handbook for creative cultural heritage cooperation projects, Report of Salzburg Research Forschungsgesellschaft m.b.H., Austria. ISBN 978-3-902448-39-2

IDEA Consult (2021) Creative FLIP so you need money guide, Brussels. Available at: https://creativesunite.eu/synm/

MacDonald S (2011) Leveraging heritage: public-private, and third-sector partnerships for the conservation of the historic urban environment. In ICOMOS 17th general assembly. 2011-11-27/2011-12-02. Paris, France

McKeever E, Jack S, Anderson A (2015) Embedded entrepreneurship in the creative re-construction of place. Journal of Business Venturing. XXX:50–65

Michiara P (2016) I patti di collaborazione e il regolamento per la cura e la rigenerazione dei beni comuni urbani. L'esperienza del Comune di Bologna. Aedon. 2/2016

OECD (2018) Job creation and local economic development 2018: preparing for the future of work. OECD Publishing

OECD (2020) Social economy and the COVID-19 crisis: current and future roles. OECD Publishing

OECD and European Commission (2022) Policy brief on making the most of the social economy's contribution to the circular economy, OECD Local Economic and Employment Development (LEED) Papers, No. 2022/01. OECD Publishing, Paris. https://doi.org/10.1787/e9eea313-en
Ost C (2016) Innovative financial approaches for culture in urban development. In UNESCO (2016) Global Report on Culture for Sustainable Urban Development
Ost C, Saleh R (2021) Cultural and creative sectors at a crossroad: from a mainstream process towards an active engagement. Built Heritage 5:14. https://doi.org/10.1186/s43238-021-00032-y. Paris, https://doi.org/10.1787/9789264305342-en
Pickerill T (2021) Investment leverage for adaptive reuse of cultural heritage. Sustainability (Switzerland) 13(9). https://doi.org/10.3390/SU13095052
Porter ME, Kramer MR (2011) Creating shared value. Harv Bus Rev 89(1–2):62–77
Rafael Chaves Ávila & José Luis Monzón Campos (2019) Recent evolutions of the social economy in the European Union, CIRIEC Working Papers 1902, CIRIEC – Université de Liège. Avalaible at https://www.eesc.europa.eu/sites/default/files/files/qe-04-17-875-en-n.pdf
Ragozino S (2016) Tools for regeneration of the urban landscape. Social enterprise as link between people and landscape. Procedia Social and Behavioral Sciences. CCXXIII:201–208
Sacco PL, Teti E (2017) Culture 3.0: A new paradigm of value creation. Economia & Management 2017:79–96
Saleh R, Ost C (2023) Innovative business model for adaptive reuse of cultural heritage in a circular economy perspective. Int J Entrep Small Bus 48(1):39. https://doi.org/10.1504/IJESB.2023.128181
Stanojev J, Gustafsson C (2022) Towards cultural heritage-based innovation strategies by evolution of the knowledge triangle of research education and business innovation. Built Herit J 2023 (Forthcoming)
Thurley S, Busquin P, Spek T, Brandt-Grau A, Clausse G, Gustafsson C, Kolar J, Lazarro E, Smith B, Mallouchou-Tufano F (2015) Getting cultural heritage to work for Europe. Report of the Horizon 2020 expert group on cultural heritage. European Commission, Brussels. https://doi.org/10.2777/745666

Part VI
Conclusions and Some Guidelines

Introduction

This section provides concluding reflections on the adaptive reuse of cultural heritage within the framework of the circular economy, summarizing key insights and lessons learned from previous chapters. It also offers practical guidelines aimed at supporting future initiatives in this field, addressing challenges such as sustainability, resource efficiency, and community engagement. By distilling the main takeaways and proposing actionable strategies, these conclusions serve as a roadmap for stakeholders looking to align cultural heritage reuse with circular economy principles, ensuring long-term success and value generation for both heritage assets and local communities.

Chapter 22
Conclusions and Some Guidelines

Luigi Fusco Girard and Antonia Gravagnuolo

Which conclusions? Which critical reflections about the CLIC results? Which recommendations?

This volume offers an overview on the theoretical framework and operational aspects of the **circular symbiotic heritage ecosystem** model (Luigi Fusco Girard, Chap. 2), focused on the relationships between cultural heritage adaptive reuse and local development in the perspective of the circular economy and circular city/region implementation.

The adoption of the circular model requires to take care of the long term,[1] of intangibles that do not have a monetary dimension, and thus it recognizes a specific weight to many benefits that are those typical of cultural heritage, that in traditional economic terms are underestimated. The circular model re-establishes "weights" with respect to conventional economics, according to ecological and human-centred values, valorising the intangible dimensions that are usually not recognized operationally in (for example) cost-benefit analyses. On the other side, the adaptive reuse of cultural heritage contributes to the realisation of the circular city model.

[1] Economic analysis, as known, is characterized by the reduction of heterogeneity/multidimensionality, and by the valorisation of "present" instead than "future", for example through the concept of the "discount rate" widely used in economic and financial analysis.

L. Fusco Girard
Emeritus Professor at University of Naples Federico II; CNR IRISS, National Research Council, Institute for Research on Innovation and Services for Development, Naples, Italy

A. Gravagnuolo (✉)
CNR ISPC Institute of Heritage Science, National Research Council, Naples, Italy
e-mail: antonia.gravagnuolo@cnr.it

L. Fusco Girard, A. Gravagnuolo (eds.), *Adaptive Reuse of Cultural Heritage*,
https://doi.org/10.1007/978-3-031-67628-4_22

1 The Contribution of the CLIC Research in the Context of Current Global Challenges: The Role of Cultural Heritage Adaptive Reuse in the Perspective of the Circular Economy

As discussed by Fusco Girard,[2] the **circular symbiotic heritage ecosystem** represents the theoretical framework developed through the CLIC research, which focuses on abandoned and underused cultural heritage identified as a potential resource for the implementation of the circular economy and circular city/region model, regenerating territorial "wasted" capitals (manufactured, social, human, environmental, financial…) and turning them from costs into drivers of enhanced social cohesion, ecological regeneration, self-sustainability, new jobs and cultural development, i.e. an engine of social, cultural and urban/rural regeneration.

Grounded on the results of previous researches, such as Cultural Heritage Counts for Europe (2015), and international guidelines and recommendations such as UNESCO Historic Urban Landscape (2011), ICOMOS Heritage Impact Assessment (2011), the ICOMOS Burra Charter (2012), and many documents representing cornerstones of the cultural heritage conservation research over the last decades, the CLIC model of the "circular symbiotic heritage ecosystem" builds a holistic (not partial) framework and actualises heritage conservation and adaptive reuse in the context of the greatest challenges of the current times as climate change, ecosystems and natural resources depletion, social fragmentation and the right of people's health and wellbeing. Recently, the contribution of cultural heritage to the "Green Deal" policies was highlighted by international organisations such as ICOMOS and Europa Nostra. The European Cultural Heritage Green Paper[3] underlines that the contribution of cultural heritage to the circular and green transition was underestimated or completely not acknowledged in the European policies. This book, which entails the CLIC research contribution, aimed to fill this gap by highlighting the many different ways in which cultural heritage can be considered as a key element of a sustainable future in Europe and beyond, supporting a just and more effective transition linking past and future, and ensuring that cultural diversity is preserved as fundamental human right.[4]

The circular model is the proposal of regeneration of a network of relationships and interdependencies, as the key for the success of adaptive reuse. Human relationships are enhanced through the presence of unique, vibrant, accessible and open places of encounter that express the cultural identity (e.g. the European "square"),

---

[2] Luigi Fusco Girard, "The circular "human-centred" adaptive reuse of cultural heritage: theoretical foundations", Chap. 2 of this book.

[3] ICOMOS and Europa Nostra (2021), European Cultural Heritage Green Paper https://www.icomos.org/en/focus/climate-change/102906-publication-of-the-european-cultural-heritage-green-paper-in-french-and-spanish

[4] See also: UNESCO Universal Declaration on Cultural Diversity, 2001. https://en.unesco.org/about-us/legal-affairs/unesco-universal-declaration-cultural-diversity

as well as through increased opportunities for participation and collaboration, promoting mutual trust, interrelation, cooperation at all levels.[5] Cultural heritage, present in almost all cities and rural areas in Europe, can largely contribute, at certain conditions, to re-build human connections, and thus social cohesion, as showed by the studies on social impacts of cultural heritage regeneration presented in this book.[6] International policy documents such as the UN Agenda 2030, the UN New Urban Agenda, the Paris Agreements, etc. have one main goal: to enhance people's wellbeing. According to the Agenda 21 of Rio de Janeiro,[7] the aim of sustainable development is the "*fulfilment of basic needs, improved living standards for all, better protected and managed ecosystems and a safer, more prosperous future*", thus "*consideration should also be given to the present concepts of economic growth and the need for new concepts of wealth and prosperity which allow higher standards of living through changed lifestyles and are less dependent on the Earth's finite resources and more in harmony with the Earth's carrying capacity*" (section 4.11). The human being is the centre of sustainable development, in harmony with Nature. Cultural heritage can provide a substantial contribution to the goals of sustainable development, and to the implementation of human rights in the territory, if the **circular symbiotic heritage ecosystem model** is implemented, contributing to more circular and "human-centred" cities and regions.

The CLIC research aimed at providing first of all a theoretical reflection on the role of cultural heritage for addressing societal challenges, as well as practical answers on whether and how the **circular symbiotic heritage ecosystem model** can be implemented in different contexts and heritage typologies: religious heritage sites, industrial heritage sites, archaeological resources, forts, castles and military sites, rural cultural landscapes, historic villages. This book comes as a useful reading, demonstrating how the adaptive reuse of cultural heritage, in a systemic perspective, has the potential to stimulate growth, sustainable development, social regeneration, welfare, jobs, income, and liveability of urban/territorial settings: to implement the circular economy model in the human-centred perspective. It also provides innovative models and a toolkit for circular financing, reusing and managing cultural heritage based on research outcomes and implementation of experimental models.

The contributions harvested in the book confirm that cultural heritage can be a significant driver for sustainable development in cities and regions. As an economic and cultural asset, if well managed cultural heritage is capable of boosting economic growth (not only through a tourism economy), enhancing urban liveability, and contributing to environmental and climate adaptability. In addition, the reuse of abandoned and underused cultural heritage and landscapes is a practical substitute

[5] Here, digital spaces become not enough to achieve a human-centred and circular perspective of sustainable development, while they can be effectively complementary and enhance connections both at local and global scale.

[6] See in particular Domaradzka et al., Chaps. 10 and 11 of this book.

[7] United Nations Conference on Environment & Development, Rio de Janerio, Brazil, 3 to 14 June 1992. https://sustainabledevelopment.un.org/outcomedocuments/agenda21

to the risk of demolition, bypassing the wasteful processes of demolition and new construction prolonging the cultural heritage lifespan and transmitting heritage values to future generations.[8] The circular model allows to acknowledge the importance of cultural heritage adaptive reuse compared to other investments, as it substantially contributes to reach societal goals under multiple, synergic perspectives (environmental, cultural, social, economic). Adaptive reuse of cultural heritage can thus be instrumental to circularize the flows of energy, raw-materials, human and cultural capital, and hence, it plays a significant role in the transition towards the circular economy. Complementary to its environmental benefit, adaptive reuse brings forth substantial economic, social and cultural advantages by reusing historic buildings, sites and landscapes attached meanings and values by a wide range of citizens and local actors, starting with the young and the elderly people.

The CLIC theoretical framework was developed through a inter- and transdisciplinary approach with researchers and practitioners from heritage conservation, cultural economics, ecological economics, social science, evaluation and decision science, financial assessment, policy making, involving diverse actors from cities and regions to discuss, test and validate the proposed model. The experts' contributions from different disciplines aim to address this gap in existing knowledge from a circular economy and sustainable development perspective, maintaining the integrity and the authenticity of the cultural assets. They introduce innovative economic, environmental and governance models and evaluation tools tested and validated for adaptive reuse within the "CLIC research".

The many examples, experimentations and reflections presented in this book focus on the main instruments that can enable the circular model in the adaptive reuse of cultural heritage:

- Innovative evaluation methods and tools for participatory decision-making processes
- Circular governance models towards heritage communities and an active role of third sector actors
- Circular business and financial models
- Circular metabolism of heritage as "living system"

The realisation of the **circular symbiotic heritage ecosystem** as "ideal" model of adaptive reuse requires a series of innovative instruments: business, governance, financing and evaluation tools. The CLIC research integrated knowledge from different disciplinary fields thanks to the contribution of high level researchers and experts, as well as practitioners, from the fields of economics, social science, ecological science, heritage conservation, etc. The research process started with the analysis of 126 practices of cultural heritage adaptive reuse, analysed under the

[8] ARUP: Circular economy in the built environment, 2016 report: https://www.arup.com/perspectives/publications/research/section/circular-economy-in-the-built-environment; Transform & Reuse: Low Carbon Futures for Existing Buildings, 2020 report: https://www.arup.com/perspectives/publications/promotional-materials/section/transform-and-resuse-low-carbon-futures-for-existing-buildings

perspective of circularity, to assess success and barrier factors. These practices were then assessed to identify exemplary cases to be investigated more in-depth (10 cases analysed through ex-post evaluation, 3 case studies analysed in-depth in their business and financing model, etc.). The practices investigated contributed to orient the identification of criteria and new indicators for evaluation, that were used as reference for the experimentation conducted in the pilot cities. In particular, in the city of Salerno the indicators were used to orient and assess the adaptive reuse alternatives, enriching the evaluation process to identify the most satisfying solution in the "wasted" area of large abandoned buildings in the historic city center.

2 Innovative Evaluation Methods and Tools for Participatory Decision-Making Processes in the Reuse of Cultural Assets

Decisions in cultural heritage adaptive reuse can be particularly complex, due to the amount of factors to be considered: tangible and intangible cultural value conservation, regulatory constraints, reduction of environmental negative impacts, divergent and often conflicting interests of stakeholders, business model viability, social acceptance, synergies with the local context, enhancement of landscape beauty, and other positive social impacts to be generated through the adaptive reuse. The "ideal" solution towards a circular symbiotic heritage ecosystem should aim to reach all circularity objectives simultaneously and realize a "living system" in synergy with the natural and urban/rural ecosystem, able to be self-sufficient from multiple points of view, e.g. economically and financially, as well as energetically.

New uses/functions of heritage buildings and sites should ensure that circularity from diverse perspectives is activated, while innovative and compatible technologies, including nature-based solutions, can largely enable the reuse of cultural heritage as auto-poietic, symbiotic living system that self-generates the resources for its functioning.

As "common good", cultural heritage adaptation should also meet the needs and desires of local communities, possibly activating them into "heritage communities" willing and able to take care of cultural heritage contributing to its conservation, maintenance, and transmission to future generations (in line with the FARO Convention, CoE 2005).

Thus, *participatory inclusive decision-making processes* in cultural heritage adaptive reuse are preferable than top-down decisions based only on expert judgements, as they ensure higher social acceptability of interventions while potentially activating the interest of local communities and stakeholders for collective care, reuse and valorisation of heritage resources in the longer term, avoiding new processes of abandonment and decay. Transparency and clarity in each phase of participatory processes is of utmost importance, as they should aim also to enhance active citizenship, promoting dialogue, cooperative relationships, collaboration, trust, and supporting the search for creative solutions. The "common good" is not given, but it needs to be "built" through dialogue and open public discussion, analysing the possible impacts and implications of alternative solutions, learning from previous experiences and evidence-based data, sharing knowledge from different backgrounds. In this context, appropriate evaluation methodologies and tools need to be employed as supporting instruments for transparent and effective participatory decision-making. There is a close relationship between the design of conservation interventions "by trial and error" (Giancarlo de Carlo 1988), evaluation and participation.

To develop **circular symbiotic heritage ecosystem** as systemic complex solutions, multiple objectives and criteria should be considered when assessing alternative solutions. Multicriteria evaluation methods were thus experimented to support decisions in the adaptive reuse of cultural heritage. The evaluation approaches used in the cultural sector are considered in general unsatisfactory.[9] In the course of the CLIC research, evaluation was applied in the different case studies at several scales, providing a replicable process to implement the circular city model. Hybrid, multidimensional, quanti-qualitative evaluation methods were tested, which can also be used in the phase of co-planning, co-design and co-management,[10] demonstrating the effectiveness of adaptive reuse in the circular model. Evaluating means interpreting a general context, foreseeing impacts before using resources, land, spaces, etc., and comparing alternatives through clear and shared criteria. By evaluating alternatives, it is possible to deduce priorities, alternatives, and new solutions, considering multiple, multidimensional and conflicting criteria/objectives. Evaluation is necessary for decision-making processes in a time of crisis, with more and more scarcity of resources and energies to improve governance, urban planning, design and management.

In the CLIC research, specific circularity objectives and related criteria were identified, allowing stakeholders and heritage managers to learn about the circular model and orient their decisions towards circularity. A set of quantitative and qualitative indicators was proposed and experimented for ex-post and ex-ante evaluation,

[9] See: UK studies on valuation of cultural heritage.

UK Government, DCMS Department (2022), Scoping culture and heritage report https://www.gov.uk/government/publications/scoping-culture-and-heritage-capital-report/scoping-culture-and-heritage-capital-report

UK Government, DCMS Department (2021), Valuing culture and heritage capital: a framework towards informing decision making https://www.gov.uk/government/publications/valuing-culture-and-heritage-capital-a-framework-towards-decision-making/valuing-culture-and-heritage-capital-a-framework-towards-informing-decision-making

Historic England (2021), Towards better valuation: The Culture and Heritage Capital approach https://historicengland.org.uk/content/docs/research/towards-better-valuation-culture-and-heritage-capital-approach/

ICCROM (2022), Valuing Cultural Capital: accounting for the hidden value of heritage (workshop) https://www.iccrom.org/news/valuing-cultural-capital-accounting-hidden-value-heritage

[10] As an example, in Italy the Law 117/2017 art. 55 foresees co-planning, co-design and co-management.

allowing best practices assessment and circular solutions design. In CLIC pilot areas, and in particular in the city of Salerno, multicriteria evaluation was used to support participatory processes of cultural heritage adaptive reuse, promoting the achievement of circularity objectives, but also fostering co-design, co-creation, co-planning with local stakeholders. The experience of Salerno showed how creative solutions can be identified through the process of participation, engagement and co-design, in a step-by-step process that started with a public consultation for "ideas", towards a circular business model workshop to work on the viability of proposed project solutions, and finally towards the identification of a "satisfying" solution which was as "near" as possible to the "ideal" circular model.

Based on the experience of CLIC, it is possible to affirm that evaluation processes are fundamental tools for new governance towards sustainability, for checking and co-designing creative and resilient solutions. New governance is based on experiences and best practices interpretation and comparison not only by experts, but also by the general public. Creative cities have to invest more and more in assessment as support for participation, and thus for decision-making.

The evaluation of the creative potential of a city is more and more required for city/region development so that the areas of strength and the ones of weakness can be properly selected.

Evaluation evolutionary processes[11] help make decisions on "if", "what", "where", "when", "for whom" and "with whom" to implement adaptive reuse creative initiatives and when to stop them. Evaluations are necessary tools in different pacts, agreements, city contracts, in participation processes, in finance and micro-credit, in taxation, in sustainability focus groups, in auditing, in choices at a strategic, tactical and management level and in general for investigations.

Evaluation is a fundamental tool for selecting innovative alternatives and for building choices in urban planning and design which can synthesize many values, and produce multiple benefits for many agents, in a win-win perspective. They are to be evaluated in their quantitative and qualitative, direct, indirect and induced impacts, in the short, medium and long term, integrating any bureaucratic or strictly economic approach.

Innovative alternatives are characterized by high uncertainty, costs and risks. Lack of knowledge is the common element in all creative choices/actions. Therefore,

[11] The evolutionary perspective should be chosen:

- when faced with the choice between what to keep and what to replace, it is necessary to move in a dynamic perspective;
- when the energy issue is fundamental to economic production;
- when there are no maximising solutions but only "satisfying" solutions;
- when the importance of fundamental values such as transparency, accountability, cooperation and trust is recognised in the production of development prosperity;
- when it is recognised that there is a general interest (common good) to be pursued and not only particular interests;
- when concepts such as self-organisation, learning capacity and regenerative capacity are introduced.

they require experimental and testing approaches in order to learn from their successes or failures and about the specific characteristics of the dynamic urban system in supporting uncertain and/or irreversible effects (critical capacity thresholds). Evaluations may suggest how to improve experiences, whether to transfer them into ordinary practices or totally change them.

An integrated assessment process does not only help to compare given and defined alternatives but it also stimulates to identify and explore alternative new solutions. So the evaluation process can become the engine of regeneration creativity. An iterative and interactive decision-making process is activated through continuous feedbacks and improvements in the level of achievement of objectives.

The "creative city" systematically collects data and information to improve knowledge for a critical judgment/assessment required in urban planning. Data, information, knowledge are to be structured in a systemic way so as to allow for comprehensive evaluations and comparisons with new ideas and their implementation and performance in satisfying needs in the material and immaterial space.

Considering that creativity is interpreted here in relation to promotion of economic, social and ecological resilience, specific indicators about density of relationships in different dimensions are required. People's involvement in reaching the common good, social inclusion, community sense, collective identity becomes relevant to reflect benefits of relations.

A particular attention should be given to the "intrinsic value" (non instrumental value) of cultural heritage[12] in the evaluation process. The intrinsic value requires attention to the natural/ecological system of a site, and to safeguard its health, on which human health itself depends, involving the enhancement of landscape quality. It contributes also to the psychological and physical well-being of people, as evidenced by several studies.[13]

Fragile or insubstantial social relationships determine a lower quality of life and generate less resilience. The process of identification, interpretation and evaluation of the intrinsic value represents a cultural and social/community construct, which can be realized with iterative and interactive participatory processes, with subsequent approximations, through iterations/interactions steps. It helps to orient the reuse of cultural assets, limiting the range of new uses/functions in the dismissed assets. These qualitative evaluations should be characterized by a level of consensus that is as high as possible in different contexts, so that they can be satisfactory, and therefore intersubjective and replicable. Once this intrinsic value has been defined, it must be placed in relation to the opportunity costs[14] that result from the preservation of this value. If the opportunity costs assessed are too high, i.e. they go beyond a certain threshold tolerable/compatible with a series of constraints that the specific context determines, the cultural site/landscape characterized by the abovementioned

[12] On the intrinsic value of cultural heritage, see: the Burra Charter (2013), Fusco Girard and Nijkamp (1997), Fusco Girard and Vecco (2021).

[13] Sacco and Teti (2017).

[14] Fusco Girard (1987).

intrinsic value will not be preserved. In other words, conservation intervention will be acceptable if the proposed changes reduce opportunity costs to a reasonably acceptable level. Of course, the threshold of tolerability/compatibility is also subject to evaluation and interpretation by the community.

3 Circular Governance Models Towards Heritage Communities and an Active Role of Third Sector Actors

The reuse of cultural assets is an engine of new employment, but specific conditions should be guaranteed. Recommendations towards a new governance for the circular reuse should be coherent with the following general points:

1. The reuse benefits are to be assessed in relation to the achievement of the global Goals of the United Nations Agenda 2030 for Sustainable Development and Sustainable Development Goals, of the UN New Urban Agenda and the Pact of Amsterdam (The Urban Agenda for the EU), that highlight the importance of the achievement of the common good. The challenge is the humanization together with reduction of climate change negative impacts. Humanization means first of all jobs for young people, inclusion for aged/poor people living in the city/territories, and wellbeing/health for all people.
2. The functional reuse of the heritage assets should be incorporated in the perspective of the circular economy, mimicking the economy of nature (the economy of living systems): the Circular economy as the Regenerative economy, as result of circular processes minimizing the consumption of resources and waste of materials. Circular adaptive reuse should contribute to the city decarbonization with specific and integrated initiatives of adaptation and mitigation;
3. The analysis of the systemic context with which cultural heritage interdepends through flow of materials, energy, goods, should be grounded and assessed through the knowledge of local metabolism: new knowledge of local level metabolism is more and more required;
4. Emphasis on the self-sustainability of heritage functional reuse (in financial and economic terms), through specific capacity to generate economic value (value generation for value capture, value sharing, and also to generate economic, social and intangible values);
5. The notion of "intrinsic value" should orient the new functions, new use values, through new decision-making support systems able to overcome the existing Heritage Impact Assessment approach proposed by ICOMOS (2011);
6. The capacity to recognize the central role of the third sector between public institutions and private organizations, to generate specific social impacts in relations to the characteristics of inclusion and cooperation that are typical of circular economy experiences (cultural heritage adaptive reuse as social investment). An important benefit coming from the reuse is the regeneration of collective memory and transforming it into a civic *hope* and a "public value";

7. The third sector can be a valuable actor to ensure that multidimensional objectives are achieved. The "gift economy", the "civil economy", the "we-economy",[15] with volunteering efforts, capacity of collaborating and creating synergies at territorial level, has already demonstrated to be a powerful engine of heritage regeneration in a circularity perspective;
8. The importance of benefits in terms of wellbeing and health coming from functions of reuse, through perception indicators about wellbeing, health and quality of life;
9. The capacity to build a local community through the adaptive reuse: to contribute to the regeneration of the social capital (social cohesion, in the fragmented society);
10. Attention to the productivity of functional reuse in all dimensions and then to the productive and generative capacity of each function (art, technologies, start-ups, etc.) with a particular attention to the productivity of creative industries activities (not ignoring circular tourism economic benefits and real estate benefits, but going beyond them);
11. The capacity to implement HUL (UNESCO Historic Urban Landscape) through circular economy, considering their reciprocal interdependences;
12. The potential key role of circular procurement in the public and private sector for the implementation of circular economy in cultural heritage adaptive reuse;
13. The availability of integrated quantitative and qualitative evaluation methods to be adapted in a spiral process, focused on the circular economy model. The evaluation method for the circular economy should consider complex values;
14. The strong correlation to be guaranteed between adaptive reuse and the general planning strategy for the city/territory;
15. The ideal adaptive reuse can be recommended. It refers to the organisation of a "**regenerative circular symbiotic ecosystem**" based on the cooperation of all stakeholders;
16. The need of new financial tools, able to leverage particularly "patient" impact investment capital promoting "blended" social/environmental/cultural and financial positive returns, also in line with the EU Taxonomy; innovative financial tools should be thus supported by adequate tools to assess the impacts and provide reliable estimations and evaluations to private and public investors.

These recommendations can specify and integrate the general proposals in Chap. 2. The existing governance structures and operational systems concerning reuse of cultural heritage and landscapes are still highly limited in the involvement of relevant stakeholders to the decision-making process. Regulatory and planning tools are not flexible enough to allow sustainable and circular transformation processes, and are restricted in the financial resources and funding arrangements that mostly rely on public funds. Therefore, to turn cultural heritage and landscapes into a resource, instead of a cost for the community, the structures of authority, institutions and financial arrangements should be adjusted. This adjustment needs to ensure larger

[15] See also the studies of Stefano Zamagni, Leonardo Becchetti.

stakeholders' involvement in decision-making, to attract private investments, and to facilitate cooperation between community actors, public institutions, property owners, informal users and producers. Specific vocational training and education programmes are required to build the necessary skills in public institutions and private organisations. In addition, to manage change, flexible, transparent and inclusive tools are required, thus leveraging the potential of cultural heritage to foster adaptive reuse practices.

The CLIC research highlighted in particular the increasing role of third sector in the management of heritage assets. In particular, some examples are interesting: the Catacombs of San Gennaro in Naples, Pakhuis de Zwijger in Amsterdam, Pfefferberg in Berlin, and many others described in the several contributions of this book. Hybrid business models related to Community cooperatives, Community enterprises, Social entrepreneurship, Civic entrepreneurship, Foundations, Civic associations, etc. can have a key role in generating financial viability (self-sustainability) as well as the "heritage community". The Third sector as a "hybrid" between public and private sector, should be taken into consideration (also according to the best practices analysed in CLIC) as relevant and promising actor to enhance cultural heritage adaptive reuse in a multidimensional circular perspective, as many best and good practices can show.

4 Good Practices in Circular Business and Financial Models

Historic urban areas in European countries have a great potential to become the new "hubs" of innovation and attractiveness, if circular solutions based on energy and digital technologies are implemented. While in other areas outside Europe (such as Korea, China, Emirates) the innovation is focused on new hyper-technological cities (e.g. Masdar city, Songdo), in European countries the historic urban areas and landscapes should be considered the centres of attractiveness, creativity and innovation, building on their unique character and new ground-breaking compatible technologies. The "European Songdo/Masdar" are the historic cities and historic villages, that can become not only digital cities/districts, but they have also a specific identity, a "soul" and a human-scale which is grounded in European historic-cultural values. Thus, the maintenance and reuse of cultural heritage and landscapes becomes a key strategy for European cities and regions attractiveness.

Innovative uses/functions that meet market demand are needed to ensure the viability of business models for reused heritage assets. Circular business models ensure that positive externalities are generated through the new uses of heritage buildings and sites. The circular business model can be seen as a responsible economic approach in which people, planet and places thrive in synergy with each other, generating win-win solutions and benefits at all levels. Here, third sector organisations and "mission-oriented" businesses can be a valid example, as social enterprises, cooperatives, foundations etc. can re-generate financial capitals through adequate revenue flows, but also generate positive impacts in the territories and at larger scale. Examples can be seen in abandoned heritage regenerated through cultural, creative and innovative enterprises as in De Hoorn in Leuven, heritage sites in which energy is generated on site through renewable sources as for example in C-Mine in Gent, heritage communities taking care of cultural heritage establishing social businesses that engage entire neighbourhoods as in the Catacombs of San Gennaro in Naples, Italy. These "best practices" showed how ambitious circularity objectives can be reached, not only representing "costs" but more exactly, "investments" as they are able to generate wealth and wellbeing for local communities.

Circular business models are also linked to management models and innovative financial mechanisms. In particular, adaptive reuse sees a large barrier in effective management once the buildings/sites are restored and made available for the new uses. Many experiences refer to the museum use/function, which after having absorbed considerable public resources, has often shown little capacity for self-financing. New uses/functions should be un-ordinary, characterized by high specificity, originality, uniqueness, in the capacity to respond to a unsatisfied demand.

One important recommendation relates to the management model, ensuring that management can be self-sufficient over time, but this strongly depends on legislation, which differs from country to country in Europe. With CLIC side-projects activated at national level (such as 'CLIC Italy'[16]), the intention should be also to verify the possibility of activating (circular) symbiosis strategies between medium-large cities and small centres/towns: that is, between city and countryside. Indicators related to cultural heritage, circular metabolism, heritage community, circular governance, circular business model can be implemented, declining them on a macro, meso and micro scale, in the perspective of the human-centred strategy.

Effective management and the capacity to develop new entrepreneurial models for the adaptive reuse of cultural heritage are key for leveraging investments in the field. According to CLIC results, it is possible to conclude that effective circular business models would be the key to unlock private, public and community investments in cultural heritage, reversing the trend of abandonment and regenerating heritage sites starting from their intrinsic values. "Impact" investors are keen to identify projects which blend financial return with high social returns, thus cultural heritage adaptive reuse can represent an interesting sector for such investors. Finally, the entrepreneurial skills in the heritage sector should be addressed, stimulating new entrepreneurial initiatives able to unlock public and private investments.

---

[16] National research project ongoing at CNR IRISS, in which the general conclusions of CLIC project for European countries are further explored under the perspective of the Italian regulatory framework.

5 Circular Metabolism of Heritage as "Living System"

The adaptive reuse of cultural heritage should be exploited as an opportunity to implement measures for decarbonisation of cities, possibly turning cultural heritage buildings and sites into "engines" of energy generation from renewable sources. The goals of decarbonisation cannot be achieved by 2030 if adequate measures to address the energy node are not taken, first of all at urban level, together with new lifestyles, choices and behaviours at all levels. This requires appropriate technologies, but also the capacity to design circular metabolic systems, also at the micro scales, that take into account synergic measures such as those for renewable energy, energy efficiency, natural lighting and ventilation, water recovery and reuse, heat reuse, waste to energy, nature-based solutions. In the design of energy systems and energy grids, it should be considered that renewable energy is a discontinuous source, and accumulation systems are fundamental to ensure the levels of performance required by demand. Thus, digital technologies become absolutely complementary to energy technologies, as smart grids and in general smart energy systems based on renewable energy sources can become efficient at the condition that a network of sensors and remote control systems are in place, to monitor energy needs, consumption and environmental parameters also for comfort and wellbeing. Moreover, digital technologies (in particular, IoT and AI) should become key to turn cultural heritage and whole historic urban areas also into energy generation sites, if we want to reach decarbonisation objectives in 2030 and 2050.

Another recommendation is that there is a growing demand for specialised professionalism in the field of crafts and traditional skills in cultural heritage, however training is not sufficiently guaranteed. Technical vocational and training schools should be established. This competence at a higher level also means specialisation and master courses on the adaptive reuse of disused and underused cultural heritage.[17]

In order to generate new value, an engine of environmental regeneration must be identified in the "circular regenerative ecosystem" that represents the energy for circularisation and the production of new values: economic efficiency, new employment, landscape quality, etc. The engine is represented by functions capable of self-healing and autopoietic. The assessment of these functions should be conducted

[17] For example, the Specialisation Course on the Adaptive Reuse of Dismissed Religious Heritage activated at the University of Naples "Federico II". Religious heritage constitutes the most frequent category of UNESCO sites, and represents a landmark that strongly characterises the urban and suburban landscape (especially in Italy and Europe); if abandoned and/or left to deteriorate, it causes damage on multiple levels (very high maintenance and management costs, reduced attractiveness of the location for activities and people, reduction of social/relational needs, etc.). Following the processes of demographic decrease and ageing, of growing urbanisation (especially in coastal areas) and with the consequent abandonment of 'inland' and mountain areas, as well as the decline of traditional religious practice, there is a growing need to identify new (non-liturgical or religious) uses for buildings once used for worship, which, however, are coherent with the 'intrinsic value' of religious heritage.

through sustainability verification, i.e. sustainability assessments for processes, products and services. This assessment and sustainability certification is growing in the current EU context, as it is in strong demand on the market and linked to the awarding of a quality mark. The above is configured as a source of financing that also benefits non-self-sustainable functions induced by adaptive reuse. Independent evaluations, *pro veritate* third party assessments, integrate the above.

6 Conclusions

The book reflects the specific implementation of the new circular economy in the space/territory. The **circular symbiotic heritage ecosystem** model is able to generate many net positive externalities to the benefit of a plurality of actors, unlike usual industrial ecosystems, and it is also a generator of relationships of mutual trust that enhance cooperation (and thus competitiveness on a higher scale). The circular model is interpreted also as a model able to foster the regeneration of social bonds, of community relationships.

Thus, the new approaches that implement the eco-social conversion of the economy through the adaptive reuse in design, planning and managing:

- are based on the ability to incorporate external effects;
- are based on the centrality of the collaborative/synergistic perspective of the different subjects involved, even institutional ones, also on the basis of new forms of agreements/contracts;
- are able to incorporate long-term impacts;
- attach importance to the use values and to intrinsic values, compared to exchange values;
- are characterized by a systemic logic, which takes into account multiple interdependencies. Are useful to avoid/minimize the different forms of underutilization of resources;
- are attentive to all technological innovation, and in particular to digital technologies to improve overall productivity;
- are attentive to the use and reuse of *local* resources (material, energy, water, human, social, ecological, etc.), reducing wastes;
- are aimed at the production of intangible services, and not only to material goods;
- are interested in ensuring long-term relationships with buyers and users;
- are interested in using fewer natural resources;
- are attentive to the flow of ecosystem services that are derived from natural resources and support human activities;
- are attentive to the circular closure of processes (as Mother Nature teaches), so that each output is as much as possible reused as input to produce other goods;
- are attentive to the well-being (and the variation of well-being) on the part of the subjects, which is also linked to qualitative and perceptual aspects.

Integration, connectivity, innovations, are the most general principles on which the recommendations here proposed are based. All these recommendations cannot go further because different regulations would require in-depth study at national level, which will be developed with specific "national CLIC projects". From a regulation perspective, it will be necessary to reshape and to operationalise all CLIC recommendations in each European Country considering the existing different rules and regulations, moving towards new specific national CLIC researches in different countries of European Union. For example, specific recommendations were provided for the construction industry in synergy with stakeholders, addressing issues with the application of the circular economy that make it difficult to apply circular virtuous practices.

As already underlined, this volume offers a comprehensive approach to the reuse of cultural heritage / landscape, moving across economic, entrepreneurial, social, cultural, environmental, geographical interdependent dimensions. Thus, the book becomes useful for academic researchers, for experts, practitioners, for public and political institutions, promoting a bridge between theory and practice. The proposal itself of the reuse of cultural assets into symbiotic circular ecosystems underlines the required systemic approach.[18]

References

Australia ICOMOS (2013) The Burra Charter: The Australia ICOMOS Charter for places of cultural significance

Council of Europe (2005) Convention on the value of cultural heritage for society (Faro Convention, 2005). Available at: https://www.coe.int/en/web/culture-and-heritage/faro-convention

De Carlo G (1988) Giancarlo De Carlo. Un Progetto per Catania. Il recupero del Monastero di San Nicolò l'Arena per l'Università. Sagep editrice.

Fusco Girard L (1987) Risorse architettoniche e culturali: valutazioni e strategie di conservazione. Franco Angeli

Fusco Girard L, Nijkamp P (1997) Le valutazioni per lo sviluppo sostenibile della città e del territorio. FrancoAngeli, Napoli

Fusco Girard L, Vecco M (2021) The "Intrinsic value" of cultural heritage as driver for circular human-centered adaptive reuse, sustainability, 13/6

ICOMOS (2011) Guidance on heritage impact assessments for cultural world heritage properties. ICOMOS, Paris

Sacco PL, Teti E (2017) Culture 3.0: a new paradigm of value creation. Econ Manag 2017:79–96. https://doi.org/10.3390/su13063231

[18] For example, the Association of Building Construction Companies in Italy, during scientific meetings and conferences analysed barriers and possible solutions to enhance reuse processes in cultural heritage works.

SPRINGER NATURE

GPSR Compliance

The European Union's (EU) General Product Safety Regulation (GPSR) is a set of rules that requires consumer products to be safe and our obligations to ensure this.

If you have any concerns about our products, you can contact us on ProductSafety@springernature.com

In case Publisher is established outside the EU, the EU authorized representative is:

Springer Nature Customer Service Center GmbH
Europaplatz 3
69115 Heidelberg, Germany

The manufacturer's authorised representative in the EU is Springer Nature Customer Service Centre GmbH, Europaplatz 3, 69115 Heidelberg, Germany. If you have any concerns regarding our products, please contact ProductSafety@springernature.com

Printed and bound by CPI Group (UK) Ltd, Croydon, CR0 4YY
01/10/2026
02225857-0017